Harnessing MicroStation® V8

Harnessing MicroStation® V8

G.V. Krishnan
James E. Taylor

THOMSON
DELMAR LEARNING   Australia • Canada • Mexico • Singapore • Spain • United Kingdom • United States

Harnessing MicroStation® V8
G.V. Krishnan and James E. Taylor

Executive Director:
Alar Elken

Executive Editor:
Sandy Clark

Acquisitions Editor:
James DeVoe

Developmental Editor:
John Fisher

Executive Marketing Manager:
Maura Theriault

Channel Manager:
Fair Huntoon

Marketing Coordinator:
Sarena Douglass

Executive Production Manager:
Mary Ellen Black

Production Manager:
Andrew Crouth

Production Editor:
Stacy Masucci

Technology Project Manager:
David Porush

Technology Project Specialist:
Kevin Smith

Editorial Assistant:
Mary Ellen Martino

COPYRIGHT 2003 by Delmar Learning, a division of Thomson Learning, Inc. Thomson Learning™ is a trademark used herein under license.

Printed in Canada
1 2 3 4 5 XX 06 05 04 03 02

For more information contact
Delmar Learning
Executive Woods
5 Maxwell Drive, PO Box 8007,
Clifton Park, NY 12065-8007
Or find us on the World Wide Web at
http://www.delmar.com

ALL RIGHTS RESERVED. No part of this work covered by the copyright hereon may be reproduced in any form or by any means—graphic, electronic, or mechanical, including photocopying, recording, taping, Web distribution, or information storage and retrieval systems—without the written permission of the publisher.

For permission to use material from the text or product, contact us by
Tel.   (800) 730-2214
Fax   (800) 730-2215
www.thomsonrights.com

Library of Congress Cataloging-in-Publication Data:

1-4018-2430-7

NOTICE TO THE READER

Publisher does not warrant or guarantee any of the products described herein or perform any independent analysis in connection with any of the product information contained herein. Publisher does not assume, and expressly disclaims, any obligation to obtain and include information other than that provided to it by the manufacturer.

The reader is expressly warned to consider and adopt all safety precautions that might be indicated by the activities herein and to avoid all potential hazards. By following the instructions contained herein, the reader willingly assumes all risks in connection with such instructions.

The publisher makes no representation or warranties of any kind, including but not limited to, the warranties of fitness for particular purpose or merchantability, nor are any such representations implied with respect to the material set forth herein, and the publisher takes no responsibility with respect to such material. The publisher shall not be liable for any special, consequential, or exemplary damages resulting, in whole or part, from the readers' use of, or reliance upon, this material.

Contents

Introduction xix

Chapter 1   Getting Started 1
   HARDWARE CONFIGURATION 2
      Processors 2
      Memory 2
      Hard Drives 2
      Video Adapter and Display 3
      Input Devices 3
   STARTING MICROSTATION V8 5
   USING DIALOG BOXES AND SETTINGS BOXES 6
      Title Bar and Menu Bar Item 6
      Label Item 8
      Edit Field 8
      List Boxes and Scroll Bars 8
      Buttons 9
      Check Boxes 9
      Option Button 10
   BEGINNING A NEW DESIGN 10
      File Names 13
      Seed Files 14
   OPENING AN EXISTING DESIGN FILE 16
      Opening a Design in Read-Only Mode 17
   MICROSTATION APPLICATION WINDOW 17
      Drop-down Menus 18
      Tool Frames and Tool Boxes 19
      Tool Settings Window 24
      Key-in Window 25
      Status Bar 26
      View Windows 27
   INPUT METHODS 28
      Keyboard 28
      Pointing Devices 28
      Cursor Menu 29
   THE DESIGN PLANE 30

WORKING UNITS	31
Custom Units	32
Advanced Units Settings	34
COORDINATE READOUT	35
CREATING MODELS	36
Create a New Model/Sheet	37
Copying an Existing Model	37
Edit Model Properties	38
Delete a Model	38
Import a Model	38
DRAWING PROPERTIES	38
SAVING CHANGES AND EXITING THE DESIGN FILE	40
Saving the Design File Settings	41
Exiting the MicroStation Program	42
GETTING HELP	42
Contents	42
Tool Index	43
MicroStation on Web	44
ENHANCEMENTS IN MICROSTATION V8	44

Chapter 2 Fundamentals I 47

PLACEMENT COMMANDS	47
Place Line	48
Place Block	50
Place Shape	52
Place Orthogonal Shape	54
Place Circle	55
Place Arc	60
MANIPULATING ELEMENTS	63
DELETE ELEMENT	64
DROP LINE STRING/SHAPE STATUS	65
PRECISION INPUT	66
Absolute Rectangular Coordinates	67
Relative Rectangular Coordinates	68
Relative Polar Coordinates	69
REVIEW QUESTIONS	74

Chapter 3 Fundamentals II 77

DRAWING TOOLS	77
The Grid System	78
Axis Lock	83
Snap Lock	84
AccuSnap	94
ELEMENT ATTRIBUTES	97
Levels	97
Element Color	103
Element Line Style	104
Element Line Weight	105

Contents

- Change Element Attributes 106
- Match Element Attributes 109
- **VIEW CONTROL** **109**
  - Update View 111
  - Controlling the Design Area Displayed in the View 111
  - Fit View 116
  - Rotate View 118
  - Pan View 121
  - View Previous 122
  - View Next 122
- **VIEW WINDOWS AND VIEW ATTRIBUTES** **123**
  - Opening and Closing View Windows 124
  - Arranging Open View Windows 126
  - Arranging Individual View Windows 131
  - Turning the View Window Scroll Bars ON and OFF 133
  - Creating and Using View Window Groups 133
  - Setting View Attributes 138
  - Saving Views 140
- **UNDO AND REDO** **146**
  - Undo Tool 146
  - Redo Tool 148
  - Things to Consider Before Undoing 148
- **REVIEW QUESTIONS** **149**

Chapter 4 Fundamentals III **151**
- **PLACEMENT TOOLS** **151**
  - Place Ellipse 152
  - Place Regular Polygon 155
  - Place Point or Stream Curve 160
  - Place Multi-line 162
- **ELEMENT MODIFICATION** **164**
  - Construct Circular Fillet 164
  - Construct Chamfer 166
  - Trim Elements 167
  - Partial Delete 177
- **ELEMENT MANIPULATION** **179**
  - Copy 179
  - Move 181
  - Move Parallel 183
  - Scale 185
  - Rotate 190
  - Mirror 194
  - Construct Array 198
- **TEXT PLACEMENT** **202**
  - Place Text 203
  - Tool Settings window 204
  - Place Text by Origin 209
  - Keyboard Shortcuts for the Text Editor window 210
- **REVIEW QUESTIONS** **213**

Chapter 5 AccuDraw and SmartLine **215**

GETTING TO KNOW ACCUDRAW **215**
- Start and Stop AccuDraw 216
- Key-In Shortcuts 217
- The AccuDraw Compass 217
- The AccuDraw Coordinates Box 220
- Recall Previous Values 222
- The Popup Calculator 223
- Smart Lock 224

CHANGE ACCUDRAW SETTINGS **224**
- Operation 225
- Display 227
- The Coordinates Tab 229
- AccuDraw Shortcuts 231

WORKING WITH ACCUDRAW **233**
- Example of Simple Placement 233
- Moving the Compass Origin 234
- Using Tentative Points with AccuDraw 234
- Rotating the AccuDraw Plane 235
- Placing Elements with AccuDraw Active 235
- Manipulating Elements with AccuDraw Active 236

PLACE SMARTLINE TOOL **237**
- The SmartLine Tool Settings Box 237
- Using SmartLine with AccuDraw 241

REVIEW QUESTIONS **243**

Chapter 6 Manipulating a Group of Elements **245**

ELEMENT SELECTION **245**
- Selecting Elements with the PowerSelector Tool 245
- Select Elements with the Element Selection Tool 251
- Consolidate Elements into a Group 252
- Ungroup Consolidated Elements 253
- Locking Selected Elements 254
- Unlock Selected Elements 254
- Drag Selected Elements to a New Position 254
- Drag an Element Handle to Change Its Shape 255
- Deleting Selected Elements 256
- Change the Attributes of Selected Elements 257
- Manipulation Tools that Recognize Selected Elements 257

FENCE MANIPULATION **258**
- Placing a Fence 259
- Fence Selection Mode 263
- Modifying a Fence's Shape or Location 266
- Manipulating Fence Contents 268
- Stretch Fence Contents 270
- Remove a Fence 271

REVIEW QUESTIONS **272**

Chapter 7 Placing Text, Data Fields and Tags ... 273

PLACE TEXT ... 274
- Fitted Method ... 274
- View Independent and Fitted VI Methods ... 275
- Above Element Method ... 275
- Below Element Method ... 278
- On Element Method ... 278
- Along Element ... 279
- Word Wrap Method ... 282

PLACE NOTE ... 283

SINGLE-CHARACTER FRACTIONS ... 285

TEXT STYLES ... 287
- Select a Text Style ... 287
- Create and Maintain Text Styles ... 287
- Text Styles Box Parts ... 288
- Create a New Style ... 289
- Modify the Values of an Existing Style ... 290
- Delete a Style ... 290
- The General Tab Settings ... 290
- The Advanced Tab Settings ... 294
- Importing Text Styles ... 294

IMPORT TEXT ... 295

TEXT MANIPULATION TOOLS ... 297
- Spell Checker ... 297
- Edit Text Elements ... 299
- Match Text Attributes ... 300
- Change Text Attributes ... 301
- Display Text Attributes ... 302
- Copy and Increment Text ... 302

TEXT NODES ... 304
- View Text Nodes ... 304
- Place Text Nodes ... 305
- Fill In Text Nodes ... 306

DATA FIELDS ... 307
- Data Field Character ... 307
- Data Field View Attribute ... 308
- Set Justification for Data Field Contents ... 308
- Fill In Data Fields ... 309
- Copy Data Fields ... 311
- Copy and Increment Data Fields ... 311
- Edit Text in a Data Field ... 312

TAGS ... 313
- Tag Terms ... 314
- Create a Tag Set and Tags ... 314
- Maintain Tag Set Definitions ... 316
- Attach Tags to Elements ... 317
- Edit Tags ... 318

Review Tags	320
Change Tags	321
Create Tag Reports	323
Tag Libraries	327
REVIEW QUESTIONS	**330**

Chapter 8 Element Modification 331

ELEMENT MODIFICATION—EXTENDING LINES 331
- Extend Element 332
- Extend Elements to Intersection 333
- Extend Element to Intersection 335

ELEMENT MODIFICATION—MODIFYING VERTICES 336
- Modify Element 336
- Delete Vertex 339
- Insert Vertex 340

ELEMENT MODIFICATION—MODIFYING ARCS 341
- Modify Arc Radius 341
- Modify Arc Angle 342
- Modify Arc Axis 343
- Create Complex Chain and Shapes 345
- Create a Complex Chain Manually 346
- Create a Complex Chain Automatically 347
- Create a Complex Shape Manually 349
- Create a Complex Shape Automatically 350
- Create a Region 352

DROP COMPLEX CHAINS AND SHAPES 358
- Drop a Complex Chain or Shape 358
- Drop Several Complex Chains or Shapes 359

DEFINE MULTI-LINES 359
- Define Line Components 361
- Start Cap or End Cap Components 363

MODIFY MULTI-LINE JOINTS 365
- Construct Closed Cross Joint 365
- Construct Open Cross Joint 367
- Construct Merged Cross Joint 368
- Construct Closed Tee Joint 369
- Construct Open Tee Joint 370
- Construct Merged Tee Joint 371
- Construct Corner Joint 372
- Cut Single Component Line 373
- Cut All Component Lines 374
- Uncut Component Lines 375
- Multi-line Partial Delete 376
- Move Multi-line Profile 376
- Edit Multi-line Cap 378

REVIEW QUESTIONS **379**

Chapter 9 Measurement and Dimensioning 381
MEASUREMENT TOOLS 381
- Measure Distance 382
- Measure Radius 385
- Measure Angle 386
- Measure Length 386
- Measure Area 387

DIMENSIONING 392
- Dimensioning Terminology 393
- Associative Dimensioning 395
- Alignment Controls 395
- Dimension Styles 396
- Linear Dimensioning 396
- Dimension Element 408
- Angular Dimensioning 409
- Dimension Radial 418
- Miscellaneous Dimensions 427
- Geometric Tolerance 433

DIMENSION SETTINGS 434
- Custom Symbols 437
- Dimension Lines 439
- Dimension with Leader 440
- Extension Lines 441
- Placement 443
- Terminators 446
- Terminator Symbols 448
- Text 449
- Tolerance 452
- Tool Settings 454
- Units 454
- Unit Format 456
- Match Dimension Settings 458
- Change Dimension 459

REVIEW QUESTIONS 460

Chapter 10 Printing 463
OVERVIEW OF THE PRINTING PROCESS 464
PRINTING FROM MICROSTATION 465
- Selecting the Area of the Design to Print 467
- Setting the Vector Output Color 467
- Selecting a Printer 468
- Setting the Printing Parameters 469
- Setting Print Attributes 470
- Creating the Print 472

SAVING PRINT CONFIGURATION 473
PEN TABLES 474
- Creating a Pen Table 475

Renaming a Pen Table Section	476
Inserting a New Pen Table Section	476
Deleting a Pen Table Section	476
Modifying a Pen Table Section	477
BATCH PRINTING	**483**
Setting Print Specifications	483
Design Files to Print	491

Chapter 11 Cells and Cell Libraries 493

CELLS	**494**
CELL LIBRARIES	**494**
Creating a New Cell Library	495
Attaching an Existing Cell Library	497
CREATING YOUR OWN CELLS	**498**
Before You Start	498
Steps for Creating Cells	500
Graphic and Point Cell Types	503
ACTIVE CELLS	**503**
Place Cells	504
Place Active Line Terminator	510
Point Placement	512
CELL SELECTOR	**520**
Opening the Cell Selector Box	520
Invoking Tools from the Cell Selector Box	520
Customizing the Cell Selector	520
Creating and Using Cell Selector Files	526
CELL HOUSEKEEPING	**528**
Identify Cell	528
Replace Cells	529
Drop Complex Status	531
Drop Fence Contents	532
Fast Cells View	533
LIBRARY HOUSEKEEPING	**533**
Edit a Cell's Name and Description	534
Delete a Cell from the Library	534
Compress the Attached Cell Library	535
Create a New Version of a Cell	535
SHARED CELLS	**536**
Turn on the Shared Cell Feature	536
Determine Which Cells Are Shared	537
Declare a Cell to Be Shared	537
Place Shared Cells	538
Turn a Shared Cell into an Unshared Cell	538
Delete the Shared Cell Copy from Your Design File	539
REVIEW QUESTIONS	**540**

Chapter 12 Patterning 541
CONTROLLING THE VIEW OF PATTERNS 542
PATTERNING COMMANDS 543
TOOL SETTINGS FOR PATTERNING 545
Spacing 545
Angle 545
Tolerance 545
Associative Pattern 545
Snappable Pattern 546
Pattern Cell 546
Scale 547
Method 547
PATTERNING METHODS 547
PLACING PATTERNS 549
Element 549
Fence 550
Intersection 550
Union 551
Difference 552
Flood 552
Points 554
USING THE AREA SETTING TO CREATE HOLES 555
Changing an Element's Area 556
DELETING PATTERNS 557
MATCHING PATTERN ATTRIBUTES 557
FILLING AN ELEMENT 558
Area Fill View Attribute 559
Changing the Fill Type of an Existing Element 560
REVIEW QUESTIONS 562

Chapter 13 Attaching References 563
OVERVIEW OF REFERENCES 564
EXAMPLES OF USING REFERENCES 565
ATTACHING A REFERENCE 566
Listing the Attached References 570
REFERENCE MANIPULATIONS 571
Selecting the Reference 571
Move Reference 572
Copy Reference Attachment 573
Scale Reference 573
Rotate Reference 575
Mirror Reference 575
Clip Reference 576
Mask Reference 577

    Delete Clipping Mask(s) 578
    Merging References into Master 579
    Reloading the Reference 579
    Detaching the Reference 580
    Exchange Reference 581
    Reference File Agent Dialog Box 581
    Update Sequence Dialog Box 582
    Hilite References 583
REVIEW QUESTIONS 584

Chapter 14   Special Features 585

GRAPHIC GROUPS 586
    Adding Elements to a Graphic Group 586
    Manipulating Elements in a Graphic Group 587
    Copying Elements in a Graphic Group 588
    Dropping Elements from a Graphic Group 588
MERGE UTILITY 589
SELECTION BY ATTRIBUTES 591
    Select By Attributes Dialog Box 591
    Selection by Level 593
    Selection by Type 593
    Selection by Symbology 593
    Controlling the Selection Mode 593
    Select By Properties Dialog Box 594
    Select By Tags Dialog Box 595
    Executing Selection Criteria into Effect 596
LEVEL SYMBOLOGY 596
CHANGING THE HIGHLIGHT AND POINTER COLOR 598
IMPORTING AND EXPORTING DRAWINGS IN OTHER FORMATS 600
    Supported Formats 600
    Opening a File of Another File Format 601
    Saving a Design File Using Another File Format 602
    Importing and Exporting Other Formats into an Open Design File 603
MANIPULATING IMAGES 604
THE ANNOTATION TOOLS 607
    The Annotate Tool Box 608
    The Drafting Tools 611
    XYZ Text Tool Box 617
DIMENSION-DRIVEN DESIGN 622
    Example of a Constrained Design 622
    Dimension-Driven Design Terms 624
    Dimension-Driven Design Tools 624
    Creating a Dimension-Driven Design 625
    Modifying a Dimension-Driven Design 637
    Creating a Dimension-Driven Cell 639
OBJECT LINKING AND EMBEDDING (OLE) 639
REVIEW QUESTIONS 644

Chapter 15   Internet Utilities 645
LAUNCHING THE WEB BROWSER AND OPENING FILES REMOTELY 646
Remote Opening of Design File 649
Remote Opening of Settings File 650
Remote Opening of Archive File 651
Remote Opening of Reference File 651
Remote Opening of Cell Library 652
PUBLISHING MICROSTATION DATA TO THE INTERNET 653
Creating HTML File from Design File Saved View 653
Creating HTML File from Design File Snapshot 654
Creating HTML File from Cell Library 655
Creating HTML File from Basic Macros 657
LINKING GEOMETRY TO INTERNET URLS 659
Attaching Link to a Geometry 659
Displaying Links 660
Connecting to URLs 660
Edit Engineering Link 661
Delete Engineering Link 662

Chapter 16   Customizing MicroStation 663
SETTINGS GROUPS 664
Selecting a Settings Group and Corresponding Component 664
Attaching a Settings Group File 666
Maintaining a Settings Group 666
LEVEL FILTERS 674
CUSTOM LINE STYLES 676
The Line Styles Settings Box 677
The Line Style Editor Settings Box 678
WORKSPACE 690
Setting Up the Active Workspace 690
Setting User Preferences 693
Working with Configuration Variables 706
Customizing the User Interface 726
FUNCTION KEYS 738
Creating and Modifying Function Key Definitions 738
INSTALLING FONTS 741
Selecting the Source Font Files 742
Selecting the Destination Fonts File 743
Importing Fonts 743
PACKAGING THE MICROSTATION ENVIRONMENT 744
SCRIPTS AND MACROS 746
Scripts 746
Macros 746

Chapter 17 3D Design and Rendering 747
WHAT IS 3D? 747
CREATING A 3D DESIGN FILE 749
VIEW ROTATION 750
- Rotate View 750
- Changing View Rotation from the View Rotation Settings Box 751
- Rotating View by Key-in 752
DESIGN CUBE 753
DISPLAY DEPTH 753
- Setting Display Depth 754
- Fitting Display Depth to Design File Elements 756
ACTIVE DEPTH 756
- Setting Active Depth 756
- Boresite Lock 758
PRECISION INPUTS 758
- Drawing Coordinate System 759
- View Coordinate System 760
AUXILIARY COORDINATE SYSTEMS (ACS) 761
- Precision Input Key-in 762
- Defining an ACS 763
- Rotating the Active ACS 765
- Moving the Active ACS 766
- Selecting the Active ACS 766
- Saving an ACS 766
3D PRIMITIVES 767
- Display Method and Surface Rule Lines 768
- Selection of Solids and Surfaces 769
- Place Slab 769
- Place Sphere 770
- Place Cylinder 771
- Place Cone 772
- Place Torus 773
- Place Wedge 774
CHANGING THE STATUS—SOLID OR SURFACE 775
USING ACCUDRAW IN 3D 776
PROJECTED SURFACES 777
EXTRUDE ALONG A PATH 778
SURFACE OF REVOLUTION 779
SHELL SOLID 780
THICKEN TO SOLID 780
PLACING 2D ELEMENTS 781
CREATING COMPOSITE SOLIDS 781
- Union Operation 781
- Intersection Operation 783
- Difference Operation 784
CHANGE NORMAL 786

MODIFY SOLID 786
REMOVE FACES 787
CUT SOLID 788
CONSTRUCT FILLET 789
CONSTRUCT CHAMFER 790
PLACING TEXT 791
FENCE MANIPULATIONS 791
CELL CREATION AND PLACEMENT 792
DIMENSIONING 792
RENDERING 792
    Setting Up Cameras 792
    Placement of Light Sources 794
    Rendering Methods 794
DRAWING COMPOSITION 796

Appendix A MicroStation Tool Boxes 797

Appendix B Key-in Commands 799

Appendix C Alternate Key-ins 813

Appendix D Primitive Commands 817

Index 829

DEDICATION

HATS OFF!

bhuvana
avinash
kavitha

dorothy

Introduction

Harnessing MicroStation V8 gives you the necessary skills to start using a powerful CADD (Computer Aided Drafting and Design) program, MicroStationV8. We have created a comprehensive book providing information, references, instructions, and exercises for people of varied skill levels, disciplines, and requirements, for applying this powerful design/drafting software.

Now in its fourth edition, *Harnessing MicroStation V8* was written and updated as a comprehensive tool for the novice and the experienced MicroStation user, both in the classroom and on the job. The book includes new features introduced in MicroStation V8.

Readers immediately gain a broad range of knowledge of the elementary CADD concepts necessary to complete a simple design. We do not believe the user should be asked to wade through all components of every tool or concept the first time that tool or concept is introduced. Therefore, we have set up the early chapters to cover and practice fundamentals as preparation for the advanced topics covered later in the book.

Harnessing MicroStation V8 is intended to be both a classroom text and a desk reference. If you are already a user of earlier versions of MicroStation, you will see an in-depth explanation provided in the corresponding chapters for all new features. The new features in MicroStation V8 give personal computer-based CADD even greater depth and breadth.

IN THIS BOOK

Chapter 1—Getting Started   The beginning of this chapter describes the hardware you need to get started with MicroStation. The balance of the chapter explains how to start MicroStation V8; the salient features of dialogs, settings boxes, and MicroStation applications windows; input methods, design planes, and working units; saving changes and exiting the design file; and a summary of the enhancements in MicroStation V8.

Chapters 2, 3, and 4—Fundamentals   These chapters introduce the basic element placement and manipulation tools needed to draw a moderately intricate design. All tools discussions are accompanied by examples. Ample exercises are designed to give students the chance to test their level of skill and understanding.

Chapter 5—AccuDraw and SmartLine   An in-depth explanation is provided for AccuDraw and SmartLine, two powerful drawing tools.

Chapter 6—Manipulating Groups of Elements Introduces the Power Selector and Fence tools that allow you to manipulate groups of elements.

Chapter 7—Placing Text, Data Fields, and Tags Provides an in-depth explanation of various methods for placing text, data fields that serve as place holders for text to be placed later, and tags that allow you to add non-graphical information to the design.

Chapter 8—Element Modification This chapter introduces various tools for modifying the elements of your design. All tools are accompanied by examples.

Chapter 9—Measurement and Dimensioning Introduces tools that allow you to display the length, angle, or area of elements; place dimensions on elements; and customize the way dimensions are placed.

Chapter 10—Plotting This chapter introduces all the features related to plotting (creating paper copies of your design).

Chapter 11—Cells and Cell Libraries Introduces the powerful set of tools available in MicroStation for creating and placing symbols—called cells—and for storing them in Cell libraries. These tools permit you to group elements under a user-determined name and perform manipulations on the group as though they were a single element.

Chapter 12—Patterning Introduces the set of tools available in MicroStation to place repeating patterns to fill regions in a design, such as the hatch lines in a cross-section.

Chapter 13—References A powerful and timesaving feature of MicroStation is its ability to view other models from an active design file (the one you currently have open for editing). MicroStation lets you display the contents of unlimited number of models from the active design file. When you view a model in this way, it is referred to as a Reference. This chapter describes all the tools used to manipulate references .

Chapter 14—Special Features MicroStation provides some special features that, though less often used than the tools described in chapters 1 through 13, add power and versatility to the MicroStation tool set. This chapter introduces several such features.

Chapter 15—Internet Utilities MicroStation provides utilities that allow you to take part in collaborative engineering projects over the World Wide Web. With these utilities you can share design information over the Web and insert engineering data directly into your design from the Web.

Chapter 16—Customizing Introduces several tools for customizing MicroStation, such as creating multi-line definitions, custom line styles, and workspaces; creating and modifying tool boxes, tool frames, and function key menus; installing fonts; and using the archive utility.

Chapter 17—3D Design and Rendering Provides an overview of the tools and specific tools available for 3D design.

Appendices provide additional valuable information to the user.

Introduction

Harnessing MicroStation V8 provides a sequence suitable for learning, ample exercises, examples, review questions, and thorough coverage of the MicroStation program, that should make it a must for multiple courses in MicroStation, as well as self-learners, everyday operators on the job, and operators aspiring to customize MicroStation.

The project exercises and drawing exercises are contained in PDF files found on the CD in the back of this book. The PDF files correspond to the material presented in Chapters 1 through 13 and 17.

STYLE CONVENTIONS

In order to make this text easier for you to use, we have adopted certain conventions that are used throughout the book:

Convention	Example
Drop-down menu names appear with the first letter capitalized.	Element drop-down menu
Tool box names appear with the first letter capitalized.	Linear Elements tool box
All interface object names are indicated in boldface.	Click the **OK** button to save the changes and close the Design File settings box. Next, select **Tools** from the main menu.
User input is indicated by boldface.	Click in the **Angle** field, key-in **180**, and press ENTER.
Instructions are indicated by italics and are enclosed in parentheses.	Enter first point: *(Place data point or key-in coordinates)*

HOW TO INVOKE TOOLS

Throughout the book, instructions for invoking tools are summarized in tabular form, similar to the example shown below:

Drop-down menu	Element > Text Styles
Place Text Settings window	Click the magnifying glass icon
Key-in Window	**textstyle dialog open** (or **texts di o**) ENTER

The left column tells you where the action will take place (such as, by selecting a drop-down menu), and the right column tells you specifically what the action is (such as, by selecting the Text Styles option from the Element drop-down menu).

EXERCISE ICONS

A special icon is used to identify step-by-step Project Exercises. Exercises that give you practice with types of drawings that are often found in a particular engineering discipline are identified by icons that indicate the discipline. Exercises that are cross-discipline—that is, the skills used in the exercise are applicable for most or all disciplines—do not have a special icon designation. The following table presents all the exercise icons:

Type of Exercise	Icon	Type of Exercise	Icon
Project Exercises		Electrical	
Mechanical		Piping	
Architectural		Civil	

ACKNOWLEDGMENTS

This book was a team effort. We are very grateful to many people who worked very hard to help create this book. We are especially grateful to the following individuals at Delmar Publishers, whose efforts made it possible to complete the project on time: Ms. Sandy Clark, Executive Editor; Mr. James DeVoe, Acquisitions Editor; Mr. John Fisher, Developmental Editor; Ms. Stacy Masucci, Production Editor; and Ms. Mary Ellen Martino, Editorial Assistant. The authors and Delmar Publishers would also like to acknowledge Mr. John Shanley and staff at Phoenix Creative Graphics.

In addition, the authors would like to acknowledge the following individuals who reviewed the previous edition of this book: Dennis C. Jackson, Malcolm A. Roberts, Jr., and Michael J. White. The authors and Delmar Publishers also gratefully acknowledge the thorough and thoughtful technical editing provided by David Newsom.

And, last but not least, special appreciation to Bentley Systems, Inc. for providing the MicroStation V8 software.

chapter 1

Getting Started

The beginning of this chapter describes the hardware you need to get started with MicroStation V8. If you need to set up the MicroStation V8 program on the computer and you are not familiar with the computer operating system (files, drives, directories, operating system commands, etc.), you may wish to review an Introduction to Windows operating system book, refer to the Installation Guide that came with the program, or consult the dealer from whom you purchased MicroStation. Once the computer is set up, you will have at your disposal a versatile design and drafting tool that continues to grow in power with each new version.

The balance of this chapter explains how to start MicroStation and gives an overview of the screen layout and the salient features of dialog and settings boxes. Detailed explanations and examples are provided for the concepts and commands throughout the chapters that follow.

At the end of the chapter, a list of enhancements for MicroStation V8 is provided. If you are already a user of MicroStation J, check the list to see the new and improved features in MicroStation V8. Throughout the book you will find in-depth explanations in the chapters corresponding to all the new features of MicroStation V8. Features introduced in MicroStation V8 give personal computer–based CADD even greater depth and breadth.

HARDWARE CONFIGURATION

The configuration of your system is a combination of the hardware and software you have assembled to create your system. Countless PC configurations are available. The goal for a new computer user should be to assemble a PC workstation that will not block future software and hardware upgrades.

To use MicroStation, your computer system must meet certain minimum requirements. Following are the minimum recommended configurations for MicroStation V8:

- Intel-compatible Pentium III or compatible processor.
- Operating System: Windows NT version 4.0, Windows 2000 Professional, Windows 98, Windows ME, or Windows XP.
- 128 MB RAM recommended, 64 MB RAM minimum.
- 300MB minimum hard disk (typical MicroStation V8 installation requires 200MB).
- Input device: mouse or tablet.
- Supported graphics card (256 or more color card recommended for rendering).
- Dual screen graphics supported.

Processors

Select the one that best suits your workload. The faster and more powerful the processor, the better MicroStation performs. Thus, the processor of choice is generally the most technologically advanced one available. With the availability of fast Pentium-based systems under $1,000, a good platform for MicroStation can be attained from a number of sources. Several manufacturers of RISC processors are introducing PC-compatible systems that can now run most Windows applications.

Memory

To run MicroStation, your computer must be equipped with at least 64MB of RAM (random access memory). Depending on your MicroStation application, 128MB of RAM or more may be required for optimal performance. The availability of low-cost RAM has made this an easy upgrade for most MicroStation users.

Hard Drives

The hard disk is the personal computer's primary data storage device. A hard disk with at least 300MB capacity is required, but a larger capacity is advised. The hard disk accesses data at a rate of 3 milliseconds to 20 milliseconds. High performance is 3 milliseconds or faster, while 20 milliseconds is considered slow. Obviously, the faster the hard-disk drive, the more productive you will be with MicroStation.

Video Adapter and Display

As with any graphics program for PC CAD systems, MicroStation requires a video adapter capable of displaying graphics information. A video adapter is a printed circuit board that plugs into the central processing unit (CPU) and generates signals to drive a monitor. MicroStation supports a number of display options, ranging from low-priced monochrome setups to high-resolution color units. Some of the video display controllers can be used in combination, giving a two-screen display.

Input Devices

MicroStation supports several input device configurations. Data may be entered through the keyboard, a mouse, or a digitizing tablet with a cursor.

Keyboard

The keyboard is one of the primary input methods. It can be used to enter commands and responses.

Mouse

A mouse is used with the keyboard as a tracking device to move the crosshairs on the screen. MicroStation supports a two-button mouse and a three-button mouse (see Figure 1–1). As you move your pointing device around on a mouse pad or other suitable surface, the cursor will mimic your movements on the screen. It may take on the form of crosshairs when you are being prompted to select a point. Each of the buttons in the mouse is programmed to serve specific functions in MicroStation. See Table 1–1 for the specific functions that are programmed for the two- and three-button mouse.

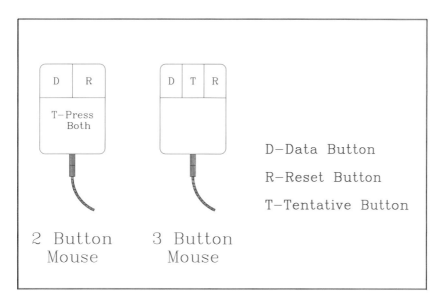

Figure 1-1 Pointing device—two-button mouse and three-button mouse

Table 1-1 Button Functions on Two- and Three-Button Mice

BUTTON FUNCTION	BUTTON POSITION	
	TWO-BUTTON MOUSE	**THREE-BUTTON MOUSE**
Data button	Left (first button)	Left (first button)
Reset button	Right (second button)	Right (third button)
Tentative button	Left and right simultaneously	Center (second button)

Digitizing Tablet

A digitizer is literally an electronic drawing board. An internal wiring system forms a grid of fine mesh, which corresponds to the coordinate points on the screen. The digitizer works with a puck. When the user moves the puck across the digitizer, crosshairs follow on the screen. The number of buttons on a puck depends on the manufacturer. A puck usually has four buttons, as shown in Figure 1-2; each of the buttons on the puck is programmed to serve specific functions in MicroStation. If there are more than four buttons on the puck, the first four are preprogrammed. For example, when you place a line, the points are selected by pressing the designated Data button in two locations. Although the user's attention is focused on the screen, the points are actually selected on the coordinates of the tablet. The coordinates from the tablet are then transmitted to the computer, which draws the image of a line on your screen.

Another advantage of a tablet is the ability to use a tablet menu for command selection. Menus have graphic representations of the commands and are taped to the digitizing tablet's surface. The menu is attached or activated with the key-in **AM=** followed by that menu's specific name. The commands are chosen by placing the puck's crosshairs over the block that represents the command you want to use and pressing the designated command button. Another powerful feature of the tablet (not related to entering commands) is that it allows you to lay a map or other picture on the tablet and trace over it with the puck. See Table 1-2 for the specific functions that are programmed for the four-button puck. Detailed explanations of the functions are given later in the chapter, in the section on Input Methods.

Table 1-2 Button Functions on the Four-Button Puck

FUNCTION	FOUR-BUTTON PUCK
Data button	Top (yellow)
Reset button	Right (green)
Tentative button	Bottom (blue)
Command button	Left (white)

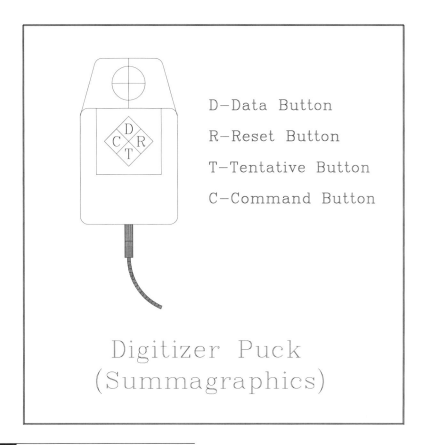

Figure 1-2 Pointing device—digitizer puck

STARTING MICROSTATION V8

Click the Start button (on Windows 98, Windows NT 4.0, Windows 2000 Professional, and Windows XP operating systems), select the MicroStation V8 program group, and then select the MicroStation V8 program. MicroStation displays the MicroStation Manager dialog box, similar to Figure 1–3.

 Note: All the screen captures shown in this textbook are taken from MicroStation V8 running in the Windows XP operating system.

Before we start a new design file, let's discuss the important features of dialog boxes and settings boxes.

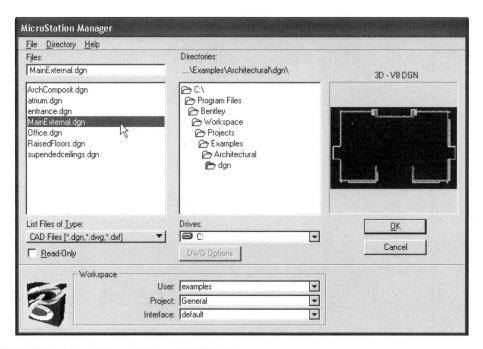

Figure 1-3 MicroStation Manager dialog box

USING DIALOG BOXES AND SETTINGS BOXES

A dialog box is a special type of window displayed by MicroStation. Dialog boxes were designed to permit the user to perform many actions easily within MicroStation. Dialog boxes force MicroStation to stop and focus on what is happening in that dialog box only. You cannot do anything else in MicroStation until you close the dialog box. The MicroStation Manager is a good example of this type of dialog box.

In addition to dialog boxes, MicroStation provides settings boxes. Several settings boxes can be left on the screen while you work in other areas of MicroStation. For instance, the Lock Toggles settings box (see Figure 1-4) can be left open as long as you need to use it. While it is open, you can turn the locks ON and OFF as you need and at the same time interact with other dialog boxes. To close the settings box, click the "X" symbol located in the top right corner of the settings box.

When you move the cursor onto a dialog box or settings box, the cursor changes to a pointer. You can use the arrow keys on your keyboard to make selections, but it is much easier with your pointing device. Another way to make selections is via the keyboard equivalents.

Title Bar and Menu Bar Item

MicroStation displays the title of the dialog box or settings box in the Title bar, as shown in Figure 1-5. Below the Title bar, MicroStation displays any available drop-down menus in the menu bar, as shown in Figure 1–5. In this case, two drop-down menus are available, **File** and **Directory**. Selecting from the list is a simple matter of moving the cursor down until the desired item is highlighted, then pressing the designated Pick (Data) button on the pointing device. If a menu item has an arrow to the right, it has a cascading submenu. To display the submenu, just click on the menu title. Menu items that include ellipses (...) display dialog boxes. To select these, just pick the menu item.

Figure 1-4 Lock Toggles settings box

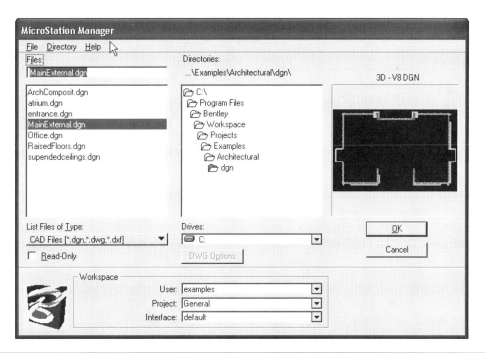

Figure 1-5 MicroStation Manager dialog box showing the Title bar and drop-down menus

Label Item

MicroStation displays the text (display only) to label the different parts of the dialog box or settings box, as shown in Figure 1–6.

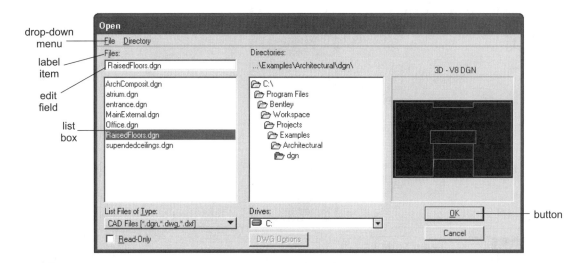

Figure 1–6 Dialog box

Edit Field

An edit field is an area that accepts one line of text entry. It normally is used to specify a name, such as a file name, including the drive and/or directory path or level name. Edit fields often make an alternative to selecting from a list of names when the desired name is not displayed in the list box. Once the correct text is keyed-in, enter it by pressing ENTER.

Moving the pointer into the edit field causes the text cursor to appear in a manner similar to the cursor in a word processor. If necessary, the text cursor, in combination with special editing keys, can help to facilitate changes to the text. You can see the text cursor and the pointer at the same time, making it possible to click the pointer on a character in the edit field and relocate the text cursor to that character. You can select a group of characters in the edit field to manipulate, for instance, or to delete, by highlighting the characters. This is accomplished by holding down the designated Pick button on your pointing device and dragging left or right. You can also use the right and left arrows on the keyboard to move the cursor right or left, respectively, across text, without affecting the text.

List Boxes and Scroll Bars

List boxes make it easy to view, select, and enter a name from a list of existing items, such as file names (see Figure 1–6). With the pointer, highlight the desired selection. The item, when clicked, will appear in the edit field. You can accept this item by clicking **OK** or by double-clicking on the item. List boxes are accompanied by scroll bars to facilitate moving long lists up and down in the list box. When you point and hold onto the slider box, you can move it up and down to cause the list to scroll. Pressing the up/down arrows cause the list to scroll up or down one item at a time.

Buttons

Actions are initiated immediately when one of the various buttons is clicked (**OK** or **Cancel**, for example). If a button (such as the **OK** button in Figure 1–6) is surrounded by a heavy line, it is the default button, and pressing ENTER is the same as clicking that button. Buttons with action that is not acceptable will be disabled; they will appear grayed out. Buttons with ellipses (...) will cause that action's own dialog box (subdialog) to appear.

Check Boxes

A button that indicates an ON or OFF setting is called a *check box*. For instance, in Figure 1–7, the Snap Lock, Graphic Group, Boresite, and Depth Lock check boxes are set to ON, and the remaining check boxes are set to OFF.

Figure 1-7 Check boxes

Option Button

A list of items is displayed when you click on the option button menu, and only one item may be selected from the list, as shown in Figure 1–8.

Figure 1-8 Option button menu

Let's get on to the business of starting a new design file in MicroStation.

BEGINNING A NEW DESIGN

To begin a new design, select the **New** tool from the **File** drop-down menu, as shown in Figure 1–9.

The New dialog box opens, as shown in Figure 1–10.

Select the appropriate seed file (more about this in the section on Seed Files), enter a name for your new design in the **Files** edit field, and click the **OK** button. The New dialog box is closed, and control is passed to the MicroStation Manager. MicroStation by default highlights the name of the file you just created in the **Files** list box.

Before you click the **OK** button to open the newly created design file, make sure the appropriate **User**, **Project**, and **Interface** are selected from the option menu located at the bottom of the MicroStation Manager dialog box.

A user is a customized workspace drafting environment that permits the user to set up MicroStation for specific purposes. You can set up as many workspaces as you need. A workspace consists of "components" and "configuration files" for both the user and the project. By default, MicroStation selects the default workspace. If necessary, you can create or modify an existing workspace. Refer to Chapter 15 for a detailed description of creating or modifying workspaces.

The selection of the project sets the location and names of data files associated with a specific design project. Refer to Chapter 16 for setting up the project. By default, MicroStation selects the default in the **Project** option menu.

The selection of the interface sets a specific look and feel of MicroStation's tools and general on-screen operation. If necessary, you can change the selection of the interface from the Interface option menu. By default, MicroStation selects the default in the **Interface** option menu. Refer to Chapter 16 for a detailed explanation on creating and modifying the MicroStation interface.

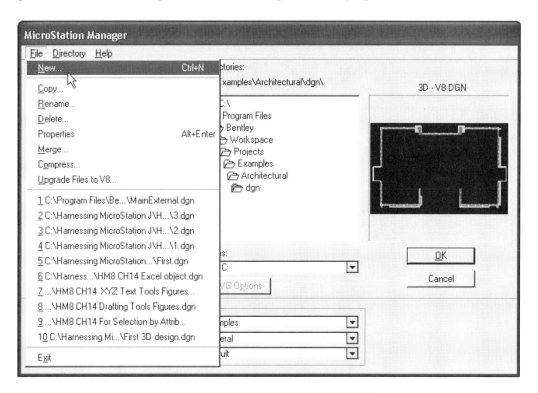

Figure 1-9 Invoking the New tool from the File drop-down menu

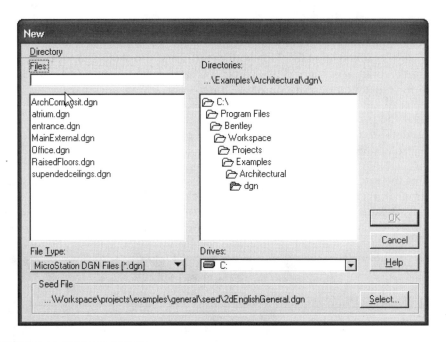

Figure 1-10 The New dialog box

To open the new design file, click the **OK** button. Your screen will look similar to the one in Figure 1–11.

 Note: If the design file name you key-in is the same as the name of an existing file name, MicroStation displays an Alert Box asking if you want to replace the existing file. Click the **OK** button to replace, or **Cancel** to reissue a new file name.

Chapter 1 Getting Started

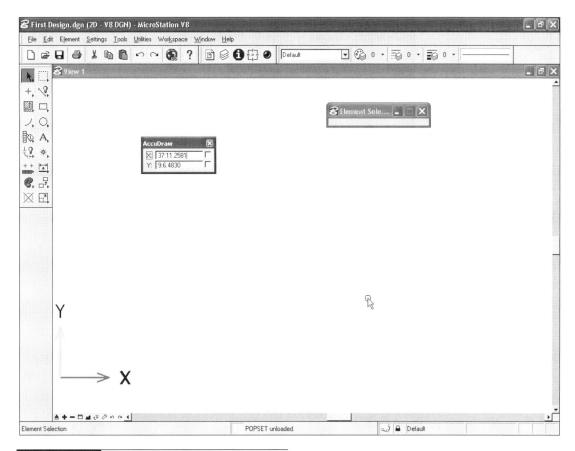

Figure 1-11 MicroStation application window

File Names

The name you enter in the **Files:** edit field will be the name of the file in which information about the design is stored. It must satisfy the requirements for file names as specified by the particular operating system on your computer.

File names and file extensions can contain up to 255 characters. Names may be made up of combinations of uppercase and lowercase letters, numbers, the underscore (_), the hyphen (-), embedded spaces, and punctuation. Valid examples include:

> this is my first design.dgn
>
> first house.dgn
>
> machine part one.dgn
>
> PART_NO5.wrk

When MicroStation prompts for a design name, just type in the file name and MicroStation will append the extension .DGN by default. For instance, if you respond to the design name as FLOOR1, then MicroStation will create a file with the file spec FLOOR1.DGN. If you need to provide a different extension, key-in the extension with the file name.

As you progress through the lessons, note how various functions ask for names of files. If MicroStation performs the file processing, it usually adds the proper default extension.

The Path

If you want to create a new design file or edit a design file that is on a drive and/or folder other than the current drive/folder, you must furnish what is called the *path* to the design file as part of the file specification you enter. Specifying a path requires that you use the correct pathfinder symbols—the colon (:) and/or the backslash (\). The drive with a letter name (usually A through E) is identified as such by a colon, and the backslashes enclose the name of the directory where the design file is (or will be) located. Examples of path/key name combinations are as follows:

a:proj1	The file proj1 in the working directory on drive A.
b:\spec\elev	The file elev in the \spec directory on drive B.
\buil\john	The file john in the \buil directory.
ACME\doors	The file doors in the working directory's ACME subdirectory.
..\PIT\flange	The file flange in the parent directory's PIT subdirectory.

Note: Instead of specifying the path as part of the design file name, select the appropriate drive and folder from the list box and then select the design file from the files list box. MicroStation Manager displays the current drive and path in the dialog box.

Seed Files

Each time you use MicroStation's Create New File utility, a copy is made of an existing "prototype," or "seed," file. If necessary, you can customize the seed file. In other words, you can control the initial "makeup" of the file. To do so, open an existing seed file, make the necessary changes in the settings of the parameters, and place elements such as title block, and so on. Whenever you start a new design file, make sure to copy the appropriate seed file.

MicroStation programs come with several seed files. Depending on the discipline, use an appropriate seed file. For instance, if you plan to work on an architecture floor plan, then use the architecture seed file (2dEnglishArch.DGN). When you open the Create Design file, MicroStation displays the name of the default seed file as shown in Figure 1–12. If necessary, you can change the default seed file by clicking the **Select** button. MicroStation displays a list of available seed files as shown in Figure 1–13. Select the one you want from the list and click the **OK** button.

Chapter 1 Getting Started

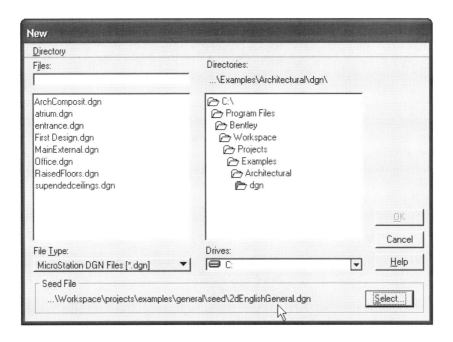

Figure 1-12 New dialog box displaying the name of the default seed file

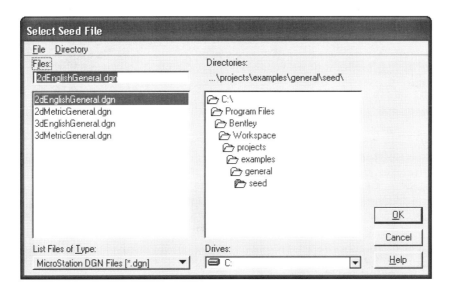

Figure 1-13 Select Seed File dialog box

OPENING AN EXISTING DESIGN FILE

Whenever you want to open an existing design file in MicroStation, simply click the name of the file in the **Files** list box item of the MicroStation Manager dialog box, and then click the **OK** button, or double-click the name of the file. If the design file is not in the current folder, change to the appropriate drive and folder from the Directories list box, then select the appropriate design file. In addition, MicroStation displays the names (including the path) of the last ten design files opened in the **File** drop-down menu, as shown in Figure 1–14. If you need to open one of these ten design files, click on the file name and MicroStation displays the design file.

 Note: You can also create a new design file or open an existing design file by selecting the New or Open tool, respectively, from the File drop-down menu located in the MicroStation Application Window.

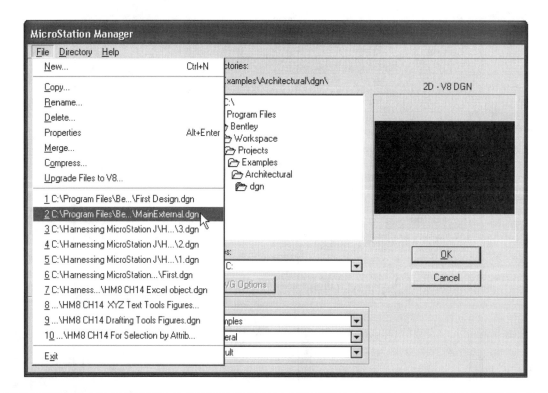

Figure 1–14 File drop-down menu from the MicroStation Manager dialog box, displaying the names of the last ten design files opened

Chapter 1 Getting Started

Opening a Design in Read-Only Mode

MicroStation allows you to open a design in read-only mode by turning ON the check box for **Read-Only**, located in the MicroStation Manager dialog box. When the **Read-Only** option is chosen, MicroStation opens the active design file in a read-only state and displays a disk icon with a large red X in the lower right corner of the application window. Any changes you make to the design will not be saved as part of the design file.

MICROSTATION APPLICATION WINDOW

The MicroStation application window consists of drop-down menus (also called *menu bars*), the Status bar, tool frames, tool boxes, the Key-in window, and view windows with the View Control bar (see Figure 1–15).

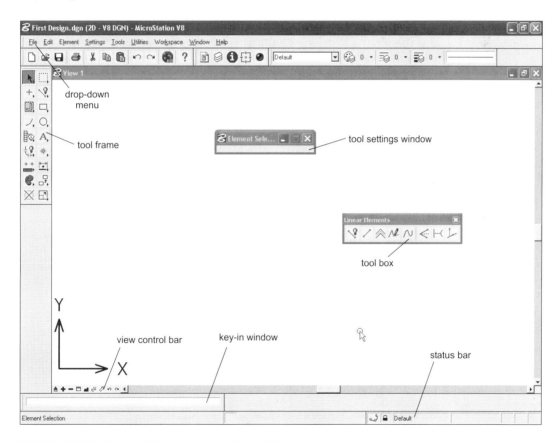

Figure 1-15 MicroStation application window

Drop-down Menus

The MicroStation Application Window has a set of drop-down menus, as shown in Figure 1–15. Several tool boxes, dialog boxes, and settings boxes are available from the drop-down menus. To select one of the drop-down menus, select the name of the drop-down menu. MicroStation displays the list of options available. Selecting from the list is a simple matter of moving the cursor down until the desired item is highlighted and then pressing the Data button on the pointing device. If a menu item has an arrow to the right, it has a cascading submenu. To display the submenu, just select the name of the submenu. Menu items that include ellipses (...) display dialog boxes. When a dialog box is displayed, no other action is allowed until that dialog box is dismissed or closed.

The **Tools** drop-down menu displays the list of available tool boxes in MicroStation. To select one of the available tool boxes, just select the name of the tool box and it will be displayed on the screen. Check marks are placed in the **Tools** menu to indicate open tool boxes. Choosing an item in the **Tools** drop-down menu toggles the state of the corresponding tool box. If the tool box is closed, it opens; if the tool box is open, it closes. You can place the tool box anywhere on the screen by dragging it with your pointing device. There is no limit to the number of the tool boxes that can be displayed on the screen.

To open multiple tool boxes at the same time, open the Tool Boxes dialog box from:

Drop-down menu	Tools > Tool Boxes

MicroStation displays the Tool Boxes dialog box, similar to Figure 1–16.

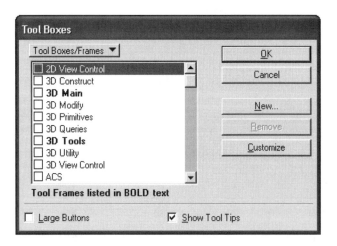

Figure 1–16 Tool Boxes dialog box

Select all the tool boxes to open by turning ON their check boxes, then click the **OK** button to close the dialog box. MicroStation displays the selected tool boxes.

If you are working with two monitors, you can drag the tool box to the second monitor. You can close the tool box by clicking on the "X" located in the top right corner of the tool box. When you open the tool box, it is displayed at the same location where it was previously open.

Tool Frames and Tool Boxes

Tool frames hold tool boxes of related tools and display the icon for the most recently invoked tool within each tool box. MicroStation provides six tool frames: Main, 3D Tools, Surface Modeling, Annotation, B-Spline Curves, and DD Design. The most important one is the Main tool frame, shown in Figure 1–17. The Main tool frame provides access to the majority of MicroStation's drawing tools, including measurement and dimensioning tools. Consider always leaving the Main tool frame open while drawing.

Figure 1-17 Main tool frame

Tool boxes consist of various tools. To access one of the tools from the tool box, click on the icon with the Data button, and the appropriate tool is invoked. To access one of the tools from a tool box in the tool frame, click, hold, and drag the pointer to the appropriate icon and release the pointer. The selected tool is invoked. If necessary, you can tear the tool box from the tool frame. To do so, select the appropriate tool box by pressing and holding the tool box, then drag it to anywhere on the screen and release it. Figure 1–18 shows the Linear Elements tool box taken out from the Main tool frame.

Resizing the Tool Box

By clicking and dragging on the edge of the tool box you can change the shape of the tool box's window. Figure 1–19 shows the resizing of the Linear Elements tool box to three different layouts. The tools contained within will move to fill the new window shape. The resized version of any given tool box will be remembered by MicroStation from session to session.

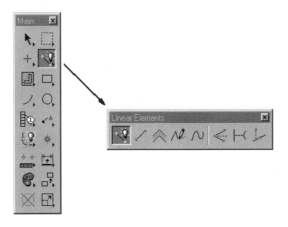

Figure 1-18 Linear Elements tool box taken out from the Main tool frame

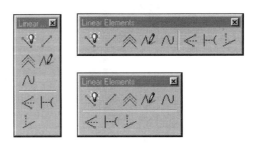

Figure 1-19 Linear Elements tool box resized to three different layouts

Docking a Tool Box

MicroStation allows you to dock a tool box to any edge of the MicroStation application window so it becomes part of the application window. Docking can happen on all four sides of the application window, and you can dock more than one tool box to each edge of the application window. Figure 1–20 shows tool boxes docked on four sides of the application window. When a tool box is docked to the top edge of the application window, it is most often referred to as a *tool bar*. For all practical purposes, a tool box and a tool bar are one and the same.

 Note: To override the docking feature, hold CTRL while dragging the tool box to the edge of the application window.

Chapter 1 Getting Started

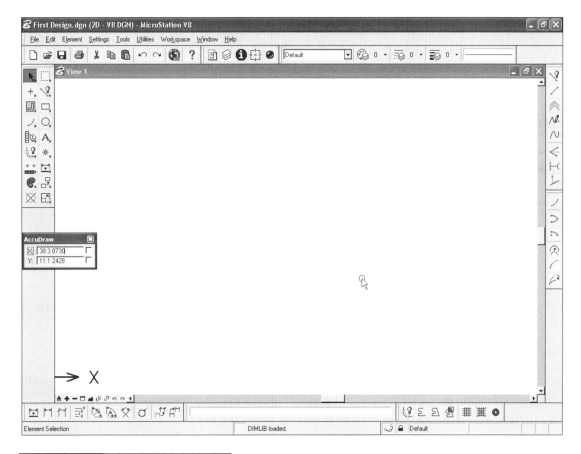

Figure 1-20 Docking of tool boxes

Undocking a Tool Box

To undock a tool box, grab part of the frame around the target tool box and drag it onto the main part of the MicroStation window. Once this is done, the tool box will return to its floating condition, or undragged.

Standard Tool Box (Also Called Standard Tool Bar)

By default, the Standard tool box (see Figure 1–21) is docked just below the drop-down menu. The Standard tool box contains icons that enable quick access to many commonly used **File** and **Edit** drop-down menu items. Table 1–3 lists the tools available in the Standard tool box and the corresponding items in the drop-down menus.

21

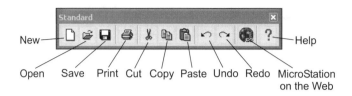

Figure 1-21 Standard tool box

Table 1-3 Tools Available in the Standard Tool Box

STANDARD TOOL BOX ITEM	DROP-DOWN MENU ITEM
New File	File > New
Open File	File > Open
Save Design	File > Save
Print	File > Print/Plot
Cut	Edit > Cut
Copy	Edit > Copy
Paste	Edit > Paste
Undo	Edit > Undo
Redo	Edit > Redo
MicroStation on the Web	————————
Help	Help > Contents

Attributes Tools Box

By default, the Attributes Tool box (see Figure 1-22) is docked along the side of the Standard tool box. The Attributes tool box contains icons that provide access to the more frequently accessed element symbology settings. Table 1-4 lists the tools available in the Attributes tool box.

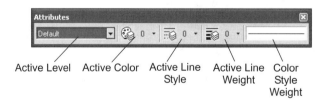

Figure 1-22 Attributes tool box

Table 1–4 Tools Available in the AttributesTool Box

ATTRIBUTES TOOL BOX ITEM	FUNCTION
Active Color	Sets the Active Color
Active Level	Sets the Active Level
Active Line Style	Sets the Active Line Style
Active Line Weight	Sets the Active Line Weight

Primary Tools Box

By default, the Primary Tools tool box (see Figure 1–23) is docked along the side of the Standard and Attributes tool boxes. The Primary Tools tool box contains icons that provide access to the more frequently used tools. Table 1–5 lists the tools available in the Primary Tools tool box.

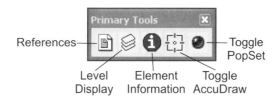

Figure 1–23 Primary Tools tool box

Table 1-5 Tools Available in the Primary Tool Box

PRIMARY TOOL BOX ITEM	FUNCTION
References	Opens the References settings box
Level Display	Opens the Level Display settings box
Element Information	Provides information about an element
Toggle AccuDraw	Opens and closes AccuDraw window
Toggle PopSet	Controls the display of Tool Settings window

Tool Tips

When you move the pointer over any icon, MicroStation displays a tool tip, as shown in Figure 1–24. In addition, MicroStation displays a brief description of the function of the tool in the Status bar. This is very helpful, especially when you are unfamiliar with which icon goes with which tool.

Figure 1-24 Invoking the Place Line tool (with the display of a tool tip) from the Linear Elements tool box

If necessary, you can disable the tool tips by turning off the option via the Help drop-down menu.

Tool Settings Window

Whenever you invoke a tool, MicroStation displays the controls required for adjusting the settings in the Tool Settings window. For example, if the Place Arc tool is selected, the **Method**, **Radius**, **Length**, **Start Angle**, and **Sweep Angle** options are displayed in the Tool Settings window, as shown in Figure 1-25. If the Tool Settings window is closed, it opens automatically when a tool with settings is selected.

Figure 1-25 Place Arc tool box and Tool Settings window

The PopSet toggle is used to automatically to prevent the display of Tool Settings window when you are done adjusting its controls. When it is enabled, the Tool Settings window will close automatically when you are done adjusting its controls. This is a great way to reclaim valuable screen "real estate" and reduce pointer movement.

Invoke the PopSet toggle from:

Primary Tools tool box	Select the Popset toggle (see Figure 1-26).
Key-in window	**popset on/off** ENTER

Figure 1-26 Invoking the PopSet toggle from Primary Tools tool box

By default, the PopSet is disabled.

Key-in Window

Key-ins are typed instructions entered into the Key-in window to control MicroStation. You can invoke any MicroStation tool by typing the name of the tool in full or in abbreviated form in the Key-in window. Open the Key-in window from:

| Tool-down menu | Utilities > Key-in |

MicroStation displays the Key-in window, similar to Figure 1–27.

As you type, the characters are matched to keywords in the list box below the Key-in window and are automatically selected in the list box. If it selects the right key-in command, press SPACE to complete the key-in, then click the Key-in button or press ENTER to enter the constructed key-in.

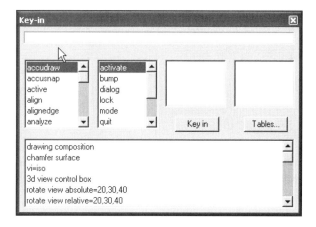

Figure 1–27 Key-in window

The list box in the Key-in window can also help you find and build key-ins. Scroll through the list of first words of key-ins in the left-most list box and select the keyword; it is then displayed in the key-in field. The subordinate, second-level keywords are shown in the key-in window's next list box. Select the desired keyword, and subsequently third-level keywords, if any, are shown. Select additional keywords, one per list box from left to right, until the desired key-in is constructed. To enter the constructed key-in, click the **Key-in** button or press ENTER.

MicroStation stores submitted key-ins in a buffer so you can recall them and, if necessary, edit them. Press UP ARROW or DOWN ARROW on your keyboard repeatedly until the desired key-in text appears in the key-in field of the Key-in window. Make any necessary changes, if any. In addition, MicroStation lists the submitted key-ins in the list box at the bottom of the Key-in window, and you can select the desired key-in from that list box.

Status Bar

The Status bar, located at the bottom of the MicroStation Application Window, as shown in Figure 1–28, displays a variety of useful information, including prompts, messages, and the name of the selected tool.

The Status bar is divided into two sections, as shown in Figure 1–28.

Figure 1-28 Status bar

Left-hand Section

The left-hand section of the Status bar shows the name of the selected tool, followed by either a "greater than" symbol (>) or a colon (:) and message text. The message text that follows ">" is the selected tool's prompts. For example, when you invoke the Place Line tool, MicroStation prompts on the left side of the Status bar as follows:

> Place Line > Enter first point

The command programs guide you step by step as you perform an operation with a tool. The text that follows the colon is a message that indicates a possible problem.

In addition, as you move the pointer on the tools in a tool box, the name of the selected tool and the associated message text are replaced with a description of the tool over which the pointer is located. This is intended as a form of online assistance.

Right-hand Section

The right-hand section of the Status bar consists of a series of fields. Following are the available fields, from left to right.

- The first field indicates the Snap Mode setting.
- The Locks icon in the second field allows you to open the Setting menu's Locks submenu. You can toggle the locks settings.
- The third field indicates the current level.
- The fourth field indicates the count of the selected elements. If this field is blank, no elements are selected.
- The fifth field indicates whether there is a fence in the design. If this field is blank, no fence is placed.
- The last field indicates whether changes to the active design file are unsaved. If the field is blank, there are no unsaved changes. If the field has a red icon with an "X" through it, the active design file is open for "read-only" access.

When you enter a tentative point or request quantitative information, the fields in this section to the right of the Snap Mode field are temporarily replaced with a single message field. To restore the fields, press the Reset button or click anywhere in the Status bar.

View Windows

MicroStation displays the elements you draw in the view windows. The portion of the design that is displaying in the view window is referred to as a *view*. With this part of the screen, all of the various commands you enter will construct your design. As you progress, you can Zoom In and Zoom Out to control the design's display. You can also move the scroll bar located both on the right side and at the bottom to pan the design. A view typically shows a portion of the design, but may show it in its entirety, as in Figure 1–29.

Eight view windows can be open (ON) at the same time, and all view windows are active. This lets you begin an operation in one view and complete it in another. You can move a view window by pressing and holding the cursor on the Title bar and dragging it to anywhere on the screen. You can resize the display window by clicking and dragging on its surrounding border, and shrink and expand the window by clicking the push buttons located at the top right corner of the window. To close the view window, click the "X" located at the top right corner of the view window.

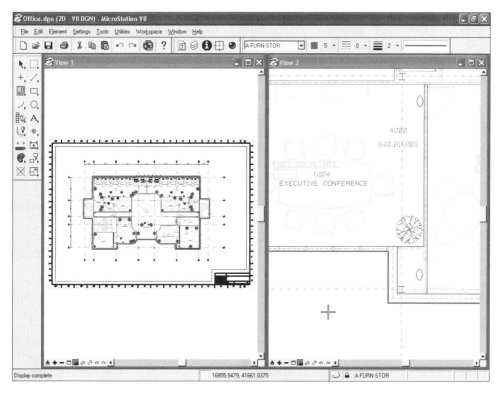

Figure 1-29 Two view windows displaying different portions of a design

MicroStation also provides a set of tools at the bottom left of the view window to control the view. View controls operate much like drawing tools; many even have "tool" settings. Detailed explanations of the View controls are provided in Chapter 3.

INPUT METHODS

Input method refers to the manner in which you tell MicroStation what tool to use and how to operate the tool. As mentioned earlier, the two most popular input devices are the mouse and the digitizing tablet, in addition to the keyboard.

Keyboard

To enter a command from the keyboard, simply key-in (type) the tool name in the Key-in window and click the **Key-in** button or press ENTER. *Key-in* is the name given to the function of providing information via the keyboard to MicroStation. MicroStation key-in language is much like plain English. For example, keying-in **PLACE LINE** selects that line command; **DELETE ELEMENT** selects the delete command, and so on. See Appendix B for a list of the key-in commands available in MicroStation.

Pointing Devices

As mentioned earlier, MicroStation supports a two-button mouse, a three-button mouse, and a digitizer. Depending on your needs and hardware, select one of the three pointing devices.

Detailed explanations of the functions that are programmed to the pointing device buttons (mouse and digitizing tablet puck) follow.

Data Button

This is the most-used button on the mouse/puck. The Data button is used to do the following:

> Select a tool from the drop-down menus and tool box.
> Define location points in the design plane.
> Identify elements that are to be manipulated.

In addition, it is used to accept tentative points, and generally tell the computer "yes" (accept) whenever it is prompted to do so. The Data button is also referred to as the *Identify button* or the *Accept button*.

Reset Button

The Reset button enables you to stop the current operation and reset MicroStation to the beginning of the current command sequence. For instance, when you are in the Place Line command, a series of lines can be drawn by using the Data button. When you are ready to stop the sequence, press the Reset button. MicroStation will stop the current operation and reset the Line sequence to the beginning. In addition, the Reset button also can reject a prompt, and generally tell the computer "no" (reject) whenever it is prompted to do so. The Reset button is also referred to as the *Reject button*.

Tentative Button

The tentative point is one of MicroStation's most powerful features. The Tentative button enables you to place a tentative (temporary) point on the screen. Once you are happy with the location of the point, accept it with the Data button. In other words, the tentative point lets you try a couple of places before actually selecting the final resting point for the data point. The Tentative button also can snap to elements at specific locations—for instance, the center and four quadrants of a circle, when the Snap lock is set to ON. For a detailed explanation, see Chapter 3.

Command Button

The Command button is only available on the digitizer tablet puck and lets you choose commands from the tablet menu. To do so, look down the tablet menu, place the puck crosshairs in the box that represents the command you want to activate, and then press the Command button. The corresponding command is invoked.

See Tables 1–1 and 1–2 for the specific functions that are programmed for the two- and the three-button mouse and the four-button digitizer puck.

Cursor Menu

MicroStation has a cursor menu that can be made to appear at the location of the cursor by pressing the designated button on your pointing device. Two cursor menus are available, one for View Control tools and another for Snap Mode options.

To invoke the View Control tools (Figure 1–30), press SHIFT plus the Reset button. The menu includes all the available tools for controlling the display of the design.

Figure 1–30 Cursor menu (SHIFT + Reset button)—View Control tools

To invoke the Snap Mode options (Figure 1–31), press SHIFT plus the Tentative button. The menu includes all the available Snap Mode options. The reason the Snap Mode options are in such ready access will become evident when you learn the significance of these functions.

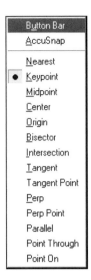

Figure 1-31 Cursor menu (SHIFT + Tentative button)—Snap Mode options

THE DESIGN PLANE

In conventional drafting, the drawing is normally done to a certain scale, such as 1' = 1"–0' or 1' = 1"–0'. But in MicroStation, you draw full scale: All lines, circles, and other elements are drawn and measured as full size. For example, if a part is 150 feet long, it is drawn as 150 feet actual size. When you are ready to plot the part, MicroStation scales the design to fit a given sheet size. Alternatively, you can specify a scale factor to plot on a given sheet size.

Whenever you start a new *two-dimensional* design, you get a design plane—the electronic equivalent of a sheet of paper on a drafting table. Unlike a sheet of drafting paper, however, the design plane (or cube in 3D) in a design file is extremely large, letting you draw your models at full scale. To draw various elements in your model, you enter data points. The center of the design plane is the global origin and is assigned coordinates (0,0). Any point to the right of the global origin has a positive *X* value; any point to the left has a negative *X* value. Any point above the global origin has a positive *Y* value; any point below has a negative *Y* value.

When you enter a data point, MicroStation saves its coordinates in IEEE64-bit floating point format. The 3D design cube is similar to the 2D design plane, but with a third axis Z (depth). Points in 2D models are stored as coordinate values expressed in the form (X,Y), while those for 3D models are stored as (X,Y, Z).

If necessary, you can change the location or coordinates of the global origin. For example, an architect might want all coordinates to be positive in value, so he or she would set the global origin at the bottom left corner of the design plane. To relocate the Global Origin, key-in **GO=0,0**

ENTER and click the Reset button. The Global Origin will be relocated to the lower left corner of the Design Plane and assigns the coordinate 0,0. To relocate the Global Origin to any location on the Design Plane, key-in **GO=0,0** ENTER and specify a data point where you want to relocate the Global Origin.

WORKING UNITS

Working units are the real-world units that you work in creating your models in a design file. Typically, the working units are defined in seed files, from which you create your design file. Normally, they will not require any adjustment.

Working Units are comprised of Master or Major units (MU) and Sub Units (SU). The Master Unit is the largest unit being used in the design, such as feet and meters. The fractional parts of a Master Unit are called Sub Units, such as inches or centimeters. Sub Units cannot be larger than Master Units. You can change your working units without affecting the size of elements in the design. For example, initially you had for feet and inches for Master units and Sub units respectively, and then you can change to Meters and Centimeters for Master units and Sub units respectively to get the Metric measurements. Existing elements and elements to be drawn will reflect the change in units. When you are inputting distances in design files, typically they are expressed in either of two forms:

- As a standard decimal number, such as 3.750
- As two numbers separated by a colon in terms of Master Units:Sub Units (MU:SU). For example, 3:4 means 3 master units and 4 sub units.

The following are the options available to key-in 3 1/8 feet when the Working Units are set as feet for Master Units and inches for Sub Units:

3.125	*(In terms of feet)*
3:1.50	*(In terms of feet and inches.)*
0:37.50	*(In terms of inches.)*
3:1 1/2	*(In terms of feet and inches in fractions.)*

Whenever you start a new design, you need not set the Working Units if you used the appropriate seed file to create the new design file. To draw an architectural floor plan, you copy 2DEnglishArch.DGN (architectural seed file); then MicroStation sets the Working Units to feet and inches for Master Units and Sub Units respectively.

If necessary, you can make changes to the current Working Units. To do so, open the Design File Settings dialog box from:

Drop-down menu	Settings > Design File

MicroStation displays the Design File Settings dialog box, similar to Figure 1–32.

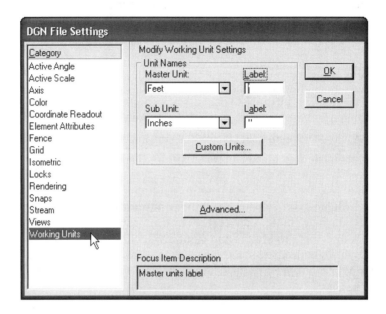

Figure 1-32 Design File Settings dialog box

Select **Working Units** in the **Category** list box. MicroStation displays the appropriate controls needed to modify the Working Units parameters (see Figure 1–32).

Select the **Master Units** from the **Master Unit** list box and appropriate label appears in the **Label** edit field. If necessary, you can change the label. Select the Sub Units from the **Sub Unit** list box and appropriate label appears in the **Label** edit field.

Custom Units

Custom Units allows you to define customized settings for Master Unit and Sub Unit. Click the **Custom Units** button in the Design File Settings dialog box to open Define Custom Units dialog box (see Figure 1–33).

Chapter 1 Getting Started

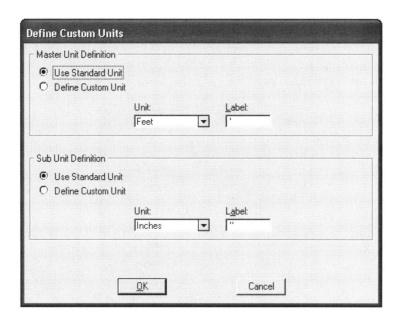

Figure 1-33 Define Custom Units dialog box

The **Master Unit Definition** section contains controls for selecting a Standard Unit or defining a Custom Unit to be your design Master Unit.

- **Use Standard Unit** provides option to select standard Master Unit settings recognized by MicroStation.
- **Define Custom Unit** creates a custom Master Unit, which is defined relative to one of the recognized Metric or English units.

The **Sub Unit Definition** section contains controls for selecting a Standard Unit or defining a Custom Units to be your design Sub Unit.

- **Use Standard Unit** provides option to select standard Sub unit settings recognized by MicroStation.
- **Define Custom Unit** create a custom Sub Unit, which is defined relative to one of the recognized Metric or English units.

After making necessary changes, click the **OK** button to close the Define Custom Units dialog box.

33

Advanced Units Settings

The Advanced Units Settings dialog box contains controls for setting the resolution of the design environment, which sets its size and accuracy. Click the **Advanced** button in the Design File Settings dialog box to open the Advanced Unit Settings dialog box (see Figure 1–34).

Figure 1-34 Advanced Unit Settings dialog box

The **Unit Type** option menu determines if the units in this design file are units of measurement or unitless representation.

- **Distance** selection set units are a unit of measurement.

- **Unitless** selection set units are other than a unit of linear measurement. For example, Longitude or Latitude.

The **Resolution** section determines the accuracy of the design plane/cube. The **Resolution** setting defines the worst case accuracy for the design environment, which occurs at the very outer limits of the (very large) working area/volume. For example, working to a "worst case" accuracy of 0.0001 meters, the size of the design plane/cube is 900 million kilometers along each axis. In almost all cases, therefore, there is no need to change the **Resolution** setting.

The **Working Areas (each axis)** section displays the length of each axis of the working environment (expressed in Miles or Kilometer when **Unit Type** is Distance) depending on the resolution. This area is recalculated, automatically, if the resolution is changed.

 Note: Do not change the **Resolution** when you have already placed elements in the design. Changing the **Resolution** alters the size of existing elements.

Once you have made necessary changes, click the **OK** button to save the changes and close the dialog box.

COORDINATE READOUT

Coordinate readout is the setting that controls the format in which MicroStation displays coordinates, distance, and angles in the status bar and dialog boxes.

If necessary, you can make changes to the Coordinate display format any time during design session. Setting the coordinate readout does not affect the accuracy of calculations, only the accuracy with which the results are displayed. To change the coordinate display format, open the Design File Settings dialog box from:

| Drop-down menu | Settings > Design File |

MicroStation displays the Design File Settings dialog box, similar to Figure 1–35.

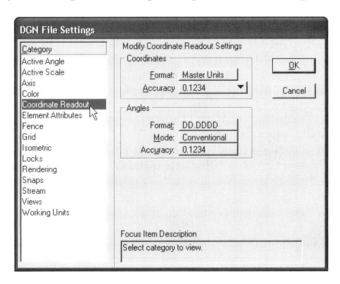

Figure 1-35 Design File Settings dialog box

Select **Coordinate Readout** in the **Category** list box. MicroStation displays the appropriate controls needed to modify the Coordinate readout settings (see Figure 1–35).

The **Coordinates** section contains controls that are used to set the manner in which coordinates are displayed.

- The **Format** option menu sets which units are displayed. Master Units selection displays only master units, Sub Units selection displays Master and Sub units in MU:SU format, and Working Units selection displays in Master, Sub and positional units in MU:SU:PU format.

- **Accuracy** sets decimal accuracy up to six decimal places for coordinates, 8 decimal places for angles or fractional accuracy to 1/2, 1/4, 1/8, 1/16, 1/32, or 1/64.

The **Angles** section contains controls that are used to set the format, direction, and accuracy of angle readout.

- **Format** sets the angle readout format.
- **Mode** sets the manner in which angles are measured.
- **Accuracy** sets decimal accuracy up to four decimal places.

Once you have made necessary changes, click the **OK** button to save the changes and close the dialog box.

CREATING MODELS

Each model in a design file is a set of design elements (such as lines, circles, etc.) that are unique to that model. A model has its own set of eight views and serves as a container for the geometry forming the model. Each model in a design is totally independent; in fact they may even have different working units. There is no limit to the number of models you can have in a design file. For example, one Design model may be used for a house plan and another for its site details. One may have feet and inches for working units, and other may have decimal feet.

MicroStation allows you to create two types of model: Design and Sheet. As mentioned earlier, Design models contain the actual designs, where as Sheet models are used to compose drawings. Sheet models are used to arrange the design elements from the Design models by referencing (refer to Chapter on References for detailed explanation), annotate and dimension them as required, add borders, title blocks etc. Printouts will normally be created from Sheet models. Only one model can be active at any time.

The Models settings box is used to create, manage and switch between models in the open DGN file. Open the Models settings box from:

| Drop-down menu | File > Models |

MicroStation displays the Model settings box, similar to Figure 1–36.

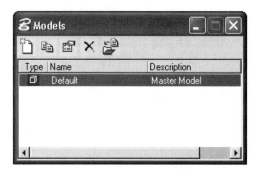

Figure 1-36 Models settings box

You can create a model as a 2D or 3D Design type, or as a Sheet. Icons at the top of the Models settings box give you access to its various functions. To switch to an active model or sheet, double-click on the name in the list box. The selected name becomes the active model and name of the active Model or Sheet is displayed as part of the View Window name.

Create a New Model/Sheet

To create a new model or sheet, click the Create a New Model icon. MicroStation opens a Create Model dialog box similar to Figure 1–37.

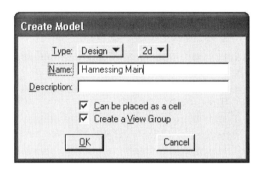

Figure 1-37 Create Model dialog box

Type option menu sets the type of model created.

- **Design** selection allows you to create a design model. When Design is chosen, a further option menu to the right lets you choose between 2D and 3D for the model.

- **Sheet** selection allows you to create a sheet model. This lets you attach references to create a drawing.

Name edit field allows you to enter a name for the model.

Description edit field allows you add a description of the model.

Can be placed as a cell check box sets whether the model can be placed as a cell in the design file.

Create a View Group check box sets whether a View Group is created with the model.

Click the **OK** button to create a new model or sheet, as the case may be, and close the dialog box.

Copying an Existing Model

To copy an existing model in the open DGN file, click the Copy a Model icon. MicroStation opens Copy Model dialog box similar to Figure 1–38.

Figure 1-38 Copy Model dialog box

Model to Copy option menu lets you choose a model contained in the open DGN file (or the selected DGN file if importing a model).

Name edit field allows you to enter a name for the copy of the model.

Description edit field allows you to enter a description of the copied model.

Click the **OK** button to create a copy of the selected model and close the dialog box.

Edit Model Properties

To edit properties of a selected model contained in the open DGN file, click the Edit Model Properties icon. MicroStation opens Model Properties dialog box and allows you to change the **Name** and **Description**. After making necessary changes, click the **OK** button to accept the changes and close the dialog box.

Delete a Model

To delete a selected model, click the Delete Model icon. MicroStation deletes the selected model.

Import a Model

MicroStation allows you to import a model into the open DGN file from another DGN or DWG file. To import a model, click the Import a Model icon. MicroStation opens the Import Model From File dialog box, which is similar to the Open dialog box. This lets you select a DGN or DWG file from which to select the model to import. After you click the **OK** button, the Copy Model dialog box opens, from which you can select a model to copy into the open DGN file.

DRAWING PROPERTIES

The Properties dialog box is used to review the active DGN file's general properties and usage statistics and change the file's design properties. Open the Properties dialog box from:

| Drop-down menu | File > Properties |

Chapter 1 Getting Started

MicroStation displays Properties dialog box similar to Figure 1-39.

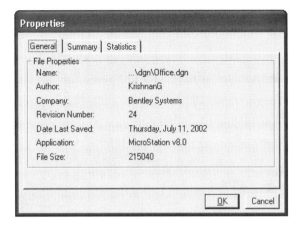

Figure 1-39 Properties dialog box

The **General** tab of the dialog box displays general information (see Figure 1-39) about the DGN file; such as name of the design file with full path, name of the person who last edited the DGN file, company of the person who last edited the DGN file, number of times the DGN file has been edited and saved, Date the last time the DGN file was edited and saved, software version on which the design was created, and size of the DGN file.

The **Summary** tab (see Figure 1-40) of the dialog box contains controls for editing a file's properties. You can store information such as the project to which the file is related, explanation of the drawing or the client to which this drawing is related, query keyword to find the file if it is in a database, comments associated with the file and the name of the manager associated with the file.

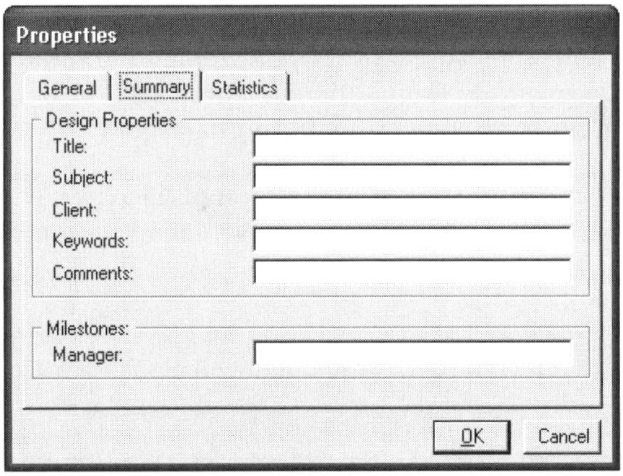

Figure 1-40 Properties dialog box (Summary tab)

The **Statistics** tab (see Figure 1–41) contains statistical information about the DGN file. Information includes total amount of time the DGN file has been edited, number of levels in the DGN file, number of levels used in the DGN file, number of models in the DGN file, number of references attached to the DGN file, and number of elements in the DGN file.

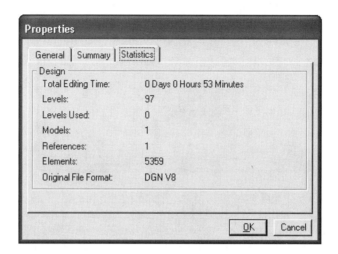

Figure 1–41 Properties dialog box (Statistics tab)

Click the **OK** button to accept the changes and close the Properties dialog box.

SAVING CHANGES AND EXITING THE DESIGN FILE

Before we discuss how to place elements in the design file, let's discuss how to save the current design file. By default, MicroStation saves all elements in your design file as you draw them. There is no separate Save command. You can get out of your design file without doing the proper exit procedure and still not lose any work. Even if there is a power failure during a design session, you will get most of the design file back without significant damage.

If necessary, you can set the Immediately **Save Design Changes** check box to OFF. Then MicroStation provides a Save tool in the **File** drop-down menu to save the design file. No other automatic save feature is provided. Whenever you want to save the design file, you have to invoke the Save tool from the **File** drop-down menu.

To change the status of the check box for **Immediately Save Design Changes**, open the Preferences dialog box from:

| Drop-down menu | Workspace > Preferences |

MicroStation displays the Preferences dialog box, similar to Figure 1–42.

Select the **Operation** option from the **Category** list box and set the check box to ON/OFF for **Immediately Save Design Changes**.

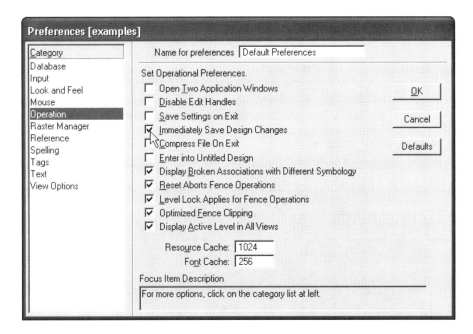

Figure 1-42 Preferences dialog box

To save the current design file to a different file name, invoke the Save As command from:

| Drop-down menu | File > Save As |

MicroStation displays the Save As dialog box. Select the folder where you want to save the design file, and key-in the name of the design file in the Files edit box. Click the **OK** button to save the file.

Saving the Design File Settings

To save the design file settings such as working units, grid spacing, or view settings between sessions, you must explicitly save the settings. To do so, invoke the Save Settings tool from:

| Drop-down menu | File > Save Settings |
| Key-in window | **file design** (or **fi**) ENTER |

MicroStation saves the current settings.

If you forget to save the settings, you will have to spend time adjusting the design and view settings to match what you had in place the last time you worked on the design.

To automatically save the settings on exiting the design file, set the check box for **Save Settings on Exit** to ON in the Preferences dialog box. By default it is set to OFF.

To change the status of the check box for **Save Settings on Exit**, open the Preferences dialog box from:

| Pull-down menu | Workspace > Preferences |

MicroStation displays the Preferences dialog box. Select the **Operation** option from the **Category** list box, and set the check box to ON/OFF for **Save Settings on Exit**.

Exiting the MicroStation Program

To exit the MicroStation program and return to the operating system, invoke the Exit tool from:

Drop-down menu	File > Exit
Key-in window	**exit** (or **exi**) ENTER

MicroStation exits the program and returns to the operating system.

If, instead, you prefer to return to the MicroStation Manager dialog box, invoke the Close tool from:

Drop-down menu	File > Close
Key-in window	**close design** (or **clo d**) ENTER

MicroStation returns to the MicroStation Manager dialog box.

GETTING HELP

When you are in a design file, MicroStation provides an online help facility available from the Help drop-down menu. Online help is provided through the Help window, by specific topics, by searching for a text string within help topic names or help articles, or by browsing key-ins.

Contents

The Contents window lists the top-level topics, as shown in Figure 1–43. To see a list of more specific subtopics related to a topic in the list, select the topic. MicroStation displays a list of subtopics, and by selecting a subtopic, MicroStation displays the available help information on that topic. In addition, you can also search for a text string by clicking the **List Topics** button and MicroStation will display all the related topics (this option available in the **Search** tab section of the Help window).

Chapter 1 Getting Started

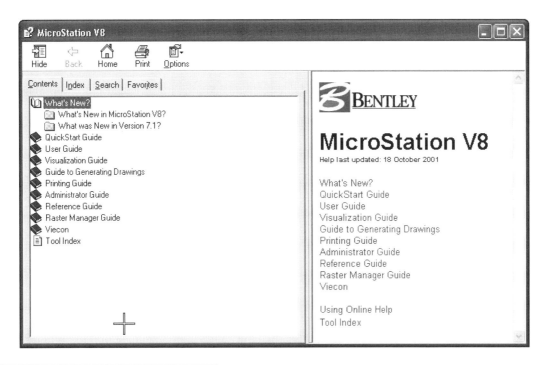

Figure 1-43 Help Contents window

You can display the Help file's previous article by clicking the **Back** button.

Tool Index

Tool Index from the Help menu opens the Tool Index window in the online help system as shown in Figure 1-44. You can either search for a tool with the **Find** form or select a tool from the alphabetical list. When the desired tool is selected, the associated tool documentation is displayed in the Help window.

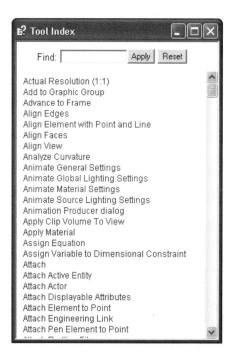

Figure 1-44 Tool Index window

MicroStation on Web

Choosing MicroStation on Web from the Help menu opens the default browser and connects to the MicroStation resource website. MicroStation on the Web offers tools, resources and information for the MicroStation user that you can use in your daily workflows, right from within your MicroStation session. It's your window to the online world of MicroStation.

ENHANCEMENTS IN MICROSTATION V8

MicroStation V8 represents a major step forward in the evolution of the entire MicroStation product line. For the first time in MicroStation's history, the fundamental file format is substantially redesigned and upgraded. The new format permits MicroStation and applications to add more intelligence to the models that users create, enabling information to be created and manipulated as components. The expanded data eliminate a number of constraints that were previously imposed on the product, and improve the reliability and robustness of MicroStation.

The features of MicroStation V8 represent the beginning of the improvements made possible by the upgraded data storage; future releases of MicroStation will take further advantage of this improved flexibility.

Following are some of the the features/benefits that were added to MicroStation V8:

- **Seamless access to compatible file formats** – New file format, which is a superset of DGN and DWG, allowing users to share data in hybrid workflows without translating files. **Models** – You can have more than one model in a DGN file. Every model has its own set of eight views. Using the Models dialog box, you can create and switch quickly between models in a DGN file.

- **Sophisticated level system** – Provides access to an unlimited number of levels that allow companies to adopt generous de facto industry-level standard level descriptions. With centralized control of level standards, users can add, delete, filter and sort levels on any attribute.

- **AccuSnap** – Provides graphical assistance for snapping to elements. It can be used by itself, or in combination with AccuDraw.

- **Location elements** – Reports visually whether an element can be selected to manipulate by the selected tool.

- **Text Enhancements** – Allows for more efficient handling of text by using True Type and AutoCAD shx fonts natively, allowing the definition of text styles, supporting masking, overline, italics, and bold, and incorporating a Quick Text editor mode.

- **Dimensioning enhancements** – Setting up alternative dimensions based on specific criteria, storing dimension styles in either the DGN file or in an external library.

- **Cells and Cell Library enhancements** – Cell names can now be up to 512 characters in length, and the size of a cell definition is only limited by the maximum size of the design file itself. You can store in a library both 2D and 3D cells. In addition, you can place a cell by True Scale option (places the cell based on the relationship between the cells's working unit's definition and that of the recipient design file).

- **References Enhancements** – Attach an unlimited number of references to the active model; in addition you can attach models within the same DGN file and as well DWG files as reference to the active model.

- **Printing enhancements** – Print dialog box is redesigned and incorporates new features related to preview and print size/scale selection.

- **Design History** – Changes made to individual drawings can be tracked as part of your default design environment. Users can record, review, and restore revisions made to a DGN file or milestones in the development of a design. This allows users to document changes, visually compare revisions, and restore discarded ideas. VBA support - Users can develop both daily automation tools and sophisticated applications, without being a professional programmer. VBA's strong inter-process capabilities allow for easier programming, better integration with other Windows applications, and faster prototyping for applications.

chapter 2

Fundamentals I

Objectives

After completing this chapter, you will be able to:

- Draw lines, blocks, shapes, circles, and arcs
- Drop blocks and shapes and delete elements
- Use Precision Input

PLACEMENT COMMANDS

MicroStation provides various tools for drawing objects. This section explains in detail the various tools for drawing lines, blocks, shapes, circles, and arcs.

Place Line

The primary drawing element is the line, and the Place Line tool enables you to draw series of lines. Invoke the Place Line tool from:

Linear Elements tool box	Select the Place Line tool (see Figure 2–1).
Key-in window	**place line** (or **pl l**) ENTER

Figure 2-1 Invoking the Place Line tool from the Linear Elements tool box

MicroStation prompts:

> Place Line > Enter first point

Specify the first point by providing a data point via your pointing device (mouse or puck) or by Precision Input coordinates (see the discussion, later in this chapter, on "Precision Input" for a more detailed explanation). After you specify the first point, MicroStation prompts:

> Place Line > Enter end point

Specify the end of the line by placing a data point via your pointing device (mouse or puck) or by Precision Input. MicroStation repeats the prompt:

> Place Line > Enter end point

Place a data point via the pointing device or by Precision Input to continue. To save time, the Place Line tool remains active and prompts for a new endpoint after each point you specify. When you have finished placing a series of lines, press the Reset button or invoke another tool to terminate the Place Line tool.

When placing data points with your pointing device to draw a series of lines, a rubber-band line is displayed between the starting point and the crosshairs. This helps you to see where the resulting line will go. In Figure 2–2 the dotted lines represent previous cursor positions. To specify the endpoint of the line, click the Data button. You can continue to place lines with the Place Line tool until you press the Reset button or select another tool.

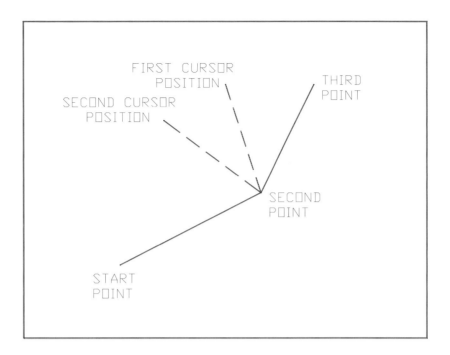

Figure 2-2 Placing data points with the cursor rather than with coordinates

Place Line to a Specified Length

To place a line to a specified length, select the Place Line tool in the Linear Elements tool box and set the check box for **Length** to ON in the Tool Settings window. Type the distance, in MU:SU:PU format, in the **Length** edit field. The prompts are similar to those for the Place Line tool, and you can place any number of line segments of specified length.

Place Line at an Angle

To place a line to a specified angle, select the Place Line tool in the Linear Elements tool box and set the check box for **Angle** to ON in the Tool Settings window. Key-in the angle in the **Angle** edit field. The prompts are similar to those for the Place Line tool, and you can place any number of line segments of a specified angle.

If necessary, you can turn both of the check boxes ON for the **Length** and **Angle** edit fields; MicroStation allows you to place a line with a specific length and constrained angle.

Place Block

MicroStation allows you to draw a rectangular block by two different methods: orthogonal and rotated.

Orthogonal Method

The Place Block tool (orthogonal) allows you to place a rectangular block by selecting two points that define the diagonal corners of the shape. Place the two diagonal corners by specifying data points via your pointing device or by keying-in *two-dimensional (2D)* coordinates (see the discussion later in this chapter on "Precision Input").

Invoke the Place Block (orthogonal) tool from:

Polygons tool box	Select the Place Block tool and Orthogonal from the Method option menu located in the Tool Settings window (see Figure 2–3).
Key-in window	**place block orthogonal** (or **pl b o**) ENTER

Figure 2–3 Invoking the Place Block (orthogonal) tool from the Polygons tool box

MicroStation prompts:

> Place Block > Enter first point *(Place a data point or key-in coordinates to define the start point of the block.)*
>
> Place Block > Enter opposite corner *(Place a data point or key-in coordinates to define the opposite corner of the block.)*

A block is a single element, and element manipulation tools such as Move, Copy, and Delete manipulate a block as one element. If necessary, you can make a block into individual line elements with the Drop Line String tool.

For example, the following command sequence shows placement of a block by placing two data points diagonally opposite to each other, as shown in Figure 2–4, using the Place Block (orthogonal) tool.

> Place Block > Enter first point *(Place a data point as shown in Figure 2–4.)*
>
> Place Block > Enter opposite corner *(Place a data point diagonally opposite to the first point.)*

Chapter 2 Fundamentals I

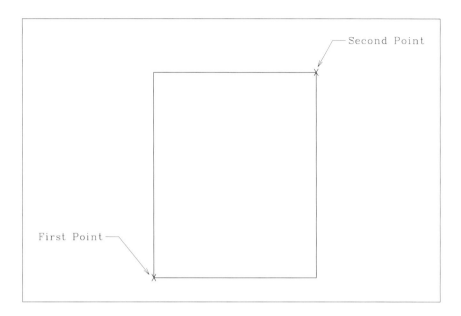

Figure 2-4 An example of placing an orthogonal block using the Place Block (orthogonal) tool

Rotated Method

The Place Block (rotated) tool allows you to place a rectangular block at any angle that is defined by the first two data points. The first data point defines the first corner of the block and the point that the block rotates around. The second data point defines the angle of the block, and the third data point, entered diagonally from the first, defines the opposite corner of the block.

Invoke the Place Block (rotated) tool:

Polygons tool box	Select the Place Block tool and Rotated from the Method option menu located in the Tool Settings window (see Figure 2–5).
Key-in window	**place block rotated** (or **pl b r**) ENTER

Figure 2-5 Invoking the Place Block (rotated) tool from the Polygons tool box

MicroStation prompts:

> Place Rotated Block > Enter first base point *(Place a data point or key-in coordinates to define the start point of the block.)*
>
> Place Rotated Block > Enter second base point *(Place a data point or key-in coordinates to define the angle of the block.)*
>
> Place Rotated Block > Enter diagonal point *(Place a data point or key-in coordinates to define the opposite corner of the block.)*

See Figure 2–6 for an example of placing a rotated block with the Place Block (rotated) tool by providing three data points.

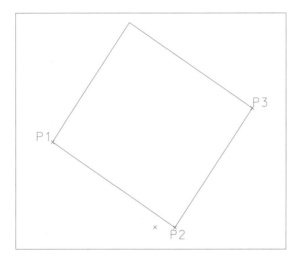

Figure 2-6 An example of placing a rotated block with the Place Block (rotated) tool

Similar to an orthogonal block, a rotated block is also a single element.

Place Shape

The Place Shape tool allows you to place a multisided shape defined by a series of data points (3 to 100) that indicates the vertices of the polygon. To complete the polygon shape, the last data point should be placed on top of the starting point. You can specify the starting point and subsequent points via Precision Input or by using your pointing device.

Invoke the Place Shape tool from:

Polygons tool box	Select the Place Shape tool (see Figure 2–7).
Key-in window	**place shape** (or **pl sh**) ENTER

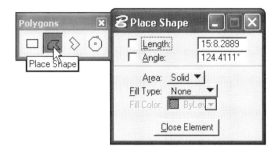

Figure 2-7 Invoking the Place Shape tool from the Polygons tool box

MicroStation prompts:

> Place Shape > Enter first point *(Place a data point or key-in coordinates to define the starting point of the shape.)*
>
> Place Shape > Enter vertex or Reset to cancel *(Place a data point or key-in coordinates to define the vertex, or press Reset button to cancel.)*

Continue placing data points. To complete the polygon shape, place the last data point on top of the starting point or click the **Close Element** button located in the Tool Settings window.

You can also draw a shape by constraining to Length and/or Angle by turning on the check boxes for **Length** and for **Angle**, appropriately located in the Tool Settings window.

See Figure 2–8 for an example of placing a closed shape with the Place Shape tool by providing six data points.

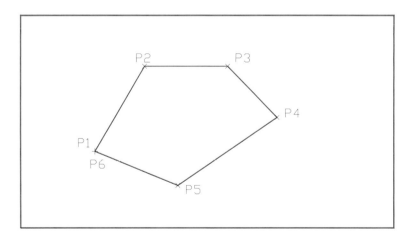

Figure 2-8 An example of placing a closed shape with the Place Shape tool

Similar to a block, a shape is also a single element. Element manipulation tools such as Move, Copy, and Delete manipulate the shape as one element. If necessary, you can make the shape into individual line elements with the Drop Line String tool.

Note: Area and Fill type options are explained in the Patterning section of Chapter 12

Place Orthogonal Shape

The Place Orthogonal Shape tool allows you to create a multisided shape that has adjacent sides at right angles. As with the Place Block (rotated) tool, the first two points define the vertices of the orthogonal shape. The additional points define the corners of the shape. To complete the polygon shape, the last data point should be placed on top of the starting point. You can specify the starting point and subsequent points with absolute or relative coordinates (see "Precision Input," later) or by using your pointing device.

Invoke the Place Orthogonal Shape tool from:

Polygons tool box	Select the Place Orthogonal Shape tool (see Figure 2–9).
Key-in window	**place shape orthogonal** (or **pl sh o**) ENTER

Figure 2-9 Invoking the Place Orthogonal Shape tool from the Polygons tool box

MicroStation prompts:

> Place Orthogonal Shape > Enter shape vertex *(Place a data point or key-in coordinates to define the start point of the shape.)*
>
> Place Orthogonal Shape > Enter shape vertex *(Place a data point or key-in coordinates to define the vertex.)*

MicroStation prompts for additional shape vertices. Continue placing data points. To complete the polygon shape, the last data point should be placed on top of the starting point.

Similar to a block, an orthogonal shape is also a single element. Element manipulation tools such as Move, Copy, and Delete manipulate the orthogonal shape as one element. If necessary, you can make the shape into individual line elements with the Drop Line String tool.

See Figure 2–10 for an example of placing an orthogonal shape with the Place Orthogonal Shape tool by providing nine data points.

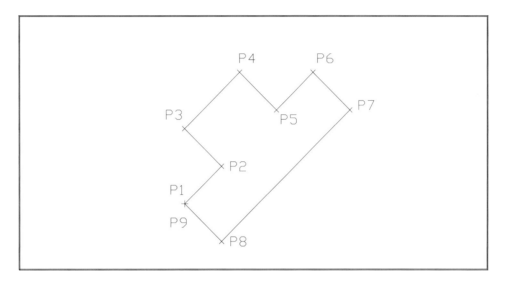

Figure 2–10 An example of placing an orthogonal shape with the Place Orthogonal Shape tool

 Note: Area and Fill type options are explained in the Patterning section of Chapter 12.

Place Circle

MicroStation offers several methods for drawing circles. These include Place Circle By Center, Place Circle By Edge, and Place Circle By Diameter. The appropriate method is selected from the **Method** option menu located in the Tool Settings window.

Place Circle By Center

With the Place Circle By Center tool, you can draw a circle by defining two points: the center point and a point on the circle.

Invoke the Place Circle By Center tool from:

Ellipses tool box	Select the Place Circle tool and Center from the Method option menu located in the Tool Settings window (see Figure 2–11).
Key-in window	**place circle center** (or **pl ci c**) ENTER

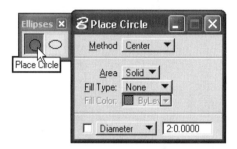

Figure 2-11 Invoking the Place Circle By Center tool from the Ellipses tool box

MicroStation prompts:

Place Circle By Center > Identify Center Point *(Place a data point or key-in coordinates to define the center of the circle.)*

Place Circle By Center > Identify Point on Circle *(Place a data point or key-in coordinates to define the edge of the circle.)*

 Note: After you place the first data point, a dynamic image of the circle drags with the screen pointer.

To save time, the Place Circle By Center tool remains active and prompts for a new center point. When you are finished placing circles, invoke another tool to terminate the Place Circle By Center tool.

For example, the following command sequence shows placement of a circle with the Place Circle By Center tool (see Figure 2–12).

Place Circle By Center > Identify Center Point *(place a data point to define the center point as shown in Figure 2-12)*

Place Circle By Center > Identify Point on Circle *(place a data point to draw a circle)*

In the last example, MicroStation used the distance between the center point and the point given on the circle for the radius of the circle.

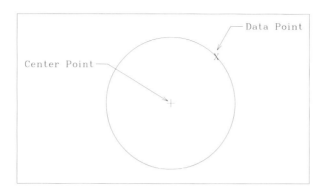

Figure 2-12 An example of placing a circle with the Place Circle By Center tool

You can also place a circle by its center by keying-in its diameter or radius. Select **Diameter** or **Radius** from the options menu located in the Tool Settings window, turn on the check box, key-in the value in working units, then press ENTER or TAB. When the desired diameter/radius is entered, a circle of that diameter/radius appears on the screen cursor. You will then be asked to identify the center point of the circle. Position your cursor where you want the center of the circle to be, and place a data point. Continue placing circles with this same diameter/radius, or press the Reset button to allow you to change the diameter/radius of the circle. If you do not wish to continue placing circles, invoke another tool to terminate the Place Circle By Center tool.

Place Circle By Edge

The Place Circle By Edge tool enables you to draw a circle by defining three data points on the circle.

Invoke the Place Circle By Edge tool from:

Ellipses tool box	Select the Place Circle tool and Edge from the Method option menu located in the Tool Settings window (see Figure 2–13).
Key-in window	**place circle edge** (or **pl ci e**) ENTER

Figure 2-13 Invoking the Place Circle By Edge tool from the Ellipses tool box

MicroStation prompts:

> Place Circle By Edge > Identify Point on Circle *(Place a data point or key-in coordinates to define the first edge point of the circle.)*
>
> Place Circle By Edge > Identify Point on Circle *(Place a data point or key-in coordinates to define the second edge point of the circle.)*
>
> Place Circle By Edge > Identify Point on Circle *(Place a data point or key-in coordinates to define the third edge point of the circle.)*

For example, the following command sequence shows placement of a circle via the Place Circle By Edge tool (see Figure 2–14).

> Place Circle By Edge > Identify Point on Circle *(Place a data point to define the first point of the circle as shown in Figure 2–14.)*
>
> Place Circle By Edge > Identify Point on Circle *(Place a data point to define the second point of the circle.)*
>
> Place Circle By Edge > Identify Point on Circle *(Place a data point to define the third point of the circle.)*

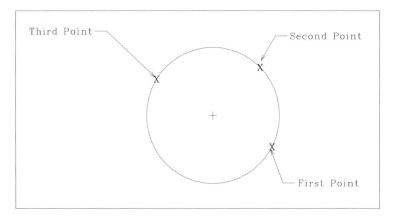

Figure 2-14 An example of placing a circle with the Place Circle By Edge tool

You can also place a circle by its edge by keying-in its diameter or radius. Select **Diameter** or **Radius** from the options menu located in the Tool Settings window, turn on the check box, key-in the value, then press ENTER or TAB. When the desired diameter/radius is specified, MicroStation prompts for two data points, instead of three, to place a circle by edge.

Place Circle By Diameter

With the Place Circle By Diameter tool, you can draw a circle by defining two data points: two endpoints of the diameter.

Invoke the Place Circle By Diameter tool from:

Ellipses tool box	Select the Place Circle tool and Diameter from the Method option menu located in the Tool Settings window (see Figure 2–15).
Key-in window	**place circle diameter** (or **pl ci d**) ENTER

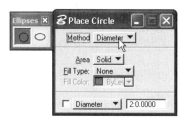

Figure 2-15 Invoking the Place Circle By Diameter tool from the Ellipses tool box

MicroStation prompts:

> Place Circle By Diameter > Enter First Point on Diameter *(Place a data point or key-in coordinates to define the first endpoint of one of its diameters.)*
>
> Place Circle By Diameter > Enter Second Point on Diameter *(Place a data point or key-in coordinates to define the second endpoint of one of its diameters.)*

For example, the following command sequence shows placement of a circle via the Place Circle By Diameter tool (see Figure 2–16).

> Place Circle By Diameter > Enter First Point on Diameter *(Place a data point to define the first point to draw a circle as shown in Figure 2–16.)*
>
> Place Circle By Diameter > Enter Second Point on Diameter *(Place a data point to define the second point to draw a circle.)*

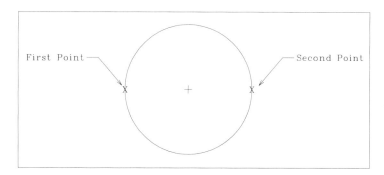

Figure 2-16 An example of placing a circle with the Place Circle By Diameter tool

Place Arc

Similar to placing circles, MicroStation offers two different methods for placing arcs: Place Arc By Center and Place Arc By Edge. Arcs can be placed clockwise or counterclockwise. The appropriate method is selected from the **Method** option menu located in the Tool Settings window.

Place Arc By Center

The Place Arc By Center tool enables you to draw an arc defined by three points: the center point, the first arc endpoint, and the second arc endpoint.

Invoke the Place Arc By Center tool from:

Arcs tool box	Select the Place Arc tool and Center from the Method option menu located in the Tool Settings window (see Figure 2–17).
Key-in window	**place arc center** (or **pl a c**) ENTER

Figure 2-17 Invoking the Place Arc By Center tool from the Arcs tool box

MicroStation prompts:

> Place Arc By Center > Identify First Arc Endpoint *(Place a data point or key-in coordinates to define the first arc endpoint.)*
>
> Place Arc By Center > Identify Arc Center *(Place a data point or key-in coordinates to define the arc center.)*
>
> Place Arc By Center > Enter point to define sweep angle *(Place a data point or key-in coordinates to define the sweep angle.)*

Note: After you place the first data point, a dynamic image of the arc drags with the screen pointer. After place the first and center points, you can place the second arc end point clockwise or counterclockwise.

For example, the following command sequence shows placement of an arc with the Place Arc By Center tool (see Figure 2–18).

> Place Arc By Center > Identify First Arc Endpoint *(Place a data point to define the first arc endpoint as shown in Figure 2–18.)*
>
> Place Arc By Center > Identify Arc Center *(Place a data point to define the arc center.)*
>
> Place Arc By Center > Enter point to define sweep angle *(Place a data point to define the arc endpoint.)*

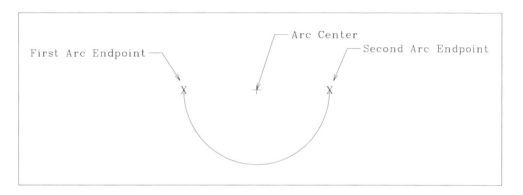

Figure 2-18 An example of placing an arc by the Place Arc By Center tool

You can also draw an arc by its center by keying-in its radius. To do so, turn on the check box for **Radius**, located in the Tool Settings window; key-in the appropriate value in the **Radius** edit field; and press ENTER or TAB. The prompts are similar to those for the Place Arc By Center tool, except the First Arc Endpoint and Second Arc Endpoint define the starting and ending directions of the arc.

You can also draw an arc by its center by keying-in its length. To do so, turn on the check box for **Length**, located in the Tool Settings window; key-in the appropriate value in the **Length** edit field; and press ENTER or TAB. MicroStation prompts for the first arc endpoint and center point of arc and draws an arc to the specified arc length.

Similarly, you can also constrain the **Start Angle** and **Sweep Angle** by keying-in appropriate angles in the respective edit fields. The MicroStation prompts depend on the number of constraints turned ON. For example, if **Radius** and **Start Angle** are preset, MicroStation prompts for the center point of the arc and the sweep angle; if **Radius**, **Start Angle**, and **Sweep Angle** are preset, MicroStation prompts only for the center of the arc.

Place Arc By Edge

The Place Arc By Edge tool allows you to draw an arc defined by three points on the arc.

Invoke the Place Arc By Edge tool from:

Arcs tool box	Select the Place Arc tool and Edge from the Method option menu located in the Tool Settings window (see Figure 2–19).
Key-in window	**place arc edge** (or **pl a e**) ENTER

Figure 2-19 Invoking the Place Arc By Edge tool from the Arcs tool box

MicroStation prompts:

> Place Arc By Edge > Identify First Arc Endpoint *(Place a data point or key-in coordinates to define the first arc endpoint.)*
>
> Place Arc By Edge > Identify Point on Arc Radius *(Place a data point or key-in coordinates to define a point on the arc radius.)*
>
> Place Arc By Edge > Identify Second Arc Endpoint *(Place a data point or key-in coordinates to define the second arc endpoint.)*

For example, the following command sequence shows the placement of an arc with the Place Arc By Edge tool (see Figure 2–20).

> Place Arc By Edge > Identify First Arc Endpoint *(Place a data point to define the first arc endpoint as shown in Figure 2–20.)*
>
> Place Arc By Edge > Identify Point on Arc Radius *(Place a data point to define the arc radius.)*
>
> Place Arc By Edge > Identify Second Arc Endpoint *(Place a data point to define the second arc endpoint.)*

You can also draw an arc by its edge by keying-in the radius. To do so, turn on the check box for **Radius**, located in the Tool Settings window; key-in the appropriate value in MU:SU:PU format in the **Radius** edit field; and press ENTER or TAB. The prompts are similar to those for the Place Arc By Edge tool, except the First Arc Endpoint and Second Arc Endpoint define the starting and ending directions of the arc.

You can also draw an arc by its edge by keying-in its length. To do so, turn on the check box for **Length**, located in the Tool Settings window; key-in the appropriate value in the **Length** edit field; and press ENTER or TAB. MicroStation prompts for the first arc endpoint and second arc endpoint and draws an arc to the specified arc length.

Similarly, you can also constrain the **Start Angle** and **Sweep Angle** by keying in appropriate angles in the respective edit fields. The MicroStation prompts depend on the number of constraints turned ON. For example, if **Radius** and **Start Angle** are preset, MicroStation prompts for the First Arc Endpoint and Second Arc Endpoint. If **Radius**, **Start Angle**, and **Sweep Angle** are preset, MicroStation prompts only for the First Arc Endpoint.

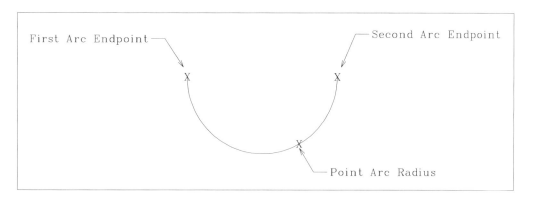

Figure 2-20 An example of placing an arc by the Place Arc By Edge tool

MANIPULATING ELEMENTS

MicroStation not only allows you to draw easily, it also allows you to manipulate the elements you have drawn.

Whenever you attempt to identify an element for manipulation that is not allowed by the selected tool (e.g., it is the wrong type for the tool, it is locked, or in a reference, etc.), MicroStation reports the reason the element cannot be selected.

MicroStation automatically locates elements as you point to them, so that you can tell if an element is acceptable without having to press the Data button. This helps you determine a tool's intent and, in some cases, saves you from entering a data point to accept the identified element.

This feature is enabled by default. To disable it, open the AccuSnap Settings box from:

| Drop-down menu | Settings > Snap > AccuSnap |

MicroStation opens the AccuSnap Settings box, as shown in Figure 2–21.

Figure 2-21 AccuSnap Settings box (General)

Set **Identify Elements Automatically** check box to OFF to disable the locating elements feature. By default, it is set to ON.

DELETE ELEMENT

Of the many manipulation tools available, the Delete Element tool is probably the one you will use most often. Everyone makes mistakes, but MicroStation makes it easy to delete them.

Invoke the Delete Element tool from:

| Main tool box | Select the Delete Element tool (see Figure 2–22). |
| Key-in window | **delete** (or **del**) ENTER |

Chapter 2 Fundamentals I

Figure 2-22 Invoking the Delete Element tool from the Main tool frame

MicroStation prompts:

> Delete Element > Identify element *(Identify the element to delete.)*

The selected element is deleted. If the **Identify Elements Automatically** check box is set to OFF (in the AccuSnap Settings box), then MicroStatation prompts:

> Delete Element > Accept/Reject (select next input) *(Click the Accept button to delete the selected element, select another element to delete, or click the Reject button to terminate the command sequence.)*

Note: The Delete Element tool deletes only one element at a time. If you need to delete a group of elements, use the Fence Delete tool, explained in Chapter 6.

DROP LINE STRING/SHAPE STATUS

The Drop Line String/Shape Status tool causes blocks and shapes to separate into a series of connected individual line elements that can be manipulated as individual elements. Once a block or shape is dropped, it behaves as if it had been drawn with a Place Line tool.

65

Invoke the Drop Line String/Shape Status tool from:

Drop tool box	Select the Drop Line String/Shape Status tool (see Figure 2–23).
Key-in window	**drop string** (or **dr st**) ENTER

Figure 2-23 Invoking the Drop Line String/Shape tool from the Drop tool box

MicroStation prompts:

> Drop Line String/Shape Status > Identify element *(Identify the block or shape to be dropped.)*

The selected element is dropped. If **Identify Elements Automatically** check box is set to OFF (in the AccuSnap Settings box), then MicroStatation prompts:

> Drop Line String/Shape Status > Accept/Reject (select next input) *(Click the Accept button to accept, select another element to drop, or click the Reject button to reject.)*

PRECISION INPUT

MicroStation allows you to draw an object at its true size and then make the border, title block, and other non-object–associated features fit the object. The completed combination is reduced (or increased) to fit the plotted sheet size you require when you are plotting.

Drawing a not-to-scale schematic does not take advantage of MicroStation's full graphics and computing potential. But even though the symbols and distances between them have no relationship to any real-life dimensions, the sheet size, text size, line widths, and other visible characteristics of the drawing must be considered to give your schematic the readability you desire. Some planning, including sizing, needs to be applied to all drawings.

When MicroStation prompts for the location of a point, instead of providing the data point with your pointing device, you can use three Precision Input commands that enable you to place data points precisely. Each of the commands allows you to key-in by coordinate, including absolute rectangular coordinates, relative rectangular coordinates, and relative polar coordinates.

The rectangular coordinates system is based on specifying a point's location by giving its distances from two intersecting perpendicular axes for *two-dimensional (2D)* points or from three intersecting perpendicular planes for *three-dimensional (3D)* points. Each data point is measured along the *X* axis (horizontal) and *Y* axis (vertical) for *2D* design and along the *X* axis, *Y* axis, and *Z* axis (toward or away from the viewer) for *3D* design. The intersection of the axes, called the *origin* (XY=0,0), divides the coordinates into four quadrants for *2D* design, as shown in Figure 2–24.

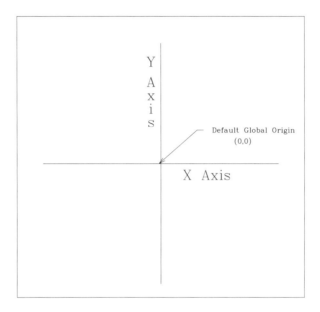

Figure 2-24 A 2D coordinate system

Absolute Rectangular Coordinates

Points are located by Absolute Rectangular Coordinates at an exact *X*, *Y* intersection on the design plane in relation to the Global Origin. By default, the Global Origin is located at the center of the design plane, as shown in Figure 2–25. The horizontal distance increases in the positive *X* direction from the origin, and the vertical distance increases in the positive *Y* direction from the origin. To enter an absolute coordinate, key-in:

       XY=<X coordinate>**,**<Y coordinate> ENTER

or

       POINT ABSOLUTE <X coordinate>**,**<Y coordinate> ENTER

The <X coordinate> and <Y coordinate> are the coordinates in working units in relation to the Global Origin. For example:

XY=2,4 ENTER

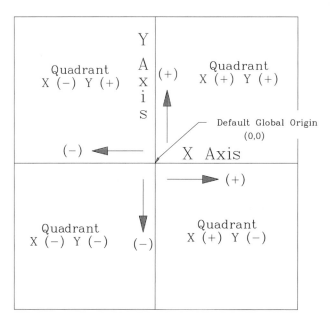

Figure 2-25 Showing Global Origin in a 2D design

The data point is located 2 master units from the origin along the X axis and 4 master units from the origin along the Y axis.

If necessary, you can relocate the Global Origin anywhere on or off the design plane. To do so, key-in **GO=MU:SU:PU,MU:SU:PU**, and MicroStation prompts:

> Global Origin > Enter monument point *(Specify a point anywhere in the design plane to define the new global origin, or click the Reset button to automatically assign the coordinates to the lower-left corner of the design plane.)*

Relative Rectangular Coordinates

Points are located by Relative Rectangular Coordinates in relation to the last specified position or point, rather than in relation to the origin. This is similar to specifying a point as an offset from the last point you entered. To enter a Relative Rectangular Coordinate, key-in:

DL=<X coordinate>**,**<Y coordinate> ENTER

or

POINT DELTA <X coordinate>**,**<Y coordinate> ENTER

The <X coordinate> and <Y coordinate> are the coordinates in relation to the last specified position or point. For example, if the last point specified was X,Y=4,4, the key-in:

DL=5,4 ENTER

is equivalent to specifying the Absolute Rectangular Coordinates X,Y=9,8 (see Figure 2–26).

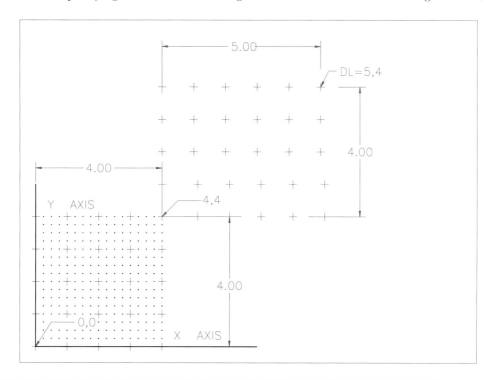

Figure 2-26 An example of placing a point by Relative Rectangular Coordinates

Relative Polar Coordinates

Relative Polar Coordinates are based on a distance from a fixed point at a given angle. In MicroStation, a Relative Polar Coordinate is determined by the distance and angle measured from the previous data point. By default, the angle is measured in a counterclockwise direction relative to the positive X axis. It is important to remember that points located by Relative Polar Coordinates are always positioned relative to the previous point, not to the Global Origin (0,0). To enter a Relative Rectangular Coordinate, key-in:

DI=<distance>,<angle> ENTER

or

POINT DISTANCE <distance>,<angle> ENTER

The <distance> and <angle> are specified in relation to the last specified position or point. The distance is specified in current working units, and the direction is specified as an angle in current angular units relative to the X axis. For example, to specify a point at a distance of 6.4 Master Units from the previous point and at an angle of 39 degrees relative to the positive X axis (see Figure 2–27), key-in:

DI=6.4,39 ENTER

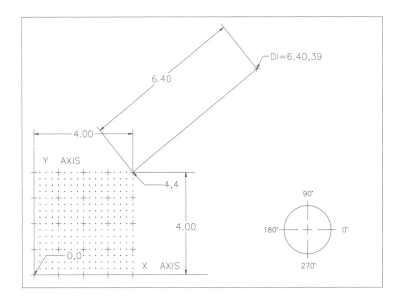

Figure 2-27 An example of placing a line by Relative Polar Coordinates

For example, the following key-ins show the placement of connected lines for the drawing shown in Figure 2–28 by the Place Line tool with absolute coordinates (see Figure 2–29):

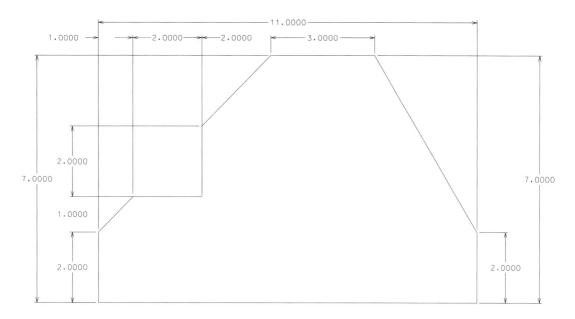

Figure 2-28 An example of placing connected lines

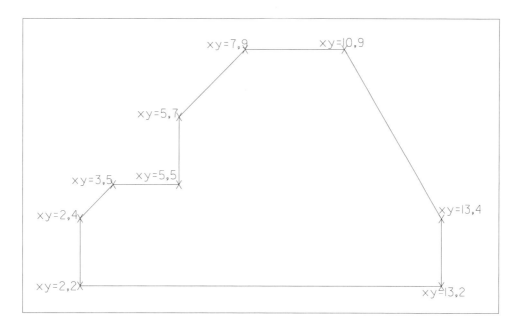

Figure 2-29 Placing connected lines using absolute coordinates

XY=2,2 ENTER
XY=2,4 ENTER
XY=3,5 ENTER
XY=5,5 ENTER
XY=5,7 ENTER
XY=7,9 ENTER
XY=10,9 ENTER
XY=13,4 ENTER
XY=13,2 ENTER
XY=2,2 ENTER
(click the Reset button to terminate the line sequence)

The following key-ins show the placement of connected lines for the drawing shown in Figure 2–28 by the Place Line tool with relative rectangular coordinates (see Figure 2–30).

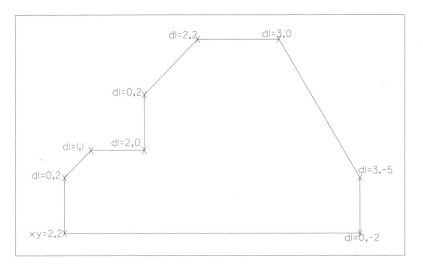

Figure 2–30 Placing connected lines using relative rectangular coordinates

XY=2,2 ENTER
DL=0,2 ENTER
DL=1,1 ENTER
DL=2,0 ENTER
DL=0,2 ENTER
DL=2,2 ENTER
DL=3,0 ENTER

DL=3,-5 ENTER
DL=0,-2 ENTER
XY=2,2 ENTER
(click the Reset button to terminate the line sequence)

The following key-ins show the placement of connected lines for the drawing shown in Figure 2–28 by the Place Line tool with relative polar coordinates and relative rectangular coordinates (see Figure 2–31):

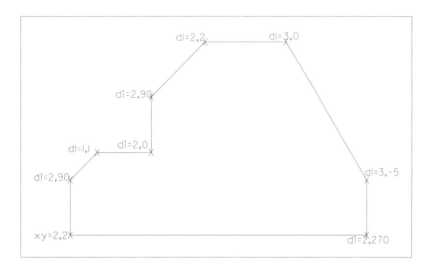

Figure 2–31 Placing connected lines using relative polar coordinates and relative rectangular coordinates

XY=2,2 ENTER
DI=2,90 ENTER
DL=1,1 ENTER
DI=2,0 ENTER
DI=2,90 ENTER
DL=2,2 ENTER
DL=3,0 ENTER
DL=3,-5 ENTER
DI=2,270 ENTER
XY=2,2 ENTER
(click the Reset button to terminate the line sequence)

Open the Exercise Manual PDF file for Chapter 2 on the accompanying CD for project and discipline specific exercises.

REVIEW QUESTIONS

Write your answers in the spaces provided.

1. How many possible positions (positional units) are there in the *X* and *Y* directions of a *2D* design file?
   _____

2. Define Master Units: _____

3. Define Sub Units: _____

4. Define Positional Units: _____

5. Name the tool that will make individual elements from a shape. _____

6. Explain briefly the differences between the absolute rectangular coordinates and relative rectangular coordinates precision key-ins.
   _____

7. Name the three key-ins that are used in Precision Input for absolute rectangular coordinates, relative rectangular coordinates, and polar relative coordinates.
   _____

8. When MicroStation displays the information regarding the size of an element or coordinates, it does so in the following format:
   _____:_____:_____

9. In a design file, if the Working Units are set up as inches, eighths, and 1600 positional units per eighth, what distance does 1:6:600 represent?
   _____

10. In a design file, if the Working Units are set up as feet, inches, and 1600 positional units per inch, what Working Units expression is equivalent to 4'–0.3200"?
    _____

11. Name the three methods by which you can place circles in MicroStation.
    _____

12. If you wish to draw a circle by specifying three known points on the circle, invoke the _____ tool.

13. The tool related to placing arcs is in the _____ tool box.

14. The Place Arc Edge tool places an arc by identifying _____ points on the arc.

15. Number of vertices allowed when you use the Place Shape tool range from _____.

16. Number of data points MicroStation prompts to draw a rotated block when you use the Place Block Rotated tool is _____.

17. Key-in to redefine the Global origin is _____.

18. What is the purpose of using the Save Settings command?
    _____

chapter 3

Fundamentals II

Objectives

After completing this chapter, you will be able to do the following:

- Use drawing tools: grid, axis, units, and tentative snap
- Control and view levels
- Set element attributes
- Match element attributes
- Use View Control: update, zoom in, zoom out, window area, fit, and pan
- Use View windows and view attributes
- Use Undo and Redo tools

DRAWING TOOLS

MicroStation provides several drawing tools to make your drafting and design layout easier.

The Grid System

The grid system is a visual tool for measuring distances precisely and placing elements accurately. The grid appears in view windows as a matrix of evenly spaced dots and crosses; it is similar to a sheet of linear graph paper. You can turn Grid display to ON or OFF as needed and change the spacing of the dots and crosses at any time. There are two parts to the grid system. The first is the Grid Reference, which appears on your screen as crosses. By default the spacing between crosses is set to one Master Unit. The second is the Grid Unit, which appears as dots on the screen. By default the spacing between dots is set to one Sub Unit. For example, if the Master Units are feet and the Sub Units are inches, the default has one foot between reference crosses and one inch between grid dots.

The grid is for visual reference only and is never plotted. The grid serves two purposes: it provides a visual indication of distances and, with Grid lock set to ON, data points placed with the Data button are forced to grid points. This is useful for keeping lines straight, ensuring that distances are exact, and making sure all elements meet.

 Note: MicroStation overrides Grid lock when you key-in the location of a point by Precision Input.

Grid Display

Grid display can be set to ON or OFF. If it is ON, you can see the Grid display on the screen. When it is set to OFF, the grid is not displayed. You change the status of the Grid display from the View Attributes setting box. Invoke the View Attributes settings box from:

| Drop-down menu | Settings > View Attributes |

MicroStation displays the View Attributes settings box as shown in Figure 3–1.

Figure 3-1 View Attributes settings box

To display the grid, turn ON the **Grid** check box. If you only want the grid displayed on a specific view window, pick the number of the window in the **View Number** drop-down menu at the top of the settings box and click the **Apply** button. If you want to turn ON the Grid display for all the open view windows, click the **ALL** button. (For a detailed explanation of Views and the View Attributes settings box, see the View Windows and View Attributes section later in this chapter.)

Note: If you turn on Grid display and do not see the grid in the view window, the window may be zoomed out too far from the design plane (the Zoom tool is discussed later in this chapter). MicroStation, by default, displays a maximum of 90 grid dots and 46 reference crosses. If the view window is zoomed out so that more than 90 dots are in the view area, MicroStation does not display the grid dots. If the zoom factor is such that more than 46 reference crosses are in the view area, the crosses are not displayed either.

Grid Spacing

The spacing between both the grid dots and the reference crosses can be changed at any time to suit your drawing needs by making appropriate changes to the Grid settings in the Design File dialog box. Invoke the Design File dialog box from:

| Drop-down menu | Settings > Design File |

MicroStation displays a Design File dialog box, similar to the one shown in Figure 3–2.

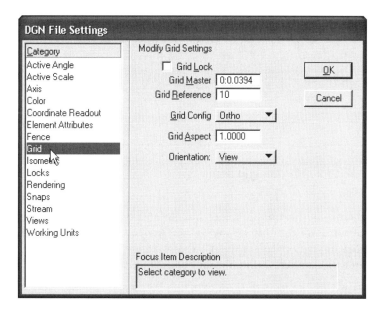

Figure 3–2 Design File dialog box

Select **Grid** from the **Category** list and MicroStation displays controls for adjusting the grid spacing. The **Grid Master** defines the distance between the grid dots and is specified in terms of the design's Master Units. Change the spacing by entering the number of Master Units or fraction of a Master Unit. The **Grid Reference** is an integer number that MicroStation multiplies by the **Grid Master** to determine the space between the reference crosses.

For example, assume that in your design the Master Units are feet and the Sub Units are inches. Setting the **Grid Master** to 0.25 and the **Grid Reference** to four puts three inches between the dots and one foot between the crosses. On the other hand, setting the **Grid Master** to 5.0 and the **Grid Reference** to 20, puts five feet between the dots and 100 feet between the crosses.

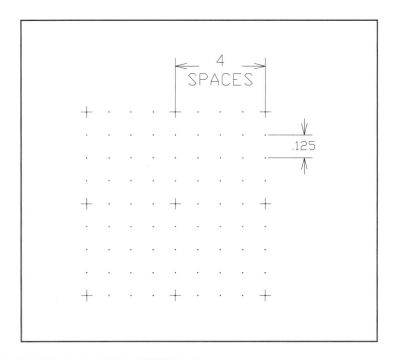

Figure 3-3 Screen display for the grid settings

You can also set the Grid Master unit and the Grid Reference by keying-in the spacing in the key-in window. To set the Grid Master unit, key-in **GU**=<distance> and press ENTER. The <distance> has to be specified in Master Units. To set the Grid Reference, key-in **GR**=<integer number> and press ENTER. The <integer number> is the number of grid units between the grid reference crosses.

To keep the grid settings in effect for future editing sessions, select **Save Settings** from the **File** drop-down menu.

Grid Configuration

MicroStation provides three choices for controlling the orientation of the Grid display: **Ortho**, **Isometric**, and **Offset**. The selection is made from the Design File dialog box under the **Grid Config** options menu. **Ortho** aligns the grid points orthogonally (the default option). **Isometric** aligns the grid points isometrically. **Offset** offsets the rows by half the distance between the horizontal grid points.

Grid Aspect Ratio (*Y/X*)

Grid Aspect allows you to set the ratio of vertical (*Y*) grid points to horizontal (*X*) grid points. The default is set at 1.000 which causes the vertical grid spacing to be identical to the horizontal spacing. If the ratio is set to 5.000, the space between rows of grid dots will be five times the space between grid columns.

Grid Orientation

Grid Orientation allows you to set the orientation of the grid in relation to the design. The **Top**, **Right**, and **Front** orientation options are only used in 3D designs and they align the grid with the 3D drawing axes. **View** and **ACS** (Auxiliary Coordinate System) are available to both 2D and 3D drawings. **View** is usually selected for 2D drawings to align the grid parallel to the plane of the X and Y axes. For more information on 3D design and ACS, see Chapter 17.

Grid Lock

The **Grid Lock** check box controls where data points are placed in the design with the pointing device. When the lock is ON, MicroStation forces all the data points to grid marks and you can only place data points on the grid with the pointing device. By setting the lock ON, you can enter points quickly with the Data button, letting MicroStation ensure that they are placed precisely. You can always override the lock by keying-in absolute or relative coordinate points.

Note: Grid Lock is effective regardless of the status of Grid display. It still locks to grid points even if you cannot see the grid in the view window.

MicroStation displays the current Grid Lock setting on the Status bar. You can toggle Grid Lock ON and OFF by clicking the lock icon on the Status bar. You can also toggle Grid Lock from two other locations:

1. The **Settings > Locks** submenu, as shown in Figure 3-4.
2. The **Settings > Locks > Toggles** settings box, as shown in Figure 3–5.

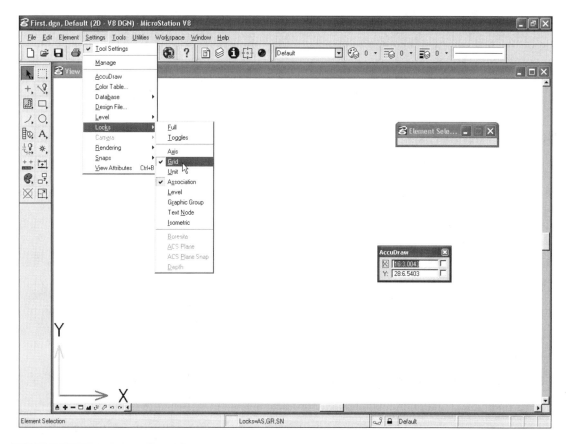

Figure 3-4 Locks submenu

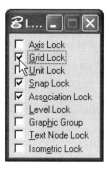

Figure 3-5 Lock Toggles settings box

To keep the grid settings (including Grid Lock) in effect for future editing sessions, choose **Save Settings** from the **File** drop-down menu.

Axis Lock

When the **Axis Lock** check box is ON, each data point is forced to lie at an incremental angle (or multiples of the angle) from the previous data point. Axis Lock settings are available in the Design File dialog box. Invoke the dialog box from:

| Drop-down menu | Settings > Design File |

MicroStation displays the Design File dialog box as shown in Figure 3–6.

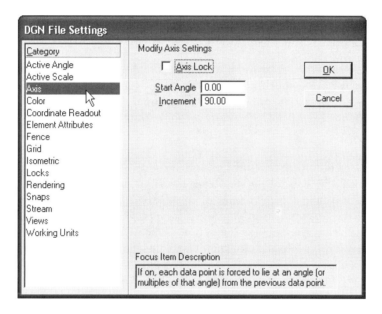

Figure 3-6 The Design File dialog box

Select **Axis** from the **Category** list and MicroStation displays controls for adjusting the Axis Lock settings. Key-in the appropriate Axis Start Angle in the **Start Angle** field and the Axis Increment angle in the **Increment** field. Turn ON the **Axis Lock** check box to constrain the data points to your settings.

For example, if you want to constrain data points to only allow drawing horizontal and vertical lines, key-in a **Start Angle** of 0 degrees and an **Increment** angle of 90 degrees. Turn ON the **Axis Lock** check box to constrain the data points.

Snap Lock

Snap Lock controls the placement of tentative snap points at specific locations on elements in the design. Tentative snapping provides a way to preview the location of a data point before it is actually entered in the design and to place data points at precise locations on existing elements.

When a tentative snap point is placed (by clicking the Tentative button), a large cross appears to identify the point. If it is snapped to an element, the element is highlighted and MicroStation displays the absolute coordinates of the point on the Status bar. If the point is at the correct location, click the Accept button (same as the Data button) to confirm it. If, however, the point is not at the correct location, move the pointer closer to the correct location and click the Tentative button again. Selecting another tentative snap point rejects the last tentative snap point and selects a new one. When you accept the tentative snap point with the Accept button, the large cross disappears and a data point is placed in the design. You can also cancel the tentative snap point by clicking the Reset button.

The pointing device buttons used to place tentative snap points vary. Pucks that are used with a digitizer board usually have a specific Tentative button. On a three-button mouse, the middle button is usually the Tentative button. Tentative snapping with a two-button mouse requires pressing both buttons simultaneously (which may take some practice to learn to do successfully).

Snap Lock Mode

If the **Snap Lock** check box is ON, you can snap to a specific point on an element, depending on the snap mode selected. For example, if you set the snap mode to **Center** and turn **Snap Lock** ON, the tentative button snaps to the center of circles and blocks, and the midpoint of lines and segments of line strings. A tentative snap point can be placed while executing any MicroStation tool that requests a data point, such as Place Line, Place Circle, Place Arc, Move, and Copy. If **Snap Lock** is OFF, tentative points do not snap to elements.

Two settings boxes are provided for selecting Snap Lock settings:

Drop-down menu	Settings > Locks > Toggles
	Settings > Locks > Full

You can toggle the **Snap Lock** check box from either the Toggles settings box or the Full settings box, as shown in Figure 3–7.

Chapter 3 Fundamentals II

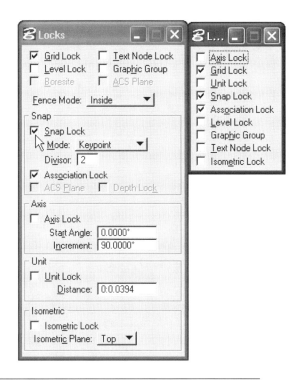

Figure 3-7 Locks Full setting box and Lock Toggles settings box

Selecting a Snap Mode

You can select an active (default) Snap mode that always stays in effect. When you occasionally need a different one, you can select an override Snap mode that applies only to the next tentative point.

To set the Snap active mode:

Drop-down menu	Settings > Locks > Full (Select the desired Snap mode from the Snap Mode option menu. See Figure 3–8.)
Drop-down menu	Settings > Snaps > Button Bar (Double-click the desired Snap mode. See Figure 3–9.)

85

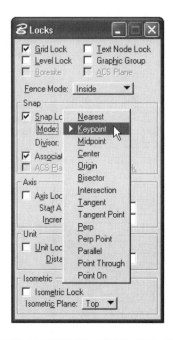

Figure 3-8 Locks settings box displaying the Snap Mode option menu.

Figure 3-9 Snap mode button bar

You can also select the default Snap mode from the **Snap Mode** pop-up menu:

- Invoke the pop-up menu from the View window by holding down SHIFT and tapping the Tentative button, as shown in Figure 3–10.
- Invoke the pop-up menu from the Status bar by positioning the pointer over the Active Snap Mode icon on the Status bar and clicking either the Data button or the Reset button.

Set the default Snap mode from the pop-up menu by positioning the pointer over the name of the desired Snap mode, holding down SHIFT, and tapping either the Data button or Reset button.

Chapter 3 Fundamentals II

Figure 3-10 The pop-up Snap Mode menu

Set a temporary Snap mode override from:

Button bar	Click the desired Snap mode.
Drop-down menu	Settings > Snaps (Choose the desired Snap mode override.)

You can also select the override Snap mode from the pop-up Snap Mode menu. Hold down SHIFT and press the Tentative button to display the pop-up menu and select the desired override snap mode from the menu.

By default, MicroStation displays the active Snap mode icon on the Status bar. MicroStation also displays a diamond-shaped object to the left of the default active Snap mode in the Snaps menu. If an override mode is selected, a square appears to the left of the default Snap active mode and a diamond shaped object appears to the left of the current override Snap mode.

Note: The number of Snap modes included in the Snaps menu varies. The menu shows only the Snap modes that are available for the active placement or manipulation command. Some commands do not support all Snap modes.

How to Use the Tentative Button

To use the Tentative button in placing and manipulating elements:

1. Select the placement or manipulation tool.
2. Select the Snap mode you want (these modes are described later).

87

3. Point to the element you want to snap to, and **click** the Tentative button.
4. Click the Data button to accept the tentative point.
5. Continue using Snap modes and tentative points, as necessary, to complete the placement or manipulation.

For example, if you want to start a line in the exact center of a block, select the Place Line tool, set the Snap Mode to **Center**, and click the Tentative button on the block. The block is highlighted and a large tentative cross appears at the exact center of the block. Place a data point to start the line at the center of the block (see Figure 3–11).

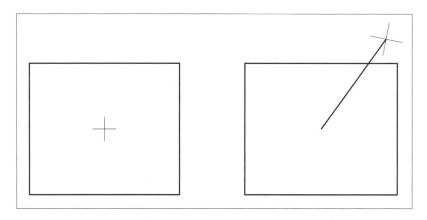

Figure 3-11 Example of using tentative snap

Keep in mind the following points when you use the Tentative button.

- You can only snap to elements when Snap Lock is ON.

- If Grid Lock is ON and the element you are trying to snap to is placed between grid points, the Tentative button may snap to grid points rather than to the element you want. If that happens, turn Grid Lock OFF.

- When the Tentative button snaps to an element, the element is highlighted and the tentative cross appears at the tentative snap point. If the cross appears but the element is not highlighted, you did not snap to the element; you may have snapped to a grid point close to the point you wanted.

- When you press the Tentative button, MicroStation starts searching for elements in the area immediately around the screen pointer. It selects elements in the order they were placed in the design. If the tentative snap point appears on the wrong element, click the Tentative button again and the next element is found; there is no need to move the screen pointer location or to press the Reset button. If the Tentative button cycles through all the elements in the area without finding the tentative snap point you want, move the screen pointer closer to the element and press the Tentative button again.

For example: you place a block, and then place a line starting very near one corner of the block. You need to snap to the end of the line for the next command, but the tentative point snaps to the corner of the block. Click the Tentative button again, and it should snap to the end of the line. If the second snap also does not find the end of the line, move the screen pointer a little closer to it and snap again.

■ You do not have to place the screen pointer exactly on the point of the element you want to snap to, just near it. In fact, to lessen the chance of snapping to the wrong element, it is best to move back along the element, away from other elements.

Types of Snap Modes

The following sections explain the available Snap modes.

Keypoint mode

Keypoint mode allows tentative points to snap to defined key points on elements. For a line, the key points are at the endpoints of the line; for a circle, they are at the center and four quadrants; for a block, they are at the four corners, and so on. See Figure 3–12 for the key point snap points for various element types. To snap to a key point on an element, position the pointer close to the key point (make sure Snap Lock is ON and the key point mode is selected) and click the Tentative button. The tentative cross appears on the element's key point and the element is highlighted. If the tentative cross appears but the element does not highlight, you have not found the element's snap point. Continue pressing the Tentative button until the correct snap point is located, then press the Accept button.

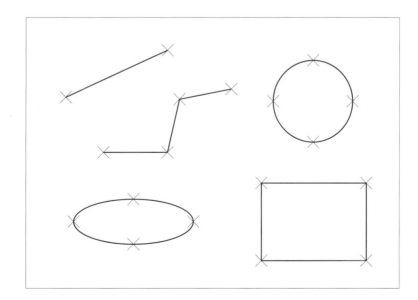

Figure 3–12 Keypoint snap locations on various element types

The Keypoint snap Divisor works with Keypoint snap mode to define additional snap points on an element. For example, setting the Key point snap Divisor to 5 divides an element into five equal divisions. Figure 3–13 shows the key point snaps for divisor values of two and five. The divisor can be set in the Full Locks settings box by keying-in the value in the Divisor edit field. You can also set the value by keying-in at the key-in window: **KY**=<number of divisors> and pressing ENTER.

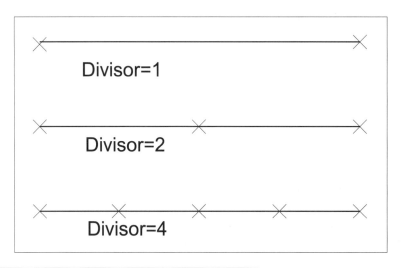

Figure 3-13 Example of various key point snap divisors

Nearest mode

When active, Nearest mode places tentative snap points on a point of an element that is closest to the pointer. This rule remains the same among all element types, except as applied to text where the only snap point is the justification point. With Nearest mode you are always certain that you can locate any point on any type of element.

To pick a specific point on an element, position the pointer close to the point you want to select, ensure that the Snap Lock is ON and the Snap mode is Nearest. When the Tentative button is pressed, the element is highlighted and the tentative cross appears at the closest point on the element. If the tentative cross appears, but the element does not highlight, you have not found the element. Continue pressing the Tentative button until a point is located, and then press the Accept button.

Midpoint mode

Midpoint mode, when active, places tentative snap points at the midpoint of an element or segment of a complex element (see Figure 3–14). The location of the midpoint varies with different types of elements:

- It bisects a line, arc, or partial ellipse.

- It bisects the selected segment of a line string, block, multi-sided shape, or regular polygon.
- It snaps to the 180-degree (9 o'clock) position of a circle or an ellipse.

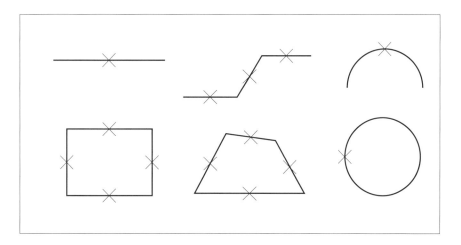

Figure 3-14 Midpoint snaps for various element types

Center mode

Center mode causes tentative snap points to snap to the center of the space occupied by an element (such as a circle, block, or arc) as shown in Figure 3–15.

Figure 3-15 Example of snapping to the center of the circle to start a line

Intersection mode

Intersection mode causes tentative snap points to snap to the intersection of two elements. To find the intersection, snap to one of the intersecting elements close to the point of intersection. One or both of the elements may be highlighted and displayed with dashes to indicate that the intersection has been found. When the intersection is found the large tentative cross appears at the intersection (see Figure 3–16).

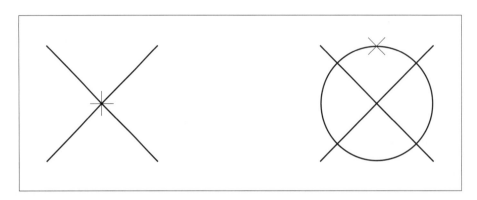

Figure 3-16 Example of snapping to the intersection of two elements for placing a circle

If the elements do not actually intersect, the tentative cross appears at the intersection of an imaginary extension of the two elements. If the two elements cannot be extended to an intersection, an error message appears in the Status bar and a tentative cross is not placed.

Through Point mode

Through Point mode causes tentative snap points to define a point on an existing element through which the element you are placing must pass (see Figure 3–17).

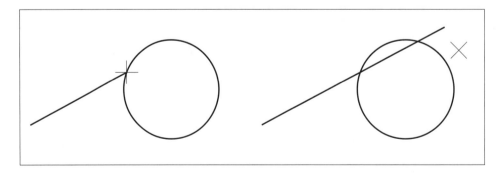

Figure 3-17 Example of snapping to a through point

Tangent mode

Tangent mode forces the element you are creating to be tangent to a nonlinear element (such as a circle, ellipse, or arc). The actual point of tangency varies, depending on how you place the element (see Figure 3–18).

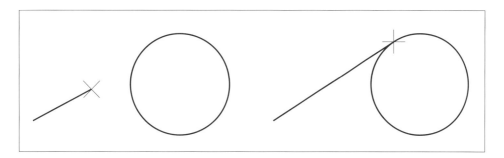

Figure 3-18 Example of snapping to a tangent point

Tangent From mode

Tangent From mode forces the element you are placing to be tangent to an existing nonlinear element (such as a circle, ellipse, or arc) at the point where you placed the tentative snap point (see Figure 3–19).

Figure 3-19 Example of snapping to a tangent on an existing nonlinear element

Origin mode

Origin mode snaps to the center of an arc, circle, origin of the text, or cell.

Bisector mode

Bisector mode sets the snap mode to bisect an element; the snap point varies with different types of elements.

Perpendicular mode

Perpendicular mode forces the element to be perpendicular to an existing element. The actual perpendicular point depends on the way the element is placed.

Perpendicular Point

Perpendicular Point forces the element to be perpendicular to an existing element at the point where you place the tentative snap point.

Parallel mode

Parallel mode forces the line or segment of the line string to be parallel to a linear element.

Point On mode

Point On mode snaps to the nearest element (after you have entered the first point of element placement) and constrains the next data point to lie on a closed element or anywhere on a linear element's line.

AccuSnap

AccuSnap improves your ability to quickly snap to elements by automatically placing tentative snap points and providing visual and audible clues to the location of the snap points. The Tentative button is not needed when AccuSnap is enabled (a timesaver for two-button mouse users).

Use AccuSnap to Snap to an Element

AccuSnap comes into play when you select a tool and move the pointer near a tentative snap point on an element in the design. AccuSnap displays a cross on the point. When you move the pointer closer to the point, AccuSnap highlights the element containing the point and changes the cross to an "X" to indicate that you are at the point. To place a data point at the tentative snap point, click the Data button, after the "X" appears. To reject it, move the pointer away from the point.

Control the Way AccuSnap Works

MicroStation provides an AccuSnap settings box that allows you to control the behavior of AccuSnap. Invoke the AccuSnap settings box from:

Drop-down menu	Settings > Snaps > AccuSnap
Pop-up menu	While holding down SHIFT, click the Tentative button and select AccuSnap from the pop-up menu.

The AccuSnap settings box is divided into three groups of settings (**General**, **Element**, and **Feel**) that are available in separate tabs. To access one of the groups, click its tab near the top of the settings box.

General Settings

The **General** tab in the AccuSnap settings box allows you to turn AccuSnap on and off (enable and disable) and control what AccuSnap does when your pointer approaches a tentative snap point on an element in the design. Figure 3-20 shows the settings box with the General tab displayed.

Figure 3-20 The General Tab on the AccuSnap settings box

If the AccuSnap check box is ON, AccuSnap is on and will automatically identify tentative snap points. If the check box is OFF, there is no AccuSnap action and you must use the Tentative button to place tentative snap points.

 Note: You can also turn AccuSnap ON and OFF from the Snap Mode toolbar. The AccuSnap button is shown in Figure 3–21. To open the toolbar, select **Settings > Snaps > Button Bar**

Figure 3-21 The AccuSnap button in the Snap Mode toolbar

Turning ON the **Show Tentative Hint** check box displays the cross when the pointer is near a tentative snap point. If the check box is OFF, AccuSnap only displays the "X" when the pointer is at a tentative snap point, there is no hint that you are getting close to a tentative point.

Turning ON the **Display Icon** check box displays a picture of the currently active snap mode's icon (example: Keypoint Snap) with the AccuSnap cross and "X." This setting provides a convenient reminder of what will happen when the tentative snap point is accepted by clicking the Data button.

The **Fixed Point for Perp./Tan. From** check box is used only with the Place SmartLine tool when Perpendicular or Tangent snap modes are in effect. The SmartLine tool is discussed in Chapter 5).

Turn ON the **Update Status Bar Coordinates** check box and AccuSnap will display the coordinate readout for each tentative snap point, in the Status bar. When this option is on you will always know where you are, on the design pane.

Turn ON the **Play Sound On Snap** check box and AccuSnap will make a clicking sound when it displays a tentative snap point (whenever the "X" appears).

Turn ON **Highlight Active Element** and AccuSnap will highlight an element as soon as it finds a tentative snap point on the element (whenever the cross appears). Turn the check box OFF so that elements are highlighted only when AccuSnap selects a tentative snap on them (when the "X" appears).

Turn ON **Identify Element Automatically** and AccuSnap will automatically identify elements, as the pointer passes over them.

The **Pop-up Info** check box controls the display of Tips that appear next to the tentative snap points that AccuSnap finds or selects. When the check box is ON, an options list is available, to the right of the check box. The options are:

- **Automatic** – causes the Tips to appear automatically.
- **Tentative** – causes the Tips to appear only when you click the Tentative button to place a tentative snap point, and then allow the pointer to hover over the point.

When the **Pop-up Info** check box is OFF, no tips appear.

Element Settings

The **Element** tab in the AccuSnap settings box provides check boxes for (B-spline) **Curves**, **Dimensions**, and **Text**. Figure 3–22 shows the settings box with the **Element** tab displayed.

Figure 3-22 The Elements Tab on the AccuSnap settings box

If the check box for B-spline **Curves**, **Dimension**, or **Text** is ON, AccuSnap snaps to that type of element. If the check box is OFF, AccuSnap does not snap to that type of element.

Note: You can override the Element settings by using the Tentative button to place tentative snap points.

Feel Settings

The **Feel** tab provides three slider bars used to control AccuSnap's sensitivity when locating and selecting tentative snap points. Figure 3–23 shows the settings box with the **Feel** tab displayed.

Figure 3-23 The Feel Tab on the AccuSnap settings box

The **Keypoint Sensitivity** slider bar adjusts how close the pointer must be to the snap point before AccuSnap will snap to it and display the "X". Moving the slider to the right (+) increases the allowable distance, while moving the slider to the left (-) reduces the allowable distance.

The **Stickiness** slider bar adjusts the sensitivity of AccuSnap to the current element when it has tentatively snapped to that element. Moving the slider to the right (+) increases the distance you can move the pointer away from the active element before AccuSnap drops it and snaps to another element. Moving the slider to the left (-) decreases the distance.

The **Snap Tolerance** slider bar adjusts how close the pointer must be to an element before you can snap a tentative point to it (the "snap tolerance"). Moving the slider to the right (+) increases the snap tolerance. Moving the distance to the left (-) decreases the snap tolerance.

ELEMENT ATTRIBUTES

There are four important attributes associated with the placement of elements: level, color, line style, and line weight.

Levels

MicroStation offers a way to group elements on levels, in a manner similar to a designer drawing different parts of a design on separate transparent sheets. By stacking the transparent sheets one on top of another, the designer can see the complete drawing but can only draw on the top sheet.

If the designer wants to show a customer only part of the design, he or she can remove from the stack, the sheets that contain parts of the design the customer does not need to see.

MicroStation has the ability to create an almost unlimited number of levels in each design file. For example, an architectural design might have the walls on one level, the dimensions on another level, electrical information on still another level, and so on. Separating parts of the design by level, allows designers to turn on only the part they need to work on and to plot parts of the design separately.

You can only draw on one level at a time (the active level), but you can turn ON or OFF the display of all other levels, in selected views. Elements on levels that are not displayed, disappear from the view and do not plot, but they are still in the design file. The name of the active level is displayed on the Status bar.

When you manipulate an element, it remains on the same level. For instance, a copy of an element stays on the same level as the original element, regardless of what level is currently active. The Change Element Attributes tool, which moves elements to different levels, is discussed later.

The Level Settings Boxes and Drop-down Menu

Two settings boxes and a drop-down menu are provided for managing levels. The Level Display settings box allows you to control which levels are displayed and sets the active level (see Figure 3–24). The Level Manager settings box allows you to create levels, set level symbology, and set the active level (See Figure 3–25).

Invoke the Level Display setting box from:

Drop-down menu	Settings > Levels > Display
Primary Toolbar	Click the Level Display icon

Invoke the Level Manager settings box from:

Drop-down menu	Settings > Levels > Manager
Status Bar	Click the name of the active level.
Level Display Settings Box	Position the pointer over a level name, click the Reset button, and, in the resulting pop-up menu, click the Level Manager option.

Chapter 3 Fundamentals II

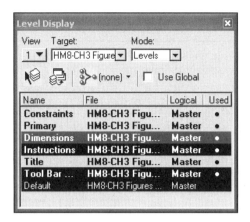

Figure 3-24 The Level Display Settings Box

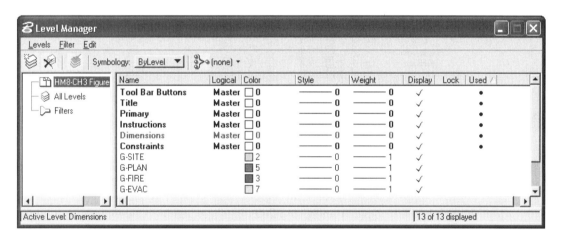

Figure 3-25 The Level Manager Settings Box

Setting the Active Level

There can be only one active level at any time, and most MicroStation placement tools place new elements on the active level. The name of the Active Level is displayed in the Status bar and is highlighted in the Level Manger and Level Display settings boxes.

You select the active level from the Level Manager or Level Display settings box by double-clicking the level's name. The selected name is highlighted in the settings boxes and displayed in the Status bar.

The Active Level can also be set from a drop-down menu and with a key-in:

Attributes Toolbar	Click the drop-down arrow on the Active Level drop-down menu to display the list of defined levels. Elect the name of the level that is to be made the active level (see Figure 3–26).
Key-in window	**LV=**<level number> ENTER

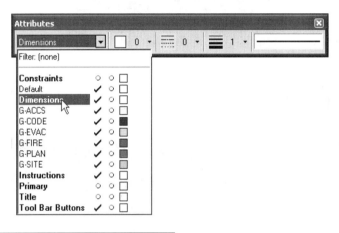

Figure 3–26 The Active Level Drop-down Menu

 Note: The active level you set remains in effect for the rest of the design session or until a new active level is selected. To save the current active level for the next design session, select **Save Settings** from the **File** drop-down menu.

Creating New Levels

New levels are created in the Level Manager settings box. First create the new level and then set the new level's properties (such as the name).

To create a new level from the Level Manager settings box:

Drop-down menu	Levels>New
Toolbar	Click the New icon
From an existing level name	Position the pointer over one of the existing level names, click the Reset button, and, in the resulting pop-up menu, select the New option.

When the new level is created, a description of its default settings is inserted at the bottom of the list of existing level names.

To change the name of the new level, click the default level name in the levels list, and type the new name. If the default name's background color does not change when you click it, click the name again before typing.

Setting Level Attributes

To the right of each named level in the Level Manager settings box are fields named **Color**, **Style**, and **Weight**. These fields allow you to specify the attributes of new elements that you place on each level. When you click one of the fields, a pop-up menu appears that contains all the settings available for that attribute. The default attribute settings are all zero (white, solid, weight zero lines).

To set the level attributes, select the **By Level** option from the **Symbology** menu in the Level Manager settings box and then use each attribute's pop-up menu to select the desired setting (For example, select the red from the **Color** menu and Center lines from the **Style** menu).

Note: Make sure the Element Attributes are set to ByLevel (can be setup through Attributes tool box), so that the attribute settings that are set for active level will be applied when elements are placed.

Changing the Properties of Existing Levels

You can change the properties of an existing level in the Level Properties

Select the level to be edited in the Level Manager settings box and invoke the Level Properties dialog box from:

Drop-down menu	Levels > Properties
From the new level name	Position the pointer over the name of the new level, click the Reset button, in the resulting pop-up menu, select the Properties option.

MicroStation will open the Level Properties dialog box, as shown in Figure 3–27. The box contains options for changing the levels name and attributes. It also provides the Description field that you can use to provide more information about the intended use of the level.

Note: The level name, color, style, and weight can also be changed by clicking each setting in the level's row in the Level Manager settings box.

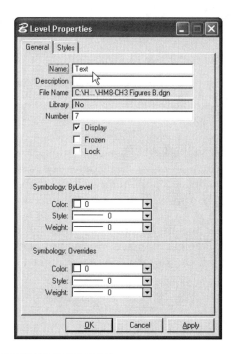

Figure 3-27 The Level Properties dialog box

Controlling the Display of Levels

The Level Display settings box controls the display of levels. You can turn display ON or OFF for one or all design views windows.

To control the display of levels, invoke the View Levels settings box from:

Drop-down menu	Settings > Level > Display
Key-in window	**ON=**<level numbers> ENTER **OFF=**<level numbers> ENTER

Displayed levels are shown in the Level Display settings box with a dark background and hidden levels are displayed with a light background.

Before changing the display of any levels, select the design view for which you want to change the level display. Click the drop-down arrow in the **View** menu and select one of the design views. (Only currently open views can be selected). Any level display changes are applied only to the selected view. To apply the changes to all the design views, turn ON the **Use Global** check box.

To change the display state of a level, click it – if display of the level was ON, clicking it turns display OFF and vise-versa. To change the display state of a contiguous set of levels, click the first or last one in the list and, while continuing to hold down the Data button, drag the pointer across all the levels. The resulting display state of all the selected levels is determined by the state of the first level you select. If the display of the first level was OFF, all selected levels are turned ON (levels that were already ON remain ON). If the display of the first level was ON, all selected levels are turned OFF.

Notes: Display of the active level cannot be turned OFF.

The active level and the level display settings for each view remain in effect until you change them or exit from MicroStation. To keep them in effect for the next editing session, select **Save Settings** from the **File** drop-down menu.

Element Color

Color helps to differentiate the various types of design elements. For example, the object being designed may be drawn using black while the dimensions are drawn using green. Before drawing elements, set the active color to ByLevel to use the color that is defined in the Level manager settings box for the Active Level. To override the color assigned to the active Level, select an alternate color from the color palette.

All new elements drawn are set to the active color. The active color remains in effect until you change it to a different color or exit the design file. To keep the active color in effect for the next editing session, select **Save Settings** from the **File** menu, after you select the active color.

Note: Element color is permanent, so the color of existing elements is not changed by changing the active color. Commands for changing an element's color are discussed later.

Set the Active Color from:

Attributes Toolbar	Click the drop-down arrow on the Color menu. Select a color from the resulting pop-up color palette or select ByLevel to use the color that is defined in the Level Manager settings box for the Active Level (see Figure 3–28). As you drag the pointer over the palette, pop-up boxes show the color numbers.
Key-in window	**CO=**<name of the color or color number from 0 to 254> ENTER

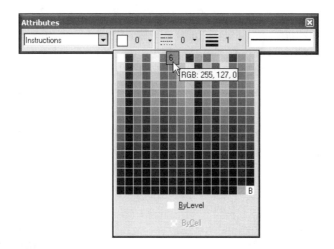

Figure 3-28 The Color options menu

Note: The actual colors shown depend on your monitor, graphics card, and what colors are defined in the MicroStation color table.

Element Line Style

Similar to color, MicroStation allows you to place elements with a specific line style such as: solid or dashed). You select the active line style from a set of eight internal styles and several custom-made line styles, or you can elect to have the line style set to the Active Level's style attribute.

All new elements drawn are set to the active line style. The active line style remains in effect until you change it or exit the design file. To keep the active line style in effect for the next editing session, select **Save Settings** from the **File** menu, after you select the active style.

Note: Element line style is permanent, so the line style of existing elements is not changed by changing the active line style. Commands for changing an element's line style are discussed later.

Set the Active Line Style from:

Attributes Toolbar	Click the drop-down arrow on the Line Style menu. Select a style from the resulting pop-up menu or select ByLevel to use the Active Level Style attribute that is set in the Level Manager settings box (see Figure 3-29).
Key-in window	**LC**=<name of the line style or line style number> ENTER

Figure 3-29 The Line Style options drop-down menu

Element Line Weight

In drafting, line width contributes to the "readability" or understanding of the design. For example, in a piping arrangement drawing, the line width for the pipe is the widest of the lines used on the drawing so that the pipe stands out from the equipment, foundations, and supports.

In MicroStation, weight refers to the width of the element and elements are placed using the active weight. You can choose from 32 line weights (numbered from 0 to 31), which is comparable to 32 different technical pens. You can also elect to have the active weight set to the Active Level Weight attribute.

All new elements you draw are set to the active weight. The active weight remains in effect until you change it or exit the design file. To keep the active line weight in effect for the next editing session, select **Save Settings** from the **File** menu, after you select the active weight.

Note: Element line weight is permanent, so the weight of existing elements is not changed by changing the active weight. Commands for changing an element's line weight are discussed later.

Set the Active Element Line Weight from:

Attributes Toolbar	Click the drop-down arrow on the Line Weight menu. Select a weight from the resulting pop-up menu, or select ByLevel to use the Active Level Weight attribute that was set in the Level Manager settings box (see Figure 3–30).
Key-in window	**WT=**<line weight number anywhere from 0 to 32> ENTER

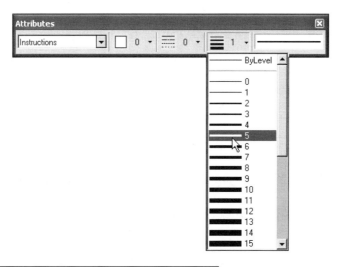

Figure 3-30 The Line Weight options drop-down menu

Change Element Attributes

The Change Element Attributes tool allows you to change the level, color, style, and weight of elements already in the design file. With this tool you select the required attributes and then select elements that will be changed to have the new attributes.Or, you can first select an element with the attributes you need, and then select the elements that will share the attributes of the first element.Invoke the Change Attributes tool from:

Change Attributes toolbox	Click the **Change Element Attributes** button.

 Note: Tool boxes are invoked from the Main tool frame. By default, the frame is located on the left side of the MicroStation desktop. Tool boxes can also be invoked by selecting **Main** from the **Tools** menu.

MicroStation displays the Tool Settings window, as shown in Figure 3–31.

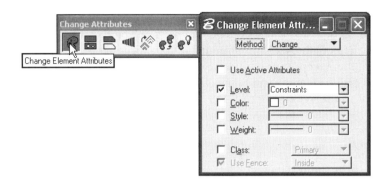

Figure 3-31 The Change Element Attributes window

Change Method

If no existing element uses the attribute settings you desire, click the drop-down arrow on the **Method** menu, at the top of the Tool Settings window, and select the **Change** option.

Change the state of an attribute by clicking the attribute's check box. For example, if you only want to change the level of an element, turn ON the **Level** check box and turn OFF the check boxes for **Color**, **Style**, **Weight**, and **Class** (Class is not discussed here).

If you want to change elements to the current active element attributes (Active Level, Active Color, Active Style, and Active Weight), turn ON the **Use Active Attributes** check box to load the Tool Settings window with the current active settings.

Note: When the **Use Active Attributes** check box is ON, any attribute setting changes you make in the Tool Settings window also change the associated Active Attribute. For example, if the Active Color is red and you change the Color on the Tool Settings window to green, the Active Color becomes green. If the check box is OFF, changes on the Tool Settings window do not change the Active Attributes.

If you do not want to use the current Active Attribute settings, turn OFF the **Use Active Attributes** check box and select the attribute settings from the drop-down menus to the right of each attribute check box.

Once the attributes are set, start selecting the elements whose attributes you want to change. MicroStation prompts:

> Change Element Attributes > Identify Element *(Identify the element whose attributes are to be changed.)*
>
> Change Element Attributes > Accept/Reject *(Click the Accept button to accept the changes for the selected element or click the Reject button to reject the element.)*

Note: If the pointer is on another element when you click the Accept button, the first element is changed, the second element is selected, and MicroStation prompts to "Accept/Reject" the second element. If the pointer is not on another element when you click the Accept button, MicroStation prompts to "Identify Element" and you can select another element to change. The Change Element Attributes tool remains active until you select another tool.

Match / Change Method

If an existing element has the attribute settings you need, start by clicking the drop-down arrow on the **Method** menu at the top of the Tool Settings window and selecting the **Match / Change** option. With this method, the attributes of the first element you select are loaded in the Tool Settings window and the following elements you select are changed to the attribute settings of the first element. Before starting to select elements, turn ON the check boxes for the attributes you want to match and change (**Level**, **Color**, **Style**, and **Weight**).

If you also want to change the Active Element Attributes to those of the first selected element, turn on the **Use Active Attributes** check box before you select the first element.

Now that you have selected the **Match / Change** option, you are ready to select the element that has the attributes you want to use and start applying those settings to other elements. MicroStation prompts:

> Match Element Attributes > Identify Element to Match *(Identify the element that has the attributes you want to match.)*
>
> Change Element Attributes > Accept/Reject Element to Match *(Click the Accept button to accept the selected element and load its attribute settings in the Tool Settings window, or click the Reject button to reject the element.)*

Note: When you accept the selected element, do not point to another element unless you accidentally selected the wrong element. If you select another element, the attribute settings of the second element will be matched.

> Change Element Attributes > Identify Element to Change *(Identify the element whose attributes are to be changed.)*
>
> Change Element Attributes > Identify Element to Change *(Click the Accept button to accept the changes for the selected element or click the Reject button to reject the element.)*

Notes: If the pointer is on another element when you click the Accept button, the first element you selected is changed and the second element is selected. If you are not pointing at another element when you click the Accept button, no element is selected. The Change Element Attributes tool remains active until you select another tool.

> To stop selecting elements and go back to the first step of matching the attributes of an existing element, double-click the Reject button. After double-clicking, MicroStation prompts you to "Identify Element to Match" and you can select another element whose attribute settings you want to use.

Match Element Attributes

The Match Element Attributes tool allows you to change the active attributes (Active Level, Active Color, Active Style, and Active Weight) to the settings of an existing element. This tool provides a quick way to return to placing elements with the same attributes as elements you previously placed in the design.

Invoke the Match Element Attributes tool from:

| Change Attributes tool box | Select the Match Element Attributes button |

MicroStation displays the Tool Settings window as shown in Figure 3–32).

Figure 3–32 The Match Element Attributes settings box

Before selecting the element, select the attributes you want to match (**Level**, **Color**, **Style**, and **Weight**) by turning on the attribute check boxes in the Tool Settings window. MicroStation prompts:

> Match Element Attributes > Identify Element *(Identify the element that has the attributes you want to match.)*
>
> Match Element Attributes > Accept/Reject (select next input) *(Click the Accept button to accept the changes for the selected element or click the Reject button to reject the changes.)*

 Note: If the pointer is on another element when you click the Accept button, the attributes are changed to those of the first element you selected, the second element is selected, and MicroStation prompts you to "Accept/Reject (select next input)." If you are not pointing at another element when you click the Accept button, no element is selected and MicroStation prompts you to "Identify Element." The Match Element Attributes tool remains active until you select another tool and the attributes will be set to the last element you accept.

VIEW CONTROL

The View Control tools allow you to select the portion of the drawing to be displayed. By allowing you to see the design in different ways, MicroStation gives you the means to draw more quickly, easily, and accurately. The View Control tools explained in this section are utility tools; they make your job easier and help you to draw more accurately but they do not make any changes to the content of the design.

Access the view control tools from:

View Control Bar	Located on the lower-left corner of each view window, as shown in Figure 3–33.
View Control tool box	Tools > View Control (see Figure 3–34)
View Control pop-up menu	While holding down SHIFT, click the Reset button to display the pop-up menu shown in Figure 3–35.

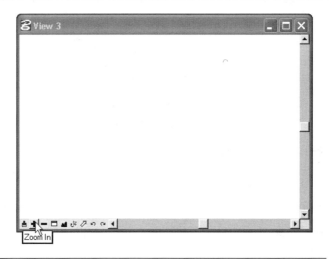

Figure 3-33 View Control Bar on the lower left corner of each view window

Figure 3-34 View Control tool box

Figure 3-35 View Control pop-up menu

If you select a View Control tool from the View Control Bar, the view whose border contains the bar is manipulated by the tool. If, you select a View Control tool from the View Control tool box or pop-up menu, MicroStation prompts you to select the view window whose view is to be manipulated. The following discussion of the view commands describes the actions required when the tools are selected from the View Control tool box.

Update View

The Update View tool instructs the computer to repaint the contents of the selected view window making sure it correctly displays the design. Use this tool when you see an incomplete image of your design in the view. For example, deleting elements can cause the false impression that there are gaps in elements that the deleted element crossed and gaps in the grid. Update the display to repaint the view and correctly display the design.

Invoke the Update View tool from:

View Control tool box	Select the Update View tool (see Figure 3–36).
Key-in window	**update view** (or **up**) ENTER

Figure 3-36 Invoking the Update View tool from the View Control tool box

MicroStation prompts:

> Update View > Select view *(Position the pointer anywhere in the view and click the Data button to update the view or click the Reset button to exit the Update View tool and return to the previously active tool.)*

Controlling the Design Area Displayed in the View

The area of the design displayed in a view can be controlled in a way similar to using a zoom lens on a camera. You can increase or decrease the viewing area by zooming in or out. When you zoom in, you view a smaller area of the design in greater detail. When you zoom out, you view a larger area of the design, in less detail. The design size is not changed, only your view of the design changes.

MicroStation provides three tools to control the area of the design that can be seen in a view window: **Zoom In**, **Zoom Out**, and **Window Area**.

Zoom In

The Zoom In tool moves the view window in closer to the design, allowing you to view a smaller area of the drawing in greater detail.

Invoke the Zoom In tool from:

View Control tool box	Select the Zoom In tool (see Figure 3–37).
Key-in window	**zoom in extended** (or **z i e**) ENTER

Figure 3-37 Invoking the Zoom In tool from the View Control tool box

The Zoom In Tool Settings window contains one field, **Zoom Ratio,** that shows the active zoom in factor. The default ratio is two, but you can set it to a range of one to 50 by typing the appropriate positive number in the field.

MicroStation prompts:

> Zoom In > Enter zoom center point *(When you move the pointer in the view, a rectangular box is displayed that indicates the new view boundary. Click the Data button to define the center of the area of the design to be displayed in the view window (see Figures 3–38a and 3–38b) or click the Reset button to exit the Zoom In tool and return to the previously active tool.)*

You can continue clicking the Accept button to zoom in closer. Zoom In remains active until you click the Reset button or invoke another tool.

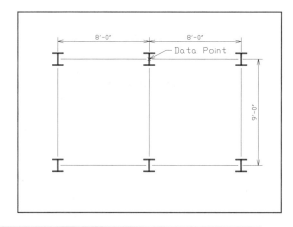

Figure 3-38a The design view before the Zoom In tool is invoked

Chapter 3 Fundamentals II

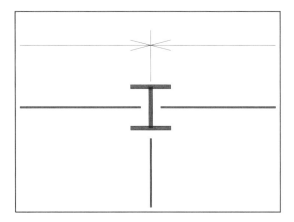

Figure 3-38b The design view after the Zoom In tool is invoked

Zoom Out

The Zoom Out tool moves the view window away from the design, allowing you to view a larger area of the drawing, but in less detail.

Invoke the Zoom Out tool from:

View Control tool box	Select the Zoom Out tool (see Figure 3–39).
Key-in window	**zoom out extended** (or **z o e**) ENTER

Figure 3-39 Invoking the Zoom Out tool from the View Control tool box

The Zoom Out Tool Settings window contains one field, **Zoom Ratio**, that shows the zoom out factor. The default ratio is two, but you can set it to a range of one to 50, by typing the appropriate positive number in the field.

MicroStation prompts:

> Zoom Out > Enter zoom center point *(Click the Data button to define the center of the design area to be displayed in the view window with decreased magnification (see Figures 3–40a and 3–40b) or click the Reset button to exit the Zoom Out tool and return to the previously active tool.)*

You continue clicking the Accept button to decrease magnification further. The Zoom Out tool remains active until you click the Reset button or invoke another tool.

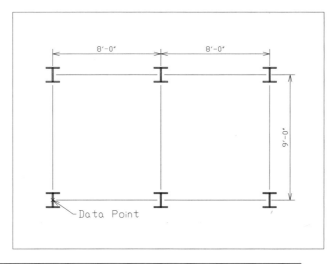

Figure 3-40a The design shown before the Zoom Out tool is invoked

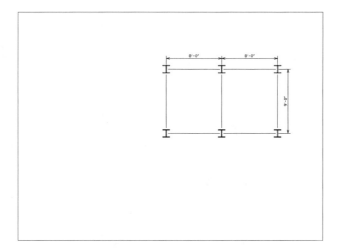

Figure 3-40b The design shown after the Zoom Out tool is invoked

Window Area

The Window Area tool allows you to specify the diagonally opposite points of a rectangle that defines the area of the design to be displayed in the view window. The center of the area selected is placed in the center of the view window and the design area inside the rectangle is enlarged to fill the view window as completely as possible.

Chapter 3 Fundamentals II

Invoke the Window Area tool from:

View Control tool box	Select the Window Area tool (see Figure 3–41).
Key-in window	**window area extended** (or **w a e**) ENTER

Figure 3-41 Invoking the Window Area tool from the View Control tool box

The Window Areas Tool Settings window contains the **Apply to Window** menu and check box. If the check box is set to ON, the window area you define is applied to the view window whose number is shown in the **Apply to Window** drop-down menu. If the check box is set to OFF, the new area is applied to the view window in which you define the area. To set the destination view window, turn ON the check box and click the drop-down arrow in the **Apply to Window** menu to display the numbers of all currently open view windows. Select the destination view window number from the list. The new area can only be applied to open view windows.

MicroStation prompts:

> Window Area > Define first corner point *(A full screen crosshair appears. Position the pointer at the point in the view window that you want to be one corner of a rectangle that defines the new displayed area and click the Data button, or click the Reset button to exit the Window Area tool and return to the previously active tool.)*
>
> Window Area > Define opposite corner point *(Position the pointer at the point in the view window that you want to be the diagonally opposite point of the rectangle and click the Data button, or click the Reset button to exit the Window Area tool and return to the previously active tool.)*

MicroStation applies the new area to destination view window (see Figures 3–42a and 3–42b). You can continue defining display areas or invoke another tool.

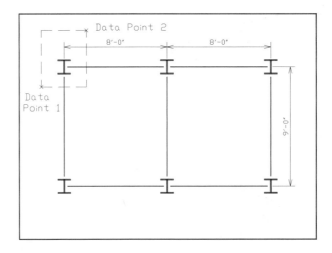

Figure 3-42a The design shown before the Window Area tool is invoked

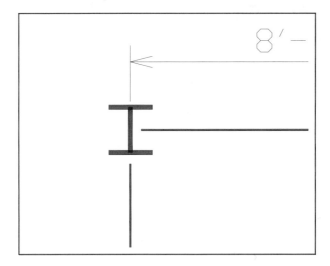

Figure 3-42b The design shown after the Window Area tool is invoked

Fit View

The Fit View tool displays the entire design in the selected view window. Invoke the Fit View tool from:

View Control tool box	Select the Fit View tool (see Figure 3–43).
Key-in window	**fit view extended** (or **fit v e**) ENTER

Chapter 3 Fundamentals II

Figure 3-43 Invoking the Fit View tool from the View Control tool box

The Fit View Tool Settings window contains an option, **Files**, that allows you to select the types of files that MicroStation fits within the selected view window. The **All** option causes MicroStation to include all types of files, including the design file, when it fits the view window. The **Active** option fits only the active design file. The **Reference** and **Raster** fit only the two types of references files. (Reference files are discussed in Chapter 13.) Click the drop-down arrow to display the file types to include in the fit and select an option by clicking on it. The default option is **All** and it is usually the selected option for fitting views.

MicroStation prompts:

> Fit View > Select view to fit *(Click the Data button anywhere in the view window to fit the entire design in the window or click the Reset button to exit the Fit View tool and return to the previously active tool.)*

MicroStation adjusts the view area and zoom factor as required to show all of the contents (elements) of the selected file types. (See Figures 3–44a and 3–44b for an example of fitting the active design file.

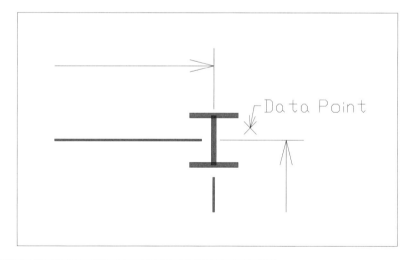

Figure 3-44a The view before the Fit View tool is invoked

117

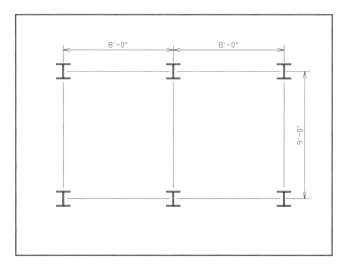

Figure 3-44b The view after the Fit View tool is invoked

Rotate View

The Rotate View tool allows you to rotate your view of the design in the view window about an imaginary Z axis that extends out from the view at a right angle to the X and Y axes. This tool rotates the X and Y axes in the view window, not the relation of the elements to the X and Y axes.

Invoke the Rotate View tool:

| View Control tool box | Select the Rotate View tool (see Figure 3–45). |

Figure 3-45 Invoking the Rotate View tool from the View Control tool box

The Rotate View Tool Settings window provides the **Method** menu that contains two rotation options. The **2 Points** option allows you to rotate the view using the Data button. The **Unrotated** option returns the view to normal rotation (X axis horizontal and Y axis vertical).

2 Points Option

To rotate the design within the view window, click the drop-down arrow in the **Method** field and select the **2 Points** option from the drop-down menu.

Chapter 3 Fundamentals II

MicroStation prompts:

> Rotate View > Define First Point *(Position the pointer in the view window at the point you want to be the center of the rotated view and click the Data button, or click the Reset button to exit the Rotate View tool and return to the previously active tool.)*
>
> Rotate View > Define X Axis of View *(Move the pointer until the view is at the rotation angle you want and click the Data button.)*
>
> Display Complete *(MicroStation rotates the design within the view window (see Figures 46a and 46b).)*

The tool is still active and your next data point will define a new center of view rotation, unless you click the Reset button or invoke another tool.

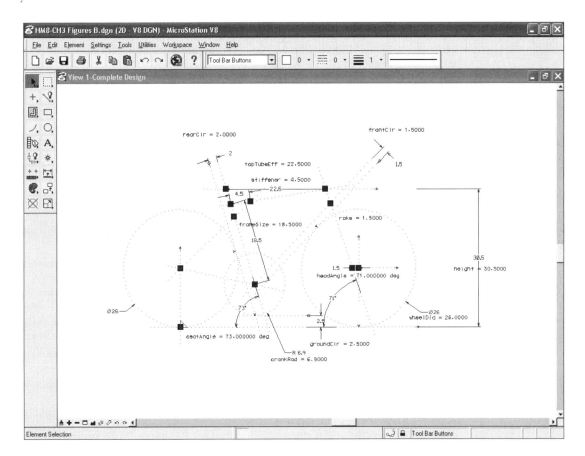

Figure 3-46a An Unrotated View Window

119

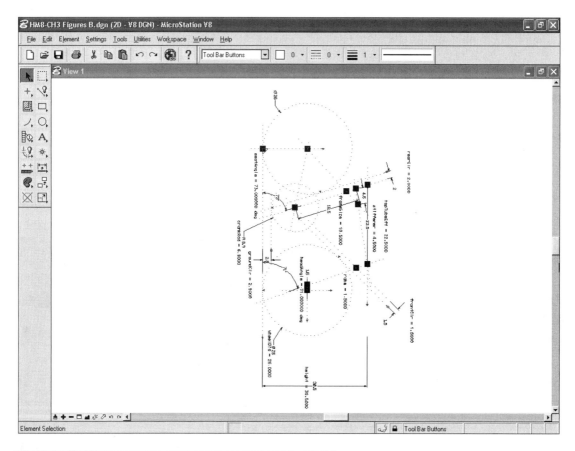

Figure 3-46b The View Window After 2 Point Rotation

Unrotated Option

To set the rotation of a view to zero degrees (display the X axis horizontally), click the drop-down arrow in the **Method** field and select the **Unrotated** option from the drop-down menu. If only one view window is open, MicroStation immediately sets the view rotation of the window to zero degrees. If more than one view window is open, MicroStation prompts:

> Top View > Select view *(Position the pointer anywhere in the view window and click the Data button, or click the Reset button to exit the Rotate View tool and return to the previously active tool.)*
>
> Display Complete *(MicroStation removed the rotation from the selected view.)*

The tool is still active and your next data point will remove rotation from the view window in which you click, unless you click the Reset button or invoke another tool.)

Pan View

The Pan View tool lets you move a view window over a part of the design that is not currently displayed in the view window. Invoke the Pan View tool from:

View Control tool box	Select the Pan View tool (see Figure 3–47).
Key-in window	**pan view** (or **pan v**) ENTER

Figure 3–47 Invoking the Pan View tool from the View Control tool box

The Pan View Tool Settings window provides the **Dynamic Display** check box that lets you control the way MicroStation pans the view. If the check box is turned OFF, you specify the pan direction and length with two data points and you do not see the panning results until after the second click. If the check box is ON, the view window follows the pointer as you drag it over the design, so you can continuously see the panning results as you move the pointer.

Panning with Dynamic Display OFF

Turn OFF the **Dynamic Display** checkbox in the Pan View Tool Settings window and MicroStation prompts:

> Pan View > Select view *(Position the pointer in the view window you want to move and click the Data button to select the view to pan and define the origin for panning, or click the Reset button to exit the Pan View tool and return to the previously active tool.)*
>
> Pan View > Define amount of panning *(Move the pointer the length you want to move the view window and in the direction you want to move it. MicroStation displays an arrow to indicate the direction and length of the pending pan movement. Click the Data button to initiate the move, or click the Reset button to exit the Pan View tool and return to the previously active tool.)*

Panning with Dynamic Display ON

Turn ON the **Dynamic Display** checkbox in the Pan View Tool Settings window and MicroStation prompts:

> Pan View > Select view *(Position the pointer in the view window you want to move and click the Data button to initiate moving the view window. Move the pointer until the part of the design you want to see is in the view window. Click the Data button again to complete the move.)*

If you click the data button after completing the move, a new move is started. To exit the Pan View tool and return to the previously active tool, click the Reset button or select another tool.

Alternately, you can press and hold down the Data button while you drag the view window until the desired part of the design is displayed. Release the button to complete the move.

View Previous

The View Previous tool displays the last displayed view. You can invoke this tool multiple times. For example, if you used the Zoom In tool two times, invoking the View Previous tool twice, moves the view window back to the zoom setting before it was zoomed in two times.

Invoke the View Previous tool:

View Control tool box	Select the View Previous tool (see Figure 3–48).
Key-in window	**view previous** (or **vi p**) ENTER

Figure 3-48 Invoking the View Previous tool from the View Control tool box

MicroStation prompts:

> View Previous > Select view (*Position the pointer anywhere in the view window you want change and click the Data button to move back to the previous view. Continue clicking the Data button until the desired previous view is displayed or there are no more previous views.*)

The View Previous tool remains active and each time you click the Data button a previous view is restored. Click the Reset button to exit the tool and return to the previously active tool, or activate another tool.

View Next

The View Next tool negates the view that was displayed by the View Previous tool. In other words, it moves forward through the previous views until the newest view is displayed again.

Invoke the View Next tool:

View Control tool box	Select the View Next tool (see Figure 3–49).
Key-in window	**view next** (or **vi n**) ENTER

Chapter 3 Fundamentals II

Figure 3-49 Invoking the View Next tool from the View Control tool box

MicroStation prompts:

> View Next > Select view *(Position the pointer anywhere in the view window you want to change and click the Data button to move forward to the next view. Continue clicking the Data button until the desired view is displayed or the newest view is reached.)*

The View Next tool remains active and each time you click the Data button a previous view is restored. Click the Reset button to exit the tool and return to the previously active tool, or activate another tool.

VIEW WINDOWS AND VIEW ATTRIBUTES

Thus far, you have been working in only one view window. That may have forced you to spend a great deal of time using the view commands to set up the view for the areas of the design that you needed to work with. MicroStation actually provides eight separate view windows that let you work in different parts of your design at the same time.

Each view window is identified by its view number (1–8). The view windows are similar to having eight zoom lens cameras that can be pointed at different parts of your design. For instance, in one view window you might display the entire drawing; in two other view windows you might be zoomed in close to different areas to show greater detail (see Figure 3–50). All eight view windows can be open at the same time on your monitor or on either monitor of a two-monitor workstation.

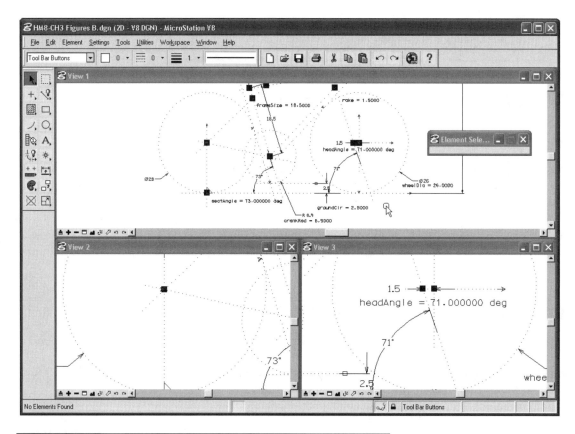

Figure 3-50 Three views showing different portions of a design

The View Control tools, which you have already been introduced to, work in any open view window. Let's look at the View tools that let you to open and close, position, and size the eight view windows.

Opening and Closing View Windows

View windows can be opened and closed from the **Windows > Views** submenu, from the View Groups settings box, and by key-in commands. To open or close one view window, select:

Drop-down menu	Window > Views > (Select a view number) (see Figure 3-51.)
Key-in window	**view off** (or **vi off**) or **view on** (or **vi on**) ENTER

Chapter 3 Fundamentals II

To quickly open or close several view windows from the View Groups settings box, select:

| Drop-down menu | Window > Views > Dialog (see Figure 3–52) |

 Note: When you select Window > Views, a submenu opens that displays the Dialog option and the numbers of the eight views. A checkmark to the left of an option in the submenu indicates that the View Groups settings box or view is already open on the MicroStation desktop. If you click an option with a checkmark, the option is turned off and vise-versa.

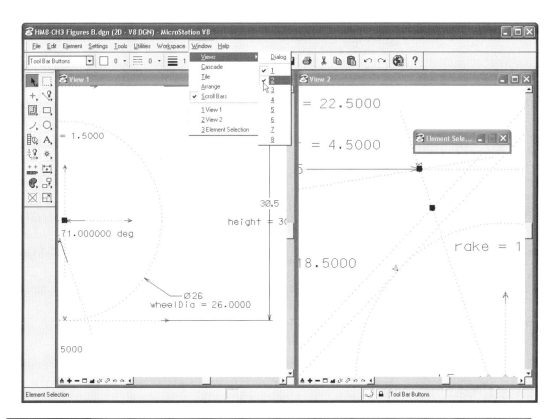

Figure 3-51 View of the Windows > Views submenu with The View Groups Dialog box and two views open

Figure 3-52 View of the View Groups settings box showing two view windows open

125

In addition to the view open and close options just discussed, most operating systems provide a tool for closing a window in the window's Control menu. Figure 3–53 shows an example of the window control drop-down menu in a view window.

Figure 3-53 View Window close option in the Microsoft Windows XP view window

Note: The view windows you open and close apply only to the current editing session. If you want the current arrangement of the view windows to be the same the next time you open the design file in MicroStation, select the **Save Settings** option from the **File** drop-down menu.

Arranging Open View Windows

The working area can become cluttered when several view windows are open and MicroStation provides three housekeeping tools for cleaning up the clutter: **Cascade**, **Tile**, and **Arrange**. The tools are provided in the Window drop-down menu, as shown in Figure 3–54.

Chapter 3 Fundamentals II

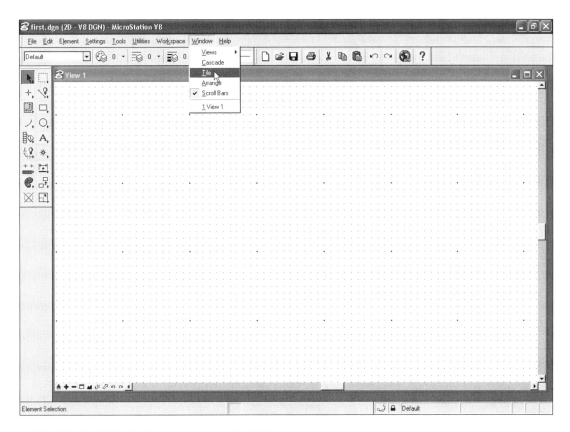

Figure 3-54 The Window drop-down menu

Cascade

The Cascade tool stacks all open view windows in numerical order, with the lowest-numbered view window on top and the other view window title bars visible behind it, as shown in Figure 3–55.

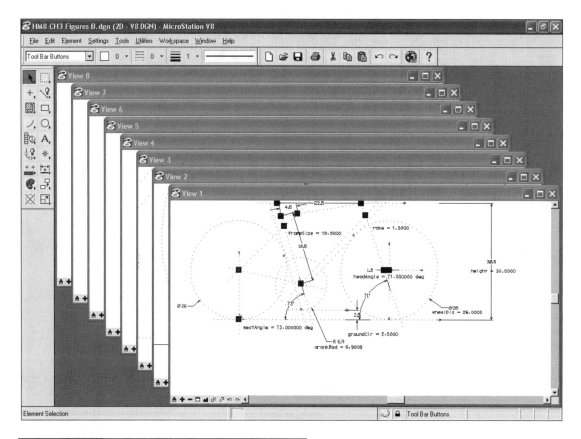

Figure 3-55 Example of cascaded view windows

To cascade the open view windows, select the Cascade tool:

Drop-down menu	Window > Cascade
Key-in window	**window cascade** (or **w c**) ENTER

The open view windows are cascaded (there are no MicroStation prompts). To work on a specific view, click its Title bar and it pops to the top of the stack.

Tile

The Tile tool arranges all open view windows side by side in a tiled fashion, with the lowest-numbered view window in the upper left, as shown in Figure 3–56.

Chapter 3 Fundamentals II

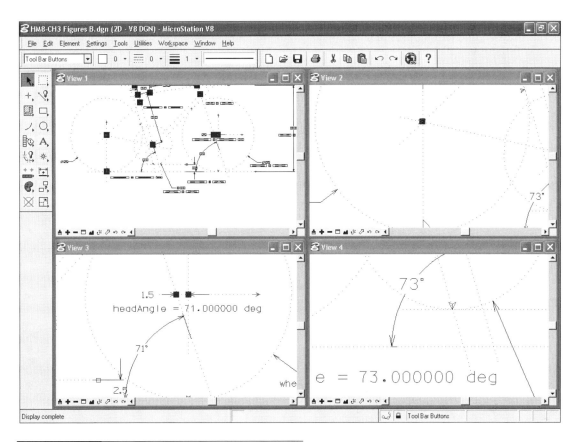

Figure 3-56 Example of four tiled view windows

To tile the open view windows, select the Tile tool:

Drop-down menu	Window > Tile
Key-in window	**window tile** (or **w t**) ENTER

The open view windows are tiled (there are no MicroStation prompts).

Note: The view arrangement commands do not place any part of the views behind tool boxes that are docked to the edges of the MicroStation desktop. For example, because the tool frame is docked to the left side of the desktop, in Figures 3–55 and 3–56, the desktop area below the tool frame is not used.

129

Arrange

The Arrange tool can size and move all open view windows as necessary to fill the MicroStation application window. The tool attempts to keep each view window as close to its original shape, size, and position as possible, as shown in Figures 3–57a and 3–57b.

To arrange the open view windows, select the Arrange tool:

Drop-down menu	Window > Arrange
Key-in window	**window arrange** (or **w arr**) ENTER

The open view windows are arranged to fill the MicroStation application window (there are no MicroStation prompts).

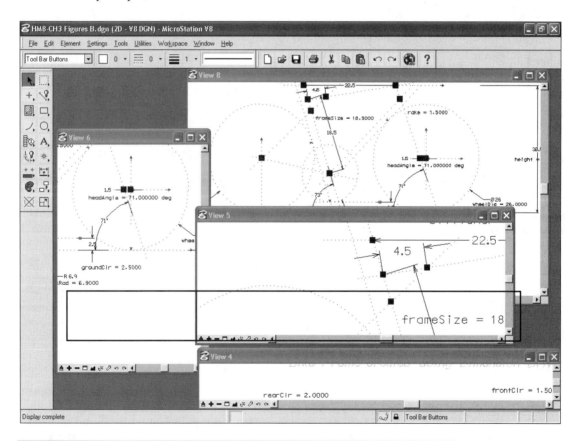

Figure 3-57a Example of four view windows before using Window>Arrange

130

Chapter 3 Fundamentals II

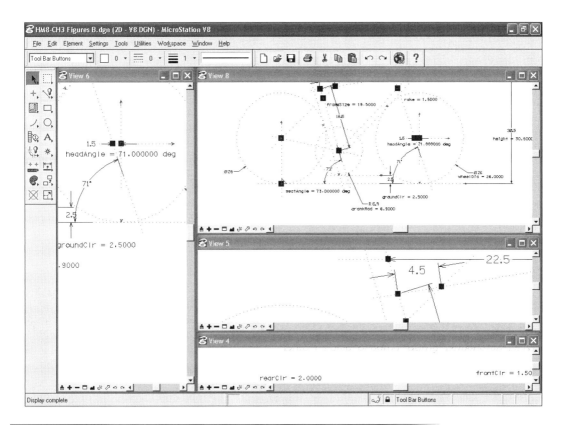

Figure 3-57b Example of four view windows after using Window>Arrange

Arranging Individual View Windows

MicroStation provides a group of tools to control the size and position of individual view windows, in addition to the tools that open and close windows. You can move a view window to a new location on the desktop, resize it, minimize it, maximize it, and pop it to the top when it is buried under a stack of other view windows.

Moving a View Window

To move a window to a new location in the MicroStation desktop:

1. Position the pointer over the view window's title bar.
2. Press and hold the Data button.
3. Drag the window to the new location.
4. Release the Data button.

The screen pointer changes to a different shape when it is over the title bar. As you drag the window, the window (or its outline) follows the screen pointer.

131

Resizing a View Window

To change the size of a window:

1. Position the pointer over the view window's border.
2. Press and hold the Data button.
3. Drag the border to a new position.
4. Release the Data button.

The screen pointer changes to a two-headed arrow when it is over a window border. As you drag the window border, the window (or its outline) follows the screen pointer.

If you grab the border on a side, you can change the size in only one dimension (height or width). If you grab the border on a corner, you can change the height and width at the same time.

Minimizing and Maximizing a View Window

View windows contain **Minimize** and **Maximize** buttons on the title bar (see Figure 3-58).

- Click the **Minimize** button to reduce the view window to its minimum possible size.
- Click the **Maximize** button to expand the view window to fill the MicroStation desktop and pop it to the top of other open view windows.

Note: The **Minimize** button reduces the window to a shortened version of the window's title bar and places it in the lower left corner of the MicroStation desktop. You may not be able to see it because the minimized window is placed behind any open windows that are located at the lower left corner.

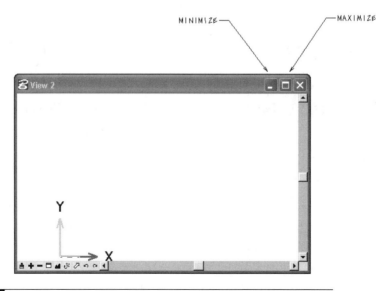

Figure 3-58 Minimize and Maximize buttons in Microsoft Windows XP

To return a view to the size it was before you clicked the Minimize or Maximize button, click the button again. For example, if you maximized a view window, click its Maximize button to return it to the size it was before being maximized.

Note: The appearance of the title bar buttons and of the window move and resize pointers varies among operating systems. For example, the resize pointer in Microsoft Windows is a two-headed arrow pointing in the directions in which the window border can be dragged.

Finding a View Window

The tools used to change the size and position of a view window may cause the window to cover all or part of other view windows, thus creating a "stack" of windows. There are two ways to bring a buried window to the top of the view window stack:

- If you can see any part of the buried view window's border or title bar, click on the title bar or border.

- The names of all open view windows are displayed at the bottom of the **Window** drop-down menu. To bring a hidden window to the top of the stack, click the **Window** menu and click on the view name in the drop-down menu. If the selected view was minimized, it is returned to the size it was before it was minimized.

Turning the View Window Scroll Bars ON and OFF

Each view window contains a horizontal and vertical scroll bar that can be used to move the view window over the design plane. These bars can be turned OFF to see a little more design plane area in each view window.

To turn view window scroll bars ON and OFF, select:

| Drop-down menu | Window > Scroll Bars |

This command toggles ON and OFF the state of the scroll bars in all open view windows every time you select it. The scroll bars are turned ON or OFF immediately (there are no MicroStation prompts).

Note: The View Control Bar in the bottom left corner of the view windows is also turned off when the scroll bars are turned OFF. To access the view control tools when scroll bars are turned off, open the **View Control** tool box from the **Tools** drop-down menu.

Creating and Using View Window Groups

MicroStation provides tools for creating and recalling named groups of view windows in a design. When a group is created, it saves the currently open view windows and the alignment of each window. When a group is selected for display, the group's view windows open on the MicroStation desktop and all other view windows close. The open windows display the same areas of the design that were showing when the group was created.

View groups are useful for setting up optimal view windows for a particular type of design. For example, a floor plan might have three groups:

- A "Rooms" group that opens all eight view windows with each view window aligned to a different room in the plan.
- A "Notes" group that opens six view windows aligned to construction notes areas on the design.
- A "Border" group that opens three view windows aligned to areas of the design border.

The same view windows can be used in more than one group. For example, when the Rooms group is selected, view one displays the living room, when the Notes group is selected, view one displays notes about electrical connections, and when the Border group is selected, view one displays the design's title block.

Selecting View Groups

A view group is selected for viewing by clicking its name in the **View Groups** drop-down menu on the Views Dialog box. This is the same settings box that we discussed earlier when we learned how to open individual views.

If the Views Dialog box is not already open on the desktop, select:

Drop-down menu	Window > Views > Dialog (a check mark next to the option name indicates that the settings box is already open on the desktop.)

To display a View Group, click the drop-down arrow on the **View Groups** menu and select a view group name from the list, as shown in Figure 3–59.

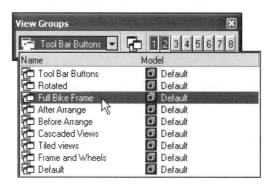

Figure 3-59 The View Groups settings box showing a typical View Groups drop-down menu

Note: If you select **File > Save Settings** after opening or closing view windows, or changing the zoom factor and alignment of view windows, the current arrangement is applied to the displayed view group. Thereafter the view group will display the saved view window settings, not its original settings. Before saving settings, display the Default view group. There is always at least a "Default" view group (but the default view group can be deleted).

Managing View Groups

MicroStation provides tools for creating, editing, and deleting view groups in the Manage View Groups settings box.

Invoke the Manage View Groups settings box from:

| View Groups settings box | Click the Manage View Groups button. |

The Manage View Groups settings box opens with all existing groups displayed (see Figure 3–60). The menu bar in the box has options to **Create View Group**, **Edit View Group Properties**, and**, Delete View Group**, and **Apply.** There is also an **Apply** button that defines an existing view group for the currently open view windows.

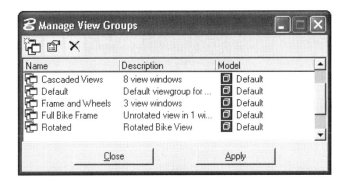

Figure 3-60 The Manage View Groups settings box showing a typical set of groups

 Note: New groups and changes to existing groups are saved automatically to the design file. It is not necessary to save settings to retain them.

The **Create View Group** tool saves the size, alignment, and zoom level of each open view window. Before creating a new group, make sure the view windows are set up the way you want them to be in the group.

To create a new view group:

| Manager View Groups settings box | Click the Create View Group button |

MicroStation displays the Create View Group settings box (see Figure 3–61). To create the new View Group, enter a group name in the **Name** field, a description in the **Description** field, and click **OK** to save the new view group. Or click **Cancel** to close the settings box without creating a new group. The new group appears in the groups list of the Manage View Groups and View Groups settings boxes.

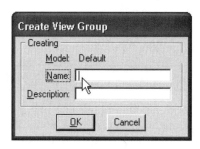

Figure 3-61 The Create View Group settings box

Edit View Group Properties

A view group's properties (**Name** and **Description**) can be changed from the View Group Properties dialog box. Invoke the dialog box from:

| Manager View Groups settings box | Select the view group to be edited and click the Edit View Group Properties button. |

MicroStation displays the View Group Properties dialog box (see Figure 3–62).

Figure 3-62 The View Group Properties dialog box

Edit the group name in the **Name** field, edit the group description in the **Description** field, and click **OK** to save the new properties, or click **Cancel** to close the dialog box without changing the properties.

Delete View Group

The third tool on the menu bar of the Manage View Groups settings box is **Delete View Group**. Use this option to remove a selected view group from the design file. Invoke the **Delete View Group** from:

| Manage View Groups settings box | Select the view group to be deleted and click the Delete View Group button. |

MicroStation displays an Alert message box (see Figure 3–63) that asks you to confirm that you want to delete the view group. Click **OK** to delete the selected view group, or click **Cancel** to close the Alert box without deleting the view group.

Figure 3-63 The Delete View Group Alert box

Apply

To change the view windows definition (number, position, content, and shape) of an existing view group, make the required changes to the view windows, open the Manage View Groups settings box, select the name of the view group you want to redefine, and click the **Apply** button. The open view windows are saved under the selected view group name.

Setting View Attributes

MicroStation provides a View Attributes settings box that contains options to control the way elements and drawing aids appear in a view window. Turning OFF certain attributes can speed up the time to update a view and reduce clutter in a view. For example, if a view window contains a large amount of patterning, turn off the display of patterns to greatly reduce the update time on a slow workstation.

Invoke the View Attributes settings box from:

| Drop-down menu | Settings > View Attributes |

MicroStation displays the View Attributes settings box as shown in Figure 3–64, each view attribute is explained in Table 3–1.

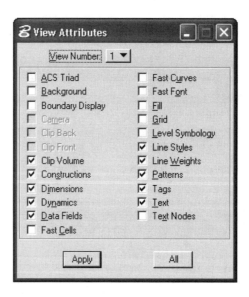

Figure 3-64 View Attributes settings box

Follow these steps to use the settings box to change view attributes:

1. Turn the appropriate attribute check box ON or OFF as required.
2. To set the attributes for all open view windows, click the **All** button.
3. To set the attributes for one view window, click the drop-down arrow in the **View Number** menu, select the number of the view from the drop-down menu, and click the **Apply** button.

Table 3-1 View Attributes

ATTRIBUTE	TURNS ON AND OFF THE DISPLAY OF...
ACS Triad	The Auxiliary Coordinate System (ACS).
Background	The background image loaded with the Active Background command.
Boundary Display	Used in 3D designs to control whether the boundaries of the clip volume are displayed for a given view, as well as reference clip boundaries.
Camera	The 3D view camera.
Clip Back	Used in 3D designs to toggle the display of elements, and parts of elements, located outside a 3D view's clipping planes. If ON, a back clipping plane is active in a view.
Clip Front	Used in 3D designs to toggle the display of elements, and parts of elements, located outside a 3D view's clipping planes. If ON, a front clipping plane is active in a view.
Clip Volume	Used in 3D designs to toggle the display of elements, and parts of elements, located outside a defined Clip Volume, for a given view. If ON, and a clip volume has been applied to the view, the view volume is restricted to the defined volume. If no clip volume has been applied to the view, it has no effect.
Constructions	Elements placed with Construction Class mode active.
Dimensions	Dimension elements.
Dynamics	Dynamic updating of elements as they are placed in the design.
Data Fields	Data Field placeholder characters.
Fast Cells	The actual cells or a box indicating the area of the design occupied by cells.
Fast Curves	The actual curve string or straight line segments indicating the vertices.
Fast Font	The actual text font for each text element or the MicroStation fast font.
Fill	The fill color in filled elements.
Grid	The grid (if the view is zoomed out far enough, the grid will be turned off even if this attribute is on).
Level Symbology	Elements according to the symbology table rather than the actual element symbology.
Line Styles	Elements with their actual line weights (when it is set to OFF, all elements are displayed with style 0, solid).

Table 3-1 View Attributes (continued)

Line Weights	Elements with their actual line weights (when it is set to OFF, all elements are displayed at line weight 0).
Patterns	Pattern elements.
Tags	The tag information for tagged elements.
Text	The display of text elements (when it is set to OFF, no text elements are displayed).
Text Nodes	Text nodes as small crosses with numeric identifiers.

Note: Changes to View Attributes setting remain in effect until they are changed or the design file is closed. To keep them in effect for future editing sessions, select **Save Settings** from the **File** drop-down menu.

Saving Views

If you regularly work in several specific areas of a design, MicroStation provides a way for you to return to those areas quickly by saving a view window's zoom factor, alignment, displayed levels, and view attributes under a user-defined name. To return to a saved view, you provide the saved view name and the number of the view window in which you want to place the saved view.

The tools for displaying and saving views are in the Saved Views settings box. Invoke the settings box from:

Drop-down menu	Utilities > Saved Views
View window control menu	Click the View Save/Recall option

MicroStation displays the Saved Views settings box as shown in Figure 3–65.

Chapter 3 Fundamentals II

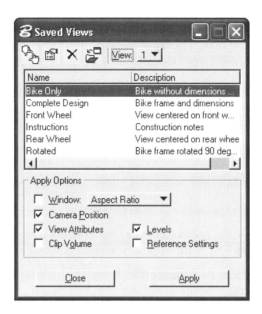

Figure 3-65 Saved Views settings box

The name and description of all saved views are displayed in the settings box under the **Name** and **Description** columns. Above the list of saved views is a menu bar containing options for managing saved views. Below the list of saved views are options that affect the way saved views are displayed.

Display a Saved View

To display a saved view using the Saved Views settings box:

1. Position the pointer over the saved view's name and click the Data button to select it.
2. On the toolbar, click the drop-down arrow next to the **View** option and select the number of the view window in which you want to display the saved view.
3. In the **Apply Options** section, choose the options you want to apply to the selected view and clear the other check boxes (See the following option descriptions).
4. Click the **Apply** button.

The saved view, whose name you selected, is loaded in the view window you selected.

You also can restore a view by keying-in at the key-in window field **vi**=<name>, and pressing ENTER. Replace <name> with the name of the view to be restored. MicroStation prompts:

> Select view *(Click the Data button in the view where you want to restore the view.)*

Apply Options

The **Apply Options** check boxes on the Saved Views settings box affect the way the saved view appears in the target view window. Following are descriptions of each option:

- Turning ON the **Window** check box enables a drop-down menu of options that control the shape and size of the selected window:

    - Select the **Aspect Ratio** option to set the height and width of the selected view window, proportionate to the shape of the window from which the view was originally saved. For example, if the original window was twice as wide as it was tall, the window in which the saved view is placed will also be twice as wide as it is tall. The longer of the selected view window's two dimensions (horizontal or vertical) is adjusted to make the window proportional.

    - Select the **Size** check box to set the selected view window the same size and shape as the window from which the view was originally saved.

    - Select the **Size and Position** option to make the selected view window match the size, shape, and location on the MicroStation desktop, of the window from which the view was originally saved.

- Turning ON the 3D **Camera Position** check box applies the camera settings, of the saved window, to the selected view window.

- Turning ON the **View Attributes** check box applies the view attributes settings, of the saved view, to the selected view.

- Turning ON the 3D **Clip Volume** check box applies the clip volume, of the saved view, to the selected view.

- Turning ON the **Levels** check box applies the displayed levels, of the saved view, to the selected view.

Note: The Active View is not impacted by the **Levels** check box. It remains at its current setting.

- Turning ON the **Reference Settings** applies the reference file view settings, of the saved view, to the selected view. (Reference files are discussed in Chapter 13.)

Save View

Before saving a view, set up a view window to display the area of the design you want to save, set the window's view attributes (**Settings > View Attributes**) to the values you need. Next, turn on the display of the levels you want (**Settings > Level > Display**), and then invoke the Save View settings box from:

Chapter 3 Fundamentals II

| Saved Views settings box | Click the Save View button (the left-most button on the tool bar that shows a pair of hands). |

MicroStation displays the Save View settings box, as shown in Figure 3–66.

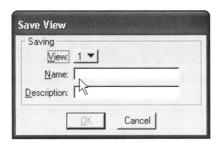

Figure 3-66 Save View settings box

In the Save View settings box:

1. Click the drop-down arrow on the **View** menu and select the number of the view window that contains the view you want to save (Only numbers for currently open views can be selected).
2. Type a name for the view in the **Name** field. The name can contain letters, numbers and the "$", ".", and "_" characters. It can be up to 511 characters long, but keep in mind that the Name field in the Saved Views settings box can only display the first 20, or so, characters of the name.
3. Optionally, type a description of the saved view in the **Description** field.
4. Click the **OK** button to save the view, or click the **Cancel** button to close the settings box without saving the view.

You also can save a view by typing in the key-in window: **sv=**<name>,<description> and pressing ENTER. Replace <name> with the name you select for the saved view and <description> with a description of the saved view. MicroStation prompts:

> Save Named View > Select view *(Position the pointer in the view window you want to save and click the Data button.)*

Edit Saved View Properties

The properties of a Saved View include the view's name and description. Invoke the Edit Properties dialog box from:

| Saved Views dialog box | Click the Edit Saved View Properties button (second button from the left on the menu bar). |

143

MicroStation displays the Edit Properties dialog box as shown in Figure 3–67.

Figure 3–67 Edit Properties dialog box

In the Edit Properties dialog box, edit the **Name** and **Description** as required. When the changes are complete, click the **OK** button to change the properties, or click **Cancel** to leave the properties unchanged.

Delete Saved View

To delete a saved view using the Saved Views settings box, click the name of the view you want to delete and invoke the **Delete Saved View** tool from:

Saved Views settings box	Click the Delete Saved View button (third button from the left on the menu bar).

MicroStation displays an Alert box that asks you to confirm the deletion of the selected saved view. Click the **OK** button to delete the selected saved view, or click **Cancel** to cancel deleting the saved view.

You also can delete a view by typing into the key-in window: **dv**=<name> and pressing ENTER. Replace <name> with the name of the saved view to be deleted. The saved view is immediately deleted.

Import Saved View

You can import saved views from other design files into your current design file. Imported views can be recalled and managed the same as any other saved views contained in the current design file. Invoke the Saved Views settings box from:

Saved Views settings box	Click the Import Saved View button (fourth button from the left on the menu bar).

MicroStation displays the Open File dialog box, as shown in Figure 3–68.

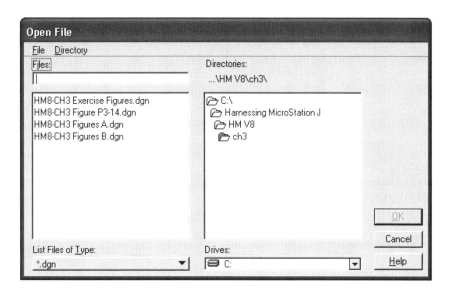

Figure 3-68 The Open File dialog box

In the **Directories** field, navigate to the folder that contains the design file with the saved views you want to import, then select the design file in the **Files** field. Click the **OK** button to continue importing views, or click the **Cancel** button to close the dialog box without importing any views.

MicroStation displays a list of the saved views that can be imported, in the Import Saved Views settings box, as shown in Figure 3–69.

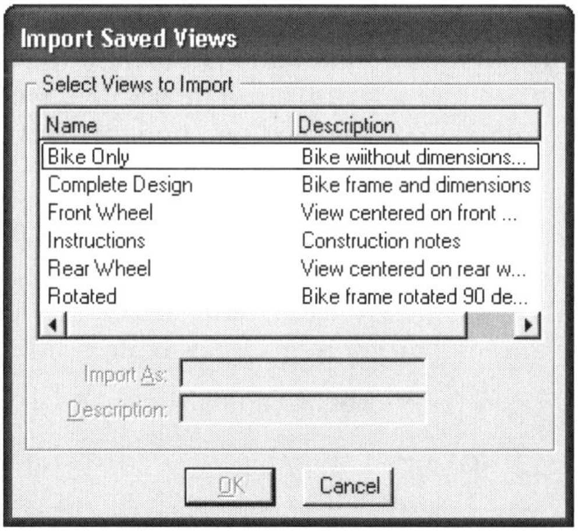

Figure 3-69 The Import Saved Views dialog box

In the Import Saved Views dialog box, the saved views are listed in the **Select Views to Import** section (If the design file does not contain any views, the list will be empty). Use the following methods to select the saved views you want to import:

- To select one saved view for import, click that saved view's name.
- To select a series of saved views, click on the name of the first view in the series, hold down SHIFT, and click on the name of the last view in the series. The selected views, and all saved views between the two selections, are selected for import.
- To select several individual saved views, click on the name of the first view, hold down CTRL while you click on the name of each subsequent saved view. Only the views you clicked will be selected for import.
- To select all saved views in the list, hold down CTRL and tap A.
- To unselect a selected saved view, hold down CTRL while you click on the view name.

Click the **OK** button to import the selected saved views, or click the **Cancel** button to cancel the import operation. After you click **OK**, the selected saved views are imported and will appear in the Saved Views settings box of the active design file.

UNDO AND REDO

The Undo tool cancels the effects of the previous commands. The Redo tool reverses the effects of the previous Undo. For example, if you place a circle in your design and then select the Undo tool, the circle is removed from the design. If you then select the Redo tool, the circle is placed back into the design.

Commands can be undone because all the steps you executed for each command are stored in an Undo buffer. The Undo tool goes to that buffer to get the information necessary to put things back the way they were before the tool was invoked. The last command you executed is the first one undone, the next-to-last command is the next one undone, and so on.

Note: The Undo tool can undo an unlimited number of commands that were executed during the current editing session. When you close a design file the Undo buffer is lost. Therefore, the next time you open the design file for editing you will not be able to undo the effects of commands issued in the previous editing session – there are other manipulation tools are used to alter changes made during previous editing sessions.

Undo Tool

The Undo tool permits you to undo the effects of the last command. To undo the last command, invoke the **Undo** tool from:

Standard tool box	Click the Undo button (see Figure 3–70).
Key-in window	**undo** (or **und**) ENTER

Figure 3-70 Invoking the Undo tool from the Standard tool box

To undo the last drawing operation, you can also select the **Undo (action)** option from the **Edit** drop-down menu. MicroStation displays the name of the last command operation that was performed, in the **Edit** drop-down menu, in place of **(action)**.

Set Mark and Undo Mark

If you are at a point in the editing session where you need to experiment with the design, but you want to be able to undo the experiment, you can place a Mark in the design before you start and then, if necessary, undo back to the Mark.

To place a Mark, invoke the **Set Mark** tool from:

Drop-down menu	Edit > Set Mark
Key-in window	**mark** (or **mar**) ENTER

To undo all the steps back to when the Mark was placed, invoke the **Undo Mark** tool from:

Drop-down menu	Edit > Undo Other > To Mark
Key-in window	**undo mark** (or **und m**) ENTER

All the commands, after the Mark was placed, are undone and the Mark is also removed.

Undo All

The **Undo All** tool lets you negate all of the drawing commands recorded in the Undo buffer. To undo all the drawing operations recorded in the Undo buffer, invoke the **Undo All** tool from:

Drop-down menu	Edit > Undo Other > All
Key-in window	**undo all** (or **und a**) ENTER

MicroStation displays an Alert box that asks you to confirm that you want to undo all changes. Click the **OK** button to undo all commands issued, or click the **Cancel** button to cancel the **Undo All** operation.

Redo Tool

Each time you invoke the **Redo** tool the effects of the last Undo are negated. You can Redo a series of Undo operations by repeatedly choosing **Redo**.

Invoke the **Redo** tool from:

Standard tool box	Select the Redo tool (see Figure 3–71).
Key-in window	**redo** ENTER

Figure 3-71 Invoking the Redo tool from the Standard tool box

Things to Consider Before Undoing

Following are two points to consider before invoking the Undo or Redo tools.

- The Undo commands back up through the Undo buffer. Sometimes, Undo is not the best way to clean up a problem. For example, if five commands ago you placed a circle you now want to remove, you must Undo the last four commands issued before you can Undo the circle placement. In this case, a better way to get rid of the circle is with the **Delete Element** tool.

- When you use one of the Undo commands, you are undoing a command, not an element. If the command manipulated multiple elements, Undo negates the manipulation of all of those elements. For example, the Fence commands can manipulate hundreds of elements at one time. If you undo **Fence Contents Delete**, you get back all the elements that the fence deleted.

Chapter 3 Fundamentals II

 Open the Exercise Manual PDF file for Chapter 3 on the accompanying CD for project and discipline specific exercises.

REVIEW QUESTIONS

Write your answers in the spaces provided.

1. Briefly explain the difference between the Zoom In and Zoom Out tools.
   _____
   _____

2. Which tool will get you closer to a portion of your design by a factor of 2?
   _____

3. Panning lets you view_____.

4. To dynamically pan in a view, select the _____ tool, turn ON the _____ check box, and press the _____ button while you move the pointer to drag the view window over the design.

5. What is the purpose of Grid lock? _____ .

6. The _____ Category in the _____ dialog box allows you to control the display of the grid.

7. The Grid Master unit defines the distance between the _____ and is specified in terms of the design's _____ _____ .

8. The Grid Reference defines the distance between the _____ .

9. To keep the grid settings in effect for future editing sessions for the current design file, select _____ from the _____ drop-down menu.

10. The Grid Aspect Ratio field in the Design File dialog box allows you to set the
    _____ .

11. Name the two settings boxes in which you can set the Snap Lock mode:
    _____ and _____ .

12. Name four of the Snap modes available in MicroStation. _____ , _____ , _____ , and _____ .

149

13. Keypoint mode allows tentative points to snap to _____ _____.on elements.

14. The Midpoint mode snaps to the _____ of a circle and an ellipse.

15. Name three ways AccuSnap can indicate that it has snapped to a point? _____ , _____ , and _____ .

16. How do you accept an AccuSnap snap point? _____ .

17. The Axis Lock forces each data point _____ .

18. List four attributes associated with the placement of elements: _____ , _____ , _____ , and _____ .

19. How many levels can be defined in a design file? _____

20. How many levels can be active at any time? _____

21. How many levels can you turn ON or OFF at any time in a specific view window? _____

22. How many colors or shades of colors are available in the color palette? _____

23. What two-letter key-in makes a color active? _____

24. How many internal Line Styles are available in MicroStation? _____

25. In MicroStation, Line Weight refers to the _____ of lines.

26. How many line weights are available in MicroStation? _____

chapter 4

Fundamentals III

Objectives

After completing this chapter, you will be able to do the following:

- Draw ellipses, polygons, point curves, curve streams, and multi-lines
- Modify elements: fillets, chamfers, trim, and partial delete
- Manipulate elements: move, copy, move and copy parallel, scale original and copy, rotate original and copy, mirror original and copy, and array
- Place text: Set text parameters and place text by origin

PLACEMENT TOOLS

In this chapter, four more placement tools are explained: Ellipse, Curve (Place Point Curve and Place Curve Stream), and Multi-line tools. This adds to the placement tools already described in Chapter 2.

Place Ellipse

MicroStation offers two methods for placing an ellipse: **Center** and **Edge**. The method is selected from the **Method** option menu in the Tool Settings window when the Place Ellipse tool is active.

Center

The **Center Method** places an ellipse by defining three points: the center, one end of the primary (major) axis, and one end of the secondary (minor) axis.

Invoke the Place Ellipse By Center tool from:

Ellipses tool box	Select the Place Ellipse tool and select Center from the Method menu on the Tool Settings window (see Figure 4–1).
Key-in window	**place ellipse center constrained** (or **pl el ce co**) ENTER

Figure 4-1 Invoke the Place Ellipse By Center and Edge tool from the Ellipses tool box

MicroStation prompts:

> Place Ellipse By Center and Edge > Identify Ellipse Center *(Place a data point or key-in coordinates to define the center of the ellipse.)*
>
> Place Ellipse By Center and Edge > Identify Ellipse Primary Radius *(Place a data point or key-in coordinates to define the one end of the primary axis.)*
>
> Place Ellipse By Center and Edge > Identify Ellipse Secondary Radius *(Place a data point or key-in coordinates to define the one end of the secondary axis.)*

For example, the following tool sequence shows how to place an ellipse by Center (see Figure 4–2).

> Place Ellipse By Center and Edge > Identify Ellipse Center *(In the Key-in window, type **XY=3,2** ENTER)*
>
> Place Ellipse By Center and Edge > Identify Ellipse Primary Radius *(In the Key-in window, type **DL=2,0** ENTER)*
>
> Place Ellipse By Center and Edge > Identify Ellipse Secondary Radius *(In the Key-in window, type **XY=3,3** ENTER)*

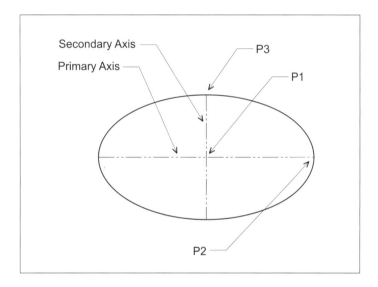

Figure 4-2 Example of placing an ellipse by means of the Place Ellipse by Center and Edge tool

Options in the Ellipse Tool Settings window allow you to constrain the **Primary** axis, **Secondary** axis, and **Rotation** angle key-in. To set a constraint, turn ON the appropriate check boxes and key-in the constraining value in the Tool Settings window. For example, to constrain the **Primary** axis, turn ON the **Primary** check box, key-in a value for the radius of the axis in the **Primary** text field and press either ENTER or TAB.

When you constrain one or more of the three values, MicroStation adjusts its prompts to handle the constraint. For example, if the **Primary** and **Secondary** radii are constrained, MicroStation prompts you to identify the ellipse center and rotation data points. If the **Primary** radius, **Secondary** radius, and the **Rotation** are all constrained, MicroStation only prompts you to identify the ellipse center point.

Place Ellipse By Edge Points

The **Edge Method** of the Place Ellipse tool (Place Ellipse By Edge Points) enables you to draw an ellipse by defining three points on the ellipse.

Invoke the Place Ellipse By Edge Points tool from:

Ellipses tool box	Select the Place Ellipse tool, and select Edge from the Method menu on the Tool Settings window (see Figure 4–3).
Key-in window	**place ellipse edge constrained** (or **pl el ed co**) ENTER

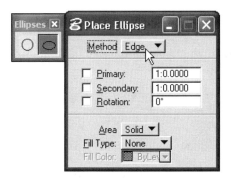

Figure 4-3 Invoke the Place Ellipse By Edge Points tool from the Ellipses tool box

MicroStation prompts:

Place Ellipse By Edge Points > Identify Point on Ellipse *(Place a data point or key-in coordinates to define the first point on the ellipse.)*

Place Ellipse By Edge Points > Identify Point on Ellipse *(Place a data point or key-in coordinates to define the second point on the ellipse.)*

Place Ellipse By Edge Points > Identify Point on Ellipse *(Place a data point or key-in coordinates to define the third point on the ellipse.)*

For example, the following tool sequence shows how to place an ellipse with the Place Ellipse by Edge Points tool (see Figure 4–4).

Place Ellipse By Edge Points > Identify Point on Ellipse *(In the Key-in window, type* **XY=1,2** ENTER*)*

Place Ellipse By Edge Points > Identify Point on Ellipse *(In the Key-in window, type* **XY=3,3** ENTER*)*

Place Ellipse By Edge Points > Identify Point on Ellipse *(In the Key-in window, type* **DL=4,0** ENTER*)*

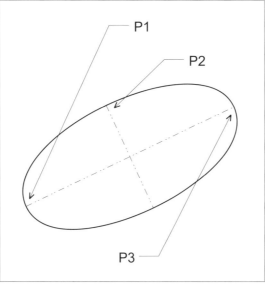

Figure 4-4 Example of placing an ellipse by edge

The Edge method provides the same placement constraints as the Center method. You can constrain the **Primary** radius, **Secondary** radius, and **Rotation**. For a discussion of the constraints see the Center method discussion.

 Note: Similar to placing arcs, you can also place half and quarter ellipses. The tools are located in the Arcs tool box.

Place Regular Polygon

With the Place Regular Polygon tool, you can place regular polygons (all edges are equal length, all vertex angles are equal) with 3 to 4,999 edges. MicroStation provides three methods for placing regular polygons: **Inscribed**, **Circumscribed**, and **By Edge**. The appropriate method is selected from the **Method** option menu on the Tool Settings window.

Inscribed Method

The **Inscribed Method** places a polygon that is inscribed inside an imaginary circle, whose diameter is equal to the distance across opposite polygon vertices (see Figure 4–5).

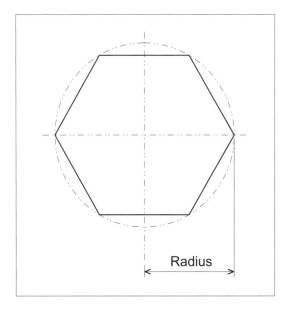

Figure 4-5 Example of displaying a polygon inscribed inside an imaginary circle

Invoke the Place Inscribed Polygon tool from:

Polygons tool box	Select the Place Regular Polygon tool and select Inscribed from the Method menu located on the Tool Settings window (see Figure 4–6).
Key-in window	**place polygon inscribed** (or **pl pol in**) ENTER

Figure 4-6 Invoke the Place Polygon Inscribed tool from the Polygons tool box

Key-in the number of polygon edges (from 3 to 4,999) in the **Edges** field on the Tool Settings window. Optionally, key-in the radius of the imaginary circle in the **Radius** edit field. If you set the **Radius** to 0, you can define the radius graphically with a data point or by key-in from the Key-in window.

MicroStation prompts:

> Place Inscribed Polygon > Enter point on axis *(Place a data point or key-in coordinates to define the center of the polygon.)*
>
> Place Inscribed Polygon > Enter first edge point *(Place a data point or key-in coordinates to define the radius of the imaginary circle, the polygon's rotation, and one vertex.)*

Circumscribed Method

The **Circumscribed Method** places a polygon circumscribed around the outside of an imaginary circle having the same diameter as the distance across the opposite polygon edges (see Figure 4–7).

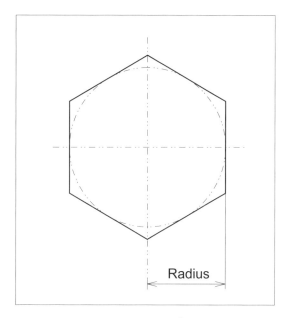

Figure 4-7 Example of displaying a polygon circumscribed around the outside of an imaginary circle

Invoke the Place Circumscribed Polygon tool:

Polygons tool box	Select the Place Regular Polygon tool and select Circumscribed from the Method option menu on the Tool Settings window (see Figure 4–8).
Key-in window	**place polygon circumscribed** (or **pl pol c**) ENTER

Key-in the number of polygon edges (from 3 to 4,999) in the **Edges** field on the Tool Settings window. Optionally, key-in the radius of the imaginary circle in the **Radius** edit field. If you set the **Radius** to 0, you can define the radius graphically with a data point or by key-in from the Key-in window.

Figure 4-8 Invoke the Place Polygon Circumscribed tool from the Polygons tool box

MicroStation prompts:

> Place Circumscribed Polygon > Enter point on axis *(Place a data point or key-in coordinates to define the center of the polygon.)*
>
> Place Circumscribed Polygon > Enter radius or point on circle *(Place a data point or key-in coordinates to define the radius of the imaginary circle, the polygon's rotation, and one vertex.)*

Edge Method

The **By Edge Method** (Place Polygon By Edge) places a polygon by defining two endpoints of a polygon edge (see Figure 4–9).

Chapter 4 Fundamentals III

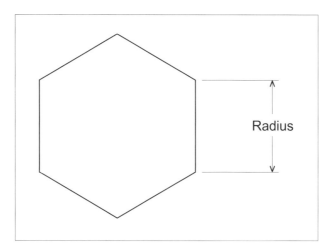

Figure 4-9 Example of a polygon placed by defining two endpoints of its side

Invoke the Place Polygon By Edge tool from:

Polygons tool box	Select the Place Regular Polygon tool and select Edge from the Method menu located in the Tool Settings window (see Figure 4–10).
Key-in window	**place polygon edge** (or **pl pol ed**) ENTER

Figure 4-10 Invoke the Place Polygon by Edge tool from the Polygons tool box

159

Key-in the number of polygon edges (from 3 to 4,999) in the Edges field on the Tool Settings window.

MicroStation prompts:

> Place Polygon by Edge > Enter first edge point *(Place a data point or key-in coordinates to define the vertex of the edge.)*
>
> Place Polygon by Edge > Enter next (CCW) edge point *(Place a data point or key-in coordinates to define the second edge point.)*

Place Point or Stream Curve

The Place Point or Stream Curve tool places a curve element by defining a series of data points through which the curve passes. The curve element can contain from 3 to 4,994 vertices. If you place more than 4,994 vertices, MicroStation starts a new Point Curve element and hooks it to the previous one as a complex chain (the elements in a chain act as if they are one element).

The **Points Method** draws the curve through data points that you define by either clicking the Data button or keying-in precision coordinates.

Invoke the Place Point Curve tool from:

Linear Elements tool box	Select the Place Point or Stream Curve tool and select Points from the Method menu on the Tool Settings window (see Figure 4–11).
Key-in window	**place curve point** (or **pl cu p**) ENTER

Figure 4-11 Invoke the Place Point Curve tool from the Linear Elements tool box

MicroStation prompts:

> Place Point Curve > Enter first point in curve string *(Place a data point or key-in coordinates to define the starting point of the curve.)*
>
> Place Point Curve > Enter point or RESET to complete *(Place data points or key-in coordinates to define the curve vertices, or click the Reset button to complete the curve.)*

At least three points are required to describe a curved element. Once you are through defining curve data points or key-in coordinates, press the Reset button to terminate the tool sequence. See Figure 4–12 for an example of placing a curve using the Place Point Curve tool by providing five data points.

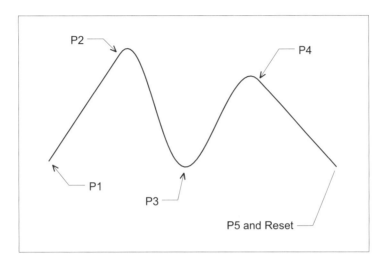

Figure 4–12 Example of placing an arc by the Place Point Curve tool

The **Stream Method** draws the curve stream by following the movement of the pointer. As you move the pointer, MicroStation defines data points based on the following streaming control settings:

- **Delta** sets the minimum distance, in working units, between points.
- **Tolerance** sets the maximum distance, in working units, between points.
- **Angle** sets the angle, in degrees, that the direction must change before a new vertex point is defined.
- **Area** sets the area that, when exceeded, causes a new vertex point to be defined.

Invoke the Place Curve Stream tool from:

Linear Elements tool box	Select the Place Point or Stream Curve tool, then select Stream from the Method option menu located in the Tool Settings window (see Figure 4–13).
Key-in window	**place curve stream** (or **pl cu st**) ENTER

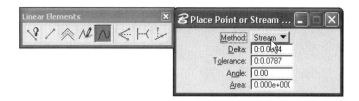

Figure 4-13 Invoke the Place Stream Curve tool from the Linear Elements tool box

MicroStation prompts:

> Place Stream Curve > Enter first point in curve string *(Place a data point or key-in coordinates to define the starting point of the curve stream.)*
>
> Place Stream Curve > Enter point or RESET to complete *(Move your cursor to define the curve stream. When you are finished, press the Reset button to complete the curve stream.)*

See Figure 4–14 for an example of placing a curve stream by means of the Place Curve Stream tool.

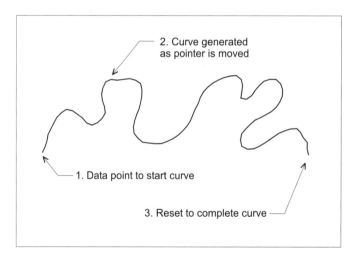

Figure 4-14 Example of placing a curve stream by means of the Place Curve Stream tool

Place Multi-line

With the Place Multi-line tool you can draw multiple parallel line segments, which are considered as one element. A multi-line can consist of as many as 16 separate, parallel lines of various line styles, weights, and colors, and the Place Multi-line tool allows you to draw a multi-line that is currently set as the active definition. If necessary, you can create or modify an existing multi-line definition with the help of the Multi-line settings box invoked from the **Element** drop-down menu. (Refer to Chapter 16 for a detailed description of creating or modifying an existing multi-line definition.)

To place a multi-line element, place a data point to identify the starting location and place additional data points to define each segment of the element. The following options in the Tool Settings window allow you to constrain each multi-line segment to a specific length and angle:

- To place segments of a specific length, turn ON the **Length** check box and key-in the length, in working units, in the **Length** text field.
- To place segments at a specific angle, turn ON the **Angle** check box and key-in the angle in the **Angle** text box. Rotation is counter-clockwise and an angle of zero draws the segment horizontally to the right.

 Note: If you set both **Length** and **Angle** to ON, each data point defines a separate, one-segment multi-line element.

Invoke the Place Multi-line tool from:

Linear Elements tool box	Select the Place Multi-line tool (see Figure 4–15).
Key-in window	**place mline constrained** (or **pl m c**) ENTER

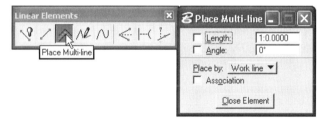

Figure 4–15 Invoke the Place Multi-line tool from the Linear Elements tool box

MicroStation prompts:

    Place Multi-line > Enter first point *(Place a data point or key-in coordinates to define the starting point of the multi-line.)*

    Place Multi-line > Enter vertex or Reset to complete *(Place a data point or key-in coordinates to define a vertex, or press the Reset button to complete.)*

 Note: MicroStation has a set of tools to edit Multi-lines called Multi-line Joints. See Chapter 8 for details about using Multi-line Joints.

ELEMENT MODIFICATION

MicroStation not only allows you to place elements easily, but also allows you to modify them as needed. This section discusses four important tools that will make your job easier: Fillet, Chamfer, Trim, and Partial Delete.

Construct Circular Fillet

The Construct Circular Fillet tool joins two elements (lines, line strings, circular arcs, circles, or shapes), two segments of a line string, or two sides of a shape with an arc of a specified radius. The arc is placed tangent to the two elements it connects.

The way MicroStation constructs a circular fillet depends on the option selected from the **Truncate** menu on the Tools Settings window.

- The **None** option places the fillet arc, but does not truncate the selected sides.
- The **Both** option places the fillet arc and truncates the elements at their point of tangency to create a smooth transition.
- The **First** option places the fillet and only truncates the first side identified.

For examples of placing a fillet using the three options, see Figure 4–16.

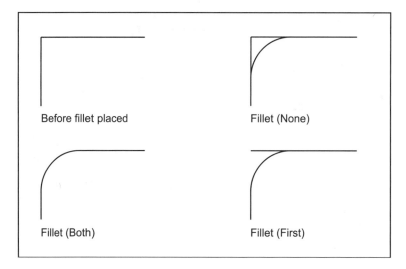

Figure 4-16 Examples of placing a fillet using the three truncation methods

Invoke the Construct Circular Fillet (no truncation) tool from:

Modify tool box	Select the Construct Circular Fillet tool. In the Tool Settings window, key-in the Radius in working units and select None from the Truncate menu (see Figure 4–17).
Key-in window	**fillet nomodify** (or **fill n**) ENTER

Figure 4–17 Invoke the Circular Fillet (no truncation) tool from the Modify tool box

MicroStation prompts:

> Circular Fillet (no truncation) > Select first segment *(Identify the first element or segment.)*
>
> Circular Fillet (no truncation) > Select second segment *(Identify the second element or segment.)*
>
> Circular Fillet (no truncation) > Accept-Initiate construction *(Click the Accept button to accept the placement of the fillet, or click the Reject button to reject the placement of the fillet.)*

Invoke the Construct Circular Fillet (Truncate Both) tool from:

Modify tool box	Select the Construct Circular Fillet tool. In the Tool Settings window, key-in the Radius in working units and select Both from the Truncate menu.
Key-in window	**fillet modify** (or **fill m**) ENTER

MicroStation prompts:

> Circular Fillet and Truncate Both > Select first segment *(Identify the first element or segment.)*
>
> Circular Fillet and Truncate Both > Select second segment *(Identify the second element or segment.)*
>
> Circular Fillet and Truncate Both > Accept-Initiate construction *(Click the Accept button to accept the placement of the fillet, or click the Reject button to reject the placement of the fillet.)*

Invoke the Construct Circular Fillet (Truncate Single) tool from:

Modify tool box	Select the Construct Circular Fillet tool. In the Tool Settings window, key-in the Radius in working units and select First from the Truncate menu.
Key-in window	**fillet single** (or **fill s**) ENTER

MicroStation prompts:

> Circular Fillet and Truncate Single > Select first segment *(Identify the first element or segment – the one that is to be truncated.)*
>
> Circular Fillet and Truncate Single > Select second segment *(Identify the second element or segment.)*
>
> Circular Fillet and Truncate Single > Accept-Initiate construction *(Click the Accept button to accept the placement of the fillet, or click the Reject button to reject the placement of the fillet.)*

 Note: The fillet appears at the location you select immediately after you select the second segment, but it is only a tentative placement. It does not become a permanent placement until you click the Accept button in response to the third prompt.

Construct Chamfer

The Construct Chamfer tool places an angled corner between two lines or between adjacent segments of a line string or shape.

The Tool Settings window provides the **Distance 1** and **Distance 2** text boxes that control the distance, in working units, that each end of the chamfer is placed from the corner. **Distance 1** is applied to the first element you select and **Distance 2** is applied to the second element. To place a 45-degree chamfer, enter equal distances in the two fields.

Invoke the Construct Chamfer tool from:

Modify tool box	Select the Construct Chamfer tool. In the Tool Settings window, key-in the appropriate Distance 1 and Distance 2, in working units (see Figure 4–18).
Key-in window	**chamfer** (or **ch**) ENTER

Figure 4–18 Invoke the Construct Chamfer tool from the Modify tool box

MicroStation prompts:

Construct Chamfer > Select first chamfer segment *(Identify the first element or segment, as shown in Figure 4–19.)*

Construct Chamfer > Select second chamfer segment *(Identify the second element or segment, as shown in Figure 4–19.)*

Construct Chamfer > Accept-Initiate construction *(Click the Accept button to accept the placement of the chamfer, or click the Reject button to reject the placement of the chamfer.)*

Note: The chamfer appears at the location you select immediately after you select the second segment, but it is only a tentative placement. It does not become a permanent placement until you click the Accept button in response to the third prompt.

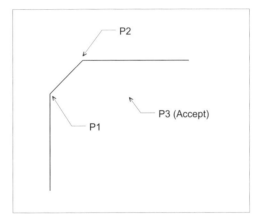

Figure 4–19 Example of placing the chamfer with the Construct Chamfer tool

Trim Elements

IntelliTrim combines into one tool the ability to **Trim** the part of selected elements that overlap a cutting element, **Extend** elements to a cutting element, or **Cut** elements into pieces. In addition, the tool provides a **Quick mode** that allows selecting one cutting element and an **Advanced mode** that allows selecting one or more cutting elements.

- As you select the elements that are to be trimmed, extended, or cut, square guide posts appear at the point where the operation will take place (such as the point to which the elements will be extended).
- Elements can be extended to the elements, such as arcs, cell headers, complex shapes, complex strings, curves, ellipses, lines, line strings, shapes, and text nodes.

- Elements can be cut or trimmed, such as arcs, b-spline curves, complex shapes, complex strings, curves, ellipses, lines, line strings, and shapes.
- Elements can be extended, such as b-spline curves, complex chains that end with a line or line string, lines, and line strings. Elements that cannot be extended will usually be deleted.

Quick Trim

When you select the **Quick Mode** and **Trim Operation**, IntelliTrim asks you to select the cutting element. Next, it asks you to draw temporary lines across all elements that are to be trimmed to the first element you selected. The parts of the elements that the temporary line touches are trimmed. Only elements that the cutting element crosses are trimmed. The cutting element is not modified.

To execute a quick trim, invoke the IntelliTrim tool from:

Modify tool box	Select the IntelliTrim Elements tool and select the Quick Mode and the Trim Operation from the Tool Settings window (see Figure 4–20).
Key-in window	**trim multi** (or **tri m**) ENTER (Set the **Mode** to **Quick** and the **Operation** to **Trim** in the Tool Settings window.)

Figure 4-20 Invoke the IntelliTrim tool from the Modify tool box, then select the Quick Mode and the Trim Operation

MicroStation prompts:

> IntelliTrim > Identify element to trim to *(Identify the cutting element.)*
>
> IntelliTrim > Create line(s) crossing the element(s) to be trimmed *(Place a data point to identify one end of the line that will define the part of the elements to be trimmed.)*
>
> IntelliTrim > Enter endpoint of the line *(Place a data point to identify the other end of the temporary line that will define the part of the elements to be trimmed. Make sure the temporary line crosses the part of the elements you want to remove.)*
>
> IntelliTrim > Create line(s) crossing the element(s) to be trimmed *(Continue identifying the start and endpoints of temporary lines until all elements to be trimmed are identified, and then click the Reset button to complete the trimming operation.)*

Chapter 4 Fundamentals III

For example, the following tool sequence shows how to use the IntelliTrim tool's **Trim Operation** with **Quick Mode** to trim the parts of a set of lines that are outside of an ellipse. The sequence of commands is illustrated in Figure 4–21.

> IntelliTrim > Identify Element *(Identify the ellipse that will be the cutting element.)*
>
> IntelliTrim > Create line(s) crossing the element(s) to be trimmed *(Place a data point on the left side of the ellipse, above the lines, to start the temporary line that will identify the lines to be trimmed on the left side of the ellipse.)*
>
> IntelliTrim > Enter endpoint of the line *(Place a data point below the lines to identify the end of the temporary line so that it passes across each of the three lines on the left side of the ellipse.)*
>
> IntelliTrim > Create line(s) crossing the element(s) to be trimmed *(Place a data point on the right side of the ellipse, above the lines, to start the temporary line that will identify the lines to be trimmed on the right side of the ellipse.)*
>
> IntelliTrim > Enter endpoint of the line *(Place a data point to identify the end of the temporary line so that it passes across each of the three lines on the right side of the ellipse.)*
>
> IntelliTrim > Create line(s) crossing the element(s) to be *trimmed (Click the Reset button to complete trimming the part of the lines outside of the ellipse.)*

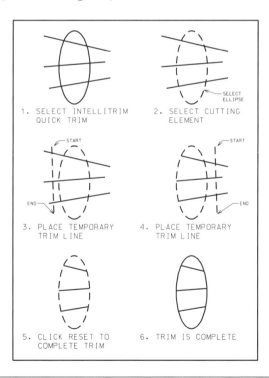

Figure 4–21 Example of trimming the parts of three lines that are outside of an ellipse

Quick Extend

When IntelliTrim is set to the **Quick Mode** and **Extend Operation**, an element to which other elements are to be extended is selected and then temporary lines are drawn across all the elements that are to be extended. If a selected element cannot be extended to actually touch the first selected element, it is not modified. If an element extends beyond the first selected element, it is not modified. If the first selected element is a closed shape, such as a circle, the extension is made to the near side of the element, unless an element already extends beyond the near side. If an element extends beyond the near side, it is extended to the far side.

To execute a quick extend, invoke the IntelliTrim tool from:

Modify tool box	Select the IntelliTrim Elements tool and select the Quick Mode and Extend Operation from the Tool Settings window (see Figure 4–22).
Key-in window	**trim multi** (or **tri m**) ENTER (Then set the **Mode** to **Quick** and the **Operation** to **Extend** in the Tool Settings window.)

Figure 4–22 Invoke the IntelliTrim tool from the Modify tool box, then select the Quick Mode and Extend Operation

MicroStation prompts:

> IntelliTrim > Identify element to extend to *(Identify the element to which the other elements are to be extended.)*
>
> IntelliTrim > Create line(s) crossing the element(s) to be extended *(Place a data point to identify the location of the start of the line that will identify the elements to be extended.)*
>
> IntelliTrim > Enter endpoint of the line *(Place a data point to identify the location of the end of the temporary line such that it extends across the elements to be extended.)*
>
> IntelliTrim > Create line(s) crossing the element(s) to be extended *(Continue identifying temporary lines until all elements to be extended are identified, and then click the Reset button to complete the extension operation.)*

For example, the following tool sequence shows how to use the IntelliTrim tool **Quick Mode Extend Operation** to extend three lines to the left side of an ellipse. The sequence of commands is illustrated in Figure 4–23.

    IntelliTrim > Identify element to extend to *(Identify the ellipse.)*

    IntelliTrim > Create line(s) crossing the element(s) to be extended *(Place a data point above the elements to be extended.)*

    IntelliTrim > Enter endpoint of the line *(Place a data point to below the lines to be extended making sure the temporary line crosses all the lines to be extended.)*

    IntelliTrim > Create line(s) crossing the element(s) to be extended *(Click the Reset button to complete extending the lines to the ellipse.)*

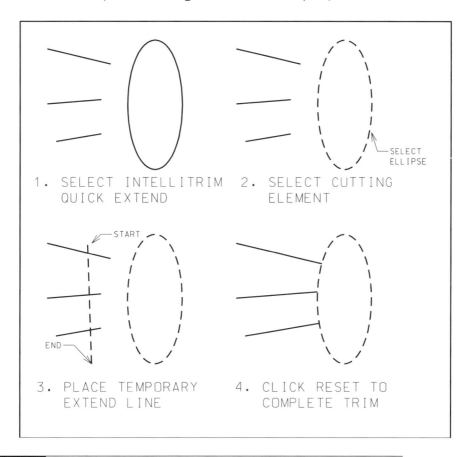

Figure 4-23 Example of extending three lines that are outside of an ellipse

Quick Cut

When IntelliTrim is set to the **Quick Mode** and **Cut Operation**, temporary lines are drawn across all elements that are to be cut. The temporary lines define the position of the cuts on each element.

The cut elements are broken into separate elements whose endpoints are where temporary lines cross the original elements. You won't be able to tell that anything has changed until you select one of the new elements (such as by selecting one of them for deletion).

To execute a quick cut, invoke the IntelliTrim tool from:

Modify tool box	Select the IntelliTrim Elements tool, then select the Quick Mode and the Cut Operation from the Tool Settings window (see Figure 4–24).
Key-in window	**trim multi** (or **tri m**) ENTER (Set the **Mode** to **Quick** and the **Operation** to **Cut** in the Tool Settings window.)

Figure 4–24 Invoke the IntelliTrim tool from the Modify tool box, then select the Quick Mode and the Cut Operation

MicroStation prompts:

> Cut elements > Create line which defines cut *(Place a data point to identify the location of the start of the line that will define the position of one of the cuts to be made on the elements it crosses.)*
>
> Cut elements > Create line which defines cut *(Place a data point to identify the location of the end of the temporary line that will define the cut points. The temporary line should cross the elements to be cut at the points where they are to be cut.)*
>
> Cut elements > Create line which defines cut *(Continue identifying temporary lines until all cut points are identified, then click the Reset button to complete the cutting operation.)*

For example, the following tool sequence shows how to use cut two segments out of an ellipse. The sequence of commands is illustrated in Figure 4–25.

> Cut elements > Create line which defines cut *(Place a data point on the left side of the ellipse to start the temporary line that will identify the top of the segment to be cut out of the ellipse.)*

Cut elements > Create line which defines cut *(Place a data point on the right side of the ellipse to identify the end of the temporary line that will identify the top of the segment to be cut out of the ellipse.)*

Cut elements > Create line which defines cut *(Place a data point on the left side of the ellipse to start the temporary line that will identify the bottom of the segment to be cut out of the ellipse.)*

Cut elements > Create line which defines cut *(Place a data point on the right side of the ellipse to identify the end of the temporary line that will identify the bottom of the segment to be cut out of the ellipse.)*

Cut elements > Create line which defines cut *(Click the Reset button to complete making the cuts in the ellipse.)*

 Note: The example procedure made two cuts each on the left and right sides of the ellipse. After the cuts were completed, the Delete Element tool was used to remove the pieces of the ellipse between the cuts as shown in Figure 4–25.

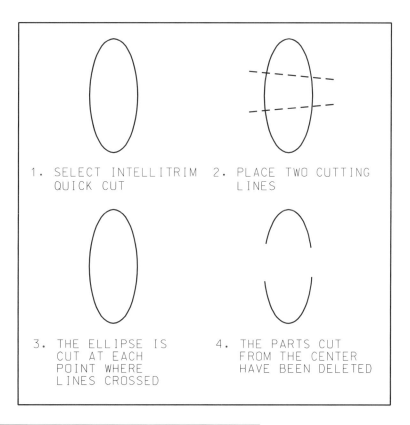

Figure 4–25 Example of making four cuts on an ellipse

Advanced Trim

When IntelliTrim is set to the **Advanced Mode** and **Trim Operation**, one or more cutting elements can be selected, and one or more elements can be trimmed to the cutting elements. Options on the Tool Settings window allow you to either **Select Elements to Trim** or **Select Cutting Elements**.

To execute an advanced trim, invoke the IntelliTrim tool:

Modify tool box	Select the IntelliTrim Elements tool and then select the Advanced Mode and the Trim Operation from the Tool Settings window. Decide which set of elements you want to start with by Clicking Select Elements to Trim or Select Cutting Elements (see Figure 4–26).
Key-in window	**trim multi** (or **tri m**) ENTER (Then set the **Mode** to **Advanced** and the **Operation** to **Trim** in the Tool Settings window.)

Figure 4-26 Invoke the IntelliTrim tool from the Modify tool box, then select the Advanced Mode and the Trim Operation

If the **Select Cutting Elements** option is selected, MicroStation prompts:

> IntelliTrim > Identify cutting elements, reset to complete step *(Select each cutting element by clicking a data point on it. When all cutting elements are selected, click the Reset button.)*
>
> IntelliTrim > Identify elements to trim/extend, reset to complete step *(Select each element to be trimmed by clicking a data point on it. When all elements to be trimmed are selected, click the Reset button.)*
>
> IntelliTrim > Enter points near portions to keep, reset to complete command *(If the wrong part of any element has been trimmed, click a data point near it to switch the part that is trimmed. When the correct part of each element is shown trimmed, click the Reset button to make the trimming operation permanent.)*

For example, the following tool sequence shows how to use the IntelliTrim tool's **Advanced Mode** and **Trim Operation** to use two orthogonal blocks as cutting elements for trimming two shapes. The sequence of commands is illustrated in Figure 4–27.

> IntelliTrim > Identify cutting elements, reset to complete step *(Select one of the orthogonal blocks by clicking a data point on it.)*
>
> IntelliTrim > Identify cutting elements, reset to complete step *(Select the other orthogonal block by clicking a data point on it, and then click the Reset button.)*
>
> IntelliTrim > Identify elements to trim, reset to complete step *(Select one of the shapes by clicking a data point on it.)*
>
> IntelliTrim > Identify elements to trim, reset to complete step *(Select the other shape by clicking a data point on it, and then click the Reset button.)*
>
> IntelliTrim > Enter points near portions to keep, reset to complete command *(If the part of the two shapes between the inner and outer blocks is tentatively removed, click the Reset button to complete the operation. If the wrong parts of the shapes are trimmed, click a data point near the trimmed parts to switch the trimming to a different part of the elements. When the correct part of each shape is selected, click the Reset button to complete the trimming operation.)*

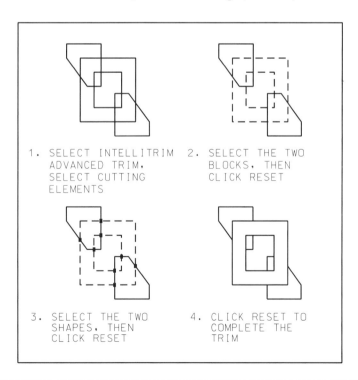

Figure 4–27 Example of using two blocks as cutting elements for trimming two shapes

Advanced Extend

When IntelliTrim is set to the **Advanced Mode** and **Extend Operation**, one or more boundary elements can be selected, and one or more elements can be extended to the boundary elements. Options on the Tool Settings window allow you to either **Select Elements to Trim** (start by selecting the elements to be extended) or **Select Cutting Elements** (start by selecting the elements to which the other elements will be extended).

To execute an advanced extend, invoke the IntelliTrim tool from:

Modify tool box	Select the IntelliTrim Elements tool, and then select the Advanced Mode and the Extend Operation from the Tool Settings window. Decide which set of elements you want to start with by Clicking Select Elements to Trim or Select Cutting Elements (see Figure 4–28).
Key-in window	**trim multi** (or **tri m**) ENTER (Then set the **Mode** to **Advanced** and the **Operation** to **Extend** in the Tool Settings window.)

Figure 4-28 Invoke the IntelliTrim tool from the Modify tool box, then select the Advanced Mode and the Extend Operation

MicroStation prompts:

> IntelliTrim > Identify cutting elements, reset to complete step *(Select each boundary element by clicking a data point on it, and then click the Reset button.)*
>
> IntelliTrim > Identify elements to trim/extend, reset to complete step *(Select each element to be extended by clicking a data point on it, and then click the Reset button.)*
>
> IntelliTrim > Enter points near portions to keep, reset to complete command *(Click the Reset button to make the extend operation permanent.)*

For example, the following tool sequence shows how to use the IntelliTrim tool's **Advanced Mode** and **Extend Operation** to extend one line to one cutting block and another line to a second cutting block. The sequence of commands is illustrated in Figure 4–29.

IntelliTrim > Identify cutting elements, reset to complete step *(Select one of the orthogonal blocks by clicking a data point on it.)*

IntelliTrim > Identify cutting elements, reset to complete step *(Select the other orthogonal block by clicking a data point on it, and then click the Reset button.)*

IntelliTrim > Identify elements to trim/extend, reset to complete step *(Select one of the lines by clicking a data point on it.)*

IntelliTrim > Identify elements to trim/extend, reset to complete step *(Select the other line by clicking a data point on it, and then click the Reset button.)*

IntelliTrim > Enter points near portions to keep, reset to complete command *(Click the Reset button to complete the trimming operation.)*

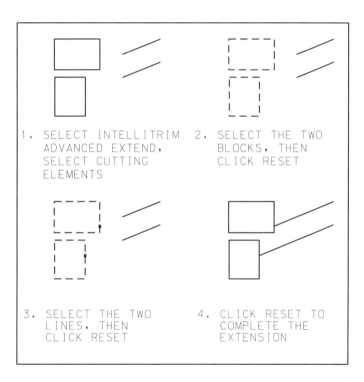

Figure 4-29 Example of using two blocks as cutting elements for extending two lines

Partial Delete

The Partial Delete tool allows you to delete part of an element. In the case of a line, line string, multi-line, curve, or arc, the Partial Delete tool removes part of the element, and the element is divided into two elements of the same type. A partially deleted ellipse or circle becomes an arc, and a shape becomes a line string. There are no Tool Settings options for this tool.

Invoke the Partial Delete tool from:

Modify tool box	Select the Partial Delete tool (see Figure 4–30.)
Key-in window	**delete partial** (or **del p**) ENTER

Figure 4–30 The Partial Delete tool

MicroStation prompts:

> Delete Part of Element > Select start pnt for partial delete *(Identify the element at the point you want to start the partial delete.)*
>
> Delete Part of Element > Select direction of partial delete *(This prompt appears only when you select a closed element, such as a circle, ellipse, or polygon. Move the pointer a short distance in the direction you want to cut, then click the data button.)*
>
> Delete Part of Element > Select end pnt for partial delete *(Place a data point or key-in coordinates to identify the location of the end point of the partial delete.)*

See Figure 4–31 for examples of the Partial Delete tool.

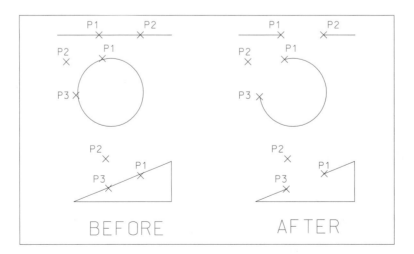

Figure 4–31 Examples of deleting part of an element with the Partial Delete tool

ELEMENT MANIPULATION

MicroStation offers two main categories of manipulation tools: single-element manipulation and multi-element manipulation. Single-element manipulation tools allow you to manipulate one element at a time and multi-element manipulation tools manipulate groups of elements. This section discusses single-element manipulation tools. Multi-element manipulations are done with Element Selection and Fence manipulation tools (see Chapter 6).

After mastering the element manipulation tools and learning when and how to apply them, you will appreciate the power and capability of MicroStation. You can draw one element and use the element manipulation tools to quickly make copies, saving you from having to draw each one separately. You will soon begin to plan ahead to utilize these powerful tools.

All the manipulation tools described here require you to identify the element to be manipulated, and then accept it. To identify an element, position the pointer on the element and click the Data button. MicroStation indicates that an element is selected by displaying it in the highlight color. If the highlighted element is the correct one, continue following the tool prompts shown in the prompt field. If the element that is highlighted is not the correct one, click the Reject button to reject the element.

Note: Be sure to check the status of your lock settings before you begin to modify your design. A good rule to follow is to turn OFF all of the locks that are not being used with the exception of the Snap Lock. It is very frustrating to try to select an element that does not lie on the grid when the Grid Lock is turned ON. The cursor bounces around from grid dot to grid dot, and it may be impossible to identify an element if it is not on the grid. It is simple to toggle the grid lock OFF quickly, identify the object, and then toggle the grid lock back ON again, if needed.

Copy

The Copy tool places a copy of the selected element at a location that you specify and leaves the original element intact. The copy is identical to the original. You can make as many copies of the original as needed. Each copy is independent of the original and can be manipulated and modified like any other element. The data point you enter to identify the element you want to copy is the (base) point on the element to which the pointer is attached. Select this point with care and use the tentative snap if you need to snap to the element at a precise location (with the appropriate snap mode selected).

When you are through placing copies of the selected element, click the Reset button to drop the element. If necessary, you can select another element to copy, or you can invoke another tool to continue working on your design file.

Invoke the Copy Element tool from:

Manipulate tool box	Select the Copy tool, then turn the Make Copy button ON in the Tool Settings window (see Figure 4–32).
Key-in window	**copy element** (or **cop el**) ENTER

Figure 4-32 The Copy tool

MicroStation prompts:

> Copy Element > Identify element *(Identify an element to copy.)*
>
> Copy Element > Enter point to define distance and direction *(Provide the location of the copy by clicking the Data button or by keying-in coordinates.)*
>
> Copy Element > Enter point to define distance and direction *(If necessary, copy to another location by providing a data point or by keying-in coordinates. When you finish placing copies, click the Reset button to terminate the tool sequence.)*

For example, the following tool sequence shows how to copy a line to the center of a circle using the Copy Element tool (see Figure 4–33).

> Copy Element > Identify element *(Identify the line by tentative snapping to the end of the line and then clicking the Data button.)*
>
> Copy Element > Enter point to define distance and direction *(Tentative snap to the center of the circle and then click the Data button.)*
>
> Copy Element > Enter point to define distance and direction *(Click the Reset button to place the element at the new location.)*

Chapter 4 Fundamentals III

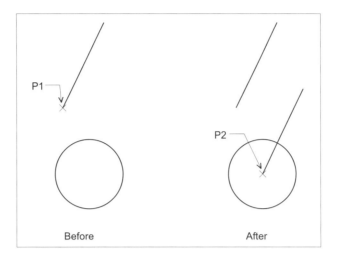

Figure 4-33 Example of copying an element by means of the Copy tool

Move

The Move tool allows you to move an element to a new location without changing its orientation or size. The data point you enter to identify the element is the (base) point on the element to which the cursor is attached. Select this point with care, and use the tentative snap if you need to snap to the element at a precise location (with the appropriate snap mode selected). Move the element to the new location and click the Data button to place it.

The element remains selected until you click the Reset button so, if you place it at the wrong location the first time, you can move it again. Once the element is in the correct location, click the Reset button to terminate the tool sequence. After clicking the Reset button, you can select another element to move, or you can invoke another tool to continue working on your design file.

Invoke the Move Element tool from:

Manipulate tool box	Select the Move tool, then turn the Make Copy button OFF in the Tool Settings window (see Figure 4–34).
Key-in window	**move element** (or **mov e**) ENTER

Figure 4-34 Invoke the Move tool from the Manipulate tool box

MicroStation prompts:

> Move Element > Identify element *(Identify an element to move.)*
>
> Move Element > Enter point to define distance and direction *(Reposition the element to its new location by providing a data point or keying-in coordinates.)*
>
> Move Element > Enter point to define distance and direction *(If necessary, move the element to another location by providing a data point or keying-in coordinates. When the element is in the correct location, click the Reset button to terminate the tool sequence.)*

For example, the following tool sequence shows how to move a line to the center of a circle using the Move Element tool (see Figure 4–35).

> Move Element > Identify element *(Identify the line by snapping to the end point of the line and clicking the Data button.)*
>
> Move Element > Enter point to define distance and direction *(Snap to the center of the circle and click the Data button.)*
>
> Move Element > Enter point to define distance and direction *(Click the Reset button to place the element at the new location.)*

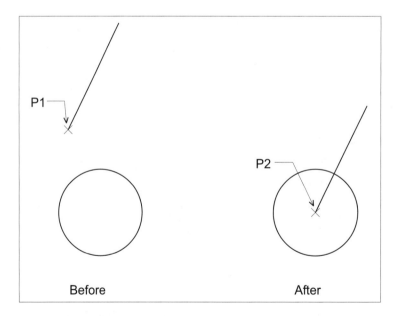

Figure 4–35 Example of moving an element

Move Parallel

The Move Parallel tool lets you move or copy an element (such as a line, line string, multi-line, circle, curve, arc, ellipse, shape, complex chain, or complex shape) parallel to the original location of the element. All points on the original element are moved an equal distance, so the effect on a closed shape such as a circle is to scale the element.

The Tool Settings window for the parallel tool provides options for controlling the manipulation results:

- If the **Make a Copy** check box is turned OFF, the new element replaces the original element. If the check box is turned ON, a parallel copy is made and the original element remains in the design.

- If the **Distance** check box is turned OFF, you graphically determine the position of the moved or copied element with a data point or precision key-in. If the check box is turned ON, the distance of the new element from the original is determined by the value you type (in working units) in the associated text box.

- The **Gap Mode** menu provides two options for controlling the shape of the corners of linear objects such as orthogonal blocks. The **Miter** option maintains a linear element's sharp corners. The **Round** option converts a linear element's sharp corners to fillets. Figure 4–36 shows the result of the **Miter** and **Round** Gap Modes on a shape that is copied parallel.

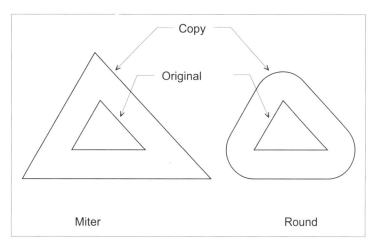

Figure 4-36 Examples of parallel copies

Invoke the Move Parallel tool from:

Manipulate tool box	Select the Move Parallel tool and select the options you need in Tool Settings window (see Figure 4–37).
Key-in window	**move parallel** (or **mov p**) ENTER **copy parallel** (or **cop p**) ENTER

Figure 4–37 The Move Parallel Tool

Move Parallel Graphically

To graphically move or copy an element parallel to its original position, turn the **Distance** check box OFF. If you want to move a copy, turn on the **Make a Copy** checkbox. If you are moving or copying a linear shape and want the corners of element to be rounded off, select the **Round** option from the **Gap Mode** menu.

MicroStation prompts:

> Move Parallel > Identify element *(Identify an element to move parallel.)*
>
> Move Parallel > Accept/Reject *(Identify the new parallel location of the element or a copy of the element by clicking a data point or by keying-in coordinates.)*
>
> Move Parallel > Accept/Reject *(If necessary, continue moving or copying the element in parallel by providing data points or by keying-in coordinates. When you are finished copying or moving the element, click the Reset button to terminate the tool sequence and place the element at the new location.)*

For example, the following sequence shows how to move line, in parallel to its original location, using the Move Parallel tool. The **Distance** and **Make Copy** check boxes are OFF. The **Gap Mode** options are not used for lines. (see Figure 4–38).

> Move Parallel by Distance > Identify element *(Identify the line.)*
>
> Move Parallel by Distance > Accept/Reject (Select next input) *(Place a data point to define the location of the parallel copy of the line.)*
>
> Move Parallel by Distance > Accept/Reject (Select next input) *(Click the Reset button to place the element at the new location.)*

Chapter 4 Fundamentals III

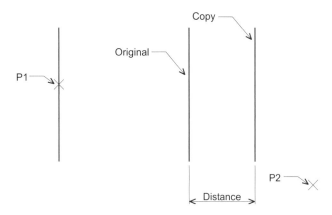

Figure 4-38 Example of moving an element by a fixed distance

Move Parallel by Keyed in Distance

You can move or copy an element parallel by a fixed distance by turning the **Distance** check box ON and entering a distance, using working units, in the associated text field. Set the **Make a Copy** check box as required and select the required **Gap Mode** option.

MicroStation prompts:

> Move Parallel by Key-in > Identify element *(Identify an element to move parallel.)*
>
> Move Parallel by Key-in > Accept/Reject (Select next input) *(Move the pointer as necessary to position the image of the parallel copy in the correct direction and click the Accept button to place the parallel element. The distance of the new element from the original is set by the value in the **Distance** field.)*
>
> Move Parallel by Key-in > Accept/Reject (Select next input) *(If necessary, continue moving or copying the element in parallel by providing data points or keying in coordinates. When you are finished copying or moving the element, click the Reset button to terminate the tool sequence and place the element at the new location.)*

Scale

The Scale tool increases or decreases the size of a selected element or a copy of the element. Options in the Tool Settings window allow you control the way the tool operates.

To scale the original element, turn the **Make Copy** check box OFF. To scale a copy of the original element (and leave the original element unchanged), turn the **Make Copy** check box ON. If you are planning to scale a multi-line and want the line offsets (distance between the lines) to also be scaled, turn the **Scale Multi-line Offsets** check box ON.

The **Method** menu allows you to scale the element graphically using the **3 Points** option or by keyed-in scale factors, using the **Active Scale** option. If you choose the **Active Scale** option, **X Scale** and **Y Scale** text fields appear on the Tool Settings window and the element is scaled using the values in the scale factor text fields. If you choose the **3 Point** method, the scale text fields disappear from the Tool Settings window and a **Proportional** check box appears. This check box forces the graphically scaled element to be proportional to the original element.

Active Scale Method

To increase the size of the element using **Active Scale**, key-in **X Scale** and **Y Scale** factors that are greater than one. To decrease the size of the element, key-in **X Scale** and **Y Scale** factors that are between 0 and 1. For example, a scale factor of 3 makes the element three times larger and a scale factor of 0.75 shrinks the selected element to three-quarters of its original size.

The scale factors can be positive or negative (-) numbers:

- Positive X and Y scale factors produce a scaled element that has the same orientation as the original element.
- A negative X and positive Y scale factor produces a scaled element that is a mirror (backward) image of the original element around the Y axis.
- A positive X and negative Y scale factor produces a scaled element that is a mirror image around the X axis.
- Negative X and Y scale factors produce a scaled element that is a mirror image around both axes. Figure 4–39 shows the effect of positive and negative scale factors.

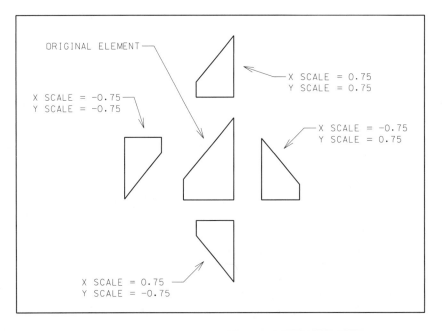

Figure 4-39 Example of the effect of positive and negative scale factors

To the right of the **X Scale** and **Y Scale** text fields is a lock symbol that can be opened and closed by clicking it with the Data button. When the lock is open, you can key-in different factors in the scale text fields. When the lock is closed, entering a new value in one field automatically changes the value of the other field to be equal.

To scale an element by keying-in the X and Y scale factors, invoke the Scale tool from:

Manipulate tool box	Select the Scale tool and in the Tool Settings window, select the Active Scale from the Method option menu. Key-in the appropriate scale factors in the X Scale and Y Scale text fields. If you want to scale a copy of the original element, turn on the MakeCopy check box (see Figure 4–40).
Key-in window	**scale original** (or **sc o**) ENTER **scale copy** (or **sc c**) ENTER

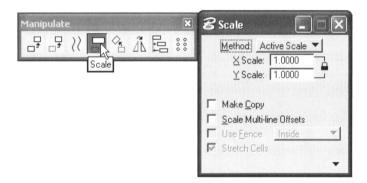

Figure 4–40 The Scale tool

MicroStation prompts:

> Scale Element > Identify element *(Identify an element to scale.)*
>
> Scale Element > Enter origin point (point to scale about) *(Identify the location of the scaled element by providing a data point or by keying-in coordinates.)*
>
> Scale Element > Enter origin point (point to scale about) *(If necessary, scale it again by providing a data point or keying-in coordinates. When you are through scaling the element, click the Reset button to terminate the tool sequence and place the scaled element.)*

 Note: If you turn the **Make Copy** check box ON, the word "Copy" is added to the MicroStation prompts. For example, "Scale Element (Copy) > Identify element.".

The following tool sequence example shows how to scale a copy of an element to half its original size (see Figure 4–41). Select Scale from the Manipulations tool box. In the Tool Settings window, set the **X Scale** and **Y Scale** to 0.5, and turn the **Make Copy** check box ON.

MicroStation prompts:

>Scale Element (Copy) > Identify element *(Identify the element by selecting a point on the bottom of the element.)*
>
>Scale Element (Copy) > Enter origin point (point to scale about) *(Move the pointer to the left in order to move the dynamic, scaled image of the element to the right, and click the Data button to place it.)*
>
>Scale Element (Copy) > Enter origin point (point to scale about) *(Click the Reset button and place the scaled element.)*

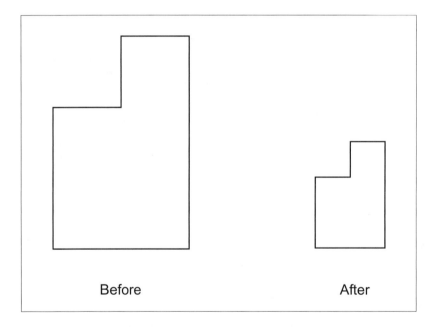

Figure 4–41 Example of scaling a copy of an element by the Active Scale tool

3-Points Method

Scaling graphically involves providing three data points or keying-in their coordinates. The scale factors are computed by dividing the distance between the first and third points by the distance between the first and second points.

Chapter 4 Fundamentals III

To scale an element graphically, invoke the Scale tool from:

Manipulate tool box	Select the Scale tool and then select 3 points from the Method option menu (see Figure 4–42).
Key-in window	**scale points original** (or **sc p o**) ENTER **scale points copy** (or **sc p c**) ENTER

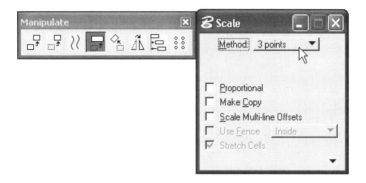

Figure 4-42 Select the Scale by 3 Points Method from the Manipulate tool box

MicroStation prompts:

>Scale Element by 3 Points > Identify element *(Identify an element to scale.)*

>Scale Element by 3 Points > Enter origin point (point to scale about) *(Place a data point or key-in coordinates to define the origin point.)*

>Scale Element by 3 Points > Enter reference point *(Place a data point or key-in coordinates to define the reference point.)*

>Scale Element by 3 Points > Enter point to define amount of scaling *(Place a data point or key-in coordinates to define the amount of scaling.)*

>Scale Element by 3 Points > Enter point to define amount of scaling *(If necessary, scale it again by providing a data point or keying-in coordinates. When you are fiinished scaling the element, click the Reset button to terminate the tool sequence and place the scaled element.)*

Depending on the relationship of the three data points, the different scale factors may be applied to the vertical and horizontal size of the element. To maintain the proportionality of the selected element, turn the **Proportional** check box ON in the Tool Settings window.

 Note: If you turn the **Make Copy** check box ON, the word "Copy" is added to the MicroStation prompts. For example, "Scale Element by 3 Points (Copy) > Identify element."

Rotate

The Rotate tool rotates a selected element, or copy of the element, about a specified pivot point. Options on the Tool Settings window allow you to change the way the tool operates.

To rotate the original element, turn the **Make Copy** check box OFF. To rotate a copy of the original element (and leave the original element unchanged), turn the **Make Copy** check box ON.

The **Method** drop-down menu provides options to rotate the element by an **Active Angle**; graphically, by defining **2 Points**; and graphically, by defining **3 Points**.. Descriptions of each method follow.

Active Angle Method

When you select the **Active Angle Method**, the text field below the **Method** menu becomes available. Key-in the rotation angle in the text box or click the scroll arrows on the right side of the text box, to select standard angles such as 180 or 270 degrees. To rotate the element or a copy of the element, by the Active Angle, identify the element and then place a data point to define the point of rotation and complete the tool sequence. The angle of rotation is counterclockwise, starting at an imaginary line from the point on the element where you originally identified it to the rotation point.

Note: MicroStation remembers the Active Angle and places it the Tool Settings window of other tools that have an **Active Angle** field, such as the Construct Array tool.

To rotate an element by the Active Angle, invoke the Rotate Element tool from:

Manipulate tool box	Select the Rotate tool. In the Tool Settings window, select Active Angle from the Method option menu, key-in the rotation angle in the text field, and, if you want to rotate a copy of the element, turn ON the Make Copy check box. (See Figure 4-43.)
Key-in window	**rotate original** (or **ro o**) ENTER. **rotate copy** (or **ro c**) ENTER

Figure 4-43 Invoke the Rotate tool from the Manipulate tool box

Chapter 4 Fundamentals III

MicroStation prompts:

> Rotate Element > Identify element *(Identify an element to rotate.)*
>
> Rotate Element > Enter pivot point (point to rotate about) *(Define the point about which you want to rotate the element by providing a data point or by keying-in coordinates.)*
>
> Rotate Element > Enter pivot point (point to rotate about) *(If necessary, rotate it again by providing a data point or keying-in coordinates. When you are through rotating the element, click the Reset button to terminate the tool sequence and place the rotated element.)*

Note: If you turn the **Make Copy** check box ON, the word "Copy" is added to the MicroStation prompts. For example, "Rotate Element (Copy) > Identify element.".

For example, the following tool sequence shows how to rotate a copy of an element to a keyed-in Active Angle of 45 degrees, using the Rotate tool (see Figure 4–44).

> Rotate Element (Copy) > Identify element *(Identify the element to rotate.)*
>
> Rotate Element (Copy) > Enter pivot point (point to rotate about) *(Reposition the rotated element to its new location by a data point.)*
>
> Rotate Element (Copy) > Enter pivot point (point to rotate about) *(Click the Reset button to place the rotated element.)*

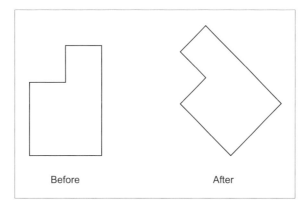

Figure 4–44 Example of rotating a copy of an element using the Active Angle Method

2 Points Method

With the **2 Points** rotation method, you identify the element with the first data point and enter a second data point to define the point about which the element is to be rotated. After you define the rotation point, a dynamic image of the element spins in a circle, around the point, as you move the pointer. The radius of the circle is equal to the distance from the selection point on the element to the rotation point. The third data point determines the angle of rotation and completes the rotation operation, and the element, or a copy of the element, is placed at the rotated position. For an example of rotating an element by 2 points, see Figure 4–45.

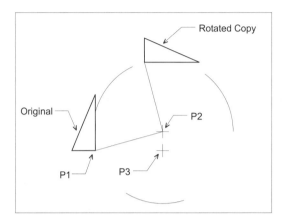

Figure 4–45 Example of rotating a copy of an element using the 2 Points Method

To rotate an element by 2 points, invoke the Rotate Element tool from

Manipulate tool box	Select the Rotate tool. From the Tool Settings window, select the 2 Points Method and if you want to rotate a copy of the original element, turn the Make Copy check box ON (see Figure 4–46).
Key-in window	**spin original** (or **sp o**) ENTER **spin copy** (or **sp c**) ENTER

Figure 4–46 Invoke the Rotate by 2 Points Method from the Manipulate tool box

MicroStation prompts:

Spin Element > Identify element *(Identify an element to rotate.)*

Spin Element > Enter pivot point (point to rotate about) *(Place a data point or key-in coordinates to define the pivot point.)*

Spin Element > Enter point to define amount of rotation *(Place a data point or key-in coordinates to define the amount of rotation and place the rotated element.)*

Spin Element > Enter point to define amount of rotation *(If necessary, rotate it again by providing a data point or keying-in coordinates. When you are finished rotating the element, click the Reset button to terminate the tool sequence and place the rotated element.)*

Note: If you turn the **Make Copy** check box ON, the word "Copy" is added to the MicroStation prompts. For example, "Spin Element (Copy) > Identify element.".

3 Points Method

With the **3 Points** rotation method, you identify the element with the first data point, enter a second data point to define the point about which the element is to be rotated, and enter a third data point to start the rotation. After the third data point, a dynamic image of the element spins around the second (rotation) data point, as you move the pointer. The fourth data point determines the angle of rotation and completes the rotation operation. The element, or a copy of the element, is placed at the rotated position.

The rotation angle is determined by the relationship of the four data points. The angle formed by data point one and data point three, with data point two as the angle vertex, is equal to the angle formed by data point three and data point four, with data point two as the angle vertex (see Figure 4–47).

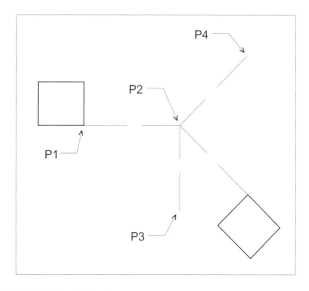

Figure 4-47 Example of rotating a copy of an element using the 3 Points Method

To rotate an element by 3 points, invoke the Rotate Element tool from:

Manipulate tool box	Select the Rotate tool. From the Tool Settings window, select the 3 points Method and if you want to rotate a copy of the original element, turn the Make Copy check box ON (see Figure 4–48).
Key-in window	**rotate points original** (or **ro p o**) ENTER **rotate points copy** (or **ro p c**) ENTER

Figure 4–48 Invoke the Rotate Element by 3 Points tool from the Manipulate tool box

MicroStation prompts:

Rotate Element by 3 Points > Identify element *(Identify an element to rotate.)*

Rotate Element by 3 Points > Enter pivot point (point to rotate about) *(Place a data point or key-in coordinates to define the pivot point.)*

Rotate Element by 3 Points > Enter point to define start of rotation *(Place a data point or key-in coordinates to define the starting point of rotation.)*

Rotate Element by 3 Points > Enter point to define amount of rotation *(Place a data point or key-in coordinates to define the amount of rotation.)*

Rotate Element by 3 Points > Enter point to define amount of rotation *(If necessary, rotate it again by providing a data point or keying-in coordinates, and/or click the Reset button to terminate the tool sequence and place the rotated element.)*

 Note: If you turn the **Make Copy** check box ON, the word "Copy" is added to the MicroStation prompts. For example, "Rotate Element (Copy) > Identify element."

Mirror

The Mirror Element tool creates a mirror (backward) image of an element. Options in the Tool Settings window allow you control the way the tool operates.

To mirror the original element, turn the **Make Copy** check box OFF. To mirror a copy of the original element (and leave the original element unchanged), turn the **Make Copy** check box

ON. If the element contains text, and you want the text mirrored, turn the **Mirror Text** check box ON. If you are mirroring a multi-line element and want the offsets mirrored, turn the **Mirror Multi-line Offsets** check box ON.

 Note: The **Mirror Text** check box is useful for mirroring groups of elements that include text elements. Group manipulation tools are discussed in Chapter 6.

The **Mirror About** menu allows you to mirror the element about a **Horizontal** axis, about **Vertical** axis, or about an imaginary **Line** that you define. Examples of the three mirror orientations are shown in Figure 4–49.

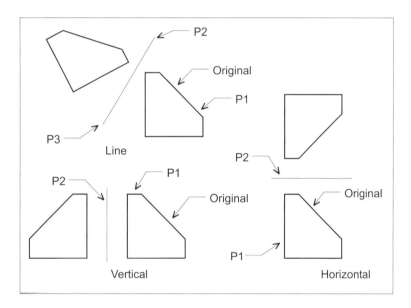

Figure 4-49 Examples of the mirroring an element using each of the Mirror About methods

Horizontal Method

To mirror an element about a horizontal axis, invoke the Mirror tool from:

Manipulate tool box	Select the Mirror tool. In the Tool Settings window, select Horizontal from the Mirror About menu and if you want to mirror a copy of the element, turn the Make Copy check box ON (see Figure 4–50).
Key-in window	**mirror original horizontal** (or **mi o h**) ENTER **mirror copy horizontal** (or **mi c h**) ENTER

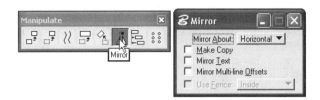

Figure 4-50 The Tool Settings window set to mirror an element about a horizontal axis

MicroStation prompts:

> Mirror Element About Horizontal (Original) > Identify element *(Identify an element to mirror.)*
>
> Mirror Element About Horizontal (Original) > Accept/Reject (Select next input) *(Identify the distance of the horizontal axis from the element by placing a data point or keying in coordinates.)*
>
> Mirror Element About Horizontal (Original) > Accept/Reject (Select next input) *(If necessary, mirror it again by providing a data point or keying-in coordinates. When you are finished mirroring the element, click the Reset button to terminate the tool sequence and place the mirrored element.)*

 Note: If you turn the **Make Copy** check box ON, the word "Original" is replaced by "Copy" in the MicroStation prompts. For example, "Mirror Element About Horizontal (Copy) > Identify element.".

Vertical Method

To mirror an element about a vertical axis, invoke the Mirror tool from:

Manipulate tool box	Select the Mirror tool. In the Tool Settings window, select Vertical from the Mirror About menu and if you want to mirror a copy of the element, turn the Make Copy check box ON.
Key-in window	**mirror original vertical** (or **mi o v**) ENTER **mirror copy vertical** (or **mi c v**) ENTER

MicroStation prompts:

> Mirror Element About Vertical (Original) > Identify element *(Identify an element to mirror.)*
>
> Mirror Element About Vertical (Original) > Accept/Reject (Select next input) *(Identify the distance of the vertical axis from the element by placing a data point or keying-in coordinates.)*

Mirror Element About Vertical (Original) > Accept/Reject (Select next input) *(If necessary, mirror it again by providing a data point or keying-in coordinates. When you are through mirroring the element, click the Reset button to terminate the tool sequence and place the mirrored element.)*

Note: If you turn the **Make Copy** check box ON, the word "Original" is replaced by "Copy" in the MicroStation prompts. For example, "Mirror Element About Vertical (Copy) > Identify element.".

Line Method

The **Mirror About Line** method allows you to mirror an element about an imaginary line whose location and angle you specify. It requires one more data point than the other mirroring methods. After selecting the element to be mirrored, you place two more data points to define the location and angle of the line about which the element is to be mirrored, as shown in Figure 5-51

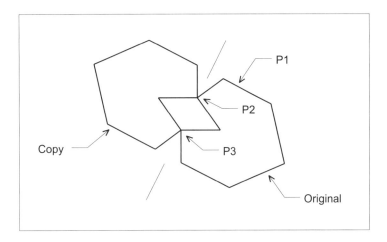

Figure 4-51 Example of mirroring a copied element about a line

To mirror an element about a line, invoke the Mirror tool from:

Manipulate tool box	Select the Mirror tool. In the Tool Settings window, select Line from the Mirror About menu and if you want to mirror a copy of the element, turn the Make Copy check box ON.
Key-in window	**mirror original line** (or **mi o l**) ENTER **mirror copy line** (or **mi c l**) ENTER

Note: If you turn the **Make Copy** check box ON, the word "Original" is replaced by "Copy" in the MicroStation prompts. For example, "Mirror Element About Line (Copy) > Identify element.".

Mirror Element About Line (Original) > Identify element *(Identify an element to mirror.)*

Mirror Element About Line (Original) > Enter 1st point on mirror line (or reject) *(Place a data point or key-in coordinates to define one end of the line about which the element will be mirrored.)*

Mirror Element About Line (Original) > Enter 2nd point on mirror line *(Place a data point or key-in coordinates to define the angle of the line about which the element will be mirrored.)*

Mirror Element About Line (Original) > Enter 2nd point on mirror line *(If necessary, mirror it again by providing a data point or keying-in coordinates. When you are finished mirroring the element, click the Reset button to terminate the tool sequence and place the mirrored element.)*

Construct Array

The Construct Array tool makes multiple copies of a selected element in a rectangular or polar array, as shown in Figure 4–52. The **Array Type** menu on the Tool Settings window allows you to select the type of array and each array type has its own set of options to control the way the array is placed.

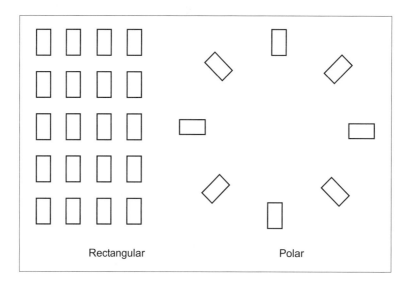

Figure 4-52 Example of a rectangular and a polar array

Rectangular Array Type

When you select **Rectangular** from the **Array Type** menu, a set of options for controlling the rectangular array structure appear in the Tool Settings tool box:

- In the **Active Angle** text field, key-in the counter clockwise rotation angle of the array rows. Alternately, you can use the up and down scroll arrows to select a standard rotation angle, such as 180 or 270 degrees.

- In the **Rows** and **Columns** text fields, key-in the number of rows and columns to include in the array. The numbers include the row and column that contains the original element, so a one by one array does not place any copies of the selected element.

- In the **Row Spacing** and **Column Spacing** text fields, key-in the row and column spacing (in working units). The spacing values can be positive or negative. Positive row spacing builds the rows upward and a negative number builds them downward. Positive column spacing builds the columns to the right and negative row spacing builds the columns to the left.

Note: The spacing is not between the copies of the element, but rather from a point on one element to the same point on adjacent elements in the array.

Invoke the Construct Array tool from:

Manipulate tool box	Select the Construct Array tool. In the Tool Setting box, select the Rectangular Array Type and specify the angle, number of columns and rows, and the row and column spacing. (See Figure 4–53).
Key-in window	**array rectangular** (or **ar r**) ENTER

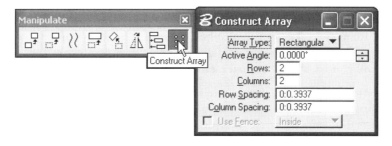

Figure 4–53 Invoke the Construct Array (Rectangular) tool from the Manipulate tool box

MicroStation prompts:

> Rectangular Array > Identify element *(Identify an element to array.)*
>
> Rectangular Array > Accept/Reject (Select the next input) *(Click the Accept button to place copies, or click the Reject button to disregard the selection.)*

Figure 4–54 shows an example of placing a rectangular array.

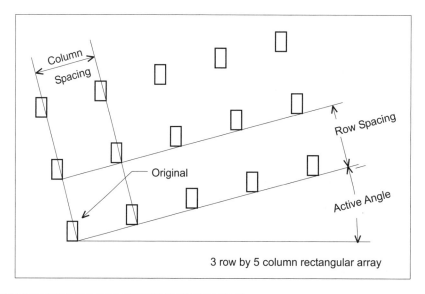

Figure 4-54 Example of placing a rectangular array

Polar Array Type

When you select **Polar** from the **Array Type** menu, a set of options for controlling the polar array structure appear in the Tool Settings window:

- In the **Items** field, key in the number of items (elements) that are to be in the array. The number includes the original element (for example, if you key in seven, the polar array will consist of the original element and six copies of the element).

- In the **Delta Angle** field, key in the degrees of rotation between each element in the polar array. The array is constructed in a counter clockwise direction and is from the point on the element where you identified it to that same point on the each copy of the element.

- Turn the **Rotate Items** check box ON, if you want each copy of the element in the array to be rotated by the **Delta Angle**. Turn the check box OFF if you want all elements in the array to be placed at the same orientation as the original element. Figure 4–55 shows difference between rotating and not rotating the array elements.

Chapter 4 Fundamentals III

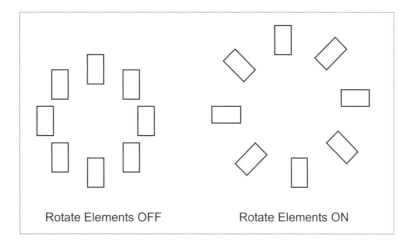

Figure 4-55 Example of rotating and not rotating the polar array elements

Invoke the Construct Array tool from:

Manipulate tool box	Select the Construct Array tool. In the Tool Setting box, select the Polar Array Type and specify the number of Items and Delta Angle (see Figure 4-56).
Key-in window	**array polar** (or **ar p**) ENTER

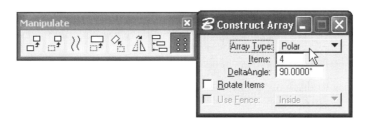

Figure 4-56 Invoke the Construct Array (Polar) tool from the Manipulate tool box

MicroStation prompts:

> Polar Array > Identify element *(Identify the element to array.)*
>
> Polar Array > Accept, select center/Reject *(Specify the center point for the array by pressing the Data button or by keying-in coordinates, or click the Reject button to cancel the selection.)*

201

The following tool sequence shows an example of using the Construct Array tool to place eight, evenly spaced bolt holes in a pipe flange (The flange and center lines were already drawn). Figure 4–57 shows flange before and after the polar array was constructed.

After selecting the tool and the **Polar Array** Type, the designer sets the **Items** to eight and the **Delta Angle** to 45 degrees. MicroStation prompts:

> Polar Array > Identify element *(Identify the shape.)*
>
> Polar Array > Accept, select center/Reject *(Place a tentative point at the center of the circle and accept it.)*

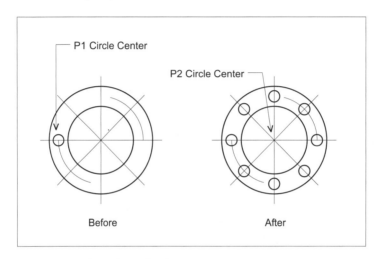

Figure 4–57 Example of placing a polar array

TEXT PLACEMENT

You have learned how to draw the geometric shapes that make up your design. Now it is time to learn how to annotate your design. When you draw by hand on paper, adding text such as design component specifications and shop and fabrication notes is a time-consuming, tedious process. MicroStation provides several text placement tools that greatly reduce the time and tedium of text placement.

MicroStation provides tools to place text, create "fill-in-the-blank" forms, import text from other applications, edit text, and change the appearance of text. In this chapter, we describe several text placement tools and settings. Additional text tools and settings are described later in Chapter 7.

You place text by typing it into the Text Editor window and then defining the location, (or locations) where the text should be placed in the design. When placed, all the keyed-in text becomes one text element called either a "text string" or a "text node." The element is a text string when all the text is in one line. The element is a text node when the text is placed as more than one line by pressing ENTER while keying in the text.

Place Text

The Place Text tool provides several text placement methods and the associated Tool Settings window provides settings that control the appearance of the text and where it is placed, in relation to the placement data point. Invoke the Place Text tool from:

Text tool box	Select the Place Text tool. MicroStation displays the Tool Settings window and the Text Editor window, as shown in Figure 4–58.
Key-in window	**place text** (or **pl tex**) ENTER

Figure 4-58 Invoking the Place Text tool from the Text tool box

MicroStation prompts:

> Place Text > Enter Text *(Make the required text settings on the Tool Settings and Text Editor windows. Type your text in the Text Editor window and place a data point in the design to define the text placement point. Continue defining points where you want to place additional copies of the text, or click the Reset button to drop the current text string. Details on these steps are provided in the following topics.)*

The Tool Settings window provides settings for controlling the way text is placed and its appearance. The Text Editor window provides the text field in which you type the text and additional settings for controlling the appearance of the text.

In the following topics, we describe the Tool Settings and Text Editor window fields, describe the process of placing text, and then describe editing features that aid in keying-in the text.

Tool Settings window

Settings on the Tool Settings window control style, placement method, angle, size, font, and justification. Additional tools can be accessed by clicking the expansion arrow in the lower right corner of the window, as shown in Figure 4–59.

Figure 4–59 Expanding the Place Text Tool Settings window

Text Style

The Text **Style** menu allows you to select a pre-defined text style. For this discussion, select the **Style (none)** option. (Text style creation is described in Chapter 7).

Method

The **Method** menu provides eight methods for placing text. For this discussion, select the first method, **By Origin**. The other methods are described in Chapter 7.

 Note: The set of options available on the Tool Settings window changes when different text placement methods are selected. The settings discussed here are the options available when the **By Origin Method** is selected.

Active Angle

The **Active Angle** field allows you to rotate the text when it is placed. The default rotation angle is zero degrees. You can key-in the degrees of rotation, or use the up and down scrolling arrows to select standard rotation angles, such as 90, 180, and 270 degrees. Enter a positive number to rotate the text in a counter-clockwise direction or a negative number to rotate the text in a clockwise direction. Figure 4–60 shows examples of various rotation angles.

Chapter 4 Fundamentals III

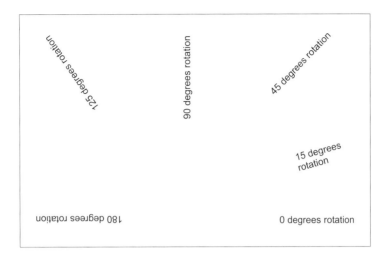

Figure 4-60 Examples of placing text at various rotation angles

 Note: MicroStation remembers the Active Angle and all tools using the Active Angle are going to use the angle you key-in here.

Line Spacing and Interchar Spacing

The **Line Spacing** and **Interchar Spacing** text fields are not available with the **Place Text By Origin Method**. Tools that use these settings are described in Chapter 7.

Height and Width

The **Height** and **Width** text fields control the size of the text when you place it in the design. Enter values in these fields in working units. For example, if the Master Unit is inches and you key-in 1.5 or 1:5 for the text height, the text height is 1.4 inches.

To the right of the **Height** and **Width** fields is a picture of a lock. If the lock is closed, the two fields will be forced to the same value: Key-in a value in one field and the other fields is automatically set equal to the value you keyed in. If the lock is open, the **Height** and **Width** can be different values.

If you are drawing an un-scaled schematic, or if you are going to print your design full size, selecting a text size is simple—key-in the actual size you want your text to be when you print it.

If you are drawing a design that must be scaled when printed, selecting a text size is a little more complicated. As mentioned earlier, in MicroStation you draw objects full size, and inform

MicroStation what scale to use when it prints the design to paper. MicroStation scales everything in the design to fit the size of paper you choose, including the text. Therefore you must scale your text by the *inverse* of the print scale so it will be the correct size on paper.

For example, if you are creating a design that will be printed at 1" = 10' and you want your text size to be 0.1 inch on paper, the text height in the design must be 1 foot. 1 inch of paper equals 10 feet, so 0.1 inch of paper equals 1 foot.

Let's put that into a formula:

> Text height in design = design scale ÷ printer scale × printed text size

(where all values are in the same measurement units, such as inches or centimeters).

Now let's try the formula for providing 1/4" tall text on paper when we print at 1/8" = 1'. Our design scale is one foot or 12 inches and our printer scale is 1/8 inch. The formula yields a text height of 24 inch text, as shown below.

> 24" = 12" ÷ 1/8" × 1/4"

Font

The **Font** menu provides character sets of different typeface designs. To display the menu, click the drop-down arrow in the right side of the field. Use the scroll bar on the right side of the menu to scroll through the font names until you find the one you want and click the name to select it, as shown in Figure 4–61.

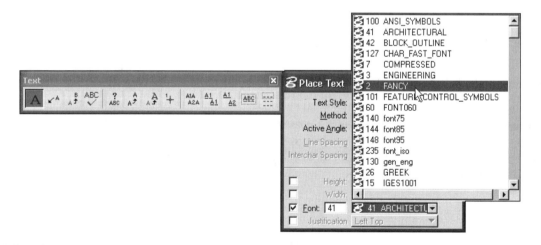

Figure 4-61 The Fonts drop-down menu

This menu can contain the names of up to 255 internal MicroStation fonts (such as Engineering or Architectural) and fonts that are in the operating system font library on your workstation (such as Arial and Times New Roman). Only a few of the internal fonts are normally installed in MicroStation. The number of system fonts varies.

 Note: This method of selecting a font does not allow you to preview the typeface the font provides. You will not see it until you place text in the design. Chapter 7 describes a method for defining sets of text styles that includes a way to preview the typeface of each font.

Justification

The **Justification** menu provides options that control where the data appears in relation to the location of the data points you define to place the text. Display the menu by clicking the drop-down arrow in the **Justification** field and then click the option, as shown in Figure 4–62.

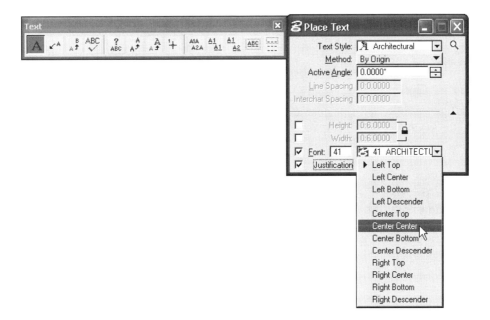

Figure 4–62 The Justification drop-down menu

The name of the justification option describes where the data point is in relation to the text, as shown in Figure 4–63.

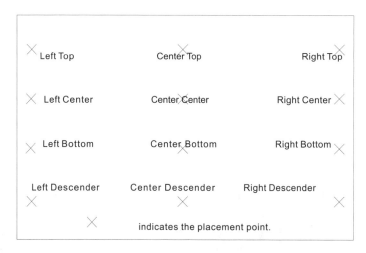

Figure 4-63 Examples of data point location for various text justifications

Text Editor Window

The Text Editor window includes a settings bar that provides additional text appearance settings and a field for keying-in the text string you need to place in the design.

Font

On the left end of the bar is the same Font menu that is available in the Tool Settings window when the Place Text tool is invoked. To change the selected font, click the drop-down arrow and select a font name from the menu.

Bold, Italics, and Underline

When these settings are available you can place text using the Bold, Italic, and Underline versions of the selected font.

To select a setting, click its icon on the settings bar. Selected icons appear to be depressed within the bar. Unselected buttons appear to be flush with the top of the bar.

If there is no version of a font for one of these settings, the button is unavailable. For example, there is no bold version of the MicroStation fonts, so when a MicroStation font is selected, the Bold button cannot be selected.

Spelling

Click the **Spelling** icon on the settings bar to invoke the MicroStation spelling check utility. Each time the utility finds a spelling error, in the text currently keyed into the text field of the Text Editor window, it displays the Spell Checker window. The Spell Checker is described in chapter 7.

Stacked Fractions

Stacked fractions are discussed in Chapter 7.

Color

When you first invoke the Place Text tool, the text color is set to the Active Color that is currently selected in the Attributes tool box. If you want to change the text color without changing the Active Color, click the drop-down arrow next to the **Color** button to display the Color palette and then select a color from the palette.

Note: When you select one of the MicroStation internal fonts, the Active Line Weight is applied to the text. If a heavy line weight is selected, the lines and curves that make up the text string are wider. True Type fonts do not use the line weight. To change the Active Line Weight, select it in the Attributes tool box.

Place Text by Origin

Now that you know how to make all the settings for controlling the appearance of the text you place in your design, you are ready to start placing text. In this chapter we describe placing text using the **By Origin Method**.

1. Invoke the Place Text tool from the Text tool box, and select **By Origin** from the **Method** menu in the Tool Settings window.
2. In the Tool Settings and Text Editor windows, select the other settings needed for the text you are about to place.
3. Click in the text field of the Text Editor window and type the text string. If you want to place the text string on multiple lines, press ENTER at each point in the text string where you want to start a new line.

Notes: You can resize the Text Editor window if you need a larger area in which to type your text. To resize it, point to the box border, press and hold the Data button while drag the box to the new size, and then release the Data button.

If you do not press ENTER, the text will wrap to a new line when you reach the right edge of the Text Editor window, but all the text will be on one line when you place it in your design.

4. After you complete typing the text, move the pointer onto the design surface and the pointer drags a dynamic image of the text.
5. To place a copy of the text string in the design, move the pointer to the location where you want to place the text and click the Data button.
6. If you want to place additional copies of the same text string, continue placing data points at each location.

7. You can edit or replace the text string at any time during the placement process by clicking in the Text Editor window and editing the text. When you move the pointer back onto the design surface, it will drag a dynamic image of the new text string.
8. You can also drop the current text string and clear the Text Editor window by clicking the Reset button.

Notes: If you need to change any of the text settings, first drop the current text string from the pointer by clicking the Reset button. Changes to settings usually do not go into effect while there is a dynamic image of a text string on the pointer.

Each copy of the text you place becomes a single element that can be manipulated like any other element. The only key point in a text string or multi-line element is the placement point.

Keyboard Shortcuts for the Text Editor window

The keyboard shortcuts listed in Table 4–1 move the text pointer to specific positions in current line and between lines when the box contains more than one line of text.

Table 4–1 Positioning the Text Cursor

PRESS:	TO MOVE THE TEXT CURSOR:
LEFT	Left one character
RIGHT	Right one character
CTRL + LEFT	Left one word
CTRL + RIGHT	Right one word
HOME	To the beginning of the current text line
END	To the end of the current text line
UP	Up to the previous line of text
DOWN	Down to the next line of text
PAGE UP	Straight up into the first text line
PAGE DOWN	Straight down into the last text line

The key-ins described in Table 4–2 delete characters from the text in the Text Editor box.

Table 4-2 Keys that Delete Text

PRESS:	TO DELETE:
Backspace	The character to the left of the text cursor
DELETE	The character to the right of the text cursor
ALT + DELETE	All characters from the text cursor to the end of the word
CTRL + DELETE	All characters from the text cursor to the end of the current word
CTRL + Backspace	All characters from the text cursor to the beginning of the current word

The key-ins described in Table 4–3 select or deselect text in the Text Editor box. Selected text is shown with a dark background, and it can be moved, copied, or deleted.

Table 4-3 Selecting Text with Key-Ins

PRESS:	TO SELECT (OR DESELECT IF ALREADY SELECTED):
SHIFT + LEFT	The character to the left of the text cursor
SHIFT + RIGHT	The character to the right of the text cursor
CTRL + SHIFT + LEFT	The characters from the text cursor to the start of a word
CTRL + SHIFT + RIGHT	The characters from the text cursor to the end of a word
SHIFT + END	The characters from the text cursor to the end of a line
SHIFT + HOME	The characters from the text cursor to the start of a line
CTRL + A	To select all text in the Text Editor box
LEFT or RIGHT	To deselect all previously selected text

The pointing device actions described in Table 4–4 select or deselect text in the Text Editor box.

Table 4–4 Selecting Text with the Pointing Device

POINTING DEVICE ACTION	RESULT
Press the Data button and drag the screen cursor across the text	Selects all the text you drag across
Double-click the Data button	Selects the word the cursor is in
Triple-click the Data button	Selects all text in the Text Editor window
Hold down SHIFT + Data button and drag across the text	Adds more text to the text already selected
Click the Data button in an area where there is no text	Deselects all previously selected text

The actions that replace, delete, and copy previously selected text are shown in Table 4–5.

Table 4–5 Replacing, Deleting, and Copying Selected Text

ACTION	RESULT
Start typing characters	Replace the selected text with the text you type
Press BackSpace	Delete all the selected text
Press DELETE	Delete all the selected text
Press CTRL + INSERT	Copy the selected text to a buffer
Press SHIFT + INSERT	Paste the previously copied or deleted text at the text cursor position

Open the Exercise Manual PDF file for Chapter 4 on the accompanying CD for project and discipline specific exercises.

REVIEW QUESTIONS

Write your answers in the spaces provided.

Write your answers in the spaces provided.

1. The Place Polygon tool places polygons that can have a maximum of _____ sides.

2. The Points Method for the Place Point or Stream Curve tool is used to place a curve string by _____.

3. The Stream Method of the Place Point or Stream Curve tool is used to place a curve string by _____.

4. The Multi-line tool allows you to place up to _____ separate lines of various _____ , _____ , and _____ with a single tool.

5. The Place Fillet tool joins two lines, adjacent segments of a line string, arcs, or circles with an _____ of a specified radius.

6. What are the three methods for constructing a fillet: _____ , _____ , and _____ .

7. The Chamfer tool allows you to draw a _____ instead of an arc.

8. The purpose of the Trim tool is to _____.

9. What is the name of the tool that deletes part of an element? _____

10. Name the two categories of manipulation tools available in MicroStation. _____ and _____ .

11. The Copy tool is similar to the Move tool, but it _____.

12. Name at least three element manipulation tools available in MicroStation: _____ , _____ , and _____

13. To rotate an element by a keyed in angle, select the _____ method.

14. The Array tool makes multiple copies of a selected element in either a _____ or _____ array.

15. List four of the settings that control the shape and size of a rectangular array: _____ , _____ , _____ , and _____ .

16. How do you display additional settings on the Place Text Tool Settings window? _____ .

17. Name three text settings that can be changed in the Place Text Tool Settings window:

    _____ , _____ , and _____ .

18. You can select a font from the _____ and _____ windows.

19. You type a new Height value in the Place Text Tool Settings window and the Width value changes to be equal to the Height. What caused that to happen?
    _____

20. If you want to place a text element that is divided into two lines of text, what must you do while typing the text?
    _____

chapter 5

AccuDraw and SmartLine

Objectives

After completing this chapter, you will be able to do the following:

- Set up AccuDraw
- Use AccuDraw to place elements with fewer data points and less typing
- Use SmartLine to draw complex models quickly with one tool

GETTING TO KNOW ACCUDRAW

AccuDraw is a powerful MicroStation feature that increases your drawing productivity by tracking what you did and anticipating what you will do next.

When AccuDraw is active, the AccuDraw coordinates box appears on the MicroStation desktop, and when you place data points, the AccuDraw compass appears on the point. The coordinates box provides a faster and easier way to enter precise coordinates than the key-in method, and the compass helps you place data points accurately when using the pointing device. The AccuDraw Settings box allows you to control the way AccuDraw interacts with the design. The boxes and compass are shown in Figure 5-1.

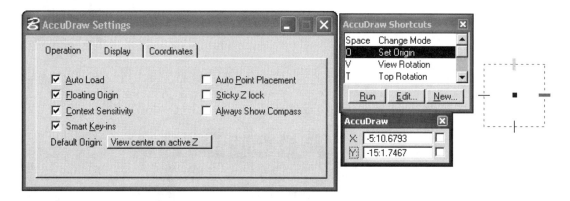

Figure 5-1 AccuDraw tools

Start and Stop AccuDraw

AccuDraw automatically starts when you open design files in MicroStation, but options are provided that allow you to stop AccuDraw while you are working in a design file.

Stop and Start AccuDraw from:

Primary tool box	Select the Start AccuDraw tool (see Figure 5-2).
Key-in window	**accudraw activate** (or **a a**) ENTER

While the pointer hovers over the Start AccuDraw tool, MicroStation prompts:

Toggle AccuDraw point input tool

Figure 5-2 Invoking the AccuDraw tool from the Primary tool box

Note: A checkbox on the AccuDraw Settings box **Operation** tab allows you to turn off starting AccuDraw automatically. The checkbox is discussed later in this chapter.

Chapter 5 AccuDraw and SmartLine

When AccuDraw is stopped, the AccuDraw coordinates box and compass disappear from the MicroStation desktop. When AccuDraw is started, the coordinates box appears and, when you define a data point in the design, the compass appears on the point.

The AccuDraw coordinates box initially appears floating, but it can be docked to the top or bottom edges of the MicroStation desktop. Figure 5–3a shows the box floating and Figure 5–3b shows it docked on the top edge of the desktop.

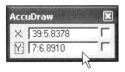

Figure 5-3a AccuDraw window shown floating in the View window

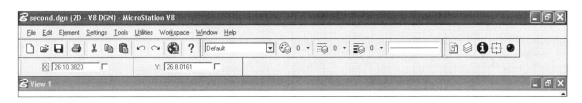

Figure 5-3b AccuDraw window docked at the top of the MicroStation application window

Key-In Shortcuts

The AccuDraw tools include one- and two-character shortcut key-ins that control AccuDraw actions and select AccuDraw features.

When focus is on the AccuDraw coordinates box, MicroStation checks all characters you key-in to see if they are AccuDraw shortcuts. Focus usually defaults to the coordinates box when AccuDraw is active.

Shortcuts are described in this chapter and the AccuDraw Shortcuts box lists all shortcuts. Instructions for opening the box and a table that lists all shortcuts are provided later in this chapter.

The AccuDraw Compass

When AccuDraw is active, the AccuDraw compass appears on data points you place in the design and on the places where you select elements for manipulation. The compass is the center of the AccuDraw plane and is your main focus for input.

Coordinate System

The AccuDraw drawing plane includes polar and rectangular coordinate systems for locating points. The systems are like those provided by the precision key-in tools, except all offsets in the

AccuDraw coordinate systems are from the AccuDraw compass origin, not the design plane's origin point. Figures 5–4a and 5–4b show the appearance of the compass for each coordinate system and the options in the AccuDraw coordinates box for each system.

Figure 5-4a Compass with the rectangular coordinate system active

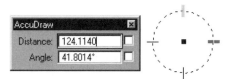

Figure 5-4b Compass with the polar coordinate system active

To switch from one coordinate system to the other, use one of these methods:

Keyboard shortcut	**Spacebar** (when focus is on the AccuDraw coordinates box)
Key-in window	**accudraw mode** (or **a m**) ENTER
AccuDraw Settings box	On the Operations tab, select Polar or Rectangular from the Type menu.

The coordinate system is switched (there are no MicroStation prompts).

Orthogonal Axes

The center of the AccuDraw design plan is indicated by a dot in the center of the compass. The orientation of the AccuDraw drawing plane X axis and Y axis is indicated by short tick marks (lines) crossing the compass rectangle or circle.

To aid in distinguishing the two axes, the positive X axis tick mark is a red line and the positive Y axis tick mark is a green line. The colors can be changed from the AccuDraw Settings box (discussed later in this chapter).

Compass Movement

The compass moves to and centers on each data point you enter. If you are drawing lines or line segments, the compass also rotates to align its X axis parallel to the angle of the last line segment drawn with the positive X axis direction pointing in the direction the line was drawn. Figure 5-5 shows an example of the Rectangular coordinate system compass at the end of a line drawn at 45 degrees or rotation.

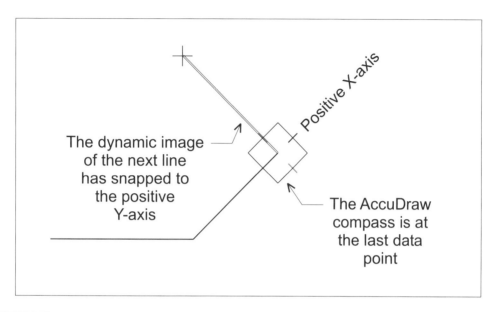

Figure 5-5 Example of the AccuDraw compass rotated to match the rotation of the last line placed in the design

Key-in any of the following one or two letter shortcuts to change the position and rotation of the compass:

- **O** (Set Origin) moves the compass from its current location to the position of the pointer.
- **T** (Top Rotation) sets the compass rotation to zero degrees (the red positive X axis tick mark points to the right).
- **RZ** (Rotate about Z) rotates the compass 90 degrees in a counterclockwise direction.
- **RQ** (Rotate Quick) allows you to dynamically set the compass rotation by moving the pointer. The compass follows the pointer by rotating around its center point until you click the Data button.

The AccuDraw Settings box has options to lock the compass at its current location so it no longer jumps to the last data point Also, the compass can be locked at its current rotation, so it does not assume the rotation angle of the last element segment placed.

Compass Indexing

As you move the pointer away from the center of the compass, AccuDraw tracks where the pointer is in relation to the compass *X* and *Y* axes. When the pointer is almost lined up with the *X* or *Y* axis, AccuDraw indexes (snaps) to the axis. Figure 5–6 shows an example of the pointer indexed to the *X* axis while placing a circle by center (as indicated by the horizontal line extending from the center of the compass).

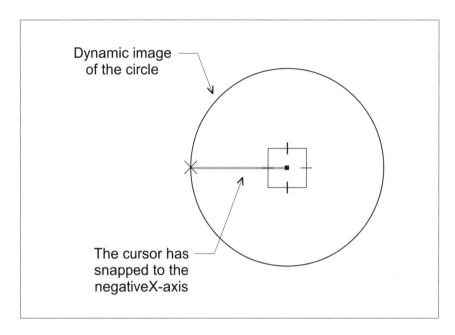

Figure 5-6 The AccuDraw compass indexed to the X axis

This makes it easy to use the pointer to place a new element in the same direction or at a right-angle to the previous element. Indexing does not lock the pointer to the axis. You can move the pointer away from the axis after it snaps to the axis.

The AccuDraw Coordinates Box

The AccuDraw coordinates box provides a place to key-in *X* and *Y* axis offsets in the Rectangular coordinate system and Distance and Angle offsets in the Polar coordinate system (see Figures 5–4a and 5–4b). This box almost completely eliminates the need to key in precision input codes such as **DL** and **DI**.

As you move the screen pointer, the coordinate box's input fields are updated automatically with the pointer's position, relative to the current AccuDraw origin.

Selecting Fields

AccuDraw usually has focus on the AccuDraw coordinates box field you are most likely to use next, so there is often no need to select the field first. Field focus is indicated by a change in the appearance of the field. If the wrong edit field is in focus, just press TAB to move the focus to the other edit field. Characters you key-in to a field replace the characters previously contained in the field.

For example, in Microsoft Windows XP, a vertical text pointer appears at the right end of the text in the field that has focus, as shown in Figure 5–7. Highlight methods vary among operating systems.

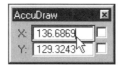

Figure 5-7 The AccuDraw coordinates box with the X axis in focus

When you key-in a value, it appears in the field that is in focus, and that field is locked until the next data point is placed. The toggle button to the right of the edit field is turned on when the field is locked.

For example, if the compass is in rectangular mode and the dynamic image is indexed to the positive *X* axis indicator, the *Y* edit field in the box contains 0 and the *X* edit field has focus. If you key-in a number on the keyboard, it is placed in the *X* edit field and the field is locked.

Accepting Field Contents

When the AccuDraw coordinates box fields contain the correct offset from the compass origin, click the Data button to place the data point at the offset values. When you place the data point, the AccuDraw edit fields are unlocked.

Locking Fields

As mentioned earlier, each edit field in the AccuDraw coordinates box has a check box that locks the field when you key-in something. The field remains locked until you place the next data point. You can also lock or unlock the fields by clicking on the field's lock check box or by keying-in a shortcut. The lock shortcut keys are as follows:

- **X** locks and unlocks the *X* edit field when the Rectangular coordinate system is active.
- **Y** locks and unlocks the *Y* edit field when the Rectangular coordinate system is active.
- **D** locks and unlocks the Distance edit field when the Polar coordinate system is active.
- **A** locks and unlocks the Angle edit field when the Polar coordinate system is active.

Negative Distances

The direction of the screen pointer from the compass origin indicates the direction of your input in the drawing plane, so there is usually no need to key-in the negative sign for distances that are to the left of or down from the AccuDraw origin. Move the screen pointer near the area where the next data point will be placed to establish the correct sign.

Recall Previous Values

AccuDraw remembers the values previously entered in the AccuDraw coordinates box. AccuDraw can use the previous distance as a hint for the next distance, and can display previously used distances and angles in the AccuDraw coordinates box.

Distance Indexing

AccuDraw remembers the linear distance between the last two data points. If the pointer is indexed to either Rectangular coordinate axis, a tick mark at the screen pointer location indicates when the linear distance from the last data point to the current pointer location is equal to the linear distance between the last two data points. For Polar coordinates, the tick mark appears at any angle of rotation. Figure 5–8 shows the tick mark at the pointer position for rectangular coordinates.

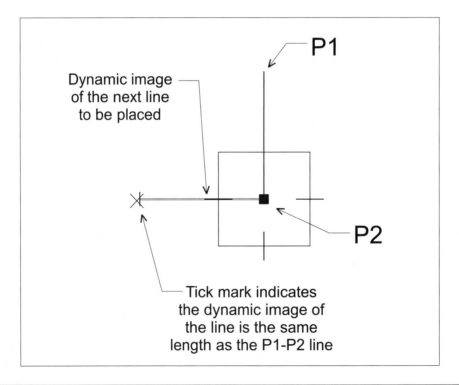

Figure 5-8 Current distance is equal to previous distance, as indicated by tick mark at pointer position

Recalling Previous Values

Each time you place a data point, AccuDraw remembers the offsets used to place that point. All distance values are stored in a buffer and all angle values are stored in a separate buffer. Press PAGE UP to load the last value used in the edit field that is in focus. Press PAGE UP again and the next-to-last value is loaded. Each time you press PAGE UP a saved value is loaded in the edit field.

All distance values are stored in the same buffer, so they can be applied to any coordinate distance field (*X* axis, *Y* axis, or Polar Distance).

The Popup Calculator

MicroStation provides a calculator that works with AccuDraw to allow you to do calculations using the values in the AccuDraw coordinates box fields. To use the calculator on the value of the field with focus:

1. Key-in a symbol for the mathematical operation you want: + (add), – (subtract), * (multiply), or / (divide).
2. Key-in a value to complete the calculation.
3. Place a data point to accept the calculation, or click the Esc key to reject the calculation.

When you key-in one of the mathematical symbols, the calculator box opens below the AccuDraw field that is in focus. The calculation result is shown at the bottom of the calculator box. Figure 5–9 shows an example of the calculator being applied to the *X* input field.

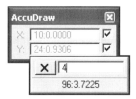

Figure 5-9 The calculator box

Replace Current Value

To perform a calculation that does not use the value in the AccuDraw field with focus, key-in an equal sign (=) to start the calculation. When you start the calculation with an equal sign, the calculation results are stored in the AccuDraw field rather than in the calculator box (see Figure 5–10).

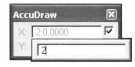

Figure 5-10 The calculator box when you start by typing an equal sign (=)

Complex Calculations

Parenthesis can be used to create complex calculations. Calculations enclosed in parenthesis are done first.

For example, keying-in the calculation: **=10*(2+3)** places 50 in the field. The two and three are added, then the result is multiplied by ten (see Figure 5–11).

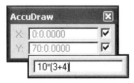

Figure 5-11 The use of parentheses in a calculation

Smart Lock

Smart Lock allows you to constrain the next data point to the nearest axis. To use Smart Lock, move the pointer near to the axis and direction you want the next data point constrained to, then press ENTER. If the rectangular compass is active, either the **X** or **Y** axis is locked. If the polar compass is active, the **Angle** is locked at 0, 90, 180, or -90 (270) degrees.

To turn off Smart Lock and remove the constraint from the next data point, press ENTER again.

CHANGE ACCUDRAW SETTINGS

The AccuDraw Settings box can be used to change the appearance of the compass and the way AccuDraw works. Options are provided on three tabs named: **Operation**, **Display**, and **Coordinates**.

Invoke the AccuDraw Settings box from:

Pull-down menu	Settings > AccuDraw
Key-in shortcut	**GS** (with focus on the AccuDraw coordinates box)
Key-in window	**accudraw dialog settings** (or **a d se**) ENTER

MicroStation opens the AccuDraw Settings box. Following are descriptions of the options available on each tab.

Operation

The **Operation** tab provides check boxes and a menu that change the way AccuDraw performs, as shown in Figure 5–12.

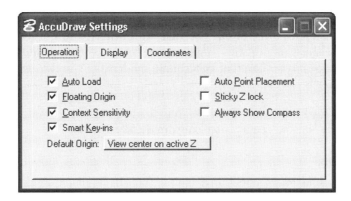

Figure 5-12 The AccuDraw Settings box Operation tab

Auto Load

If the Auto Load check box is ON, AccuDraw starts automatically when you open the design file in MicroStation.

To stop AccuDraw from starting automatically, turn this check box OFF, or use the Toggle AccuDraw button on the Primary Tool box to stop AccuDraw and then select **File > Save Settings**. This stops AccuDraw from starting automatically only in the current design file.

Floating Origin

When the **Floating Origin** check box is ON, the AccuDraw compass always moves to the location of last data point. If the check box is turned OFF, the compass stays at its current location and does not jump to the location of new data points.

Context Sensitivity

When the **Context Sensitivity** check box is ON, tools can override AccuDraw default behavior to insure smoother operation of AccuDraw with the tool. For example, when lines or line segments are placed, the AccuDraw compass rotates to the angle of the last placed line or line segment. If the last line or line segment was rotated 45 degrees, the rectangular coordinate compass is rotated 45 degrees.

If the check box is turned OFF, the AccuDraw tools do not override AccuDraw default behavior. For example, if the check is OFF, the compass does not rotate to match the rotation of the last line or line segment. If the compass is rotated when the checkbox is turned OFF, it remains at that rotation.

Smart Key-ins

When the **Smart Key-ins** check box is ON, AccuDraw interprets the numbers you type into the coordinate fields as positive or negative, depending on the pointer position in relation to the compass. If the compass is in rectangular coordinate mode, focus is also switched to the field of the axis that the pointer is closest to.

If the check box is OFF, you must key-in a dash to indicate a negative distance and you must press TAB to switch focus between the rectangular coordinate axis fields.

Auto Point Placement

When the **Auto Point Placement** check box is ON, AccuDraw places data points automatically when you have locked both the X and Y values, or when you have locked one or the other while the pointer is indexed to zero.

Sticky Z Lock

The **Sticky Z Lock** check box is used in 3D designs to keep the Z axis locked when data points are placed (Locks normally turn off when a data point is placed). Keeping the Z axis locked makes it easy for designers to draw in one plane.

Always Show Compass

If the **Always Show Compass** check box is ON, turning on AccuDraw makes the compass appear at the location of the last data point, in the current operation. For example, if you invoke the Copy tool, select an element to copy, and then turn on AccuDraw, the compass appears at the point on the element where you selected it.

If the check box is OFF, the compass does not appear until you place the next data point. In the case of the Copy operation, the compass does not appear until you define the location of the copy by placing a data point.

Default Origin

The **Default Origin** is a menu with three options for determining where AccuDraw starts operation in a 3D design:

- **View Center on Active Z** starts with origin at the center of the view and at the Active Z depth of the view.
- **Global Origin** starts with the origin at the Global Origin of the design file.
- **Global Origin on Active Z** with the origin at the Global Origin and at the Active Z depth of the view.

Display

The **Display** tab contains menus and check boxes that control the compass colors and the display of AccuDraw information, as shown in Figure 5–12.

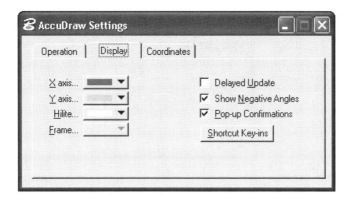

Figure 5-13 The AccuDraw Settings box Display tab

The Color Menus

Four color menus allow you to change the colors used for the AccuDraw compass. When you click one of the options, a menu with ten colors appears allowing you to select one of the colors.

- **X axis** sets the color of the positive X axis indicator.
- **Y axis** sets the color of the positive Y axis indicator.
- **Hilite** sets the color of the negative axis indicators and the compass center indicator.
- **Frame** sets the color of the frame (the dashed rectangle or circle).

Delayed Update

If the **Delayed Update** check box is OFF, the fields in the AccuDraw coordinates box are updated continuously as you move the pointer. If the check box is ON, the fields are only update when the pointer is not moving.

Show Negative Angles

The **Show Negative Angles** check box controls the way angles are displayed in the coordinates box when the Polar coordinate system is active. With the check box ON, AccuDraw displays negative angles, in the range of 0 to -180 degrees, when the pointer is below the X axis and positive numbers, in the range of 0 to 180 degrees, when the pointer is above the X axis.

When the check box is OFF, AccuDraw displays positive numbers in the range of 0 to 360 degrees.

Pop-up Confirmations

If the **Pop-up Confirmations** check box is ON, AccuDraw displays pop-up boxes to confirm keyed-in shortcuts. If you key-in a one-character shortcut, the name of the shortcut appears below the AccuDraw coordinates box. If you key in a two-character shortcut, a list of shortcuts starting with the first character appears (see Figure 5–14). If the check box is OFF, no messages appear.

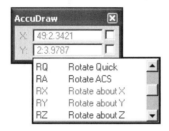

Figure 5-14 The pop-up box that appears when the first character of a two-letter shortcut is pressed

Shortcut Key-ins

Click the **Shortcut Key-ins** button to open the AccuDraw Shortcuts box, as shown in Figure 5–15.

Figure 5-15 The AccuDraw Shortcuts box

The AccuDraw Shortcuts box displays all defined AccuDraw keyboard shortcuts. To invoke a shortcut from the box, select the shortcut and click the **Run** button. The box is rather small when initially opened, but you can drag down the box's bottom border to increase its size.

The **Edit** button opens the Edit Shortcut box where you can change the shortcut's name, description, and the MicroStation command it executes. To edit a shortcut, select it in the AccuDraw Shortcuts box and click the **Edit** button. The Edit Shortcut box opens displaying the selected shortcut's current settings, as shown in Figure 5–16. After you complete editing the shortcut, click the **OK** button to close the box and apply the changes, or click the **Cancel** button to close the box without saving the changes.

Chapter 5 AccuDraw and SmartLine

Figure 5-16 The Edit Shortcut box

The **New** button opens the New Shortcut box in which you can create your own shortcut by entering a one- or two-character shortcut, a shortcut description, and the command to be executed by the shortcut. Shortcuts provide a quick way to invoke commonly used tools, such as Place Line and Copy Element. For example, Figure 5–17 shows the New Shortcut box with a new Place Line shortcut. Click the **OK** button to close the box and create your new shortcut, or click the **Cancel** button to close the box without creating the shortcut.

 Note: AccuDraw can support up to 400 keyboard shortcuts.

Figure 5-17 The New Shortcut box

The Coordinates Tab

The **Coordinates** tab contains menus and check boxes that change the way the AccuDraw coordinate system works, as shown in Figure 5–18.

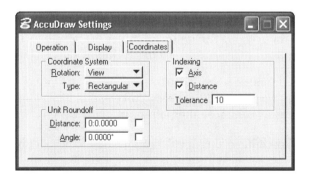

Figure 5-18 The AccuDraw Settings box Coordinates tab

Rotation

The Coordinate System **Rotation** menu options allow you to align the compass to the following settings:

- **Top**—Aligns the compass to the top view in *three dimensions*, or the current view axes in *two dimensions* (same as View).
- **Front**—Aligns the compass to the front view in *3D* only.
- **Side**—Aligns the compass to the side view in *3D* only.
- **View**—Aligns the compass to the current view axes.
- **Auxiliary**—Aligns the compass to the last defined auxiliary coordinate system.
- **Context**—A temporary orientation affected by several factors, including the RQ keyboard shortcut.

Type

The Coordinate System **Type** menu options allow you to switch the compass between the **Polar** and **Rectangular** coordinate systems (the same as pressing the Spacebar when focus is on the AccuDraw coordinates box).

Distance

The Unit Roundoff **Distance** text field and checkbox allow you to force distances, defined by moving the pointer, to be limited to the roundoff value or multiples of the value. To lock a roundoff distance value, turn the **Distance** checkbox ON and key-in the roundoff value in the **Distance** text field. For example, a **Distance** roundoff value of 0.25 Master Units (0.25, 0.5, 0.71, 1.0, and so forth) forces the line length to change in increments of 0.25 Master Units, as you move the pointer in the design.

The roundoff value applies to the distance, when AccuDraw is using the polar coordinate system, and the X and Y axis values, when AccuDraw is using the rectangular coordinate system.

Angle

The Unit Roundoff **Angle** text field and lock allow you to force angles, defined by moving the pointer, to be limited to the roundoff value or multiples of the value. To lock a roundoff angle value, turn the **Angle** check box ON and key-in the roundoff value in the **Angle** text field.

The roundoff value only applies when AccuDraw is using the polar coordinate system. For example, an **Angle** roundoff value of 30 degrees, forces all angles to increments of 30 degrees (0, 30, 60, 90, and so forth) as you move the pointer in the design.

 Note: Values keyed-in to the AccuDraw coordinates box edit fields will override the roundoff values.

Axis

The Indexing **Axis** check box turns compass indexing ON and OFF. If the check box is ON, moving the pointer close to one of the axis marks on the AccuDraw compass causes the dynamic image, or the element you are placing or manipulating, to snap to the mark. If the checkbox is OFF, no snapping occurs.

Distance

The Indexing **Distance** check box turns distance indexing ON and OFF. When the check box is ON, AccuDraw places a tick mark on the end of the dynamic image of the line you are placing, when the line is snapped to a compass axis mark and is the same length as the previously placed line. If the check box is OFF, there is no indication of equal length lines.

The Indexing **Tolerance** text field allows you to control how close you must be to the Axis and Distance indexing points before AccuDraw triggers indexing. The **Tolerance** is in screen pixels and can be in the range of 1 to 99. A pixel is the smallest addressable unit on a display screen. The higher the pixel resolution, the more information can be displayed. For example, a resolution of 1024x768 means the screen is divided into 1024 columns and 768 rows of pixels.

AccuDraw Shortcuts

As was discussed earlier in this chapter, AccuDraw includes one- and two-character keyboard shortcuts that allow you to change the action AccuDraw is about to perform. A shortcut is invoked by typing the shortcut while focus is in the AccuDraw coordinates box. Table 5–1 lists the shortcuts available in *2D* designs (additional shortcuts are available for *3D* designs).

Note: AccuDraw can support up to 400 keyboard shortcuts.

Table 5–1 AccuDraw Shortcuts for *2D* design.

KEY-IN	ACCUDRAW DIRECTIVE
?	Open the AccuDraw Shortcuts box.
~	Bumps the selected item in the top menu in the Tool Settings window.
Spacebar	Toggle between the Rectangular and Polar coordinate compass.
M	Open or set focus to the Data Point Key-in box that can be used for precision input (*XY=*, DL=, DI=, etc.).
O	Move the origin point to the current screen pointer location.
P	Open or set focus to the Data Point Key-in window that can be used for precision input (*XY=*, DL=, DI=, etc.).

Table 5-1 AccuDraw Shortcuts for *2D* design. (continued)

Q	Close the AccuDraw tool.
G + A	Get a saved ACS (Auxiliary Coordinate System)
G + K	Open, or move focus, to the Key-in window (same as choosing Key-in from the Utility menu).
G + T	Open or set focus to the AccuDraw Tool Settings window.
W + A	Save the drawing plane alignment as an ACS.
LOCKS	**ACCUDRAW DIRECTIVE**
ENTER	Turn Smart Lock ON and OFF.
X	Turn the Rectangular coordinate X value lock ON and OFF.
Y	Turn the Rectangular coordinate Y value lock ON and OFF.
D	Turn the Polar coordinate Distance value lock ON and OFF.
A	Turn the Polar coordinate Angle value lock ON and OFF.
L	Turn the Index lock ON and OFF. If the Index lock is OFF, the only way to index the pointer position to an axis is to use Smart Lock.
ROTATE THE DRAWING PLANE	**ACCUDRAW DIRECTIVE**
B	Rotate the compass between its current rotation and the drawing's top view. It is a toggle switch that alternates between rotating the compass between the two positions.
T	Rotate the drawing plane to align with the top view.
V	Rotate the drawing plane to align with the view (its normal rotation).
R + Q	Temporarily rotate the drawing plane about the compass origin point. The lock is turned OFF after the next data point.
R + A	Permanently rotate the drawing plane. It stays active after the current tool terminates.
R + Z	Rotate the drawing plane 90 degrees about its Z axis. (In a *2D* drawing, the Z axis is perpendicular to the drawing plane.)

SNAP MODES	ACCUDRAW DIRECTIVE
C	Activate Center snap mode.
I	Activate Intersect snap mode.
N	Activate Nearest snap mode.
K	Open the Keypoint Snap divisor setting box so the snap divisor can be set.

WORKING WITH ACCUDRAW

You've seen the parts of AccuDraw and how to change the way it functions. Now, let's look at how you can use it with various tools available in MicroStation.

Example of Simple Placement

This example shows how AccuDraw can be used to draw the arrowhead shown in Figure 5–19.

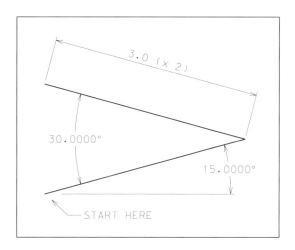

Figure 5-19 Simple Arrowhead drawn with AccuDraw

1. Open the AccuDraw Settings box, click the **Operation** tab and turn ON the **Floating Origin** and **Context Sensivity** check boxes.
2. Invoke AccuDraw and the Place Line tool.
3. Place a data point to define the lower left end of the arrowhead.
4. If the AccuDraw Rectangular coordinates compass appears, press the **Spacebar** to switch to the Polar coordinates compass.

5. Drag the pointer a short distance to the right and upwards.
6. If the AccuDraw coordinates box is focused on the **Angle** text field, click TAB to focus on the **Distance** text field.
7. Key-in **3** to place and lock the line length in the **Distance** text field.
8. Press TAB to move focus to the **Angle** edit field, and key-in **15** to place and lock the rotation angle in the **Angle** text field.
9. Click the data button to draw the bottom half of the arrowhead.
10. Drag the pointer to the left and up until the distance indexing tick mark appears at the screen pointer, and press **X** to lock that length in the **Distance** text field (the length should be 3).
11. If the AccuDraw window is not focused on the **Angle** text field, click TAB to focus on it.
12. Key-in **30** to place, and lock the angle of the second line in the **Angle** text field.
13. Click the data button to draw the top half of the arrowhead.

Moving the Compass Origin

In almost all uses of AccuDraw, the compass origin is on the previous data point, but it can be moved without placing a data point. To relocate the compass origin, key in **O**. If a tentative snap point is defined, the compass jumps to the snap point. Otherwise, it jumps to the current pointer location.

Using Tentative Points with AccuDraw

Tentative points can be used with the compass to place elements in precise relationships to other elements. For example, to start a line 2 Master Units to the right of the corner of an existing element, do the following:

1. Invoke the Place Line tool.
2. Tentative snap to the corner of the existing element from which the offset is to be measured.
3. Key-in the letter **O** to move the compass to the tentative point.
4. If the Polar coordinate system is not active, press the **Spacebar** to switch to it. (This example uses Polar coordinates, but it can be done in Rectangular coordinates as well.)
5. In the AccuDraw coordinates box, key-in **2** in the **Distance** field and **0** in the **Angle** field.
6. Click the Data button and the first point of the line is placed 2 units to the right of the tentative point.
7. Finish drawing the line.

Note: AccuDraw provides keyboard shortcuts for selecting some of the tentative snap modes, these are listed in Table 5–1.

Chapter 5 AccuDraw and SmartLine

Rotating the AccuDraw Plane

In *2D* designs, the AccuDraw plane can be rotated about the Z axis (which is perpendicular to the 2D plane) any time the compass is visible. To rotate the plane, key-in one of the two-letter rotation shortcuts (see Table **5–1**).

 Note: The compass rotates automatically to the same angle as the previously placed line segment when the context sensitivity check box is set to ON. The rotation shortcuts allow you to override that rotation.

Placing Elements with AccuDraw Active

Here are two examples of using AccuDraw to enhance placing elements in a design.

Ellipse

The following example uses the Place Ellipse tool with the **Center Method** and AccuDraw to place an ellipse with a major axis 8 Master Units long, rotated 30 degrees, and a minor axis 4 Master Units long. You start by placing a data point to define one end of the major axis and then use the AccuDraw coordinates box to complete the ellipse

1. Invoke the Place Ellipse tool and select **Center** from the **Method** menu.
2. Place the first data point to define one end of the major axis.
3. If the rectangular compass is active, press the **Spacebar** to switch to the polar compass.
4. In the AccuDraw coordinates box, set the **Distance** to **4** and the **Angle** to **30**.
5. Click the Data button to define the major ellipse axis.
6. In the AccuDraw window, key-in **2** in the **Distance** field.
7. Click the Data button to complete the ellipse by defining its minor axis.

Block

The following example uses the Place Block tool with the **Rotated Method** and AccuDraw to place a 3 x 5 block, rotated 15 degrees. You start by placing a data point to define one corner of the block and then use the AccuDraw coordinates box to define the rotation angle and place the block.

1. Invoke the Place Block tool and select **Rotated** from the **Method** menu.
2. Place a data point to define the lower left corner of the rotated block and move the pointer slightly above and to the right of the data point
3. If the rectangular compass is active, press the **Spacebar** to switch to the polar compass.

4. In the AccuDraw coordinates box, key-in **3** in the **Distance** field and **15** in the **Angle** field.
5. Click the Data button to define the bottom edge of the rotated block (the compass switches to rectangular coordinates) and move the pointer up a short distance.
6. In the AccuDraw coordinates box, key-in **5** in the **Y** field.
7. Click the Data button to complete the rotated block.

Manipulating Elements with AccuDraw Active

AccuDraw also enhances the manipulation of elements. Here is an example that uses AccuDraw and AccuSnap to place two copies of the lower left block shown in Figure 5–20.

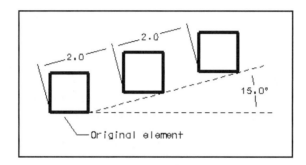

Figure 5-20 Example of using AccuDraw with the Copy Element tool

If AccuDraw is not active, invoke it.

1. Invoke the Copy Element tool and turn the **Make Copy** check box ON.
2. Open the AccuSnap Settings box, turn the **Enable AccuSnap** check box ON, and use the other check boxes to set AccuSnap operation as desired.
3. Open the AccuDraw Settings box, set the Coordinate System **Type** to **Polar**, turn the Unit Roundoff **Distance** check box ON, key-in **2** in the **Distance** text field. Turn the Unit Roundoff **Angle** check box ON and key in **15** in the **Angle** field.
4. Select the lower right corner of the block by moving the pointer toward the corner until AccuSnap snaps to the corner and then click the Data button to accept it.
5. Drag the pointer up and to the right until the AccuDraw coordinates box shows an **Angle** of **15** and a **Distance** of **2**.
6. Click the Data button to place the copy.
7. Repeat steps 5 and 6 for the second copy.
8. Click the Reset button to drop the element.

PLACE SMARTLINE TOOL

The SmartLine tool places a chain of connected lines and arcs as individual elements, or as a line string, shape, complex chain, or complex shape. The vertexes between segments can be a sharp point, a tangent arc (rounded), or a chamfer.

SmartLine, in combination with AccuDraw and AccuSnap, provide a powerful toolset for creating designs in MicroStation..

Invoke SmartLine from:

Linear Elements tool bar	Select the Place SmartLine tool (see Figure5–21) and make the initial tool settings in the Tool Settings window (see the next topic).
Key-in window	**place smartline** (or **pl sm**) ENTER

Figure 5-21 Place SmartLine tool icon in the Linear Elements tool box

MicroStation prompts:

> Place SmartLine > Enter first vertex *(Place the first data point to start the SmartLine.)*
>
> Place SmartLine > Enter next vertex or reset to complete *(Place data points to define SmartLine segments and change tool settings as required to complete the design. When you have complete drawing all segments, click the Reset button to end the SmartLine.)*Place SmartLine > Enter next vertex or reset to complete (Place the remaining data points. Change the Segment, Vertex Type, and Rounding Radius or Chamfer Offset settings as required between points.)

The SmartLine Tool Settings Box

When the SmartLine tool is active, the options available in the Tool Settings window vary depending on what options are selected and what you are currently doing with the tool. Following is a discussion of the options and their impact on SmartLine operation.

Choosing the Type of Segment to Place

The **Segment Type** menu allows you to choose between placing **Lines** and **Arcs**. When you choose to place **Lines**, the operation is similar to the Place Line tool. When you place **Arcs**, the operation is similar to the Place Arcs tool with the **Center Method** selected. You can switch between the two segment types while the SmartLine tool is active, as shown in Figure 5–22.

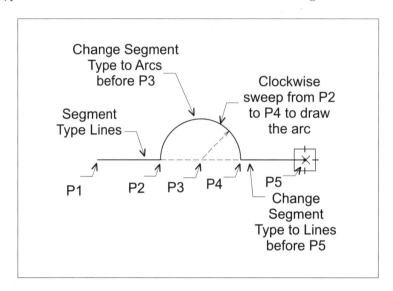

Figure 5-22 Using the SmartLine tool to place lines and arcs

When you switch to placing arcs, the next data point defines the center of the arc. As you move the pointer around the center point, a dynamic image shows how the arch will be drawn when you place the next data point. You control the direction in which the arc sweeps, by the direction in which you move the pointer, after the center point is defined.

Controlling the Shape of Line Segment Edges

You can choose a **Vertex Type** to change the way the intersections of the segments are drawn:

- Choose **Sharp** to join the two line segments at the data point and form a sharp edge.

- Choose **Rounded** to round off the intersection of the two line segments with an arc and form a smoothly curved edge. Set the radius of the rounding arcs by keying-in a radius in the **Rounding Radius** field. A rounded vertex can be placed when joining line segments or a line and arc segment.

- Choose **Chamfered** to cut off the edge formed by the intersection of the two line segments, forming a chamfered edge. To set the amount that is cut off, enter a distance in the **Chamfer Offset** field. The offset value is cut from each line segment. Chamfered edges can only be created when joining two line segments. If one of the segments is an arc, a sharp corner is created.

Note: SmartLine only creates the Rounded or Chamfered vertex if the adjacent segments are long enough to contain them. If the segments are too short, a sharp vertex is placed.

Figure 5–23 shows examples of vertex types.

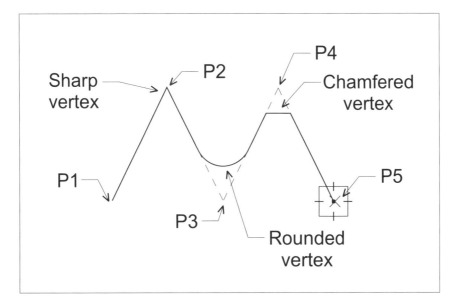

Figure 5-23 The SmartLine tool vertex types

Joining Elements

The **Join Elements** check box controls the way the SmartLine is placed.

If the check box is ON, all the SmartLine segments you place, before clicking the Reset button, are a single element.

If the check box is OFF, each SmartLine segment you place is a separate element. If you choose to place rounded or chamfered corners for line segments, the arc and line corners are also separate elements.

Closing SmartLines

If the **Join Elements** check box is ON, the SmartLines you place can be closed by moving the pointer over the first data point of the SmartLine, snapping to the point and then clicking the Data button. When MicroStation has snapped to the starting point, the Tool Settings window expands to show the **Join Element** check box and other close options, as shown in Figure 5–24.

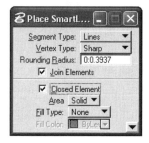

Figure 5-24 The Expanded SmartLine Tool Settings window

The additional options in the Tool Settings window control the completion of the SmartLine and the type of element it becomes when you complete it. If you want to change these options, you must click the Tentative button to snap to the starting point. Otherwise, the extra options disappear as soon as you move the pointer away from the starting point. The AccuSnap snap, alone, does not make the options "stick."

If the **Closed Element** check box is ON, the SmartLine will be completed as a closed element when you click the Data button to accept the tentative snap. If you turn the check box OFF, the SmartLine will not be completed and you must click the Reset button when you are through drawing it.

If the check box is ON, the **Area**, **Fill Type**, and **Fill Color** options allow you to fill the interior of the closed element with color or a pattern. The fill and patterning tools are discussed in Chapter 12.

 Note: If the **Closed Element** check box is OFF when you signal closing a SmartLine element by tentative snapping to the starting point of the element, the only additional option that appears in the Tool Settings window is the **Closed Element** checkbox. If you turn the check box ON, the additional options will appear.

SmartLine Placement Settings

On the bottom right corner of the SmartLine Tool Settings window is down-pointing arrowhead which opens the **SmartLine Placement Settings** box, shown in Figure 5-25. The box contains two check boxes that change the way SmartLine operates.

Figure 5-25 The SmartLine Placement Settings box

240

Chapter 5 AccuDraw and SmartLine

If the **Rotate AccuDraw to Segments** check box is ON, the AccuDraw compass is rotated so that its X axis is in line with the angle of the segment to which it is attached. If the check box is OFF, the compass rotation is always zero, with its X axis parallel to the design plane's X axis. This check box overrides rotations in the AccuDraw Settings box.

If the **Always Start in Line Mode** checkbox is ON, SmartLine always starts with the **Segment Type** set to **Lines** when you select it from the Linear Elements toolbox This occurs even if you left it set for Arcs the last time you used it.

Using SmartLine with AccuDraw

This example illustrates the use of SmartLine and AccuDraw by creating the simple design shown in Figure 5–26. The letters in the figure point to the locations of all data points. It assumes the design file has its Master and Sub Units both set to inches and the active view window is set to display an area of about six by eight inches.

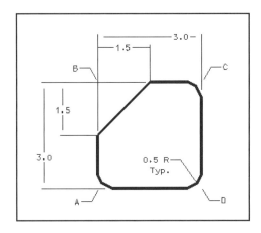

Figure 5–26 Example of a drawing made with the SmartLine tool

1. If AccuDraw is not active, invoke the AccuDraw tool.
2. Open the AccuDraw Settings box.
3. On the **Coordinates** tab, lock the **Distance** roundoff value at **0.5** inch, lock the **Angle** roundoff value at **90** degrees, select the **Polar** Coordinate System **Type**, and close the AccuDraw Settings box.
4. Invoke SmartLine from the Lines tool box.
5. In the Tool Settings window, select the **Lines Segment Type**, select the **Chamfered Vertex Type**, and key-in a **Chamfer Offset** of **1.5** inches.
6. Define a point near the bottom left of the view window to start the object (Point A).

Note: Because the AccuDraw distance roundoff value also forces the first data point to a multiple of the roundoff value on the design plane grid, the point may not be placed exactly where you click the Data button.

7. Move the pointer up until the AccuDraw coordinates box **Distance** field displays 3 inches and click the Data button (Point B).
8. Move the pointer to the right until the AccuDraw coordinates box **Distance** field displays 3 inches and click the Data button (Point C).

Note: The 1.5 inch offset chamfer does not appear until there is room for it on the line segment (when the pointer has moved at least 1.5 inches to the right).

9. In the Tool Settings window, select the **Rounded Vertex Type** and set the **Rounding Radius** to **0.5** inch.
10. Move the pointer down until the AccuDraw coordinates box **Distance** field displays 3 inches and click the Data button (Point D).
11. Slide the pointer to the left until the AccuDraw coordinates box **Distance** field displays 3 inches and touches the starting point of the SmartLine (Point A), then click the Tentative button .
12. In the Tool Settings window, if the **Closed Element** check box is off, turn it ON.
13. Click the Data button to close the element and complete the SmartLine operation.

The completed element should be identical to the element shown in Figure 5–26 (without the dimensions and letters).

Chapter 5 AccuDraw and SmartLine

 Open the Exercise Manual PDF file for Chapter 5 on the accompanying CD for project and discipline specific exercises.

REVIEW QUESTIONS

Write your answers in the spaces provided.

1. How do you activate AccuDraw?

2. Name the two coordinate systems you can use with AccuDraw.

3. Name three settings you can adjust in the AccuDraw Settings box **Operation** tab.

4. What is the purpose of rounding off AccuDraw distances and angles?

5. How do you recall the previous values in the AccuDraw coordinates box?

6. What is the shortcut key-in that will move the compass origin from the previous data point?

7. Explain briefly the benefits of manipulating elements with AccuDraw active.

8. Explain the difference between the Place Line tool and the Place SmartLine tool.

9. Name the two segment types you can place with the SmartLine tool.

10. What can cause SmartLine to place a sharp vertex when you have SmartLine set to place 1.5 inch radius rounded vertexes between line segments?

chapter 6

Manipulating a Group of Elements

Objectives

After completing this chapter, you will be able to do the following:

- Select elements with the PowerSelector tool and manipulate them
- Place fences and manipulate fence contents

ELEMENT SELECTION

While you were practicing the element manipulation tools described in the preceding chapters, did little squares occasionally appear on the corners of one of your elements? Did they make the tools act differently from the way the book said they would? This chapter turns those "handles" from a nuisance into a useful feature by showing you how the PowerSelector and Element Selector tools provide a powerful new way to manipulate elements.

Selecting Elements with the PowerSelector Tool

In the previous chapters you first invoked an element manipulation tool, and then identified the element you wanted to manipulate. With the PowerSelector Tool, you select or deselect one element, several elements, or all elements in the design. You can then manipulate all elements in the "selection set."

By default, when an element is selected, "handles" appear on the element. When you select more than one element, then all of the selected elements are highlighted. See Figure 6–1 for examples of the handles on different element types.

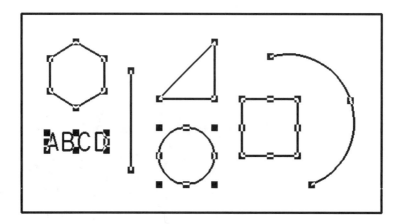

Figure 6-1 Examples of the handles on different element types

Invoke the PowerSelector tool from:

Element Selection tool box	Select the PowerSelector tool, then pick a selection Method and Mode from the Tool Settings window (see Figure 6–2).
Key-in window	**powerselector** (or **pow**) ENTER

Figure 6-2 Invoking the PowerSelector tool from the Main tool frame

The PowerSelector action and prompt depend on the position of the cursor and the **Method** and **Mode** combination selected. If the cursor is on one of the buttons in the Tool Settings window, the prompt displays the name of the button and the keyboard shortcuts that will activate the button's function. If the cursor is on the View Window, the prompt tells you what selection **Method** and **Mode** combination is active.

The **Method** buttons control the way elements are selected or deselected. Table 6–1 describes the four Methods. The first three **Mode** buttons determine if the Method buttons select or deselect elements. The fourth Mode button clears the current selection set and starts a new set. The fifth button toggles between selecting all elements and deselecting all elements. Table 6–2 describes the five modes.

Table 6–1 The PowerSelector Methods

Method	Action
Individual	Select or deselect individual elements by clicking on them or by dragging a block around them.
	Activate this Method by clicking on the first Method button from the left, or by pressing either the **Q** or **U** key.
Block	Select or deselect groups of elements by dragging a block around them.
	Activate this Method by clicking on the second Method button, or by pressing either the **W** or **I** key.
Shape	Select or deselect elements by defining a multi-sided, closed shape that can have up to 101 vertices.
	Activate this Method by clicking on the third Method button, or by pressing either the **E** or **O** key.
Line	Select or deselect elements by drawing a line. All elements the line touches are included.
	Activate this Method by clicking on the fourth Method button, or by pressing either the **R** or **P** key.

Table 6–2 The PowerSelector Modes

Mode	Action
Add	Causes the active PowerSelector Method to add elements to the current selection set.
	Activate this Mode by clicking the first Mode button from the left, or by pressing either the **A** or **J** key.
Subtract	Causes the active PowerSelector Method to remove elements from the current selection set.
	Activate this Mode by clicking the second Mode button, or by pressing either the **S** or **K** key.

Table 6-2 *(continued)*

Mode	Action
Invert	Causes the active PowerSelector Method to add previously unselected elements to the selection, and to remove previously selected elements from the selection set.
	Activate this Mode by clicking the third Mode button, or by pressing either the **D** or **L** key.
New	Clears the current selection set and starts a new set.
	Activate this Mode by clicking the fourth Mode button, or by pressing either the **F** or **semicolon (;)** key.
Select All or Clear	Is a check box that works independently of the active PowerSelector Method. If there are any selected elements, this button Clears all element selections. The result is no selected elements in the design. If there are no selected elements, this button Selects All elements in the design.
	Activate this Mode by clicking the fifth Mode button, or by pressing either the **G** or **Comma (,)** key.

Select Elements

The following discussion provides an example of selecting elements using the Block Method.

Invoke the PowerSelector tool from:

Main tool frame	Select the PowerSelector tool, then select the Block Method and the Invert Mode from the Tool Settings window (see Figure 6–3).
Key-in window	**powerselector** (or **pow**) ENTER (Activate PowerSelector) **powerselector area block** (or **pow ar b**) ENTER (Selects Block Method) **powerselector mode invert** (or **pow m i**) ENTER (Selects Invert Mode)

Figure 6-3 Invoking the PowerSelector tool and setting the options to select elements

Chapter 6 Manipulating a Group of Elements

MicroStation prompts (when you move the cursor off of the Tool Settings window):

> PowerSelector > Place Shape for elements to invert in set *(Place a data point to define one corner of the selection block, as shown in Figure 6–4.)*
>
> PowerSelector > Place Shape for elements to invert in set *(Place a data point to define the diagonally opposite corner of the selection block. The elements are selected, as shown in Figure 6–5.)*

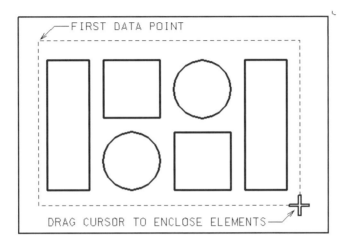

Figure 6-4 Define one corner of the selection block

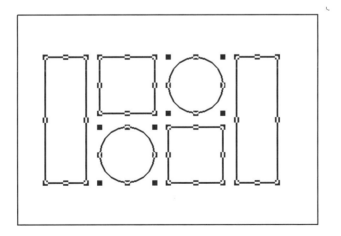

Figure 6-5 Elements inside the block are selected.

Remove Elements from a Selection Set

The following discussion provides an example removing selected elements from a selection set. The example removes the selected circles from the set that was selected in the previous example.

Invoke the PowerSelector tool from:

Tool Selection Window	Select the Individual Method from the Tool Settings window (see Figure 6–6).
Key-in window	**powerselector area individual** (or **pow ar i**) ENTER (Only works when PowerSelector tool is active.)

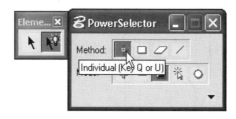

Figure 6-6 Selecting the PowerSelector tool's Individual Method in the Tool Settings window

MicroStation prompts (when you move the cursor off of the Tool Settings window):

> PowerSelector > Identify element to invert in set *(Place a data point on one of the circles to remove it from the selection set.)*
>
> PowerSelector > Identify element to invert in set *(Place a data point on the other circle to remove it from the selection set. The handles are removed from the circles, but the handles remain on the other elements, as shown in Figure 6–7.)*

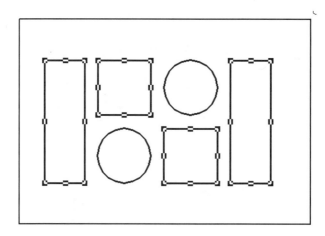

Figure 6-7 Circles are removed from the selection group

Selecting Elements by Attributes

Attribute tabs in the PowerSelector settings window let you select elements by one or more attributes — level, color, style, weight, type, or class. Alternatively, when you select elements graphically, the active set of attributes displays as a highlighted group at the top of each tab list box.

Clicking the Show Selection Information arrow expands the Tool Settings window to reveal the Attribute tabs as shown in Figure 6–8. Choose the desired Attribute tab as the selection criterion. In the Attribute list box, click the attributes to be included in the selection criterion.

If elements exist with these attributes in the design file, the attributes are highlighted and displayed at the top of the list box and the elements are selected. If you select an attribute that does not coincide with an element in the design file, the attribute is not highlighted.

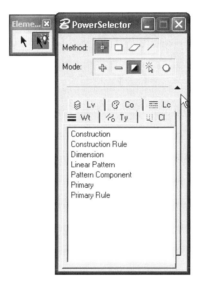

Figure 6-8 Power Selector – Attribute tabs

Select Elements with the Element Selection Tool

The Element Selection tool box also contains a tool named Element Selection. The PowerSelector tool almost makes the Element Selection tool obsolete, but there are a few situations where it is still useful.

The Element Selection tool is similar to the Individual **Method** and Invert **Mode** of the PowerSelector tool, in that you can select an individual element by clicking on it, or several elements by dragging a block around them. If you click on, or drag a block around previously selected elements, those elements are unselected.

The action of the Element Selection tool is different than PowerSelector when you want to add additional elements to the selection set or remove elements from the set. To add elements, hold down the CTRL key while you select the additional elements. To remove elements from the selection, hold down the CTRL key while you click on, or drag, a block around each element to be removed from the set. To remove all elements from the set, click the Data button without holding down the CTRL key, and without touching any elements.

Invoke the Element Selection tool from:

Element Selection tool box	Select the Element Selection tool (see Figure 6–8).
Key-in window	**choose element** (or **cho e**) ENTER

Figure 6–9 Invoking the Element Selection tool from the Element Selection tool box

The screen pointer changes to an arrow with a circle, similar to the one shown in Figure 6–10, and MicroStation prompts:

> Element Selection *(Select elements using the methods discussed in the previous paragraphs.)*

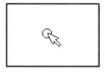

Figure 6–10 Screen pointer shape when the Element Selection tool is active

Consolidate Elements into a Group

The MicroStation Group option consolidates all selected elements (those that have selection handles) into a permanent group that acts like a single element when manipulated. The group has handles on its boundary, as shown in Figure 6–11.

Chapter 6   Manipulating a Group of Elements

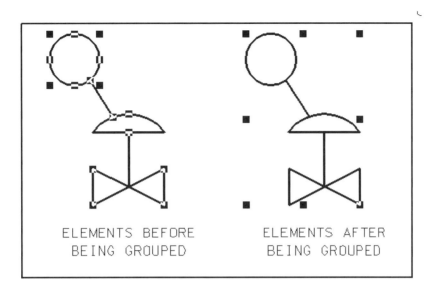

Figure 6-11   Example of elements before and after being grouped

To create a group, select the elements, then invoke the Group tool from:

Drop-down menu	Edit > Group
Key-in window	**group selection** (or **gr s**) ENTER

The selected elements are immediately consolidated into a group with one set of handles on the group boundary (there are no MicroStation prompts).

 Note: The Group tool is dimmed in the Edit drop-down menu if no elements are selected.

Ungroup Consolidated Elements

The MicroStation Ungroup option permanently ungroups all consolidated elements that have handles. The ungrouped elements can be manipulated separately. When consolidated elements are ungrouped, the group boundary handles are replaced by sets of handles on each element.

To ungroup a group, select the group (or groups) of elements, then invoke the Ungroup tool from:

Drop-down menu	Edit > Ungroup
Key-in window	**ungroup** (or **ung**) ENTER

The selected group (or groups) immediately returns into separate elements (there are no MicroStation prompts).

Note: The Ungroup tool is dimmed in the Edit drop-down menu if no elements are selected.

Locking Selected Elements

The MicroStation Lock option locks selected elements to prevent them from being manipulated. If you attempt to select a locked element for manipulation, MicroStation ignores the element and displays the message "Element is locked" in the Status bar. Locking is an easy way to protect completed parts of a design from accidental changes.

To lock elements, select the elements, then invoke the Lock tool from:

Drop-down menu	Edit > Lock
Key-in window	**change lock** (or **chan lo**) ENTER

The selected elements are immediately locked and protected from manipulation (there are no MicroStation prompts).

Note: The Lock tool is dimmed in the Edit drop-down menu if no elements are selected.

Unlock Selected Elements

Locked elements can be unlocked with the MicroStation Unlock option. After unlocking, the elements can be manipulated.

To unlock elements, select the group (or groups) of, then invoke the Unlock tool from:

Drop-down menu	Edit > Unlock
Key-in window	**change unlock** (or **chan u**) ENTER

The selected elements are immediately unlocked (there are no MicroStation prompts).

Note: The Unlock tool is dimmed in the Edit drop-down menu if no elements are selected.

Drag Selected Elements to a New Position

When the Element Selection tool is active, selected elements can be dragged (moved) to a new location in the design plane. You do not need to select the Move tool to move them.

Chapter 6 Manipulating a Group of Elements

1. Select the element or elements to be moved.
2. If PowerSelector is active, switch to the Element Selection tool.
3. Press and hold the Data button anywhere on one of the selected element outlines (but *not* on a handle).
4. Drag the elements to the new location.
5. Release the Data button to place the elements at the new location.

The elements are placed at the new location and removed from the original location, as shown in Figure 6–12 (there are no MicroStation prompts).

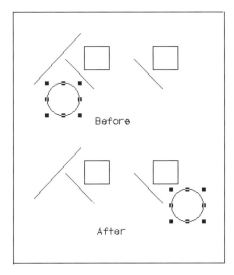

Figure 6-12 Example of dragging a selected element to a new location

 Note: The PowerSelector tool cannot be active when you want to drag elements to a new location, because it will capture the Data button press, and assume you are adding elements to the selection set or removing them from the set.

Drag an Element Handle to Change Its Shape

When the Element Selection tool is active, the geometric shape of an element can be changed by dragging one of the element's handles to a new location in the design—you do not need to use the Modify Element tool.

1. Select the element to be modified.
2. If PowerSelector is active, switch to the Element Selection tool.

3. Press and hold the Data button on the element handle to be modified.
4. Drag the handle to the new shape.
5. Release the Data button to complete the modification.

The element's shape is changed (there are no MicroStation prompts).

See Figure 6–13 for an example of modifying an element's shape.

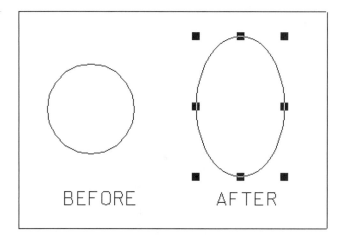

Figure 6-13 Example of modifying an element's shape

 Notes: The PowerSelector tool cannot be active when you want to drag elements to a new location, because it will capture the Data button press, and assume you are adding elements to the selection set or removing them from the set.

You can only modify the geometric shape of one element by dragging its handle. If you want to modify the shape of several elements proportionally, first select and group them.

Deleting Selected Elements

All selected elements can quickly be deleted by pressing DELETE on the computer keyboard or by selecting the Delete tool in the Main tool frame. To delete elements:

1. Select the elements to be deleted.
2. Click DELETE or select the Delete tool from the Main tool frame.

All selected elements are deleted (there are no MicroStation prompts).

Note: On some systems, the Backspace key will also delete selected elements.

Change the Attributes of Selected Elements

The element attributes (color, level, style, and weight) of selected elements can be changed simply by selecting the desired attribute in the Attributes tool box. For example, to change the attributes of a group of elements:

1. Select the elements whose attributes are to be changed.
2. From the Attributes tool box, select the desired attributes (e.g., change the line weight from 2 to 4 and the color from green to red).

The selected elements are immediately changed to the new attribute each time an attribute is changed (there are no MicroStation prompts).

Note: This only works when you select new attributes in the Attributes tool bar. If you use the Element Attributes dialog box to change attributes, the attributes of the selected elements are not changed.

Manipulation Tools that Recognize Selected Elements

Several MicroStation manipulation tools work with selected elements. The tools that recognize selected elements are the Array, Copy, Delete, Mirror, Move, Rotate, Scale, and Change Element attributes.

These manipulation tools affect all selected elements as if they were one element, and the tools exit after completing the requested change.

To manipulate the elements, first select them, and then invoke the appropriate manipulation tool. Because the prompts are slightly different than those you see when you manipulate individual elements, you should always read the MicroStation tool prompts in the Status bar. For example: the Move and Copy tools still require two data points, but they define the relative distance to move or copy the selected elements. In other words, the new location of the elements has the same relationship to the second data point as the original elements had to the first data point (see Figure 6–14).

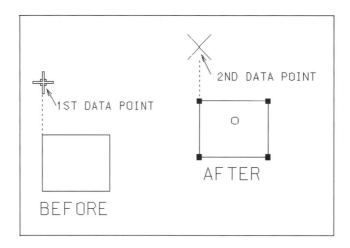

Figure 6-14 Example of using the Move tool with a selected element

FENCE MANIPULATION

The Fence manipulation tools provide another way to manipulate sets of elements. A fence is placed around the elements to be manipulated, and then the fence contents tools can manipulate all elements in the fence. Only one fence at a time can be placed in the design plane, and it remains active until either a new fence is placed or the design file is closed.

The Fence tool box is opened from the Main tool frame, as shown in Figure 6-15.

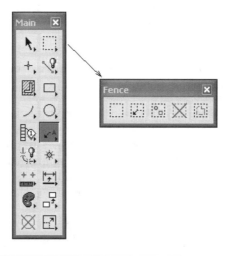

Figure 6-15 Fence tool box location in the Main tool frame

Placing a Fence

Six types of fences can be placed in the design:

- A block defined by diagonally opposite data points
- A shape defined by a series of vertex data points
- A circle defined by center and edge data points
- An existing closed element
- A fence defined by the contents of the selected View
- A fence defined by the contents of the active design file

Fence Block

The Block Fence Type places the fence as an orthogonal block.

Invoke the Place Fence Block tool from:

Fence tool bar	Select the Place Fence tool, then select Block from the Fence Type option menu in the Tool Settings window (see Figure 6–16).
Key-in window	**place fence block** (or **pl f b**) ENTER

Figure 6–16 Invoking the Place Fence Block tool from the Fence tool bar

MicroStation prompts:

> Place Fence Block > Enter first point *(Define one corner of the block in the design plane.)*
>
> Place Fence Block > Enter opposite corner *(Define the diagonally opposite corner of the block in the design plane.)*

Shape Fence

The Shape Fence Type places the fence as a closed multi-sided shape that can have up to 5000 vertexes.

Invoke the Place Fence Shape tool from:

Fence tool bar	Select the Place Fence tool, then select Shape from the Fence Type option menu in the Tool Settings window (see Figure 6–17).
Key-in window	**place fence shape** (or **pl f s**) ENTER

Figure 6-17 Invoking the Place Fence Shape tool from the Fence tool bar

MicroStation prompts:

> Place Fence Shape > Enter Fence Points *(Define the location of each fence vertex in the design plane.)*

To complete the fence shape, either place the last vertex data point on top of the first fence data point or click the **Close Shape** button in the Tool Settings window, as shown in Figure 6–17.

Fence Circle

The Circle Fence Type places the fence as a circle that you place by defining the center of a point on the circumference of the circle.

Invoke the Place Fence Circle tool from:

Fence tool bar	Select the Place Fence tool, then select Circle from the Fence Type option menu in the Tool Settings window (see Figure 6–18)
Key-in window	**place fence circle** (or **pl f c**) ENTER

Figure 6-18 Invoking the Place Fence Circle tool from the Fence tool bar

MicroStation prompts:

> Place Fence Circle > Enter circle center *(Define the location of the fence center in the design plane.)*
>
> Place Fence Circle > Enter edge point *(Define a data point on the circumference of the fence circle.)*

Fence From Element

The Element Fence Type places the fence on top of a selected closed element. A closed element is a circle, ellipse, shape, or complex shape.

Invoke the Place Fence From Element tool from:

Fence tool bar	Select the Place Fence tool, then select Element from the Fence Type option menu in the Tool Settings window (see Figure 6–19).
Key-in window	**place fence from shape** (or **pl f e**) ENTER

Figure 6–19 Invoking the Place Fence Element tool from the Fence tool bar

MicroStation prompts:

> Create Fence From Element > Identify element *(Select the closed element whose outline will define the fence.)*
>
> Create Fence From Element > (Accept/Reject) Shape Element *(Click anywhere in the design plane to accept the element.)*

Fence From View

The From View Fence Type places a fence block that is the size of the selected View Window.

Invoke the Place Fence From View tool from:

Fence tool bar	Select the Place Fence tool, then select From View from the Fence Type option menu in the Tool Settings Window (see Figure 6–20)
Key-in window	**place fence view** (or **pl f v**) ENTER

Figure 6-20 Invoking the Place Fence From View tool from the Fence tool bar

MicroStation prompts:

>Create Fence From View > Select View *(Place a data point anywhere on the view to place a fence in that view.)*

MicroStation places a fence block the exact size of the view window. The fence outline appears along the edges of the view window.

Fence From Active Design File

The From Design File Fence Type places a fence block that encloses all elements in the design file.

Invoke the Place Fence From Active Design File tool from:

Fence tool bar	Select the Place Fence tool, then select From Design File from the Fence Type option menu in the Tool Settings window (see Figure 6–21).
Key-in window	**place fence active** (or **pl f a**) ENTER

Figure 6-21 Invoking the Place Fence From Active Design File tool from the Fence tool bar

MicroStation prompts:

>Create Fence From Active Design File > Select View *(Click anywhere in the view window to place the fence.)*
>
>Create Fence From Active Design File > Fence placed - <Reset> to place again. *(Proceed with using the fence contents manipulation commands.)*

Chapter 6 Manipulating a Group of Elements

Fence Selection Mode

Before you manipulate the contents of a fence, select the **Fence Mode** to control which elements are actually manipulated by the fence. For example, you can elect to manipulate only elements that are completely inside the fence, or only elements that are completely outside the fence.

Six Fence Mode options are available from the Tool Settings window when a Fence tool is active (see Figure 6–22). The Fence Mode options are also available in the **Settings>Locks>Full** dialog box.

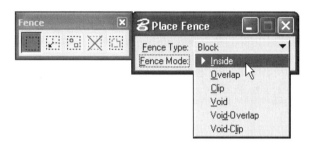

Figure 6-22 Fence Mode options in the Tool Settings window

Following are descriptions of each of the six fence modes.

Inside Mode

Inside Fence Mode limits manipulation to the elements completely inside the fence. For example, the Delete Fence Contents tool deletes circles A, B, and D, shown in Figure 6–23a.

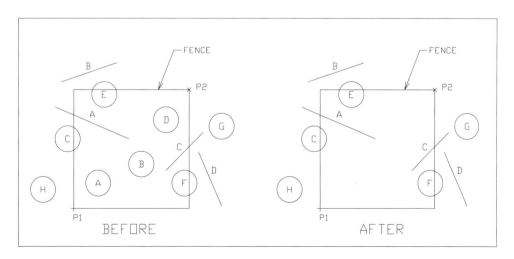

Figure 6-23a Example of deleting the fence contents with Inside mode selected

263

Overlap Mode

Overlap Fence Mode limits manipulation to the elements that are inside and overlapping the fence. For example, the Delete Fence Contents tool deletes circles A, B, C, D, E, and F, and lines A and C, shown in Figure 6–23b.

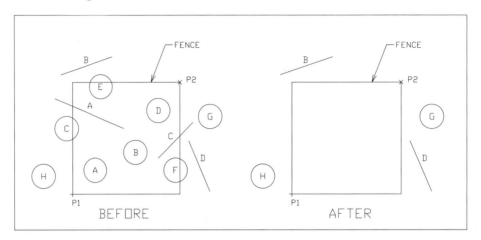

Figure 6-23b Example of deleting the fence contents with Overlap mode selected

Clip Mode

Clip Fence Mode limits manipulation to elements inside the fence and the inside part of elements overlapping the fence. Elements overlapping the fence are clipped at the fence boundary. For example, the Delete Fence Contents tool deletes circles A, B, and D, part of circles C, E, and F, and parts of lines A and C, shown in Figure 6–23c.

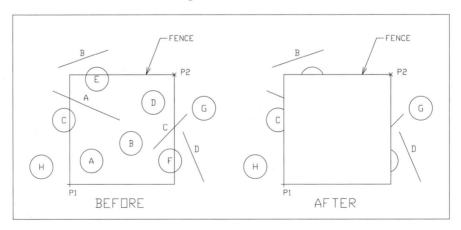

Figure 6-23c Example of deleting the fence contents with Clip mode selected

Void Mode

Void Fence Mode limits manipulation to elements completely outside the fence. For example, the Delete Fence Contents tool deletes circles G and H and lines B and D, shown in Figure 6–23d.

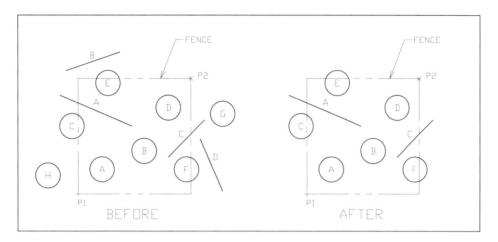

Figure 6-23d Example of deleting the fence contents with Void mode selected

Void-Overlap Mode

Void-Overlap Fence Mode limits manipulation to elements outside and overlapping the fence. For example, the Delete Fence Contents tool deletes circles C, E, F, G, and H, and lines A, B, C, and D, as shown in Figure 6–23e.

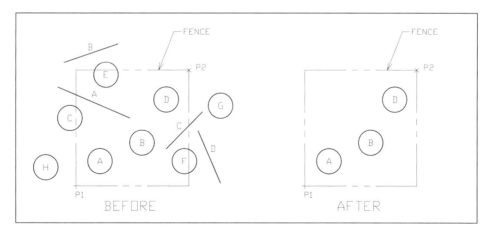

Figure 6-23e Example of deleting the fence contents with Void-Overlap mode selected

Void-Clip Mode

Void-Clip Fence Mode limits manipulation to elements outside the fence and the parts of the overlapping elements that are outside the fence. Elements are clipped at the fence boundary. For example, the Delete Fence Contents tool deletes circles G and H, lines B and D, parts of circles C, E, and F, and parts of lines A and C, as shown in Figure 6–23f.

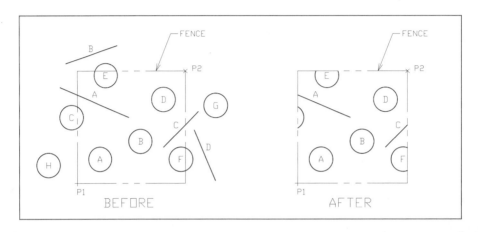

Figure 6-23f Example of deleting the fence contents with Void-Clip mode selected

Modifying a Fence's Shape or Location

You just finished placing a complicated fence shape and there, sitting outside the fence, is an element that should be inside the fence. There is no need to place the fence again; the Modify Fence tool can modify a fence vertex or move the fence to a new location.

Modifying Fence Shape

To modify a fence, invoke the Modify Fence Vertex tool from:

Fence tool box	Select the Modify Fence tool, then select Vertex from the Modify Mode option menu in the Tool Settings window (see Figure 6–24).
Key-in window	**modify fence** (or **modi f**) ENTER

Figure 6-24 Invoking the Modify Fence Vertex tool from the Fence tool box

MicroStation prompts:

> Modify Fence Vertex > Identify vertex *(Click the Data button on the fence outline near the vertex to be modified, drag the vertex to the new position, and click the Data button again. Click the Reset button to complete the modification.)*

See Figure 6–25 for an example of modifying a fence.

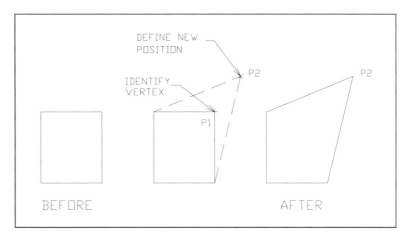

Figure 6-25 Example of modifying a fence vertex

 Note: After the first data point is placed, a dynamic image of the fence drags with the screen pointer.

Moving a Fence

To move a fence to a new location in the design plane, invoke the Move Fence tool from:

Fence tool box	Select the Modify Fence tool, then select Position from the Modify Mode option menu in the Tool Settings window (see Figure 6–26).
Key-in window	**move fence** (or **mov f**) ENTER

Figure 6-26 Invoking the Modify Fence tool (Move Position) from the Fence tool box

MicroStation prompts:

> Move Fence Block/Shape > Define origin *(Click the Data button in the design plane to identify the relative starting position of the move.)*
>
> Move Fence Block/Shape > Define distance *(Click the Data button in the design plane to identify the relative position to which the fence is to be moved. Click the Reset button to complete the move.)*

The fence is moved a distance equal to the distance between the two data points, and the relationship of the final fence position to data point two is the same as the original fence position was to data point one.

 Note: The Modify Fence tool modifies only the shape or position of the fence, not the contents of the fence. The Modify Fence Contents tools are used to modify the elements contained by the fence.

Manipulating Fence Contents

After you place a fence and select the appropriate fence selection mode, you are ready to manipulate the contents of the fence. There are fence contents manipulation equivalents for each of the element manipulation tools discussed in the previous chapters. The only difference is that you do not have to select the elements to manipulate them—the fence does that for you.

The Copy, Move, Scale Rotate, Mirror, and Array tools in the Manipulate tool box can be switched between element manipulation and fence contents manipulation. The **Use Fence** check box in the Tool Settings window determines which type of manipulation is done. If the button is set to ON, the fence contents are manipulated; if it is set to OFF, individual elements are manipulated. Figure 6–27 shows the Manipulate tool box with the Copy tool selected and the Use Fence check box set to ON in the Tool Settings window.

Similarly, when you invoke the Change Element Attributes tool from the Change Attributes tool box, a check box in the Tool Settings window can switch between element or fence manipulation.

Figure 6–27 Invoking the Copy Element tool from the Manipulate tool box with the Use Fence check box set to ON

In addition, MicroStation provides three fence contents manipulation tools:

Chapter 6 Manipulating a Group of Elements

- The Manipulate Fence Contents tool, invoked from the Fence tool box, has an **Operation** option menu in the Tool Settings window from which you can select the type of manipulation needed (see Figure 6–28).

- The Delete Fence Contents tool, invoked from the Fence tool box, deletes all elements within the contents of the fence.

- The Drop Fence Contents tool, invoked from the Fence tool box, drops all complex elements within the contents of the fence.

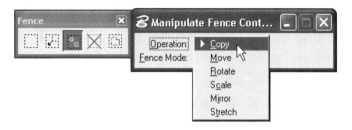

Figure 6-28 The Manipulate Fence Contents tool's Operation option menu

For example, to move the contents of the fence to a new location in the design plane, select the Manipulate Fence Contents tool from:

Fence tool box	Select the Manipulate Fence Contents tool, then select Move from the Operation menu in the Tool Settings window (see Figure 6–29).
Key-in window	**fence move** (or **f mo**) ENTER

Figure 6-29 Invoking the Manipulate Fence Contents Move tool from the Fence tool box

MicroStation prompts:

> Move Fence Contents > Define origin *(Locate the starting position of the move in the design plane.)*
>
> Move Fence Contents > Define distance *(Locate the final destination in the design plane.)*

269

Stretch Fence Contents

The Fence Stretch tool allows you to stretch the contents of a fence. There is no equivalent element manipulation tool for this command. Note the following, when using the Fence Stretch tool:

- Line, Line String, Multi-line, Curve String, Shape, Polygon, Arc, and Cell (see Chapter 10) elements that overlap the fence are stretched.
- Circle and Ellipse elements that overlap the fence are ignored.
- Elements completely inside the fence are moved.

The Stretch tool only checks to see if an inside mode (Inside, Overlap, or Clip) is active, or if a void mode (Void, Void-Overlap, or Void-Clip) is active. If an inside mode is active, the element vertexes inside the fence are stretched. If a void mode is active, the element vertexes outside the fence are stretched.

See Figure 6–30 for an example of stretching the contents of a fence when one of the Inside Modes is active.

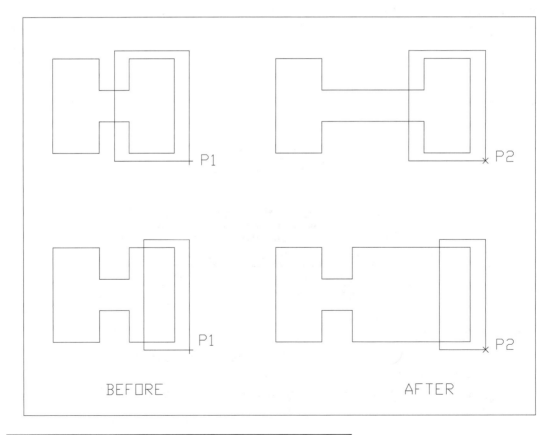

Figure 6-30 Example of stretching the contents of a fence

Chapter 6 Manipulating a Group of Elements

To stretch a group of elements, place a fence that overlaps the elements you want to stretch, then select the Fence Stretch tool from:

Fence tool box	Select the Manipulate Fence Contents tool and Stretch from the Operations option menu (see Figure 6–31).
Key-in window	**fence stretch** (or **f st**) ENTER

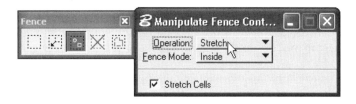

Figure 6–31 Invoking the Manipulate Fence Contents Stretch tool from the Fence tool box

MicroStation prompts:

> Fence Stretch > Define origin *(Locate a relative point in the design plane to start the stretch.)*
>
> Fence Stretch > Define distance *(Locate a relative point in the design plane to end the stretch.)*

The new fence location has the same relationship to the second data point as the original fence location did to the first data point.

Remove a Fence

To remove a fence, invoke the Place Fence tool and the existing fence will be removed. If you do not want to place another fence, select another tool and continue working. There is no separate tool to remove a fence.

Note: Always remove the fence after you are finished with it to protect against accidental fence contents manipulation. For example, if a fence is defined and you accidentally select the tool to delete the fence contents while thinking you selected the delete element tool, you could delete the contents of the fence.

Open the Exercise Manual PDF file for Chapter 6 on the accompanying CD for project and discipline specific exercises.

REVIEW QUESTIONS

Write your answers in the spaces provided.

1. The PowerSelector tool, _____ Method, and _____ Mode allow you to select all elements by dragging a block around them.

2. The PowerSelector tool Mode that allows selecting and unselecting elements at one time is the _____ mode.

3. The PowerSelector tool's Individual Method and Add Mode can be turned on by pressing two keyboard shortcuts. What two keys will turn them on? _____.

4. You can only change the shape of an element by dragging one of its handles when the _____ tool is active.

5. Briefly explain the purpose of locking individual elements.
   _____
   _____

6. List the six fence selection modes available in MicroStation.
   _____
   _____

7. Explain briefly the difference between the Overlap and Void-Overlap mode.
   _____
   _____

8. The Fence Stretch tool will stretch an arc. TRUE or FALSE

9. The element manipulation tools will manipulate the fence contents when the _____ button is turned ON.

10. How do you remove a Fence?
    _____
    _____

chapter 7

Placing Text, Data Fields and Tags

Objectives

After completing this chapter, you will be able to do the following:

- Place single-character fractions
- Use several tools to place text elements
- Import text from other computer applications
- Edit the content of text elements
- Manipulate the attributes of text elements
- Place notes in the design
- Create and use "fill-in-the-blanks" Text Node and Data Field elements
- Place and manage Tags

PLACE TEXT

In Chapter 4 you learned how to use the Place Text tool's **By Origin** method to place text strings in a design and how to use settings in the Tool Settings window to control the size and shape of the text strings. In this section you will learn how to use the other placement methods that the Place Text tool offers.

Fitted Method

The Place Text tool's **Fitted** method places the text you key-in the Text Editor box between two data points that you define. The text is placed parallel to an imaginary line from data point 1 to date point 2 and is scaled to completely fill the space between the two data points, as shown in Figure 7–1.

Before using the tool, select the text **Font** and **Justification** from the Tool Settings window. The top, center, and bottom justification options control where the text is placed within the space between the two data points. For example, if you select **Left Bottom**, **Center Bottom**, or **Right Bottom**, the two data points will be below the text string. The text in Figure 7–1 was placed using one of the bottom justification options.

 Note: The Tool Settings window may not initially display all text tool setting fields. If you do not see all the fields, click the down-pointing arrow in the lower right corner of the window to display the additional fields.

Once placed, the text string is a normal element and all the manipulation commands can be applied to it. The line string has one keypoint for tentative snapping at the location of the first data point.

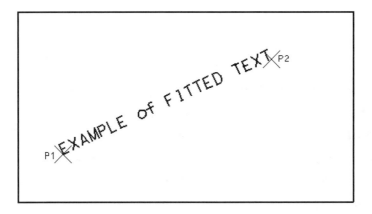

Figure 7–1 Example of placing fitted text

Invoke the Place Text tool from:

Text tool box	Select the Place Text tool. In the Tool Settings window, select Fitted from the Method menu, select a Font, and select a Justification (see Figure 7–2).
Key-in window	**place text fitted** (or **pl tex f**) ENTER

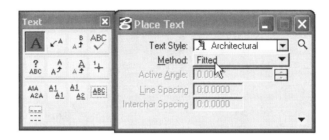

Figure 7–2 Invoking the Place Text and selecting the Fitted method

MicroStation prompts:

Place Fitted Text > Enter text *(Type the text in the Text Editor box.)*

Place Fitted Text > Enter more text or position start point *(Place a data point to define the starting point of the text.)*

Place Fitted Text > Enter more text or position end point *(Place a data point to define the ending point of the text.)*

Place Fitted Text > Enter more text or position start point *(Continue placing fitted copies of the text, key-in different text in the Text Editor box, or click the Reset button to drop the text.)*

View Independent and Fitted VI Methods

The **View Independent** method places text in the same way as the **Place Text** method, and the **Fitted VI** method places text in the same way as **Fitted**. Both of these methods are used in 3D designs to place text that can be read regardless of the design orientation. For example, if text is placed in the top view of the design, it can still be read when viewed in the left, right, and bottom views.

Above Element Method

The Place Text tool's **Above Element** method places a text string above a linear element that you select. You key-in the text in the Text Editor box and then identify the linear element at the point where you want to place the text. The text string is placed parallel to and above the selected element, as shown in Figure 7–3.

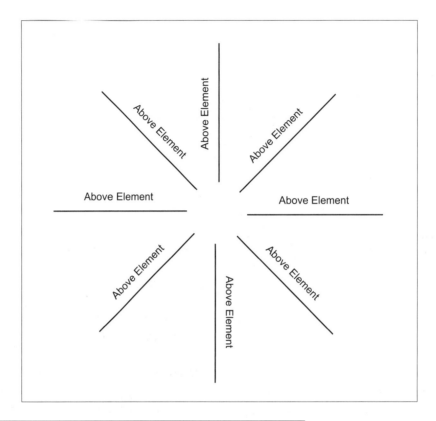

Figure 7–3 Examples of placing text above a linear element

 Note: The "above" side of the line or line segment depends on the line's rotation. The first data point that placed the line is the line's origin point. The lines in Figure 7-3 were all drawn from the center out.

The **Font**, **Height**, and **Width** fields in the Tool Settings window determine the text style and size. The **Line Spacing** field sets the space between the bottom of the text string and the selected element in working units. For example, a **Line Spacing** of zero causes the text to be placed touching the selected element, and a **Line Spacing** of 3.0000 causes the text to be placed three Master Units from the element.

The left, center and right **Justification** selections determine where the text string is placed in relation to the point on the element where you selected it. For example, if any of the left settings are selected (**Left Top**, **Left Center**, or **Left Bottom**) is selected, the text is placed such that the selection point on the element is to the left of the text string.

Once placed, the text string is a normal element and all the manipulation commands can be applied to it. The line string has one keypoint for tentative snapping at the left, center, or right of the element depending on the selected **Justification**. The text string has no connection to the element it was placed above. If the element is deleted or moved, the text stays at the same location.

Invoke the Place Text tool from:

Text tool box	Select the Place Text tool. In the Tool Settings window, select the Above Element from the Method menu (see Figure 7–4).
Key-in window	**place text above** (or **pla tex a**) ENTER

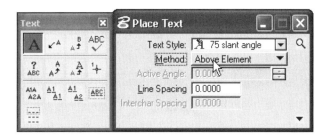

Figure 7–4 Invoking the Place Text and selecting the Above Element method

MicroStation Prompts:

> Place Text Above > Enter text *(Type the text in the Text Editor box.)*
>
> Place Text Above > Identify element *(Identify the element at the point where the text is to be placed. A dynamic image of the text appears above the element.)*
>
> Place Text Above > Accept/Reject (Select next input) *(Click the Data button to accept the element and place the text, or click the Reset button to reject it.)*

 Note: If the Text Editor box contains text when the **Above Element** method is selected, MicroStation skips the "Enter text" prompt and goes straight to asking you to identify the element.

Below Element Method

The Place Text tool's **Below Element** method works the same way as the **Above Element** method, except that it places the text string below the element, as shown in Figure 7–5.

Text tool box	Select the Place Text tool. In the Tool Settings window, select Below Element from the Method menu.
Key-in window	**place text below** (or **pla tex b**) ENTER

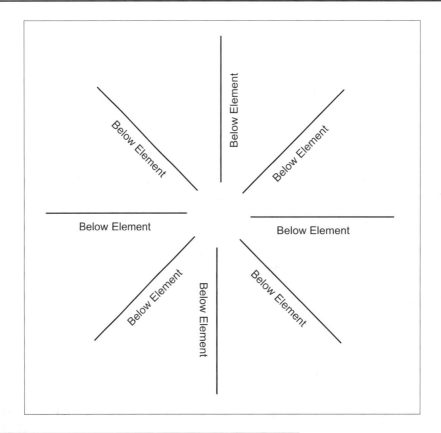

Figure 7-5 Examples of placing text below a linear element

On Element Method

The Place Text tool's **On Element** method works in the same way as the **Above Element** and **Below Element** methods, except that it cuts a hole in the linear element that is slightly wider than the text string and places the text string in the hole, as shown in Figure 7–6.

The left, center, and right **Justification** options determine where the hole is cut on the line or line segment. For example, if **Left Top**, **Left Center**, or **Left Bottom** is selected, the hole will be cut such that the data point is near the left end of the hole and to the left of the text string.

Text tool box	Select the Place Text tool. In the Tool Settings window, select the On Element from the Method menu.
Key-in window	**place text on** (or **pla tex o**) ENTER

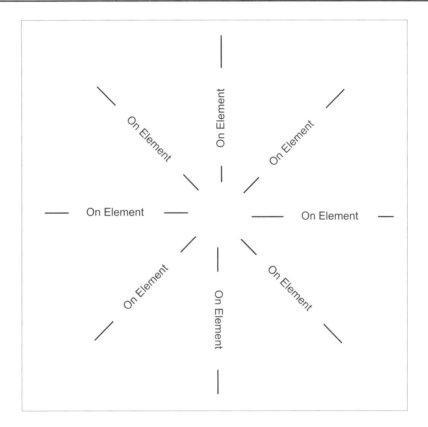

Figure 7-6 Examples of placing text on a linear element

Along Element

The Place Text tool's **Along Element** method places text that follows the contour of an arc, a circle, or a curve string, and can bend around the vertex of a linear multi-segment element. Figure 7–7 shows examples of placing text along elements.

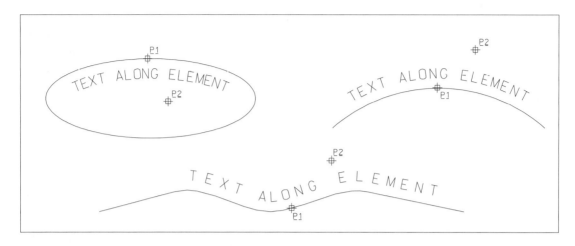

Figure 7-7 Examples of placing text along elements

The **Font**, **Height**, and **Width** fields in the Tool Settings window determine the text style and size. The left, center, and right text **Justification** options determine where the text lines up in relation to the point where the element is identified. The examples in Figure 7–7 were placed with one of the center justifications.

Text elements are linear, so to make the text follow the contour of a curving element, the **Along Element** method places each character as a separate element. To compensate for tight curves, the **Interchar Spacing** field can be used to increase the space between the characters in the string and the **Line Spacing** field can be used to move the text away from the element.

The characters in the string placed along the element are grouped together in a "complex shape" and they act as if they were one element and can be edited with the Edit Text tool that is discussed later. The complex shape has one keypoint for tentative snapping at the left, center, or right of the element depending on the selected **Justification**. The text string has no connection to the element it was placed along. If the line is deleted or moved, the text stays at the same location.

Invoke the Place Text tool from:

Text tool box	Select the Place Text tool. In the Tool Settings window, select the Along Element from the Method menu (see Figure 7–8).
Key-in window	**place text along** (or **pl tex al**) ENTER

Chapter 7 Placing Text, Data Fields, and Tags

Figure 7-8 Invoking the Place Text tool and selecting the Along Element method

MicroStation prompts:

Place Text Along > Enter text *(Type the text in the Text Editor box.)*

Place Text Along > Identify element, text location *(Define the point on the element where the text is to be placed. Dynamic images of the text appear both above and below the element, as shown in Figure 7–9, and you must select the one to place.)*

Place Text Along > Accept, select text above/below *(Click the Data button on the side of the element where you want the text placed, or click the Reset button to reject it.)*

The text on the side of the element where you accepted the text is placed in the design, and the dynamic image on the other side disappears.

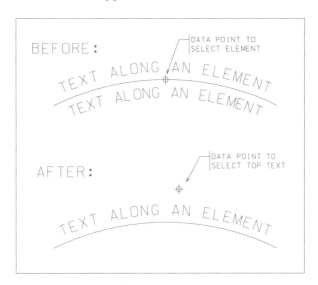

Figure 7-9 Example of placing text along an element

Note: If you insert line breaks in the text string, MicroStation removes them before placing the text along the selected element.

281

Word Wrap Method

The Place Text tool's **Word Wrap** method places a text string within the area of a temporary box using word wrapping as necessary to stay within the box width. You start the tool sequence by placing two data points that define the diagonally opposite points of the box. The dynamic image of a box is displayed in the design using a dashed line style.

After defining the box, you key-in the text in the Text Editor box. As you type, a dynamic image of the text appears in the box starting the corner of the box defined by the box to the next line. Wrapping occurs at word breaks. If the text string is too long to fit within the box, the lines wrap to the area outside the box.

When you complete keying-in the text, place a third data point anywhere in the design to complete the tool sequence and place the text in the design. The third data point also removes the box. To terminate the sequence without placing the text, click the **Reset** button. When you reset the sequence, the dynamic image of the box and text should disappear. If they do not, update the view.

The location of the two data points that define the box is important. If the first data point is to the right of the second data point, the text is placed as a mirror image. If the first data point is below the second data point, the text is placed upside down. Figure 7–10 shows examples of placing with different data point relationships. In each example the text is shown before the acceptance (third) data point is placed.

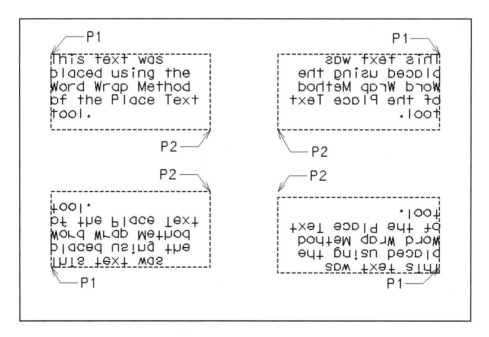

Figure 7-10 Examples of placing text using the Word Wrap method

Chapter 7 Placing Text, Data Fields, and Tags

The **Font**, **Height**, and **Width** fields determine the text style and size. The left, center, and right **Justification** options determine how the lines of text are placed within the area of the box. For example, if you select **Left Top**, **Left Center**, or **Left Bottom**, the left margin of the text is smooth. **Line Spacing** in the Text Style dialog box determines the space between the wrapped lines of text.

Invoke the Place Text tool from:

Text tool box	Select the Place Text tool. In the Tool Settings window, select Word Wrap from the Method menu (see Figure 7–11).

 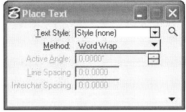

Figure 7–11 Invoking the Place Text tool and selecting the Word Wrap method

MicroStation prompts:

> Place Word Wrapped Text > Place first corner point *(Place a Data point to define one corner of the box.)*
>
> Place Word Wrapped Text > Place second corner point *(Place a Data point to define the diagonally opposite corner of the box.)*
>
> Place Word Wrapped Text *(Key-in the text string in the Text Editor box.)*
>
> Place Word Wrapped Text > Enter data point to accept text *(Click the Reset button to place the text or click the Reset button to terminate the tool without placing the text.)*
>
> Place Word Wrapped Text > Place first corner point *(Place a Data point to define one corner of the box or select another tool.)*

PLACE NOTE

The Place Note tool allows you to place a text string at the end of an arrow.

In the Tool Settings window:

- The **Text Frame** menu allows you to choose **None** to just place the text at the end of the arrow, **Line** to place a vertical line between the text and the arrow, or **Box** to enclose the text in a box.

- When you type a multi-line note, the **Justification** menu allows you to control which side of the text has a smooth margin. The **Left** option pushes the text against the left margin, the **Right** option pushes the text against the right margin, and **Dynamic** pushes the text to the side next to the arrow.
- Turn the **Generate Leader** check box ON to place a short horizontal line at the end of the arrow or OFF to place the arrow without a horizontal line segment.
- Turn the **Association** check box ON to associate the arrow with another element. To associate the arrow, tentative snap to the other element and then click the **Data** button to accept the snap point. If you move or scale an element that has a note arrow associated with it, the arrowhead moves with the element, but the text remains at the place where you initially placed it.
- Set the size of the text in the note by entering values in the **Height** and **Width** fields.

Figure 7–12 shows examples of notes placed with different tool settings.

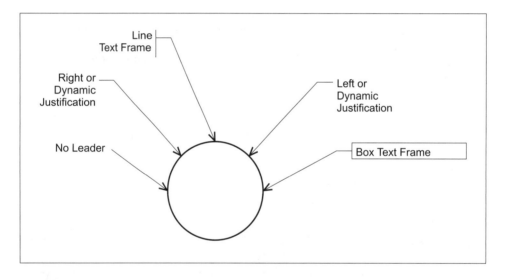

Figure 7–12 Examples of notes placed with various tool settings

Invoke the Place Note tool from:

Text tool box	Select the Place Note tool. In the Tool Settings window, make the selections for the note layout you need (see Figure 7–13).
Key-in window	**place note** (or **pl not**) ENTER

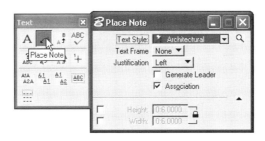

Figure 7-13 Invoking the Place Note tool from the Text tool box

MicroStation prompts:

> Place Note > Define start point *(Type the note text in the Tool Editor box and then define the point where the leader arrowhead is to be placed.)*
>
> Place Note > Define next point, or <Reset> to abort *(Define the point where the note text is to be placed.)*
>
> Place Note > Define next point, or <Reset> to abort *(Continue placing copies of the note or click the Reset button to drop the note and clear the Text Editor box.)*

SINGLE-CHARACTER FRACTIONS

Natural fractions are several characters long, which can take up a lot of space in the design. For example, 9/16 is four characters long. To reduce the space required for natural fractions, MicroStation provides "stacked fractions" in several fonts that place natural fractions as one, slightly taller than normal, character.

The Text Editor box provides the **Stacked Fractions** menu for selecting the way in which stacked fractions are placed in the design. The menu options allow you to choose to align stacked fractions such that the bottom of the fraction is in line with the bottom of the text string, the fraction is vertically centered within the text string, or the top of the fraction is in line with the top of the text string. Figure 7-14 shows the menu open in the Text Editor box.

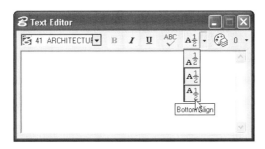

Figure 7-14 The Stacked Fractions option on the Text Editor box

Note: The single-character fractions option appears only in the word processor type of Text Editor box. If you do not see the option on the box, change to the word processor type, by selecting **Workspace > Preferences** and then selecting **Word Processor** from the **Text Editor Style** drop-down menu and clicking the **OK** button.

Figure 7–15 shows an example of natural fractions placed both as separate characters and as one-character stacked fractions.

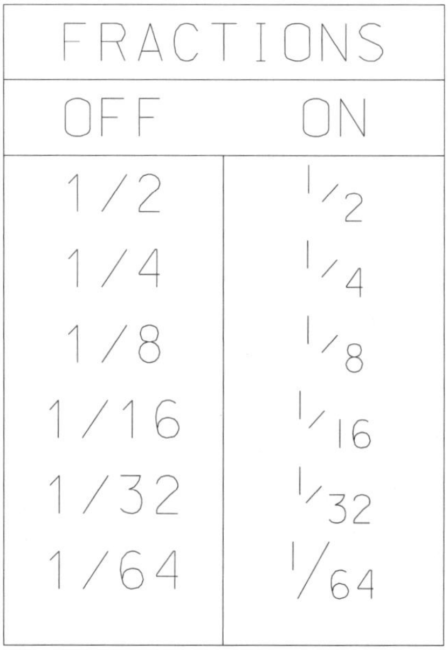

FRACTIONS	
OFF	ON
1/2	¹⁄₂
1/4	¹⁄₄
1/8	¹⁄₈
1/16	¹⁄₁₆
1/32	¹⁄₃₂
1/64	¹⁄₆₄

Figure 7-15 Examples of natural fractions

Invoke one-character natural fractions placement mode from:

Text Editor box	Click the Stacked Fractions option.

Things to Keep in Mind About Natural Fractions

- If a font does not include stacked fractions, the fractions are placed as separate characters.

- If you include a natural fraction in a string of text, there must be a space character before and after the fraction in order for MicroStation to recognize it as a natural fraction.

- One-character fractions take up slightly more horizontal space than single characters, so you may need to insert extra space characters to keep the fraction from running into the characters before and after it.

TEXT STYLES

With the text placement methods you learned thus far, changing the style of text you are placing requires making several setting changes in the Tool Settings window. For example, the switch from placing the names of the rooms in a house floor plan to placing text for the details may require changing the text font, size, and justification. On a complex drawing, you may be working with several different styles, and changing the tool settings every time you need to place a different type of text is time consuming.

To eliminate the need to constantly change the text tool settings, MicroStation provides options for building a library of text styles. With text styles you can define and save all the settings for each type of text you need to place. You can change placing one type of text to another simply by selecting the text style for the type of text you need.

Select a Text Style

A **Text Style** drop-down menu is available in the Tool Settings window for all text placement tools. The menu lists the names of all text styles in the design file's text styles library. Figure 7–16 shows a typical **Text Style** drop-down menu.

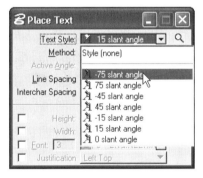

Figure 7-16 The Text Style menu in the Place Text Tool Settings window

Note: When a text style is selected, check boxes appear next to the settings in the expanded part of the Tool Settings window for the Place Text and Place Note tools. If you set a check box to ON, it overrides the selected text style for the setting associated with the check box, and the setting can be changed in the Tool Settings window.

Create and Maintain Text Styles

Text Styles are created and maintained in the Text Styles box

Invoke the Text Style box from:

Drop-down menu	Element > Text Styles
Place Text Tool Settings window	Click the magnifying glass symbol
Key-in window	**textstyle dialog open** (or **texts di o**) ENTER

Figure 7–17 shows a typical Text Styles box, and the following discussion describes how to use the box to create and maintain text styles.

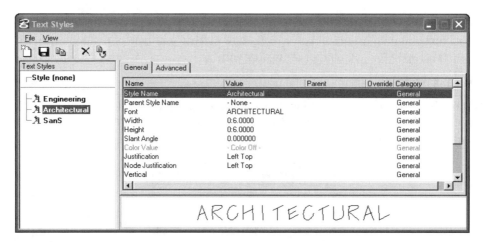

Figure 7-17 The Text Styles box

Text Styles Box Parts

The Text Styles box contains the following parts:

Menu and Button Bars

The menu bar and a button bar provide options that initiate actions such as saving a new style.

Text Style Names

The **Text Styles** area on the left side of the settings box below the bars displays all styles currently defined in the design file. To select a style for editing, click its name in this area. If no styles are defined, **Style (none)** appears in the field.

Text Style Settings

The area to the right of the **Text Styles** area lists the settings for the selected style or the default settings if there are no defined styles. There are two groups of settings, **General** and **Advanced**, that are accessed by clicking the tabs at the top of the area.

The most commonly used settings are on the **General** tab. The **Advanced** tab contains the same settings as the **General** tab and other less frequently used settings.

Preview Area

The bottom area of the Text Styles box displays the name of the text style using the style's settings to provide a preview of the appearance of text when placed with the selected text style. If this area is not visible in the Text Style box, display it by selecting the **Preview** option from the Text Style box's **View** drop-down menu.

Create a New Style

To create a new style in your design file, do the following:

1. Select **New** from the **File** drop-down menu in the Text Styles box. MicroStation creates a new text style that appears in the **Text Styles** field with "Untitled" as the style name. The new style's settings are set to default values.

2. To change the style's name, click **Untitled** in the **Text Styles** area to open the default name for editing, as shown in Figure 7–18. Type the style's name and click in another value field to complete the change. You may have to click the field two times to open it for editing.

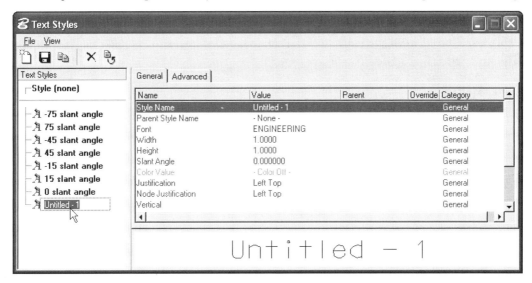

Figure 7–18 Selecting a new style's default name for editing

Note: The style's name is also the first setting under the **General** and **Advanced** tabs, and the name can also be edited from those locations.

3. For each setting that you need to change, go to the setting's line under the **General** or **Advanced** tab, click the item's **Value** field, and make the required changes. Some values (such as **Width**) are a text field in which you type a new value. Other values (such as **Font**) open drop-down menus from which you select a value. A few values (such as **Italics**) are check boxes that you turn on and off by clicking them. The settings on the tabs are described later.

4. After you complete entering the new style's values, save the style by selecting **Save** from the Text Styles box's **File** drop-down menu.

 Note: If you define a new style while the Place Text or Place Note tools are active, the new style may not appear in the **Text Style** menu. To make it appear, select another tool and then select the Place Text or Place Note tool again.

Modify the Values of an Existing Style

To modify the settings of an existing text style in your design file, do the following:

1. Select the style in the **Text Styles** area of the Text Styles box. The selected style's settings are displayed on the **General** and **Advanced** tabs.
2. To change the value of a setting, click the settings **Value** on the **General** or **Advanced** tab and make the required changes.
3. After you complete making the required changes, save the style by selecting **Save** from the Text Styles box's **File** drop-down menu.

Delete a Style

To delete a style from your design file, do the following:

1. Select the style name in the **Text Styles** field of the Text Styles box.
2. Select **Delete** from the Text Styles box's **File** drop-down menu.
3. In the Alert box, click the **Yes** button to delete the text style or click the **No** button to cancel deleting the text style.

The General Tab Settings

Following are descriptions of the text style settings that are available on the **General** tab. These settings are also available on the **Advanced** tab.

Style Name

The **Style Name** value is the name of the text style currently displayed in the Text Styles box.

Parent Style Name

A text style can be created as the "child" of another "parent" text style. The child takes on the setting values of the parent, but setting values can be changed in the child after it is created. The ability to create a hierarchical structure of parent and children is useful for ensuring uniform use of values for all text in a design. For example, the only value changes in the children might be the text size for different uses of the text (dimensions, construction notes, and so forth).

The procedure for creating a child text style is the same as for creating a parent text style, except you must select the parent style name. To select the parent after creating the new style, click in the **Parent Style Name** setting's **Value** field to open a menu that lists all defined text styles. Select the name of the parent from the menu. Any text style can be selected as the parent, even if it is the child of another text style.

Font

The **Font** setting provides a drop-down menu that lists all fonts available in the current design, as shown in Figure 7–19. The font you select from the menu is applied to all text placed using the current text style.

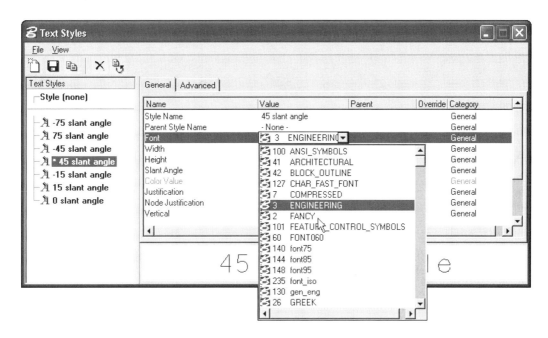

Figure 7-19 The Fonts drop-down menu

Width and Height

The **Width** and **Height** settings allow you to specify the size of text placed using the current style.

Slant Angle

The **Slant Angle** setting allows you to specify a rotation angle for text placed using the current style. The **Slant Angle** is applied to each character within text strings placed using the current text style, as shown in Figure 7–20. A positive angle slants the text counterclockwise and a negative angle slants it clockwise.

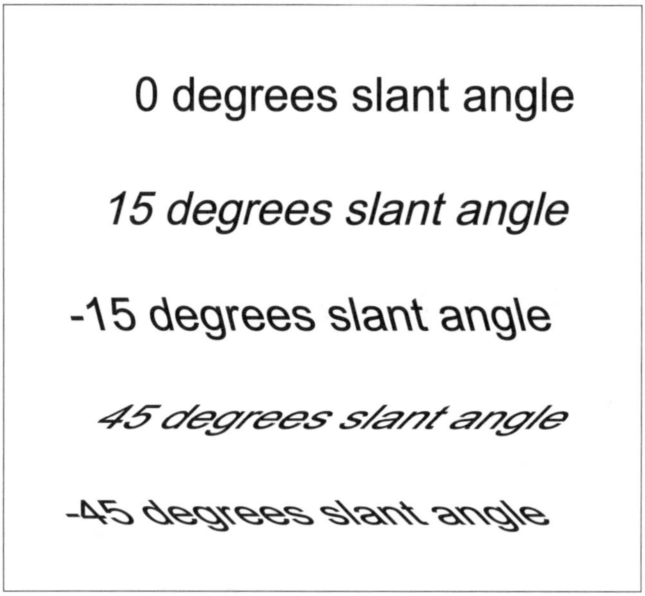

Figure 7-20 Examples of placing text with various slant angles

Color Value

The **Color Value** setting provides a palette of colors that can be used to set the color of text placed using the current text style. This setting is available only when the **Color** check box (located near the bottom of the settings area) is ON.

- If the **Color** check box is OFF, the design's Active Color (set in the Attributes tool box) is applied to new text and the **Color Value** line is not available for editing.

- If the **Color** check box is ON, the **Color Value** palette is available and its color is applied to new text. To change the color value, click the **Value** field of the **Color Value** setting and select a color from the **Color Value** palette.

Justification

The **Justification** setting provides a drop-down menu from which you can select the relationship of the placement data point to the text string when you use the Place Text tool. As was discussed in Chapter 4, the name of the option describes where the data point will be in relation to the text. For example, if the **Left Top Justification** is chosen, the placement data point is at the top left corner of the test string.

Node Justification

The **Node Justification** drop-down menu sets the justification for importing text and is discussed later in this chapter.

The Check Boxes

At the bottom of the list of settings on the **General** tab is a set of check boxes titled **Vertical**, **Underline**, **Overline**, **Color**, **Italics**, and **Bold**. Turning these check boxes ON causes additional formatting to be applied to the text strings you place in the design as shown in Figure 7–21. The indication that a check box is on is a check mark in the check box's **Value** field.

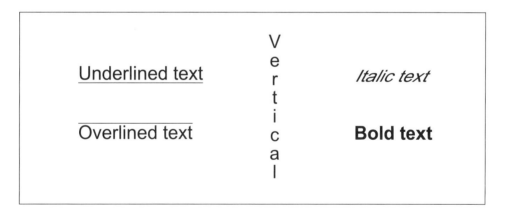

Figure 7-21 Examples of the effect of the style check boxes

If the **Vertical** check box is ON, the **Underline** and **Overline** check boxes are ignored and the text is placed without underlining or overlining.

As was described earlier, turning the **Color** check box ON enables the **Color Value** palette from which you can select the color that is always used with the current text style.

The **Bold** check box is available only when a TrueType **Font** is selected.

The Advanced Tab Settings

The **Advanced** tab provides settings that allow greater control over text formatting. With these settings you can do things such as change the style of the lines used for underlining and overlining. In addition to the advanced settings, this tab also contains the same settings as the General tab.

Importing Text Styles

In addition to creating text styles in the active design file, you can import all the text styles from another design file. The imported text styles are added to the text styles library for your design file.

Invoke the Textstyle Import tool from:

| Text Styles box | File > Import |

MicroStation opens the Textstyle Import box as shown in Figure 7–22.

1. In the **Directories** field, navigate to the folder that contains the file with text styles to import.
2. Select the file's name in the **Files** field.
3. Click **OK** to import the text styles or click **Cancel** to close the Textstyles Import box without importing anything.

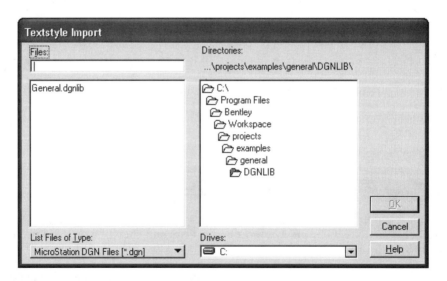

Figure 7-22 The Textstyle Import box

IMPORT TEXT

The Import Text tool allows you to import text from a file created by another computer application. The file to be imported must contain only unformatted (ASCII) text. For example, the Microsoft Word for Windows word processing application has an export option in its Save As option that creates an unformatted text file with line breaks.

When the text is imported into the design file, each line of text is placed as a separate text element.

Invoke the Import Text tool from:

Drop-down menu	File > Import > Text
Key-in window	**include** (or **in**) ENTER

MicroStation displays the Include Text File dialog box, as shown in Figure 7–23. Find and select the file that contains the text to be imported and click on the **OK** button.

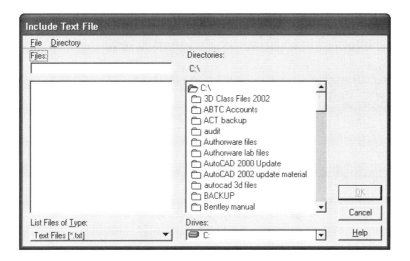

Figure 7–23 Include Text File dialog box

MicroStation prompts:

> Import Text File > Identify upper left of text block *(Define the point in the drawing plane where the text is to be placed.)*

The text is placed, starting at the location of the data point. There is no dynamic image of the text before you place the data point.

Text Attribute Settings

The text file can contain MicroStation element and text attribute setting key-ins to control the way the text appears when placed.

The Rules for Adding Attribute Setting Tools to a Text File:

- Standard MicroStation attribute key-ins are used.
- Each key-in must be preceded by a period.
- The key-in must be the only thing on the line.
- The settings act on the text that follows them in the file.
- The key-ins are not placed in the imported text string.
- If an attribute setting is not included in the file, the drawing's current active setting applies.
- The settings in the imported file become the drawing's active settings after the text is imported.

Table 7–1 lists useful key-ins for importing text. Figure 7–24 shows an example of a text file that contains key-ins.

Table 7–1 Element Attribute Key-ins

KEY-IN	SETS THE ACTIVE . . .
.AA=	Rotation angle in degrees
.CO=	Color name or number
.FT=	Font name or number
.LS=	Line Spacing in working units
.LV=	Level name
.TH=	Text Height in working units
.TW=	Text Width in working units
.TX=	Text Size (sets height and width equal) in working units
.WT=	Weight number
Indent #	Indents each following line of text with "#" number of spaces

```
.AA=5
.LV=Dimensions
.FT=Architectural
.CO=Red
.WT=1
.TH=1.5
.TW=1.0
.LS=0.8
When this text is imported, it will be placed at an angle of
5 degrees on the Dimensions level using using the Architectural
font, the color red, line weight one, and a text size of 1.5 x 1.0
master units.
.Indent=5
NOTE: This note will be indented five spaces.
```

Figure 7-24 An example of a text file with MicroStation key-ins

TEXT MANIPULATION TOOLS

The manipulation tools (Move, Copy, and Rotate, among others) manipulate the text element but not the text itself. Changes to the text in the element are handled by a set of text manipulation tools that include editing the text, setting the active text settings to match an existing text element, changing a text element's settings to match the current active text settings, copying and incrementing numbers in text elements, and displaying the text settings used to place a text element.

Spell Checker

The Spell Checker tool compares the words in a selected text element with the MicroStation dictionary and suggests possible corrections for words it does not find in the dictionary. You can select one of the selected words or type your own correction.

Invoke the Spell Checker tool from:

Text tool box	Select the Spell Checker tool. (see Figure 7–25).
Key-in window	**spellcheck** (or **sp**) ENTER

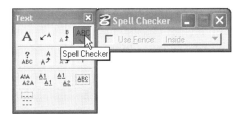

Figure 7-25 Invoking the Spell Checker tool from the Text tool box

MicroStation prompts:

> Spell Checker > Identify element *(Click the Data button on the text element whose spelling you want to check. When the text string is highlighted, click the Data button again to accept the text element.)*

If the selected element contains words that are not in the MicroStation dictionary, the Spell Checker box opens, as shown in Figure 7–26.

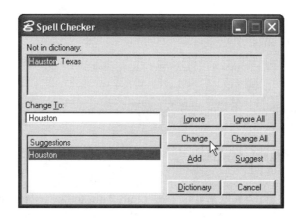

Figure 7-26 The Spell Checker box

The selected text string is displayed in the **Not in dictionary** field and the first misspelled word is highlighted. The word that MicroStation thinks is the best replacement word is displayed in the **Change To** field, and alternate spellings are displayed in the **Suggestions** field. If you click on one of the words in the **Suggestions** field, the word is placed in the **Change To** field.

- To change the misspelled word to the **Change To** word, click the **Change** button, or to change all occurrences of the word in the selected text element to the **Change To** word, click the **Change All** button.

- To ignore the misspelled word and continue checking for misspellings in the selected text element, click the **Ignore** button, or to ignore all occurrences of the word in the selected text element, click the **Ignore All** button.

- If you don't think any of the suggested spellings is the correct word, ask for more suggestions by clicking the **Suggest** button. If MicroStation has any additional suggestions, they appear in the **Suggestions** field.

- If the word is actually spelled correctly, you can add it to the dictionary by clicking the **Add** button.

- To close the Spell Checker box without checking any more words, click the **Cancel** button.

Chapter 7 Placing Text, Data Fields, and Tags

When there are no more misspelled words in the selected text string, the Spell Checker box closes.

 Note: The **Dictionary** button displays the Edit User Dictionary settings box. This box provides options that allow you to add and remove dictionary words and define special actions for words.

Edit Text Elements

The Edit Text tool provides a way to change the text in an existing text element.

Invoke the Edit Text tool from:

Text tool box	Select the Edit Text tool (see Figure 7–27).
Key-in window	**edit text** (or **edi te**) ENTER

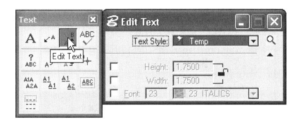

Figure 7-27 Invoking the Edit Text tool from the Text tool box

MicroStation prompts:

> Edit Text > Identify element *(Select the text element to be edited.)*
>
> Edit Text > Accept/Reject (Select next input) *(Click the Data button again to accept the text and place it in the Text Editor box for editing.)*
>
> Edit Text > Accept/Reject (Select next input) *(Make the required changes to the text in the Text Editor box and click the Data button somewhere outside of the Text Editor box to replace the text in the design with your edited text, or click the Reset button to leave the text unchanged and clear the Text Editor box.)*

 Note: Refer to the discussion of "Place Text Tools" in Chapter 4 for notes on keyboard shortcuts you can use in the Text Editor box.

The Edit Text Tool Settings window allows you to change the **Text Style**, **Height**, **Width**, and **Font** settings while editing the selected text. By default, only the **Text Style** menu is visible in the window. Click the **Expand** arrow on the right side of the window to display the additional settings. If the arrow is pointing up, the window has already been expanded and the arrow's name is **Collapse**.

To change formatting for the text, select all of the text string in the Text Editor box and then make each required setting change. The appearance of the selected text changes in the Text Editor box as you change the settings. Accept the changes by clicking the Data button outside of the box.

You can also add text to a string using different settings than the original text. Select the desired options and key-in the new text. For example, if you change the text height and width settings and key-in new text at the end of the text string in the Text Editor window, the new text is placed at the new text size and the original text remains it the same size. The text string is still one element, even though it contains text of two different sizes.

The method of selecting text and changing its formatting only works when the selected text has uniform formatting. For example, if the selected text has two different font values, it cannot be changed to new settings. You must select the part of the text that has one font, make the changes, and then select and change the part that has the other font.

Match Text Attributes

The Match Text Attributes tool sets the active text attributes to match those of an existing text element. The tool changes the active font, text size, line spacing, and text justification to the settings of the selected text element, and all text placed afterwards uses the new active settings.

Invoke the Match Text Attributes tool from:

Text tool box	Select the Match Text Attributes tool (see Figure 7–28).
Key-in window	**match text** (or **mat t**) ENTER

Figure 7-28 Invoking the Match Text Attributes tool from the Text tool box

MicroStation prompts:

Match Text Attributes > Identify text element *(Identify the text element to match.)*

Match Text Attributes > Accept/Reject (Select next input) *(Click the Data button to set the active settings to selected text element, or click the Reset button to reject it.)*

MicroStation sets the new active text attributes and displays them on the Status bar. To make these changes permanent, select **File > Save Settings**.

Change Text Attributes

The Change Text Attributes tool changes the attributes of an existing text element.

Invoke the Change Text Attributes tool from:

Text tool box	Select the Change Text Attributes tool (see Figure 7–29). In the Tool Settings window, turn ON the check box for each attribute that needs to be changed and select or type the required attributes in the option menus and edit fields of the attributes you turn on.
Key-in window	**modify text** (or **modi te**) ENTER

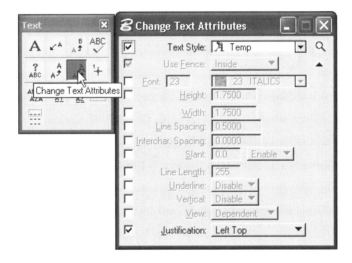

Figure 7–29 Invoking the Change Text Attributes tool from the Text tool box

MicroStation prompts:

> Change Text Attributes > Identify text *(Select text element.)*
>
> Change Text Attributes > Accept/Reject (Select next input) *(Click the Data button to change the attributes of the selected text element, or click the Reset button to reject it. This data point can also select another text element at the same time it accepts the current element.)*

Display Text Attributes

Display Text Attributes is an information-only tool that displays the attributes that were used to place an existing text element.

Invoke the Display Text Attributes tool from:

Text tool box	Select the Display Text Attributes tool (see Figure 7–30).
Key-in window	**identify text** (or **i t**) ENTER

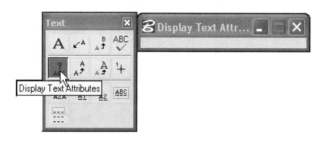

Figure 7–30 Invoking the Display Text Attributes tool from the Text tool box

MicroStation prompts:

Display Text Attributes > Identify text *(Identify the text element.)*

MicroStation displays the attributes in the left side of the Status bar. If desired, you can select another text element. The text attributes displayed in the Status bar are different for one-line text elements and multi-line text elements (Text Nodes):

- *For one-line text elements*, the displayed attributes include the text height and width, the level the element is on, and the font number.

- *For multi-line text elements*, the displayed attributes include the Text Node number, the maximum characters per line, the line spacing, the level the element is on, and the font number.

Copy and Increment Text

Annotating a series of objects with an incremented identification (such as P100, P101, P102) would be a tedious job without the Copy/Increment Text tool. This tool copies and increments numbers in text strings. To make incremented copies, select the element to be copied and incremented and place data points at each location where an incremented copy is to be placed.

Chapter 7 Placing Text, Data Fields, and Tags

The **Tag Increment** text field in the Tool Settings window allows you to set a positive or negative increment value. For example, an increment value of 10 causes each copy to be 10 greater than the previous one. A value of –10 causes each new copy to be 10 less than the previous one.

Only the numeric portion of a text string is incremented, and, if the string contains more than one numeric portion separated by nonnumeric characters, only the rightmost numeric portion is incremented. For example, only the 30 in the string P100-30 is incremented (P100-31, P100-32, P100-33, and so on).

To place a series of incremented text strings, place the starting text string and invoke Copy/Increment Text from:

Text tool box	Select the Copy/Increment Text tool and, optionally, set the Tag Increment value in the Tool Settings window (see Figure 7–31).
Key-in window	**increment text** (or **incr t**) ENTER

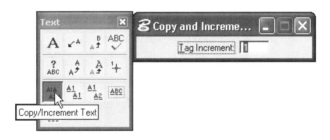

Figure 7–31 Invoking the Copy/Increment Text tool from the Text tool box

MicroStation prompts:

    Copy and Increment Text > Identify element *(Identify the text string to be copied and incremented.)*

    Copy and Increment Text > Accept/Reject (Select next input) *(Define the location of each incremented copy, or press the Reset button to reject the copy.)*

 Note: The Copy/Increment tool only accepts single-line text strings that contain numbers. If you attempt to select a string that does not contain numbers, or a multi-line string, the element is not accepted.

TEXT NODES

The Text Node tool provides a way to reserve space in a design where text is to be placed later. Once a Text Node is placed, it saves the active element and text attribute settings. When text is added to the node at a later time, the text takes on those settings.

Nodes are most often used in "fill-in-the-blank" forms that can be inserted in a design and filled in with information specific to the design. A common example is a title block form that has all the required fields held with Text Nodes. Use of the form provides a standard title block layout for all designs.

View Text Nodes

The visual indication of a text node is a unique identification number and a cross indicating the node origin point. Figure 7–32 shows examples of the way empty and filled-in Text Nodes appear in a design.

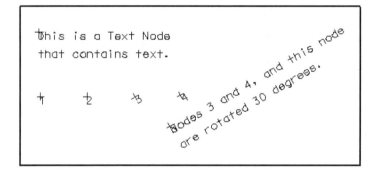

Figure 7-32 Examples of Text Node indicators

The Text Node view attribute controls the display and printing of Text Node indicators for selected views. You can change the display of Text Node view attribute from the View Attributes settings box (invoked from the **Settings** drop-down menu). The Text Node check box is at the bottom of the right column in the dialog box, as shown in Figure 7–33.

Chapter 7 Placing Text, Data Fields, and Tags

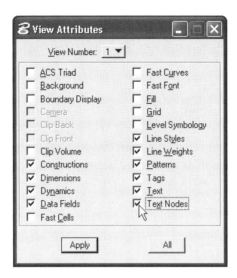

Figure 7-33 The View Attributes box

 Note: A tool is provided that places empty text nodes in the design, but multi-line text strings are also text nodes. When a multi-line text string is placed in the design, a text node is assigned to the string. If the **Text Nodes** view attribute is ON, the node cross and number appears with the text element at the justification point.

Place Text Nodes

To place empty text nodes, invoke the Place Text Node tool from:

Text tool box	Select the Place Text Node tool. In the Tool Settings window, set the Active Angle to control the angle at which the node is placed (see Figure 7-34).
Key-in window	**place node** (or **pla n**) ENTER

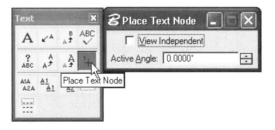

Figure 7-34 Invoking the Place Text Node tool from the Text tool box

305

MicroStation prompts:

> Place Text Node > Enter text node origin *(Define the origin point for each node that is to be placed.)*

 Note: If the **Text Node**s view attribute is turned OFF, nothing appears to happen when the Place Text Node tool is invoked to place empty nodes. Turn ON the **Text Nodes** view attribute to see the results of placing empty nodes.

The Place Text Node settings include a **View Independent** check box that applies to 3D designs. If the check box is ON, MicroStation prompts "Place View Independent Text Node," but there is no difference in the way the nodes are placed in a 2D design.

Fill In Text Nodes

Text can only be placed on empty text nodes when the Text Node Lock check box is ON.

Turn on the Text Node Lock from:

Settings > Locks	Select the Text Node option in the submenu.
Settings > Locks > Full	Turn the Text Node Lock check box ON.
Settings > Locks > Toggles	Turn the Text Node Lock check box ON as shown in Figure 7–35.
Key-in window	**lock textnode on** (or **lock t on**) ENTER

Figure 7–35 Setting the Text Node Lock check box to ON.

After turning the **Text Node Lock** ON, use the Place Text tool's **By Origin** method to place text on empty text nodes.

MicroStation prompts:

> Place Text > Enter text *(Enter the text in the Text Editor box and select the Text Node the text is to be placed on.)*
>
> Place Text > Enter more characters or position text *(Click a second data button to accept the previous node. This click can also select another node. You can also enter more text in the Text Editor box before clicking again, or click the Reset button to clear the Text Editor box.)*

The text is placed using the element and text attributes that were in effect when the node was created.

Note: If the selection or acceptance points are placed in an empty space or on a node that already contains text, MicroStation displays the message "Text node not found" on the Status bar. When the **Text Node Lock** is set to ON, you can only place text on empty Text Nodes.

DATA FIELDS

Data Fields are similar to Text Nodes in that they create placeholders for text that will be filled in later. Data Fields, though, are more powerful than Text Nodes because there are tools available to automate filling in the fields.

A common use for Data Fields is to provide placeholders for descriptive text in cells. For example, a control valve cell might contain Data Fields for the valve type, size, and identification code. Cells are discussed in Chapter 11.

Data Field Character

Data Fields are created by typing a contiguous string of underscores in a text string. For example, "_____" is a five-character Enter Data field. Any of the Place Text tool methods can be used to create the data fields.

You can only key-in as many characters in a Data Field as there are underscores, so when creating the field you must anticipate the number of text characters that will be placed in the field. The Edit Text tool can be used to add or remove underscores in Data Fields that have already been placed in the design.

A text string can contain more than one data field. For example, the string, "Pump ____ - __" contains two data fields. If you insert a line break in a string of underscores, there will be a separate data field on each line.

Note: The underscore is the default Data Field character, but you can change the reserved character from the MicroStation Preferences settings box (Preferences are discussed in Chapter 16).

Data Field View Attribute

The View Attributes box includes a **Data Field**s check box for turning ON and OFF the display of the Data Field underscores for a selected view, as shown in Figure 7–36.

Figure 7-36 View Attributes box

When the **Data Fields** View Attributes check box is:

- OFF—the underscores disappear from the selected view. If a field has been filled in, the fill-in text is still visible.

- ON—the underscores are visible in the selected view, and they print. If the data fields contain characters, the underscores appear at the bottom of each fill-in character.

Set Justification for Data Field Contents

The Data Field contents can be justified Left, Right, or Center within the field when there are fewer characters in the field than underscores. Data Field content justification is different from text justification and is applied to the Data Field only *after* the text string is placed.

Invoke the Data Field justification tool from:

Key-in window	**justify left** (or **ju l**) ENTER
	justify center (or **ju c**) ENTER
	justify right (or **ju r**) ENTER

The MicroStation prompt shown here appears when Center justification is selected (The MicroStation prompt includes the selected justification):

Center Justify Enter Data Field > Identify element *(Click on the Data Field to be justified.)*

No acceptance is required for this tool. The field is justified as soon as it is selected.

Note: If the justified Data Field contains text, the position of that text does not change. If you replace the existing text after changing the justification, the new text takes on the new justification.

Fill In Data Fields

MicroStation provides two tools to fill in text in Data Fields. The Fill In Single Enter-Data Field tool allows you to place text by identifying a specific Data Field. The Auto Fill In Enter Data Fields tool prompts you to select a specific view and then MicroStation finds and lets you fill in each empty Data Field in the view in the order they were created.

Fill In Single Enter Data Field Tool

Invoke the Fill In Single Enter Data Field tool from:

Text tool box	Select the Fill In Single Enter Data Field tool (see Figure 7–37).
Key-in window	**edit single** (or **edi s**) ENTER

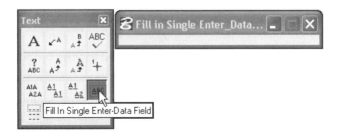

Figure 7–37 Invoking the Fill In Single Enter Data Field tool from the Text tool box

MicroStation prompts:

> Fill in Single Enter_Data Field > Identify element *(Identify the Data Field.)*

The Text Editor box opens when you identify a field. Type the text in the Text Editor box and press ENTER to place the text in the selected Data Field. The text placed in the data field is also displayed on the status bar. You can continue by identifying additional Data Fields.

 Note: The Data Field tools open the dialog box type of Text Editor box. The dialog box type does not have the formatting tools on the menu bar that the word processing type has.

Auto Fill In Enter Data Fields Tool

Invoke the Auto Fill In Enter Data Fields tool from:

Text tool box	Select the Auto Fill In Enter Data Fields tool (see Figure 7–38).
Key-in window	**edit auto** (or **edi au**) ENTER

Figure 7–38 Invoking the Auto Fill In Enter Data Fields tool from the Text tool box

MicroStation prompts:

> Auto Fill in Enter Data Fields > Select view *(Click the Data button anywhere in the view containing the fields to be filled in.)*
> Auto Fill in Enter Data Fields > <CR> to fill in or DATA for next field *(Type the text in the Text Editor box and press ENTER, or click the Data button to skip the field.)*

The tool continues through the view selecting empty Data Fields in the order they were created. It skips Data Fields that already contain text. If there are no empty data fields in the view, nothing happens when you select the view.

Copy Data Fields

To copy the contents of one Data Field to another Data Field, invoke the Copy Enter Data Field tool from:

Text tool box	Select the Copy Enter Data Field tool (see Figure 7–39).
Key-in window	**copy ed** (or **cop e**) ENTER

Figure 7–39 Invoking the Copy Enter Data Field tool from the Text tool box

MicroStation prompts:

> Copy Enter Data Field > Select enter data field to copy *(Select the Data Field containing the text to be copied and then click in each Data Field to which the text is to be copied.)*

When you select the Data Field that contains the text to be copied, MicroStation displays the selected text in the Status bar. This is useful for making sure you have the correct field when the view window is zoomed out so far that you cannot read the text.

Copy and Increment Data Fields

The Copy and Increment Enter Data Field tool copies the text from a filled-in Data Field and increments the numeric portion of the text before placing it in the next empty Data Field that you select.

The number in the **Tag Increment** field in the Tool Settings window is added to the number you are copying each time you select a destination field. A positive increment value increases the number and a negative increment value decreases it. For example, an increment value of 10 causes each copy to be 10 greater than the previous one and a value of –10 causes each copy to be 10 less than the previous one.

To copy the contents of one Data Field to another Data Field and increment the numeric portion of the copy, invoke the Copy/Increment Enter Data Fields tool from:

Text tool box	Select the Copy/Increment Enter Data Fields tool (see Figure 7–40), and, optionally, set the Tag Increment value in the Tool Settings window.
Key-in window	**increment ed** (or **incr e**) ENTER

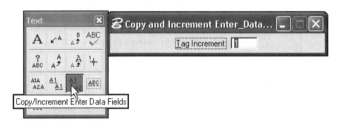

Figure 7-40 Invoking the Copy and Increment Enter Data Field tool from the Text tool box

MicroStation prompts:

> Copy and Increment Enter Data Field > Select enter data field to copy *(Select the Data Field containing the text to be copied, then click in each Data Field to which the text is to be copied and incremented.)*

Edit Text in a Data Field

The number of underscore characters in an existing Data Field can be changed by editing the text string in the Text Editor box.

In the Text Editor box the underscores are represented by spaces enclosed in pairs of angle brackets. For example, the string "PUMP __-__" will appear as "Pump << >>-<< >>" in the Text Editor box.

- To shorten a Data Field, remove spaces from between the angle brackets.
- To lengthen a Data Field, insert spaces (or underscores) between the angle brackets.
- To delete a Data Field completely, delete the angle brackets and the spaces between them.
- To insert a new Data Field:
    1. Position the cursor at the insertion point in the text string.
    2. Type a pair of left angle brackets (<<).
    3. Type the spaces (or underscores) to define the length of the Data Field.
    4. Type a pair of right angle brackets (>>).

TAGS

Engineering drawings have long served to convey more than just how a model looks. Drawings must tell fabricators how to construct the design. This nongraphical information includes such things as the construction material, how many to make, colors, where to obtain materials, and what finishes to apply to the surface. When models were created on paper, painstaking work was required to extract lists of this information from the drawings. A major innovation of CAD models is the ability to automate the creation of such lists.

MicroStation can provide this automation by attaching "Tags" to objects. Any element, or element group, can be tagged with descriptive information, and tag reports can be generated. For example, each electrical fixture in an architectural floor plan can be tagged with its rating, order number, price, and project name. An estimator can extract a fixture tag report from the design and insert the resulting data in a spreadsheet to obtain the total project cost for electrical fixtures. A purchasing agent can use the tag reports from several projects to order fixtures and take advantage of quantity discounts. Receiving clerks can employ the order numbers and project names to route the received fixtures to the correct projects.

MicroStation's Tag tools are helpful when the tagging requirements are fairly simple and when the project must import or export drawings from other CAD packages that store nongraphical data inside their design files. For complex tagging requirements, MicroStation supports connections to databases (which is beyond the scope of this text book).

Tags can be placed on any element in a design file. Figure 7–41 shows an example of tags. In the example, the tags are assigned to a small point (actually a short line) in each tract of land in a plot plan. The points are given just to provide an element to hook the tag to. Each "Tract" tag set includes the tract identification number, the purchase status, and the tract size.

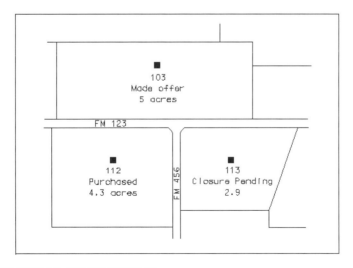

Figure 7–41 Tags displayed in a plot plan

Tag Terms

Adding tags to a design requires an understanding of tag terminology. Table 7–2 defines important tagging terms.

Table 7–2 Tagging Terms

TERM	DEFINITION	EXAMPLES
Tag set	A set of associated tags.	Separate tag sets for doors, windows, and electrical fixtures.
Tag	Nongraphical attributes that may be attached to graphical elements.	Part number, size, material of construction, vendor, price.
Tag report template	A file that specifies the tag set and the set's member tags to include on each line of reports that use the template.	For the fixtures set, report the part number, rating, price, and project name of report. One tag set per each tagged element.
Tag report	A list of all tags based on a tag report template.	F300-2, 220V, $300, New ABC, Inc. building.
Tag set library	Files containing tag sets exported from design files.	A library of architectural tag definitions for use in multiple sets.

 Note: If you delete an element with attached tags, the tag attachments are also deleted.

Create a Tag Set and Tags

The first step in creating tags in a design is to create the tag set and define the tags in the set.

Open the Tag Sets settings box from:

Drop-down menu	Element > Tags > Define
Key-in window	**mdl load tags define** (or **md l tags define**) ENTER

MicroStation displays the Tag Sets settings box as shown in Figure 7–42. All defined tag sets are displayed on the left side of the window, and the tag names for the selected tag set are displayed on the right side of the window.

Chapter 7 Placing Text, Data Fields, and Tags

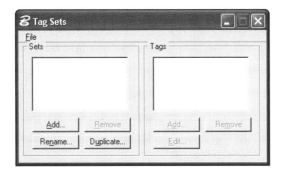

Figure 7-42 Tag Sets settings box

To create a new tag set name, click the **Add** button in the **Sets** area of the Tag Sets settings box. MicroStation displays the Tag Set Name dialog box, as shown in Figure 7–43.

Figure 7-43 Tag Set Name dialog box

Key-in the Tag Set name in the **Name** field of the Tag Set Name dialog box and click the **OK** button to create the new Tag Set. The new Tag Set name appears in the **Sets** field in the Tag Sets settings box.

To create a new tag under a specific tag set, select the Tag Set name in **Sets** list and click the **Add** button from the right side of the settings box under the **Tags** names. MicroStation displays the Define Tag dialog box, as shown in Figure 7–44.

Figure 7-44 Define Tag dialog box

315

Key-in the appropriate information in the fields provided in the Define Tag dialog box and click the **OK** button to create the tag, or click the **Cancel** button to close the dialog box without creating the tag. If you want to clear all fields in the dialog box and start over, click the **Reset** button.

Refer to Table 7–3 for a detailed explanation of the available fields in the Define Tag dialog box.

Table 7-3 Tag Attributes

ATTRIBUTE	DESCRIPTION
Tag Name	The name of the tag.
Prompt	A 32-character-maximum text string that will serve to tell the user what the tag is for when it is assigned to an element.
Type	A menu from which you can select the type of information that will be placed in the tag: Character—a text string Integer—a whole number Real—a number with a fractional part
Variable	A check box that, when OFF, prevents the tag value from being changed with the Edit Tags tool (discussed later). If ON, the value can be edited.
Default	A check box that, when OFF, uses the default tag value and prevents the tag value from being changed with the Edit Tags tool. If ON, the check box uses the default but allows editing of the value.
Default Tag Value	The default value for the tag when it is assigned to an element. The default value can be overridden.
Display Tag	Controls how the tags are displayed in the views and what can be done to them.

Maintain Tag Set Definitions

The Tag Sets settings box provides options for maintaining existing tag and tag set definitions. The options include:

- **Remove**—remove (delete) the selected tag set (with all its tags) or tag. A confirmation window opens, and you initiate the removal by clicking the **OK** button.

- **Rename**—change a tag set's name. It opens the Tag Set Name window, in which you type the new tag set name.

Chapter 7 Placing Text, Data Fields, and Tags

- **Duplicate**—create a duplicate copy of a tag set. It opens the Tag Set Name window, in which you type the name to use for the duplicate tag set.
- **Edit**—edit the attributes of the selected Tag. It opens the Define Tag window, in which you to edit the tag attributes.

Attach Tags to Elements

To assign a tag to an element, invoke the Attach Tags tool from:

Tags tool box	Select the Attach Tags tool and select a Tag Sets name from the Tool Settings window (see Figure 7–45).

Figure 7–45 Invoking the Attach Tags tool from the Tags tool box

MicroStation prompts:

    Attach Tags > Identify element *(Select a Tag Set from the Tool Settings window and identify the element to which tags are to be attached.)*

    Attach Tags > Accept/Reject (Select next input) *(Click the Accept button to accept the selected element, and, if required, select the next element to which tags will be attached after the current one is completed.)*

If the selected Tag Set contains Tags, the Attach Tags window opens when the element is accepted. Figure 7–46 provides an example of the window displaying the tag names for a "windows" tag set; Table 7–4 describes each field.

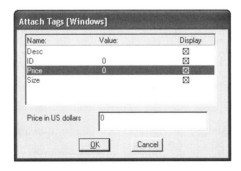

Figure 7–46 Typical Attach Tags dialog box

Table 7-4 The Fields in the Attach Tags Window

FIELD	DESCRIPTION
Name	The name of each tag in the set.
Value	The default value, if any, of each tag in the set.
Display	A check box that can be used to turn ON or OFF the display of each tag.
Prompt	The input prompt for the selected tag (in Figure 7-46, the prompt for the "Price" tag is for "Price in US dollars").
Value Field	Next to the prompt is an input field where you enter the values for selected tags. If a tag has a default value, the value appears here when the tag is selected.

For each tag you want to place on the selected element:

1. Turn the **Display** check box ON or OFF. If the check box is ON, the tag is visible in the design when placed. If check box is OFF, the tag will not be visible in the design but is included in tag reports.
2. Type the selected tag's value in the entry field, unless the tag has an acceptable default value.

   When all the tags that you want to place on the selected element are ready, click the **OK** button to close the window and place the tags on the element, or click the **Cancel** button to discard the changes.

If you click **OK** and there are selected tags with the **Display** check box ON, MicroStation prompts:

> Attach Tags > Place Tag *(Click the Data button at the location in the design where you want to place the tags.)*

The tag values are placed using the active element and text attribute settings. If no tags are to be displayed, there is no prompt for placing the tags.

Edit Tags

To make changes to the tag values attached to an element, invoke the Edit Tags tool from:

Tags tool box	Select the Edit Tags tool (see Figure 7-47).
Key-in window	**edit tags** (or **edi t**) ENTER

Figure 7-47 Invoking the Edit Tags tool from the Tags tool box

Chapter 7 Placing Text, Data Fields, and Tags

MicroStation prompts:

> Edit Tags > Identify element *(Select the element containing tags to be edited.)*
>
> Edit Tags > Accept/Reject (Select next input) *(Click the Accept button to accept the selected element.)*

 Note: If more than one tag set is attached to the selected element, the Edit Tags dialog box appears and lists all tag sets attached to the selected element (as shown in Figure 7–48) and the following prompt appears. Otherwise, the Edit Tags [tag set name] dialog box appears, as shown in Figure 7–49.

> Edit Tags > Select Tag to Edit *(If, the Edit Tags dialog box appears, select one of the tag sets from the dialog box and click the **OK** button.)*

Figure 7-48 Typical Edit Tags window

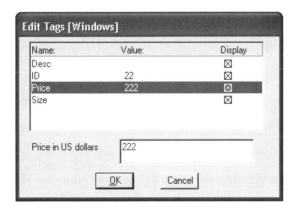

Figure 7-49 Typical Edit Tags [tag set name] window

The Edit Tags [tag set name] settings box shows all tags in the selected tag set. For each tag in the dialog box, you can turn the **Display** check box ON or OFF and change the tag's value. After you complete the required tag changes, click the **OK** button to apply the changes to the tags for the selected element or click the **Cancel** button to discard the changes.

Note: Tags that were created with the Variable option OFF cannot be edited.

Review Tags

The Review Tags tool allows you to select an element, and view the tags attached to the element. The tool works the same as the Edit Tags tool, except that it allows you only to view the tags, not edit them.

Invoke the Review Tags tool from:

Tags tool box	Select the Review Tags tool (see Figure 7–50).
Key-in window	**review tags** (or **rev t**) ENTER

Figure 7-50 Invoking the Review Tags tool from the Tags tool box

MicroStation prompts:

> Review Tags > Identify element *(Identify the element whose tags you want to review.)*
> Review Tags > Accept/Reject (Select next input) *(Click the Accept button to accept the selected element or click the Reset button to reject the element.)*

Note: If more than one tag set is attached to the selected element, the Review Tags dialog box appears and lists all tag sets attached to the selected element, and the following prompt appears. Otherwise, the Edit Tags [tag set name] dialog box appears. Except for the box name, these two boxes are identical to the Edit Tag tool boxes.

> Review Tags > Select Tag to review *(If, the Review Tags dialog box appears, select one of the tag sets from the dialog box and click the **OK** button.)*

The Review Tags [tag set name] settings box shows all tags in the selected tag set. When you are through reviewing the tags, click the **OK** button to close the dialog box.

Change Tags

The Change Tags tool allows you to change the values assigned to tags in the design. For example, you can change every "Door Material" tag with the value "Pine," to "White Oak."

Invoke the Change Tags tool from:

Tags tool box	Select the Change Tags tool and select required change options in the Tool Settings window (see Figure 7–51).
Key-in window	**change tags** (or **chan t**) ENTER

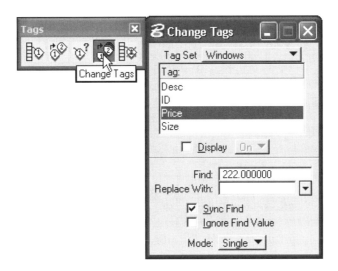

Figure 7-51 Invoking the Change Tags tool from the Tags tool box

When you invoke the Change Tags tool, the Tool Settings window contains several fields for controlling the way the tag changes are handled. Table 7–5 describes each option.

Table 7-5 The Fields in the Change Tags Window

FIELD	DESCRIPTION
Tag Set	An options menu that allows you to select the Tag Set that contains the Tag you need to change.
Tag	Lists all Tags for the selected Tag Set. You pick the Tag to be changed from this list.
Display	If the **Display** check box is ON, you can turn ON or OFF the display of the selected Tag.
Find	Provides an edit field in which you can type the value you want to change.
Replace With	Provides an edit field in which you can type the new value to assign to the selected Tag.
Sync Find	Provides a check box that enables and disables the function of the **Find** field.
Ignore Find Value	Provides control over the way the tool handles the contents of the **Find** edit field. If this check box is ON, the contents of the **Find** edit field are ignored and all occurrences of the selected Tag are changed to the contents of the **Replace With** edit field. If this check box is OFF, only occurrences of the selected Tags that contain the text in the **Find** edit field are changed to the contents of the **Replace With** edit field.
Mode	This option menu controls how the tool searches for occurrences of the selected Tag. If the Mode is **Single**, each Tag must be selected before its value can be changed. If the Mode is **Fence**, all occurrences of the selected Tag inside the fence are changed (a fence must be defined before this Mode can be used). If the Mode is **All**, all occurrences of the selected Tag in the design are changed to the new value.

For example, all Material Tags in the Doors Tag Set with a value of "Pine" need to be changed to "White Oak." The following steps explain how to do it.

1. Invoke Change Tags from the Tags tool box.
2. In the Tool Settings window, enter the following settings:
    - In the **Tag Set** menu, select the Doors Tag Set.
    - In the **Tag** list, select the Material Tag.
    - Turn the **Display** check box ON and select **On** from the **Display** menu so you can see all occurrences of the Material Tag in the design after the tool action is completed.

Chapter 7 Placing Text, Data Fields, and Tags

- In the **Find** edit field, type **Pine**.
- In the **Replace With** edit field, type **White Oak**.
- Turn on the **Sync Find** check box to enable searching for Material Tags with values equal to the contents of the **Find** edit field.
- Turn off the **Ignore Find Value** check box to make the tool use the contents of the **Find** edit field for the search.
- Set the **Mode** to **All** to cause the tool to find all occurrences of the Materials Tag in the design (MicroStation prompts, "Change Tag > Accept/Reject entire design file").

3. Click a data button anywhere in any open View Window to initiate the change.

Create Tag Reports

As mentioned earlier, MicroStation creates a tag report that lists the nongraphical information required to fabricate the model created in the project's design files.

Creating a report is a two-step process. First, a tags template is generated to control what is included in each type of report. Second, a report is generated based on the tag template.

Generating a Tags Template

A tag template defines the data columns in a tag report. Each column is either a tag from one tag set or an element attribute. To create a tag template, open the Generate Template settings box from:

Drop-down menu	Element > Tags > Generate Templates
Key-in window	**mdl load tags template** (or **md l tags template**) ENTER

MicroStation displays the General Templates settings box, as shown in Figure 7–52; Table 7–6 describes the fields in the settings box.

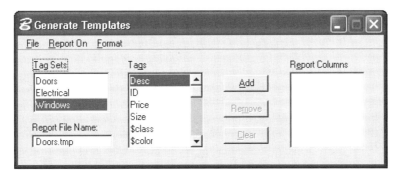

Figure 7-52 Typical Generate Templates settings box

Table 7-6 Generate Templates Settings Box Fields

FIELD	DESCRIPTION
Tag Sets	Lists the tag sets attached to the design file. Select the one for which the template is to be created.
Tags	Lists all tags in the selected tag set and all element attributes that can be included in the report. The names in the list that start with a dollar sign ($) are element attributes.
Report Columns	Lists the tags and element attributes that have been selected for inclusion in the report. Each name will be the name of a column in the report, and the order in the window determines the column order in the report.
Report File Name	Key-in an eight-character-maximum report file name, and, optionally, a three-character-maximum file extension. This name is used for the report files generated from this template.
Report On Menu	Select the type of elements to include in the report: *Tagged elements*—Include only tagged elements in the report. *All elements*—Include all elements, both tagged and untagged. If all elements are included but no element attribute columns are included, the untagged elements will show up in the report as empty rows.
File Menu	*Open*—Displays the Open Template dialog box, from which you can select an existing template file to open. *Save*—Saves the template information to the same file that was previously opened or saved as. *Save As*—Opens the Save Template As dialog box, from which you can save the template information to any directory path with a file name that you supply in the window. Use this tool to save new templates. Both the Open and Save Template As windows display the default template files directory path the first time they are opened.

Following are the steps to create a new template in the Generate Templates settings box.

1. Select a Tag Set from the **Tag Sets** menu.
2. Key-in the file name for the template in the **Report File Name** field.

Chapter 7 Placing Text, Data Fields, and Tags

3. For each Tag or element attribute that you want to include in the report, select the name in the **Tags** column and click the **Add** button.
4. If you add a name by mistake, select it in the **Report Columns** list and click the **Remove** button.
5. When all report columns are created, select **File > Save** to save the new template using the name currently in the **Report File Name** field.

Generating a Tags Report

Tag reports list tags and element data information based on Tags Templates. To generate a tag report, open the Generate Reports dialog box from:

Drop-down menu	Element > Tags > Generate Reports
Key-in window	**mdl load tags report** (or **md l tags report**) ENTER

MicroStation displays the Generate Reports dialog box, as shown in Figure 7–53; Table 7–7 describes the fields in the dialog box.

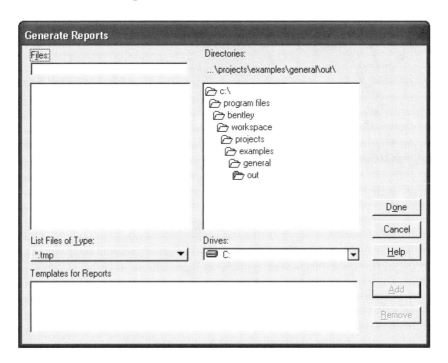

Figure 7-53 The Generate Reports dialog box

Table 7-7 Generate Reports Dialog Box Fields

FIELD	DESCRIPTION
Files	Lists the existing tag template files in the directory path shown in Directories.
Directories	Lists the directory path to the templates, and is used to change the path. It opens initially displaying the MicroStation default templates path.
List Files of Type	Provides an option menu for selecting the type of template files to display. It has two options: *.tmp—List files with the extension "tmp" (the default templates file extension). All files [*.*]—List all files in the directory path.
Drives	Provides a menu for selecting the letter of the disk drive containing the template files.
Templates for Reports	Displays the templates that have been selected for the report. Separate report files are generated for each template in this list.

Following are the steps to generate a report from the Generate Templates dialog box:

1. If required, change the directory path and file type to display the required templates.
2. For each report to be generated, click on the required template file name in the **Files** list and click the **Add** button. The file specification for each selected template appears in the **Templates for Reports** list.
3. If a mistake was made in selecting a template, select it in the **Templates for Reports** list and click the **Remove** button to remove it. The **Remove** button is then dimmed unless a template is selected.
4. After all required templates are selected, click the **Done** button to generate the reports.

Tag reports are created using the template's file name and the "rpt" extension. The reports are stored in MicroStation's default reports path:

Program Files\Bentley\Workspace\projects\untitled\out

The folder under which this path is found varies, depending on how MicroStation was installed and your computer's operating system. The complete path is defined in the MS_TAGREPORTS configuration variable.

Accessing the Reports

Tag reports are in ASCII files (also called "flat files") that can be accessed in several ways. For example:

- View and print a report with one of the operating system's text viewers, such as NotePad in Microsoft Windows.

- Import a report into another application, such as Excel, the Microsoft Windows spreadsheet application.

Tag Libraries

MicroStation provides tools that allow you to collect the tag sets from several design files into a tag library. You can then import tag sets from the library into other design files. This allows you to quickly obtain the tag sets you need without having to go to several other design files. Tools for creating tag libraries, exporting sets to existing tag libraries, and importing tag sets from a tag library are available from the Tag Sets settings box's **File** menu.

Open the Tag Sets settings box from:

Drop-down menu	Element > Tags > Define
Key-in window	**mdl load tags define** (or **md l tags define**) ENTER

Create a Tag Set Library

Following are the steps to create a tag library:

1. Invoke the Tag Sets settings box.
2. Select a tag set in the **Sets** list box.
3. From the settings box's menu bar, select **File > Export > Create Tag Library** to open the Export Tag Library dialog box, as shown in Figure 7–54.
4. In the **Directories** field, locate and select the folder in which you want to place the new tag set library.
5. Key-in the library file name in the **Files** field.
6. Click the **OK** button.

MicroStation creates a tag library file and exports the tag set you selected to the new library.

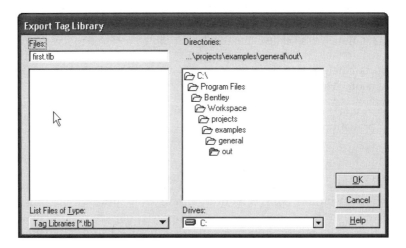

Figure 7-54 The Export Tag Library dialog box

Append a Tag Set to an Existing Library

Following are the steps to export a tag set from your design to an existing tag library:

1. Invoke the Tag Sets settings box.
2. Select the tag set to be exported from the **Sets** list box.
3. From the settings box's menu bar, select **File > Export > Append to Tag Library** to open the Export Tag Library dialog box, as shown in Figure 7-54.
4. In the **Directories** field, locate and select the folder that contains the tag library.
5. In the **Files** field, find and select the name of the library to which you want to export the selected tag set.
6. Click the **OK** button.

MicroStation appends the selected tag set to the tag library file.

 Note: If the library contains a tag set with the same name as the tag set you selected, a message box appears with a question asking if you want to replace the existing set in the library. Click the **OK** button to replace it or click the **Cancel** button to save the set already in the library.

Import a Tag Set from a Library

Following are the steps to import a tag set from a selected library:

1. Invoke the Tag Sets settings box.
2. From the settings box's menu bar, select **File > Import > From Tag Library** to open the Open Tag Library dialog box.

3. In the **Directories** field, locate and select the folder that contains the tag library.
4. In the **Files** field, find and select the tag library that contains the tag set you want to import.
5. Click the **OK** button to open the Import Sets dialog box.
6. Find and select the tag set you want to import.
7. Click the **OK** button.

MicroStation imports the selected tag set to your design file.

Note: If the design file contains a tag set with the same name as the tag set you selected, a message box appears with a question asking if you want to replace the existing set in the design file. Click the **OK** button to replace it or click the **Cancel** button to save the set already in the design.

Open the Exercise Manual PDF file for Chapter 7 on the accompanying CD for project and discipline specific exercises.

REVIEW QUESTIONS

Write your answers in the spaces provided.

1. An option to control the presentation of stacked fractions is available on the menu bar of the _____ box.

2. The Place Fitted Text tool fits the text between two _____.

3. When you place a text string above a line with the Place Text Above tool, the distance between the line and the text is controlled by the Text Style _____ setting.

4. Explain briefly why you might use Intercharacter Spacing attribute in placing text along a curving element.
   _____

5. Under what circumstance might you use the Match Text Attributes tool?
   _____
   _____

6. What is the Tag Increment for? _____

7. To determine the text attributes of an existing text element in a design file, invoke the _____ tool.

8. Explain briefly the purpose of placing nodes in a design file.
   _____
   _____

9. Explain briefly the purpose of defining tags.
   _____
   _____

10. List the steps involved in generating a template and report file.
    _____
    _____

chapter 8

Element Modification

Objectives

After completing this chapter, you will be able to:

- Extend Elements
- Modify vertices and arcs
- Modify elements with AccuDraw
- Create complex shapes and chains
- Create multi-line profiles
- Modify multi-line joints

ELEMENT MODIFICATION—EXTENDING LINES

MicroStation allows you not only to place elements easily, but also to modify them as needed. Three tools that are helpful for cleaning up the intersections of elements are available in

MicroStation. The tools—Extend Element, Extend Elements to Intersection, and Extend Element to Intersection—are available from the Modify Element tool box.

Extend Element

The Extend Line functions to extend or shorten a line, line string, or multi-line via a graphically defined length (with a data point) or via a keyed-in distance. Figure 8–1 shows examples of extending lines.

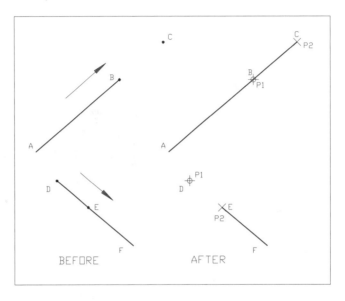

Figure 8-1 Examples of extending and shortening an element graphically

Extend Element Graphically

To extend an element graphically, invoke the Extend Element tool from:

Modify tool box	Select the Extend Line tool (see Figure 8–2).
Key-in window	**extend line** (or **ext l**) ENTER

Figure 8-2 Invoking the Extend Line tool from the Modify Element tool box

MicroStation prompts:

> Extend Line > Identify element *(Identify the element near the end to be extended or shortened.)*
>
> Extend Line > Accept or Reject (Select next input) *(Drag the element to the new length and click the Data button to accept, or click the Reject button to disregard the modification.)*

Extend Line by Key-in

To extend an element by keying-in the distance, invoke the Extend Element tool from:

Modify tool box	Select the Extend Element tool. Set the Distance check box to ON and key-in the distance in the Tool Settings window (see Figure 8–3).
Key-in window	**extend line keyin** (or **ext l k**) ENTER

Figure 8–3 Invoking the Extend Element tool via key-in from the Modify Element tool box

MicroStation prompts:

> Extend Line by Key-in > Identify element *(Identify the element near the end to be extended or shortened.)*
>
> Extend Line by Key-in > Accept/Reject (Select next input) *(Click the Data button again anywhere in the design plane to accept the extension.)*

 Note: To shorten the element, key-in a negative value in the Distance edit field.

Extend Elements to Intersection

Two elements can be extended or shortened to create a clean intersection between the two. Elements that can be extended to a common intersection with each other are lines, line strings, arcs, half ellipses, and quarter ellipses. Figure 8–4 shows several examples of possible extensions to an intersection.

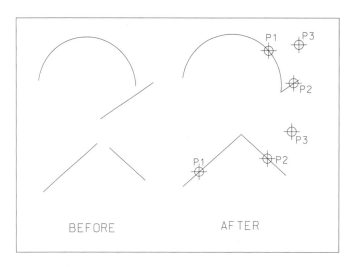

Figure 8-4 Examples of extending elements to a common intersection

To extend two elements to their common intersection, invoke the Extend two Elements to Intersection tool from:

Modify tool box	Select the Extend Elements to Intersection tool (see Figure 8–5).
Key-in window	**extend line 2** (or **ext l 2**) ENTER

Figure 8-5 Invoking the Extend Elements to Intersection tool from the Modify Element tool box

MicroStation prompts:

> Extend 2 Elements to Intersection > Select first element for extension *(Identify one of the two elements.)*
>
> Extend 2 Elements to Intersection > Select element for intersection *(Identify the second element.)*
>
> Extend 2 Elements to Intersection > Accept/Initiate Intersection *(Place a data point anywhere in the view to initiate the intersection.)*

Dynamic update shows the intersection as soon as you select the second element, but the intersection is not actually created until you accept it by clicking the Data button a third time.

Chapter 8 Element Modification

Note: If an element overlaps the intersection, select it on the part you want to keep. The part of the element beyond the intersection is deleted. If dynamic update shows the wrong part of the element deleted, click the Reset button to back up and try again.

Extend Element to Intersection

The Extend Element to Intersection tool serves to change the endpoint of the first selected line to extend to the second selected line, line string, shape, circle, or arc. Elements that can be extended are lines, line strings, arcs, half ellipses, and quarter ellipses. Figure 8–6 shows several examples of possible extensions to intersection.

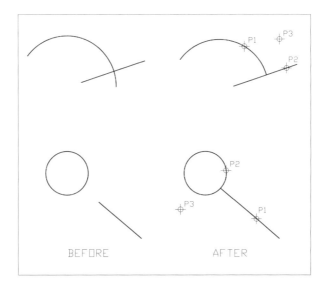

Figure 8-6 Examples of extending an element to an intersection

To extend an element to its intersection with another element, invoke the Extend Element to Intersection tool from:

Modify tool box	Select the Extend Element to Intersection tool (see Figure 8–7).
Key-in window	**extend line intersection** (or **ext l in**) ENTER

Figure 8-7 Invoking the Extend Element to Intersection tool from the Modify Element tool box

335

MicroStation prompts:

> Extend Element to Intersection > Select first element for extension *(Identify the element to extend.)*
>
> Extend Element to Intersection > Select element for intersection *(Identify the element to which the first element will be extended.)*
>
> Extend Element to Intersection > Accept/Initiate Intersection *(Place a data point anywhere in the view to initiate the intersection.)*

Dynamic update shows the intersection as soon as you select the second element, but the intersection is not actually created until you accept it by clicking the Data button a third time.

 Note: If the element to be extended overlaps the intersection, select it on the part you want to keep. The part of the element beyond the intersection is deleted. If dynamic update shows the wrong part of the element deleted, click the Reset button to back up and try again.

ELEMENT MODIFICATION—MODIFYING VERTICES

Several tools are available to modify the geometric shape of elements by moving, deleting, or inserting vertices. For example, you can change the size of a block by grabbing and moving one of the vertices of the block, or you can turn the block into a triangle by deleting one of the vertices.

Modify Element

The Modify Element tool can modify the geometric shape of any type of element except text elements. Here are the types of modifications it can make:

- Move a vertex or segment of a line, line string, multi-line, curve, B-spline control polygon, shape, complex chain, or complex shape
- Scale a block about the opposite vertex
- Modify rounded segments of complex chains and complex shapes created with the Place SmartLine tool while preserving their tangency
- Change rounded segments of complex chains and complex shapes to sharp, and vice versa
- Scale a circular arc while maintaining its sweep angle (use the Modify Arc Angle tool to change the sweep angle of an arc)
- Change a circle's radius or the length of one axis of an ellipse (if the ellipse axes are made equal, the ellipse becomes a circle and only the radius can be modified after that)
- Move dimension text or modify the extension line length of a dimension element

Typical element modifications are shown in Figure 8–8.

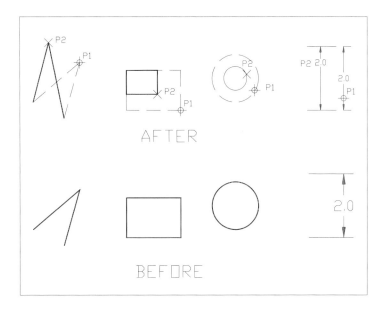

Figure 8-8 Examples of element modifications

Invoke the Modify Element tool from:

Modify tool box	Select the Modify Element tool (see Figure 8–9).
Key-in window	**modify element** (or **modi e**) ENTER

Figure 8-9 Invoking the Modify Element tool from Modify tool box

MicroStation prompts:

> Modify Element > Identify element *(Identify the element to be modified.)*
>
> Modify Element > Accept/Reject (Select next input) *(Move the selection point to the desired new location and place a data point to complete the modification, or click the Reset button to deselect the element.)*

When a vertex is selected for modification, the Tool Settings window presents options for modifying the shape of the vertex. Figure 8–10 shows a typical Modify Element tool settings window when a vertex is selected. The settings are described in Table 8–1.

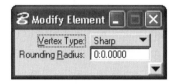

Figure 8-10 Modify Element tool settings window

Table 8-1 Modify Element Tool Settings for Vertices

SETTING	EFFECT OF SETTING
Vertex Type	Set the shape of each vertex to one of these types: **Sharp** **Rounded** (a Fillet) **Chamfered**
Rounding Radius	Enter the rounded vertex radius in working units (MU:SU:PU). **Note:** The "Rounding Radius" prompt appears only when the Vertex Type is "Rounded."
Chamfer Offset	Enter the offset of each end of the chamfer from the vertex point. Each offset is equal. **Note:** The "Chamfer Offset" prompt appears only when the Vertex Type is "Chamfer."
Orthogonal	If an orthogonal vertex is identified, turn this ON to maintain the orthogonal shape of the vertex.

Note: If you select a segment near its center, the segment is moved. If you select it near a vertex, that vertex is moved.

Using Modify Element and AccuDraw Together

Turn on AccuDraw before invoking the Modify Element tool to benefit from the extra drawing aids AccuDraw provides. These aids make Modify Element a more efficient tool. For example, the angle can be locked for a line to make it easy to adjust only the line length, or the length can be locked to make it easy to change only the rotation angle.

Delete Vertex

The Delete Vertex tool removes a vertex from a shape, line string, or curve string. Figure 8–11 shows an example of deleting a vertex from a line string.

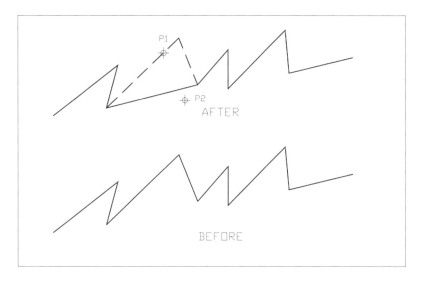

Figure 8–11 Example of deleting a vertex from a line string

To delete a vertex from an element, invoke the Delete Vertex tool from:

Modify tool box	Select the Delete Vertex tool (see Figure 8–12).
Key-in window	**Delete Vertex** (or **del v**) ENTER

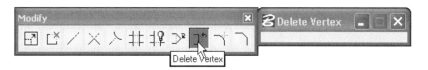

Figure 8–12 Invoking the Delete Vertex tool from the Modify Element tool box

MicroStation prompts:

>Delete Vertex > Identify element *(Identify the element near the vertex you want to delete.)*

>Delete Vertex > Accept/Reject (Select next input) *(Click the Data button to accept the deleted vertex, or click the Reject button to disregard the modification.)*

When you select the vertex to delete, dynamic update shows the element without the vertex, but it is not actually removed until you click the Data button again. The second data point can also select another vertex to delete.

 Note: If the element has only the minimum number of vertices required to define that type of element, you cannot delete a vertex from it. The tool indicates it is deleting the vertex, but nothing is deleted. For example, a minimum of three vertices is required to define a shape.

Insert Vertex

The Insert Vertex tool inserts a new vertex into a shape, line string, or curve string. Figure 8–13 shows an example of inserting a vertex for a line string.

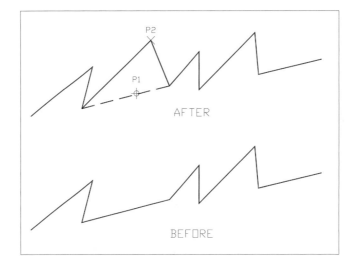

Figure 8-13 Example of inserting a vertex

To insert a vertex from an element, invoke the Insert Vertex tool from:

Modify tool box	Select the Insert Vertex tool (see Figure 8–14).
Key-in window	**insert vertex** (or **ins v**) ENTER

Figure 8-14 Invoking the Insert Vertex tool from the Modify Element tool box

Chapter 8 Element Modification

MicroStation prompts:

> Insert Vertex > Identify element *(Identify the element at the point where you want the vertex inserted.)*
>
> Insert Vertex > Accept/Reject (Select next input) *(Drag the new vertex to where you want it in the design plane and click the Data button to insert it, or click the Reject button to reject your selection.)*

When you select the element at the point where you want the new vertex inserted, a dynamic image of the new vertex follows the screen pointer until you place the second data point where you want the vertex located. The second data point causes the new vertex to be inserted, and dynamic update continues dragging the new vertices. Click the Reset button or select another MicroStation tool once you are through with the modification.

ELEMENT MODIFICATION—MODIFYING ARCS

After you place an arc, you can modify its radius, sweep angle, and axis. The tools are available in the Arcs tool box, or you can key-in the tools at the key-in window.

Modify Arc Radius

The Modify Arc Radius tool changes the length of the radius of the selected arc. Figure 8–15 shows examples of modifying an arc radius.

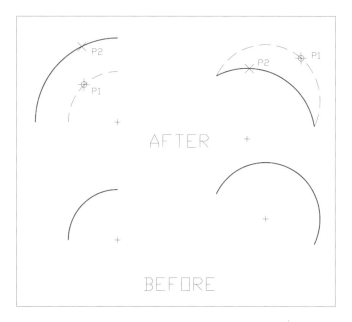

Figure 8-15 Examples of modifying an arc radius

To modify the radius of an arc, invoke the Modify Arc Radius tool from:

Arcs tool box	Select the Modify Arc Radius tool (see Figure 8–16).
Key-in window	**modify arc radius** (or **modi a r**) ENTER

Figure 8-16 Invoking the Modify Arc Radius tool from the Arcs tool box

MicroStation prompts:

Modify Arc Radius > Identify element *(Identify the arc to modify.)*

Modify Arc Radius > Accept/Reject (Select next input) *(Reposition the arc and click the Data button to place it, or click the Reject button to reject the modification.)*

Dynamic update shows the arc following the screen pointer after you select the arc. The arc is actually modified once you place the second data point, after which dynamic update continues to drag the arc. Click the Reset button or select another MicroStation tool once you are through with your modifications.

Modify Arc Angle

The Modify Arc Angle tool increases or decreases the sweep angle of the selected arc. Figure 8–17 shows examples of modifying an arc angle.

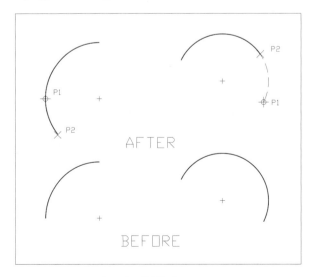

Figure 8-17 Examples of modifying an arc angle

Chapter 8 Element Modification

To modify the sweep angle of an arc, invoke the Modify Arc Angle tool from:

Arcs tool box	Select the Modify Arc Angle tool (see Figure 8–18).
Key-in window	**modify arc angle** (or **modi a a**) ENTER

Figure 8–18 Invoking the Modify Arc Angle tool from the Arcs tool box

MicroStation prompts:

> Modify Arc Angle > Identify element *(Identify the arc near the end whose sweep angle you want to modify.)*
>
> Modify Arc Angle > Accept/Reject (Select next input) *(Reposition the end of the arc and click the Data button to place it, or click the Reject button to reject the modification.)*

Dynamic update shows the arc following the screen pointer after you select the arc. The arc sweep angle is actually modified once you place the second data point, after which dynamic update continues to drag the arc. Click the Reset button or select another MicroStation tool once you are through with the modification.

 Note: If you drag the sweep angle around until the arc appears to be a circle, it is still an arc. Arcs that look like circles can be confusing later when using tools like patterning. If the arc should have been a circle, place a circle instead and delete the arc.

Modify Arc Axis

The Modify Arc Axis tool changes the major or minor axis radius of the selected arc. Figure 8–19 shows an example of modifying an arc axis.

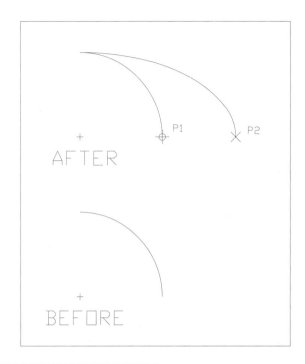

Figure 8-19 Example of modifying an arc axis

To modify an arc axis, invoke the Modify Arc Axis tool from:

Arcs tool box	Select the Modify Arc Axis tool (see Figure 8–20).
Key-in window	**modify arc axis** (or **modi a ax**) ENTER

Figure 8-20 Invoking the Modify Arc Axis tool from the Arcs tool box

MicroStation prompts:

> Modify Arc Axis > Identify element *(Identify the arc whose axis has to be modified.)*
> Modify Arc Axis > Accept/Reject (Select next input) *(Reposition the axis of the arc and click the Data button to place it, or click the Reject button to reject the modification.)*

Dynamic update shows the arc following the screen pointer after you select the arc. The arc is actually modified once you place the second data point, after which dynamic update continues to drag the arc. Click the Reset button or select another MicroStation tool once you are through with the modification.

 Note: After an arc's axis has been changed, the Modify Arc Radius tool can no longer be used on the element.

Create Complex Chain and Shapes

The Create Complex Chain and Create Complex Shape tools turn groups of connected elements into one complex element. A complex shape is a closed element (you could say it "holds water"), and a complex chain is an open element (the "water" can flow out between the two ends of the chain). The element manipulation tools treat the elements in a complex group as one element.

When you create a complex chain or shape from separate elements, the elements take on the current active element attributes (all the available settings in the Element Attributes settings box), and any gaps between the elements are closed. You can key-in the Maximum allowable gap in the Max Gap edit field. Figure 8–21 shows a group of individual elements before and after being turned into a complex shape.

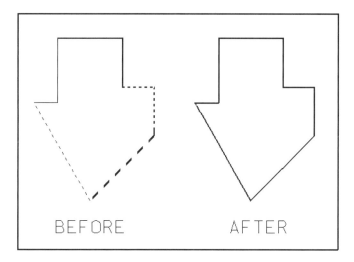

Figure 8–21 Example of a shape being turned into a complex shape

You can create a complex chain or shape manually by selecting each element to be included, or automatically by letting MicroStation find each element. If you want the elements that make up the complex chain or shape to be individual elements again, you can drop them with the Drop Complex tool (the dropped elements keep the attributes of the complex shape).

A quick way to check to see if you really created a complex group from the elements is to apply one of the element manipulation tools to it. If the elements are complex, dynamic update shows an image of all elements following the screen pointer, and the element type in the Status bar tells you it is either a Complex Chain or a Complex Shape.

Note: It is easy to make a mistake in creating complex chains and shapes, and only experience will make you competent. If you goof while creating a complex chain or shape, undo it and try again.

Create a Complex Chain Manually

When you create a complex chain manually, you must select and accept, in order, each of the elements to be included in the chain. Any gaps between the elements are closed, and the complex chain takes on the current active element attributes.

To create a complex chain manually, invoke the Create Complex Chain tool from:

Groups tool box	Select the Create Complex Chain tool. Select Manual from the Method option menu in the Tool Settings window (see Figure 8–22).
Key-in window	**create chain** (or **cr ch**) ENTER

Figure 8–22 Invoking the Create Complex Chain tool from the Groups tool box

MicroStation prompts:

> Create Complex Chain > Identify element *(Identify the first element to include in the complex chain.)*
>
> Create Complex Chain > Accept/Reject (Select next input) *(Identify the next element to include in the complex chain and continue selecting elements in order until all elements are selected. When all elements are selected and accepted, click the Reset button to create the complex chain.)*

If the **Simplify Geometry** check box is set to ON, then connected lines are added as line strings. If you identify only connected lines, the tool produces a primitive line string element rather than a complex chain. Figure 8–23 provides an example of creating a complex chain manually.

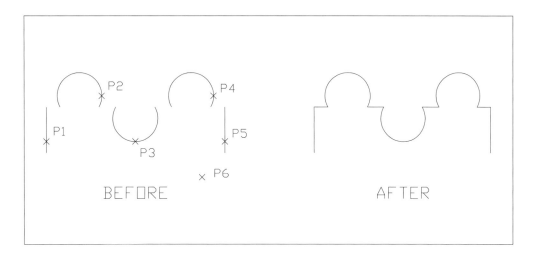

Figure 8-23 Example of creating a complex chain manually

 Note: Sometimes when you update a view, the lines closing the gaps between elements in the complex chain may disappear. They are still there—update the view again, and they should reappear.

Create a Complex Chain Automatically

To create a complex chain automatically, start by selecting and accepting the first element in the chain. After that, MicroStation finds and highlights more elements, in series, and you must accept or reject each one.

The automatic version of the tool also allows you to specify a maximum gap that tells MicroStation how far away, in working units, from the end of the previous element it can search for another element. If the tolerance is set to zero, the next element must touch the last selected one before MicroStation finds it.

If there are two or more possible elements at a junction, MicroStation tells you that there is a fork in the path and selects one of the possible elements. You can either accept the selected element or reject it and have MicroStation highlight another possible element in the fork.

 Note: If the complex chain contains many elements, and there are not very many forks, the automatic method is probably faster than the manual method. If there are many fork points, creating the chain manually may go faster.

To create a complex chain automatically, invoke the Create Complex Chain command from:

Groups tool box	Select the Create Complex Chain tool. Select Automatic from the Method option menu in the Tool Settings window (see Figure 8–24).
Key-in window	**create chain automatic** (or **cre ch a**) ENTER

Figure 8–24 Invoking the Create Complex Chain Automatic tool from the Groups tool box

MicroStation prompts:

> Automatic Create Complex Chain > Identify element *(Identify the first element to include in the complex chain.)*
>
> Automatic Create Complex Chain > Accept/Reject (Select next input) *(Move the screen pointer in the direction you want the search to go and click the Data button to accept the first element, or click the Reject button to reject it and start over.)*

If there are no forks in the path from the previous element, MicroStation prompts:

> Automatic Create Complex Chain > Accept Chain Element *(Click the Data button to accept the element and continue the search, or click the Reset button to complete the chain with the previous element.)*

If there is a fork in the path from the previous element, MicroStation prompts:

> Automatic Create Complex Chain > FORK—Accept or reject to See Alternate *(To accept the fork element MicroStation selected, click the Data button; or click the Reject button to disregard the current selection and select another fork element.)*

The process continues until MicroStation cannot find another element to add or until you reject a selection when there is no fork in the path. You *cannot* end a search at a fork point. If the **Simplify Geometry** check box is set to ON, then connected lines are added as line strings. If you identify only connected lines, the tool produces a primitive line string element, rather than a complex chain. Figure 8–25 shows an example of creating a complex chain automatically.

Chapter 8 Element Modification

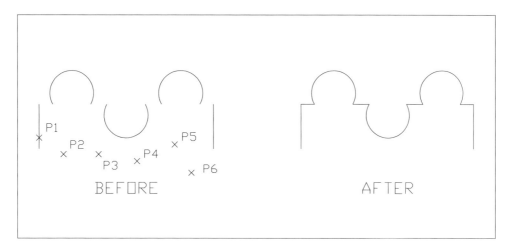

Figure 8–25 Example of creating a complex chain automatically

Create a Complex Shape Manually

To create a complex shape manually, you must select and accept, in order, each of the elements to be included in the shape. Any gaps between the elements are closed, and the complex chain takes on the current active element attributes.

To create a complex shape manually, invoke the Create Complex Shape tool from:

Groups tool box	Select the Create Complex Shape tool. Select Manual from the Method option menu Tool Settings window (see Figure 8–26).
Key-in window	**create shape** (or **cr s**) ENTER

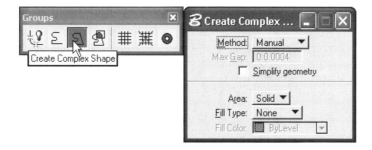

Figure 8–26 Invoking the Create Complex Shape tool from the Groups tool box

349

MicroStation prompts:

> Create Complex Shape > Identify element *(Identify the first element to include in the complex shape.)*
>
> Create Complex Shape > Accept/Reject (Select next input) *(Identify the next element to include in the complex shape and continue selecting elements in order until all elements are selected. When all elements are selected and the shape appears closed, click the Accept button to create the complex shape.)*

If the **Simplify Geometry** check box is set to ON, then connected lines are added as line strings. If you identify only connected lines, the tool produces a primitive line string element, rather than a complex chain. Figure 8–27 provides an example of how to create a complex shape manually.

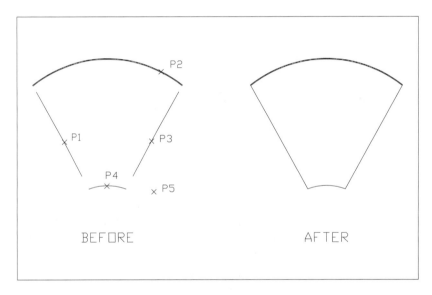

Figure 8-27 Example of creating a complex shape manually

Create a Complex Shape Automatically

To create a complex shape automatically, start by selecting and accepting the first element in the shape. After that, MicroStation finds and highlights more elements, in series, and you must accept or reject each one.

If there are two or more possible elements at a junction, MicroStation tells you there is a fork in the path and picks one of the possible elements. You can either accept or reject the element and have MicroStation highlight another possible element.

Chapter 8 Element Modification

The automatic version of the tool also allows you to specify a maximum gap that tells MicroStation how far away, in working units, from the end of the previous element it can search for another element. If the tolerance is set to zero, the next element must touch the last selected one before MicroStation finds it.

 Note: If the complex shape contains many elements, and there are not very many forks, the automatic method is probably faster than the manual method. If there are many fork points, creating the shape manually may go faster.

To create a complex shape automatically, invoke the Create Complex Shape tool from:

Groups tool box	Select the Create Complex Shape tool. Select Automatic from the Method option menu, and if necessary, key-in the gap distance in the Max Gap edit field in the Tool Settings window (see Figure 8–28).
Key-in window	**create shape automatic** (or **cr s a**) ENTER

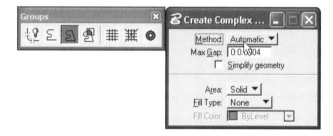

Figure 8-28 Invoking the Create Complex Shape Automatic tool from the Groups tool box

MicroStation prompts:

> Automatic Create Complex Shape > Identify element *(Identify the first element to include in the complex shape.)*
>
> Automatic Create Complex Shape > Accept/Reject (Select next input) *(Move the screen pointer in the direction you want the search to go and click the Data button to accept the first element, or click the Reject button to reject it and start over.)*

If there are no forks in the path from the previous element, MicroStation prompts:

> Automatic Create Complex Shape > Accept chain Element *(Click the Data button to accept the element and continue the search, or click the Reset button to complete the chain with the previous element.)*

351

If there is a fork in the path from the previous element, MicroStation prompts:

> Automatic Create Complex Chain > FORK—Accept or reject to See Alternate *(To accept the fork element MicroStation selected, click the Data button; click the Reject button to disregard the current selection and select another fork element.)*

The process continues until MicroStation cannot find another element to add or until you reject a selection when there is no fork in the path. You *cannot* end a search at a fork point. If you press the Reset button when another element is highlighted, that element is not used in the shape. If **Simplify Geometry** check box is set to ON, then connected lines are added as line strings. If you identify only connected lines, the tool produces a primitive line string element rather than a complex chain. Figure 8–29 shows an example of creating a complex shape automatically.

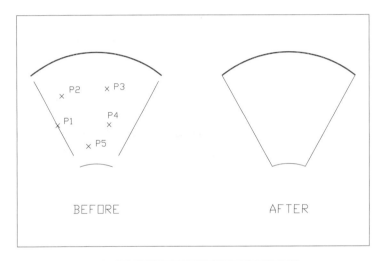

Figure 8-29 Example of creating a complex shape automatically

Create a Region

The Create Region tool creates a complex shape, similar to Complex Shape tools. You can create a complex shape from either of the following:

- The union, intersection, or difference between two or more closed elements
- A region bounded by elements that have endpoints that are closed together by the Maximum Gap

A **Keep Original** check box controls the handling of the original elements. If the check box is set to OFF, the original elements are deleted and only the complex region remains. If the check box is set to ON, the original elements remain in the design.

Create a Complex Shape from Element Intersection

The Intersection option allows you to create a complex shape from a composite area formed from the area that is common to two closed elements.

Chapter 8 Element Modification

To create a complex shape from the intersection of two overlapping circles as shown in Figure 8–30, invoke the Create Region tool from:

Groups tool box	Select the Create Region tool. Select Intersection from the Method option menu, and, if you want to keep the original elements, set the Keep Original check box to ON (see Figure 8–31).
Key-in window	**create region intersection** (or **cr r in**) ENTER

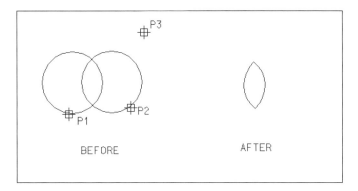

Figure 8-30 Example of creating a complex shape with the Intersection option

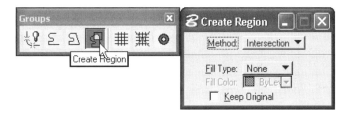

Figure 8-31 Invoking the Create Region tool with Intersection method from the Groups tool box.

MicroStation prompts:

Create Region From Element Intersection > Identify element *(Identify one of the two circles.)*

Create Region From Element Intersection > Accept/Reject (Select next input) *(Identify the second circle.)*

Create Region From Element Intersection > Accept/Reject (Select next input) *(Click the Data button again to accept the selection of the second circle.)*

Create Region From Element Intersection > Identify additional/Reset to complete *(Click the Reset button to create the complex shape.)*

Create a Complex Shape from Element Union

The Union option allows you to create a complex shape from a composite area formed in such a way that there is no duplication between two closed elements. The total resulting area can be equal to or less than the sum of the areas in the original closed elements.

To create a complex shape from the union of two overlapping circles as shown in Figure 8–32, invoke the Create Region tool from:

Groups tool box	Select the Create Region tool. Select Union from the Method option menu, and, if you want to keep the original elements, set the Keep Original check box to ON (see Figure 8–33).
Key-in window	**create region union** (or **cr r u**) ENTER

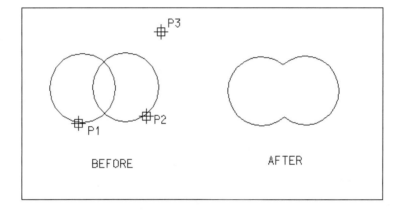

Figure 8–32 Example of creating a complex shape with the Union option

Figure 8–33 Invoking the Create Region tool with Union method from Groups tool box.

Chapter 8 Element Modification

MicroStation prompts:

> Create Region From Element Union > Identify element *(Identify one of the two circles.)*
>
> Create Region From Element Union > Accept/Reject (Select next input) *(Identify the second circle.)*
>
> Create Region From Element Union > Accept/Reject (Select next input) *(Click the Data button again to accept the selection of the second circle.)*
>
> Create Region From Element Union > Identify additional/Reset to complete *(Click the Reset button to create the complex shape.)*

Create a Complex Shape from Element Difference

The Difference option allows you to create a complex shape from a closed element after removing from it any area it has in common with a second element.

To create a complex shape from the difference of two overlapping circles as shown in Figure 8–34, invoke the Create Region tool from:

Groups tool box	Select the Create Region tool. Select Difference from the Method option menu, and, if you want to keep the original elements, set the Keep Original check box to ON (see Figure 8–35).
Key-in window	**create region difference** (or **cr r d**) ENTER

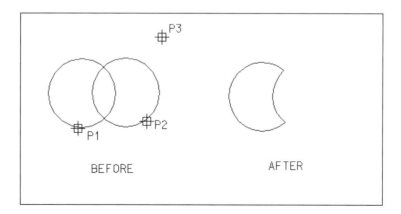

Figure 8–34 Example of creating a complex shape with the Difference option

Figure 8-35 Invoking the Create Region tool with Difference method from the Groups tool box

MicroStation prompts:

> Create Region From Element Difference > Identify element *(Identify the circle on the left.)*
>
> Create Region From Element Difference > Accept/Reject (Select next input) *(Identify the circle on the right.)*
>
> Create Region From Element Difference > Accept/Reject (Select next input) *(Click the Data button again to accept the selection of the second circle.)*
>
> Create Region From Element Difference > Identify additional/Reset to complete *(Click the Reset button to create the complex shape.)*

Create a Complex Shape from an Enclosed Area

The Flood option allows you to create a complex shape made up of one or more elements. MicroStation prompts you to pick a point inside the closed area. When you place the first data point inside the area, MicroStation searches for the elements that enclose the area, highlights pieces of the elements as it finds them, and then creates the complex shape.

You can specify a maximum gap between elements. If the gap is zero, all elements must touch. If the gap is greater than zero, the enclosed area will only be found if all gaps between elements are less than the gap value.

To create a complex shape from an enclosed area as shown in Figure 8–36, invoke the Create Region tool from:

Groups tool box	Select the Create Region tool. Select Flood from the Method option menu. If necessary, set the KeepOriginal check box to ON to keep the original elements and enter a value in the Max Gap field to specify how far apart the elements can be (see Figure 8–37).
Key-in window	**create region flood** (or **cr r f**) ENTER

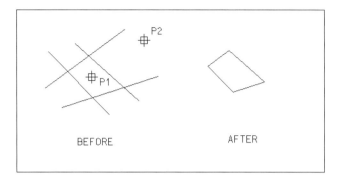

Figure 8-36 Example of creating a complex shape with the Flood option

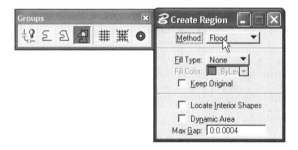

Figure 8-37 Invoking the Create Region tool with Flood method from Groups tool box

MicroStation prompts:

> Create Region From Area Enclosing Point > Enter data point inside area *(Place a data point inside the enclosed area.)*
>
> Create Region From Area Enclosing Point > Accept-Create Region *(Click the Data button to accept the complex shape.)*

If the **Locate Interior Shapes** check box set to ON, then closed elements inside the selected area are included as part of the new complex shape. If the **Dynamic Area** check box set to ON, then the region to be created displays dynamically, as you move the screen pointer over the shapes.

 Note: If the tool does not find an enclosing area, it displays the following message in the status bar: "Error – No enclosing region found."

DROP COMPLEX CHAINS AND SHAPES

If you want to return the elements in a complex chain or shape to individual elements, you can drop their complex status. The Drop Complex Status tool drops an individual complex group. The Fence Drop Complex Status tool drops all complex groups within the boundary of a fence.

Dropped complex elements return to being individual elements, but they keep the element attributes (color, weight, etc.) of the complex shape. The elements do not return to their original attributes.

Drop a Complex Chain or Shape

To drop a complex chain or shape, invoke the Drop Element tool from:

Groups tool box	Select the Drop Element tool. Set the Complex check box to ON, and turn OFF the other check boxes (see Figure 8–38).
Key-in window	**drop element** (or **dr e**) ENTER

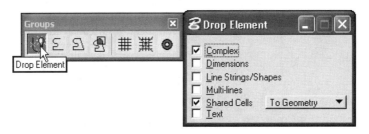

Figure 8-38 Invoking the Drop Element tool from the Groups tool box

MicroStation prompts:

> Drop Complex Status > Identify element *(Identify the complex chain or shape to drop into individual elements.)*
>
> Drop Complex Status > Accept/Reject (select next input) *(Click the Data button to accept the change in status from complex element into individual elements, or click the Reject button to reject the change in the status.)*

 Note: If there were gaps between the elements that make up the complex shape or chain, the gaps will return after the complex shape or chain is dropped. You may not see the gaps until you repaint the view window.

Drop Several Complex Chains or Shapes

The Drop Complex Status of Fence Contents tool breaks all the complex elements enclosed in a fence into separate elements. Before you invoke the tool, place a fence that encloses all the complex elements you want to drop.

Invoke the Drop Fence Contents tool from:

Fence tool box	Select the Drop Fence Contents tool. Select the desired Fence Mode from the Tool Settings window (see Figure 8–39).
Key-in window	**fence drop** (or **fe dr**) ENTER

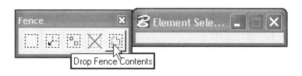

Figure 8–39 Invoking the Drop Fence Contents tool from the Fence tool box

MicroStation prompts:

> Drop Complex Status of Fence Contents > Accept/Reject Fence Contents *(Click the Data button to accept the change in status from complex element into individual elements, or click the Reject button to disregard the change in the status.)*

DEFINE MULTI-LINES

In Chapter 4 we learned how to place multi-lines using the active multi-line profile (usually three parallel lines with a dashed center line). Now we are going to learn how to customize the active multi-line profiles.

MicroStation provides the Multi-lines settings box for defining a new multi-line profile. Table 8–2 describes each multi-line component that can be defined in the settings box, and Figure 8–40 shows examples of typical multi-line profiles.

Table 8-2 The Multi-line components

Component	Description
Line	The parallel lines placed by the multi-line. You can define from one to sixteen lines components and set unique attributes for each one (color, weight, style, level).
Start Cap	An element that can be placed across the start of the multi-line to close it. The cap can be an arc or line completely across the start end, or arc across each pair of line components. You can also specify the attributes of the start cap.
End Cap	An element that can be placed across the end of the multi-line to close it. The cap can be an arc or line completely across the start end, or arc across each pair of line components. You can also specify the attributes of the start cap.
Joint	A line that can be placed at each multi-line joint (vertex). You can specify the attributes of the joint.

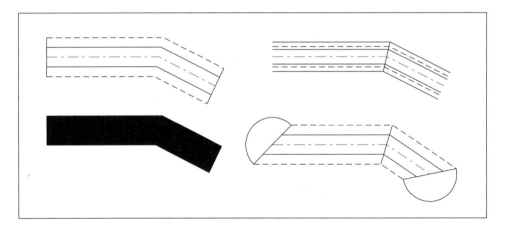

Figure 8-40 Examples of typical multi-line definitions

Invoke the Multi-Lines settings box:

Drop-down menu	Element > Multi-lines
Key-in window	**dialog multiline open** (or **di mu o**) ENTER

MicroStation displays the Multi-lines settings box as shown in Figure 8–41.

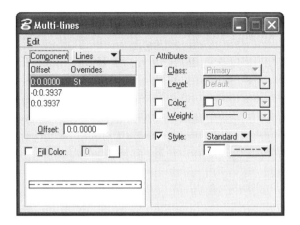

Figure 8-41 Multi-lines settings box

Note: The menu options displayed in the Multi-lines settings box depend on what component option is selected.

Define Line Components

To edit the line components in the active multi-line profile, select **Lines** from the **Component** options menu in the Multi-lines settings box. When you add a new line component to a multi-line definition, you must specify the line's offset from the placement point (data point in the design) and its attributes.

The multi-line placement point is called the "working line" and is the zero offset point in the set of parallel line components. The working line may not be an actual line. For example, if three lines are defined with offsets of –2, 0, and +2 respectively, the line with 0 offset is the working line. If two lines are defined with offsets of –2 and +2 respectively, the working line is at 0 offset (half-way in between the two parallel lines), but there is no line at the 0 offset position.

The Multi-line settings box provides several options for viewing and maintaining the line components of a multi-line (see Figure 8–41):

- The Offset/Overrides list box shows the offset (in working units) of each defined line component and the line attributes that are being controlled from this settings box. It also allows you to select the line you need to edit or delete.

- The current definition area at the bottom left of the settings box provides a graphic picture of the current multi-line definition.

- The **Edit** drop-down menu in the settings box's menu bar, provides options to **Insert**, **Delete**, and **Duplicate** line components.

- The **Offset** edit field provides a place to type in the offset of the selected line component from the working line.
- The Attributes area provides options for setting the selected line component's attributes.

Add a New Line

To add a line to the active multi-line definition, invoke the **Insert** option from:

| Drop-down menu (Multi-lines settings box) | Edit > Insert |

MicroStation adds the new line above the selected line in the Components list box.

Delete a Line

To delete a component line from the active multi-line definition, select the component line from the Components list box, then invoke the **Delete** option from:

| Drop-down menu (Multi-lines settings box) | Edit > Delete |

MicroStation deletes the selected component line from the multi-line definition.

Create a Duplicate of an Existing Line

To add a component line with the same attributes as the currently selected, invoke the **Duplicate** option from:

| Drop-down menu (Multi-lines settings box) | Edit > Duplicate |

MicroStation adds a component line with the same attributes as the selected line.

Set a Line's Offset

The **Offset** edit field sets the distance in Working Units (MU:SU:PU) from the working line to the selected component line. A positive offset places the line above the working line, and a negative offset places it below the working line. A zero offset places the line on the working line.

To enter the offset, select the line from the Offset/Overrides list, then type the offset in the **Offset** edit field and press the Tab key. Each line offset should be unique. If two lines have the same offset, they will be placed one on top of the other.

Set a Line's Attributes

When you create a new line component, it defaults to using the current active attributes for the design. You can override the active attributes by turning on the check boxes in the Attributes area of the settings box. When an attribute check box is on, that attribute is saved with the multi-line definition.

To set the attributes for an individual component line, select the component line in the Offset/Overrides list box, then set the appropriate attribute options that are located in the top right side of the Multi-lines settings box. If the check box to the left of an attribute option is:

- ON, the line will always be placed using that attribute setting.
- OFF, the line will always be placed using the design attribute value in effect when the multi-line line is placed. When the check box is set to OFF, the options are dimmed and cannot be accessed.

Figure 8–42 shows attributes being set as part of the multi-line definition.

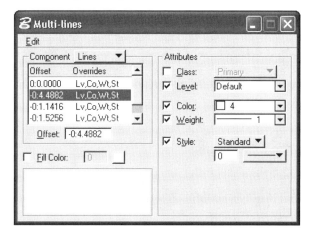

Figure 8–42 Example of line component attributes set as part of the multi-line definition

Control Multi-line Fill

The **Fill** color check box controls the color of the background area within the multi-line's parallel lines.

Set the check box to ON and enter a color number in the edit field to fill the multi-line with the selected color when it is placed. Turn it OFF to have an unfilled multi-line.

You will only see filled multi-lines if the Fill view attribute is set to ON (**Settings** > **View Attributes**).

Start Cap or End Cap Components

The Start Cap and End Cap options specify the appearance of the start and end of the multi-line. To define the start and end caps, select either **Start Cap** or **End Cap** from the **Component** options menu. Figure 8–40 shows examples of different start and end cap arrangements. Figure 8–43 shows the dialog box when the **Start Cap** option is selected.

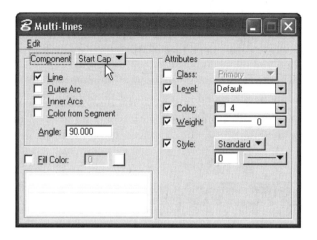

Figure 8–43 Start Cap selection displaying the controls for specifying the appearance of the start cap

The following Start Cap and End Cap options are available in the settings box:

- The **Line** check box controls the display of the cap as a line.
- The **Outer Arc** check box controls the display of the cap as an arc connecting the outermost component lines.
- The **Inner Arcs** check box controls the display of the cap, with arcs connecting pairs of inside component lines. If there is an even number of inside component lines, all are connected; if there is an odd number of inside component lines (three or more), the middle line is not connected.
- The **Angle** edit field sets the cap's line angle in degrees.

To set the attributes for a cap, set the appropriate attribute options that are located in the top right side of the Multi-lines settings box. If the check box to the left of an attribute option is:

- ON, the cap will always be placed using that attribute setting.
- OFF, the cap will always be placed using the design attribute value in effect when the multi-line line is placed. When the check box is set to OFF, the options are dimmed and cannot be accessed.

Joints Component

To display a joint line at each multi-line vertex, select **Joints** from the **Component** options menu. One of the multi-line examples in Figure 8–40 has joints displayed. Figure 8–44 shows the **Joints** options in the settings box.

Chapter 8 Element Modification

The **Display Joints** check box controls the placement of joint lines in multi-line elements. If it is set to ON, the joint lines are displayed at vertices.

To set the attributes for displayed joints, set the appropriate attribute options that are located in the top right side of the Multi-lines settings box. If the check box to the left of an attribute option is:

- ON, the joints will always be placed using that attribute setting.
- OFF, the joints will always be placed using the design attribute value in effect when the multi-line line is placed. When the check box is set to OFF, the options are dimmed and cannot be accessed.

Note: Manage Groups setting box allows you to save the newly created multi-line definition. Refer to Chapter 14 for a detailed explanation of the various options available.

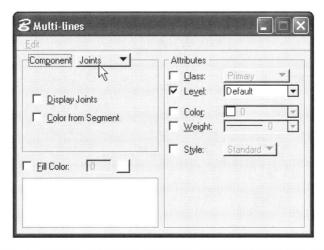

Figure 8-44 The Joints option selection displaying the controls for specifying the appearance of joints

MODIFY MULTI-LINE JOINTS

When multi-lines cross over other multi-lines, the intersection formed by the crossing needs to be cleaned up. The Multi-Line Joints tool box provides several tools for cleaning up such intersections, and for cutting holes in multi-lines.

Construct Closed Cross Joint

The Construct Closed Cross Joint tool cuts all lines that make up the first multi-line selected at the point where it crosses the second multi-line, as shown in Figure 8–45.

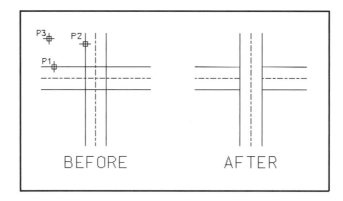

Figure 8-45 Example of using a Construct Closed Cross Joint tool

Invoke the Construct Closed Cross Joint tool from:

Multi-line Joints tool box	Select the Construct Closed Cross Joint tool (see Figure 8–46).
Key-in window	**join cross closed** (or **jo cr c**) ENTER

Figure 8-46 Invoking the Construct Closed Cross Joint tool from the Multi-line Joints tool box

MicroStation prompts:

> Construct Closed Cross Joint > Identify element *(Identify the multi-line P1 as shown in Figure 8–45.)*
>
> Construct Closed Cross Joint > Identify element *(Identify the multi-line P2 as shown in Figure 8–45.)*
>
> Construct Closed Cross Joint > Identify element *(Click the Data button anywhere in the view to initiate cleaning up of the intersection, or click the Reject button to reject the change.)*

 Note: For each Multi-Line Joint tool, dynamic update shows the intersection cleaned up after the second data point, but it does not become permanent until you provide the third data point.

Construct Open Cross Joint

The Construct Open Cross Joint tool cuts all lines that make up the first multi-line you select and cuts only the outside line of the second multi-line, as shown in Figure 8–47.

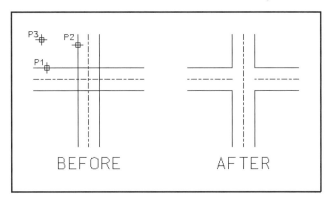

Figure 8–47 Example of using a Construct Open Cross Joint tool

Invoke the Construct Open Cross Joint tool from:

Multi-line Joints tool box	Select the Construct Open Cross Joint tool (see Figure 8–48).
Key-in window	**join cross open** (or **jo cr o**) ENTER

Figure 8–48 Invoking the Construct Open Cross Joint tool from the Multi-line Joints tool box

MicroStation prompts:

> Construct Open Cross Joint > Identify element *(Identify the multi-line P1 as shown in Figure 8–47.)*
>
> Construct Open Cross Joint > Identify element *(Identify the multi-line P2 as shown in Figure 8–47.)*
>
> Construct Open Cross Joint > Identify element *(Click the Data button anywhere in the view to initiate cleaning up of the intersection, or click the Reject button to reject the change.)*

Construct Merged Cross Joint

The Construct Merged Cross Joint tool cuts all lines that make up each of the intersecting multi-line you select, except the center lines, as shown in Figure 8–49. If there are no center lines, all lines in each multi-line are cut.

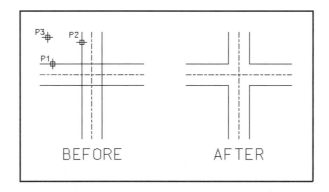

Figure 8-49 Example of using a Construct Merged Cross Joint tool

Invoke the Construct Merged Cross Joint tool from:

Multi-line Joints tool box	Select the Construct Merged Cross Joint tool (see Figure 8–50).
Key-in window	**join cross merge** (or **jo cr m**) ENTER

Figure 8-50 Invoking the Construct Merged Cross Joint tool from the Multi-line Joints tool box

MicroStation prompts:

Construct Merged Cross Joint > Identify element *(Identify the multi-line P1 as shown in Figure 8–49.)*

Construct Merged Cross Joint > Identify element *(Identify the multi-line P2 as shown in Figure 8–49.)*

Construct Merged Cross Joint > Identify element *(Click the Data button anywhere in the view to initiate cleaning up of the intersection, or click the Reject button to reject the change.)*

Construct Closed Tee Joint

The Construct Closed Tee Joint tool extends or shortens the first multi-line you identify to its intersection with the second multi-line. The first multi-line ends at the near side of the intersecting multi-line, which is left intact, as shown in Figure 8–51.

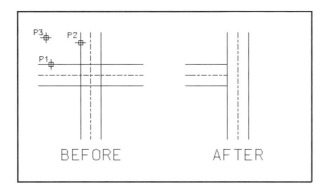

Figure 8–51 Example of using a Construct Closed Tee Joint tool

Invoke the Construct Closed Tee Joint tool from:

Multi-line Joints tool box	Select the Construct Closed Tee Joint tool (see Figure 8–52).
Key-in window	**join tee closed** (or **jo t c**) ENTER

Figure 8–52 Invoking the Construct Closed Tee Joint tool from the Multi-line Joints tool box

MicroStation prompts:

> Construct Closed Tee Joint > Identify element *(Identify the multi-line P1 as shown in Figure 8–51.)*
>
> Construct Closed Tee Joint > Identify element *(Identify the multi-line P2 as shown in Figure 8–51.)*
>
> Construct Closed Tee Joint > Identify element *(Click the Data button anywhere in the view to initiate cleaning up of the intersection, or click the Reject button to reject the change.)*

Construct Open Tee Joint

The Construct Open Tee Joint tool is similar to the Closed Tee Joint tool, except it leaves an open end at the intersecting multi-line, as shown in Figure 8–53.

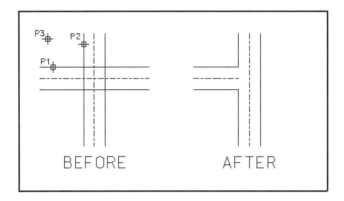

Figure 8–53 Example of using a Construct Open Tee Cross Joint tool

Invoke the Construct Open Tee Joint tool from:

Multi-line Joints tool box	Select the Construct Open Tee Joint tool (see Figure 8–54).
Key-in window	**join tee open** (or **jo t o**) ENTER

Figure 8–54 Invoking the Construct Open Tee Joint tool from the Multi-line Joints tool box

MicroStation prompts:

> Construct Open Tee Joint > Identify element *(Identify the multi-line P1 as shown in Figure 8–53.)*
>
> Construct Open Tee Joint > Identify element *(Identify the multi-line P2 as shown in Figure 8–53.)*
>
> Construct Open Tee Joint > Identify element *(Click the Data button anywhere in the view to initiate cleaning up of the intersection, or click the Reject button to reject the change.)*

Construct Merged Tee Joint

The Construct Merged Tee Joint tool is similar to the Open Tee Joint tool, except the center line of the first multi-line is extended to the center line of the intersecting multi-line, as shown in Figure 8–55.

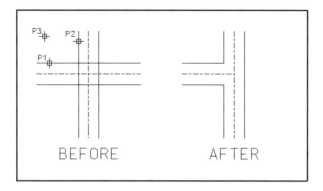

Figure 8–55 Example of using a Construct Merged Tee Joint tool

Invoke the Construct Merged Tee Joint tool from:

Multi-line Joints tool box	Select the Construct Merged Tee Joint tool (see Figure 8–56).
Key-in window	**join tee merge** (or **jo t m**) ENTER

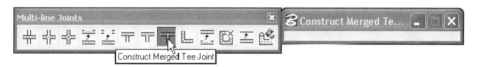

Figure 8–56 Invoking the Construct Merged Tee Joint tool from the Multi-line Joints tool box

MicroStation prompts:

> Construct Merged Tee Joint > Identify element *(Identify the multi-line P1 as shown in Figure 8–55.)*
>
> Construct Merged Tee Joint > Identify element *(Identify the multi-line P2 as shown in Figure 8–55.)*
>
> Construct Merged Tee Joint > Identify element *(Click the Data button anywhere in the view to initiate cleaning up of the intersection, or click the Reject button to reject the change.)*

Construct Corner Joint

The Construct Corner Joint tool lengthens or shortens each of the two multi-lines you select as necessary to create a clean intersection, as shown in Figure 8–57.

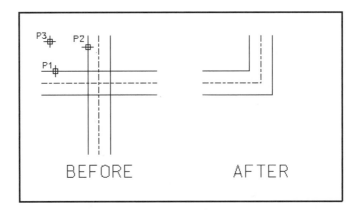

Figure 8–57 Example of using a Construct Corner Joint tool

Invoke the Construct Corner Joint tool from:

Multi-line Joints tool box	Select the Construct Corner Joint tool (see Figure 8–58).
Key-in window	**join corner** (or **jo c**) ENTER

Figure 8–58 Invoking the Construct Corner Joint tool from the Multi-line Joints tool box

MicroStation prompts:

Construct Joint > Identify element *(Identify the multi-line P1 as shown in Figure 8–57.)*

Construct Joint > Identify element *(Identify the multi-line P2 as shown in Figure 8–57.)*

Construct Joint > Identify element *(Click the Data button anywhere in the view to initiate cleaning up of the intersection, or click the Reject button to reject the change.)*

Cut Single Component Line

The Cut Single Component Line tool cuts a hole in the line you select in a multi-line from the first data point to the second data point, as shown in Figure 8–59.

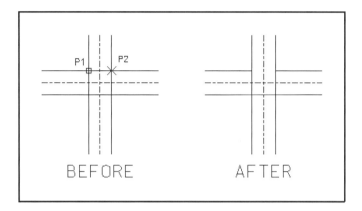

Figure 8–59 Example of using a Cut Single Component Line tool

Invoke the Cut Single Component Line tool from:

Multi-line Joints tool box	Select the Cut Single Component Line tool (see Figure 8–60.)
Key-in window	**cut single** (or **cu s**) ENTER

Figure 8–60 Invoking the Cut Single Component Line tool from the Multi-line Joints tool box

MicroStation prompts:

> Cut Single Component Line > Identify element *(Identify the multi-line at P1 as shown in Figure 8–59.)*
>
> Cut Single Component Line *(Identify the multi-line at P2 as shown in Figure 8–59 to remove the portion of the line, or click the Reset button to reject the change.)*

Cut All Component Lines

The Cut All Component Lines tool cuts a hole in the multi-line you select from the first data point to the second data point, as shown in Figure 8–61. The multi-line is still an element after this tool cuts a hole in it.

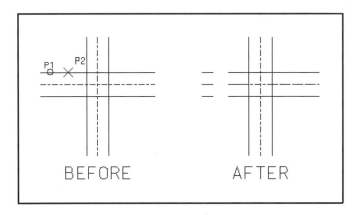

Figure 8–61 Example of using a Cut All Component Lines tool

Invoke the Cut All Component Lines tool from:

Multi-line Joints tool box	Select the Cut All Component Lines tool (see Figure 8–62.)
Key-in window	**cut all** (or **cu a**) ENTER

Figure 8–62 Invoking the Cut All Component Lines tool from the Multi-line Joints tool box

MicroStation prompts:

>Cut All Component Lines > Identify element *(Identify the multi-line at P1 as shown in Figure 8–61.)*
>
>Cut All Component Lines *(Identify the multi-line at P2 as shown in Figure 8–61 to remove the portion of the multi-line, or click the Reset button to reject the change.)*

Uncut Component Lines

The Uncut Component Lines tool provides a special undo tool for multi-lines. With it you can undo a cut in one line of a multi-line. Identify one end of the cut with a Keypoint snap, then click the Data button to accept the snap, and click the Data button a second time to initiate removing the cut, as shown in Figure 8–63.

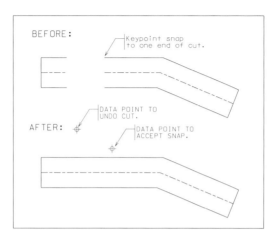

Figure 8-63 Example of using an Uncut Component Lines tool

Invoke the Uncut Component Lines tool from:

Multi-line Joints tool box	Select the Uncut Component Lines tool (see Figure 8–64).
Key-in window	**uncut** (or **un**) ENTER

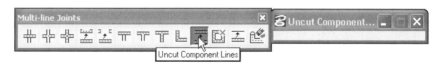

Figure 8-64 Invoking the Uncut Component Lines tool from the Multi-line Joints tool box

MicroStation prompts:

> Uncut Component Lines > Identify element *(Keypoint snap to one end of the cut, as shown in Figure 8–63.)*
>
> Uncut Component Lines *(Place a data point to accept the Keypoint snap, or click the Reset button to reject the change.)*
>
> Uncut Component Lines *(Place a second data point to undo the cut, or click the Reset button to reject the change.)*

Multi-line Partial Delete

The Multi-line Partial Delete tool enables you to reduce the length of a multi-line element by selecting one end of it, or break it into two separate elements by cutting a hole in it. This tool also allows you to control cap placement on each end of the hold, using one of the following four options in the **Cap Mode** option menu:

- **None** No caps are created, similar to the Cut All Component Lines tool.
- **Current** Uses the start cap and end cap definitions that were in effect when the multi-line was originally placed.
- **Active** Uses the active multi-line profile start cap and end cap definitions.
- **Joint** Place a 90-degree joint line on each end of the hole.

Invoke the Multi-line Partial Delete tool from:

Multi-line Joints tool box	Select the Multi-line Partial Delete tool, then select one of the four options from the Cap Mode option menu (see Figure 8–65).
Key-in window	**mline partial delete** (or **ml p d**) ENTER

Figure 8-65 Invoking the Multi-line Partial Delete tool from the Multi-line Joints tool box

MicroStation prompts:

Multi-line Partial Delete > Identify multi-line at start of delete *(Identify the multi-line at one end of the part to delete.)*

Multi-line Partial Delete > Define length of delete *(Place a data point to define the length of the delete, or click the Reset button to reject the change.)*

Move Multi-line Profile

The Move Multi-line Profile tool will move an individual component line of a multi-line, or reposition the working line of a multi-line without moving any component lines.

Move Component

The **Component Move** option allows you to move a selected line component in a multi-line element. The line can only be moved in parallel to the other lines in the multi-line element.

To move a line component, invoke the Move Multi-line Profile from:

Multi-line Joints tool box	Select the Move Multi-line Profile tool. Select Component from the Move option menu in the Tool settings Window (see Figure 8–66).
Key-in window	**mline edit profile** (or **ml e p**) ENTER

Figure 8-66 Invoking the Move Multi-line Profile tool from the Multi-line Joints tool box

MicroStation prompts:

> Move Multi-line Profile > Identify multi-line component to move *(Identify the component to move.)*
>
> Move Multi-line Profile > Define component position (reset to reject element) *(Place a Data point to reposition the component, or click the Reset button to reject the move.)*

Move Working Line

The **Workline Move** option allows you move the location of the insertion point (working line) in a multi-line element. This option has the effect of changing the zero Offset position. You usually will not see any change in the multi-line after using this tool.

The result of a working line move becomes apparent when you do element manipulations on the multi-line element. For example, if you use the Modify Element tool to move one end of the multi-line element, it will pivot about the new working line position.

To move the working line, invoke the Move Multi-line Profile from:

Multi-line Joints tool box	Select the Move Multi-line Profile tool. Select Workline from the Move option menu in the Tool settings Window (see Figure 8–66).
Key-in window	**mline edit profile** (or **ml e p**) ENTER

MicroStation prompts:

> Move Multi-line Profile > Identify multi-line profile *(Identify the multi-line.)*
>
> Move Multi-line Profile > Define new workline (reset to reject element) *(Place a Data point to reposition the working line, or click the Reset button to reject the move.)*

Edit Multi-line Cap

The Edit Multi-line Cap tool changes the end cap of a multi-line. MicroStation provides four options under the **Cap Mode** option menu:

- **None** Removes any end caps. The effect is the same as with the Cut All Component Lines tool.
- **Current** Does not change the end cap; enabled only when Adjust Angle is turned ON.
- **Active** Uses the active multi-line definitions for the end cap.
- **Joint** Uses the identified multi-line's joint definition instead of the end cap definition; ensures the end cap will always be 90 degrees.

Invoke the Edit Multi-line Cap tool from:

Multi-line Joints tool box	Select the Edit Multi-line Cap tool. Select one of the four options from the Cap Mode option menu in the Tool Settings window (see Figure 8–67).
Key-in window	**mline edit cap** (or **ml e c**) ENTER

Figure 8-67 Invoking the Edit Multi-line Cap tool from the Multi-line Joints tool box

MicroStation prompts:

Edit Multi-line Cap > Identify multi-line near the end cap to modify *(Identify the multi-line near the end cap.)*

Edit Multi-line Cap > Data to change end cap (reset to reject element) *(Place a data point to change the end cap, or click the Reject button to reject the change.)*

Chapter 8 Element Modification

 Open the Exercise Manual PDF file for Chapter 8 on the accompanying CD for project and discipline specific exercises.

REVIEW QUESTIONS

Write your answers in the spaces provided.

1. Explain briefly the options available with the Extend Line tool.
   _____
   _____

2. List the element types that can be modified by means of the Modify Element tool.
   _____
   _____

3. List the tools available to modify an arc.
   _____
   _____

4. Explain the difference between creating a chain manually and creating one automatically.
   _____
   _____

5. To drop a complex chain, invoke the _____ tool.

6. Explain the difference between the Construct Closed Cross Joint tool and the Construct Merged Cross Joint tool.
   _____
   _____

7. Explain the difference between the Construct Closed Tee Joint tool and the Construct Merged Tee Joint tool.
   _____
   _____

8. Give the steps involved in moving the Multi-line profile.
   _____
   _____

chapter 9

Measurement and Dimensioning

Objectives

After completing this chapter, you will be able to do the following:

- Use the measurement tools, such as Measure Distance, Measure Radius, Measure Angle, Measure Length, and Measure Area
- Use the dimensioning tools for linear, angular, and radial measurement
- Create and modify dimension styles

MEASUREMENT TOOLS

"Is that line really 12 feet long?" "What's the radius of that circle?" "What is the surface area of that foundation?" MicroStation can answer these questions with the measurement tools.

The measurement tools do nothing to your design. They just display distances, areas, and angles in the Status bar/Tool Settings window.

All measurement tools are available from the Measure tool box, shown in Figure 9–1.

Figure 9-1 Measure tool box

Measure Distance

MicroStation provides four distance measurement options. Distance options include measuring the distance between points you define, the distance along an element between points you define, the perpendicular distance from an element, and the minimum distance between two elements.

Measure Distance Between Points

This tool measures the cumulative straight-line distance from the first data point, through successive data points, to the last data point you define.

Invoke the Measure Distance Between Points tool from:

Measure tool box	Select the Measure Distance tool. In the Tool Settings window select Between Points from the Distance option menu (see Figure 9–2).
Key-in window	**measure distance points** (or **me dist po**) ENTER

Figure 9-2 Invoking the Measure Distance tool and selecting Between Points method

MicroStation prompts:

> Measure Distance Between Points > Enter start point *(Place a data point to start the measurement.)*
>
> Measure Distance Between Points > Define distance to measure *(Place a data point to define the distance to measure and continue placing data points to define additional measurement segments. Click the Reset button to terminate the tool sequence.)*

As you place each data point, the cumulative linear distance between points is displayed in the Status bar/Tool Settings window.

Chapter 9 Measurement and Dimensioning

Measure Distance Along Element

This tool measures the cumulative distance along an element, from the data point that selects it through successive data points on the element to the last data point you define.

Invoke the Measure Distance Along Element tool from:

Measure tool box	Select the Measure Distance tool. In the Tool Settings window select Along Element from the Distance option menu (see Figure 9–3).
Key-in window	**measure distance along** (or **me dist a**) ENTER

Figure 9-3 Invoking the Measure Distance tool and selecting Along Element method

MicroStation prompts:

> Measure Distance Along Element > Identify Element @ first point *(Place a data point on the element to start the measurement.)*
>
> Measure Distance Along Element > Enter end point *(Place a data point on the element to end the measurement.)*
>
> Measure Distance Along Element > Measure more points/Reset to select *(Continue placing data points on the element to obtain the cumulative measurement from the first data point through the succeeding points, or click the Reset button to terminate the measurement.)*

As you place each data point after the first one, the cumulative distance along the element is displayed in the Status bar/Tool Settings window.

Measure Distance Perpendicular From Element

This tool measures the perpendicular distance from a point to an element.

Invoke the Measure Distance Perpendicular From Element tool from:

Measure tool box	Select the Measure Distance tool. In the Tool Settings window select Perpendicular from the Distance option menu (see Figure 9–4).
Key-in window	**measure distance perpendicular** (or **me dist p**) ENTER

Figure 9-4 Invoking the Measure Distance tool and selecting Perpendicular From Element method

MicroStation prompts:

> Measure Distance Perpendicular From Element > Enter start point *(Identify the element from which to measure the perpendicular distance.)*
>
> Measure Distance Perpendicular From Element > Measure more points/Reset to select *(Place a data point to measure the perpendicular distance to the element.)*
>
> Measure Distance Perpendicular From Element > Measure more points/Reset to select *(Continue placing data points to obtain additional perpendicular measurements from the element, or click the Reset button to terminate the measurement.)*

A line from the element to the cursor shows the location of the calculation. The line is a temporary image that disappears when you click the Reset button or select another tool.

Note: If you place the measurement end point beyond the end of a linear element (such as a line or a box), the perpendicular is calculated from an imaginary extension of the measured element.

Measure Minimum Distance Between Elements

This tool measures the minimum distance between two elements.

Invoke the Measure Minimum Distance Between Elements tool from:

Measure tool box	Select the Measure Distance tool. In the Tool Settings window select Minimum Between from the Distance option menu (see Figure 9–5).
Key-in window	**measure distance minimum** (or **me dist m**) ENTER

Chapter 9 Measurement and Dimensioning

Figure 9-5 Invoking the Measure Distance tool and selecting Minimum Between Elements method

MicroStation prompts:

> Measure Minimum Distance Between Elements > Identify first element *(Identify the first element.)*
>
> Measure Minimum Distance Between Elements > Accept, Identify 2nd element/Reject *(Identify the second element, or click the Reject button to start all over again.)*
>
> Measure Minimum Distance Between Elements > > Identify first element *(Identify another element to continue.)*

After you place the second data point, a line appears in the design to indicate where the minimum distance is, and the minimum distance is displayed in the Status bar/Tool Settings window. The line is a temporary image that disappears when you click the Reset button or select another tool.

Measure Radius

The Measure Radius tool displays the radius of arcs, circles, partial ellipses, and ellipses in the Status bar/Tool Settings window.

Invoke the Measure Radius tool from:

Measure tool box	Select the Measure Radius tool (see Figure 9–6).
Key-in window	**measure radius** (or **me r**) ENTER

Figure 9-6 Invoking the Measure Radius tool from the Measure tool box

MicroStation prompts:

>Measure Radius > Identify element *(Identify the element to measure the radius.)*
>
>Measure Radius > Accept, Initiate Measurement *(Click the Data button to accept the element and initiate the measurement.)*

After the second data point is placed, the element's radius is displayed, in the current Working Units, in the Status bar/Tool Settings window. If the element you are measuring is an ellipse or a partial ellipse, the major axis and minor axis radii are displayed.

Measure Angle

The Measure Angle Between Lines tool measures the minimum angle formed by two elements.

Invoke the Measure Angle Between Lines tool from:

Measure tool box	Select the Measure Angle tool (see Figure 9–7).
Key-in window	**measure angle** (or **me a**) ENTER

Figure 9–7 Invoking the Measure Angle tool from the Measure tool box

MicroStation prompts:

>Measure Angle Between Lines > Identify first element *(Identify the first element.)*
>
>Measure Angle Between Lines > Accept, Identify 2nd element/Reject *(Identify the second element to measure the angle.)*
>
>Measure Angle Between Lines > Accept, Initiate Measurement *(Click the Data button to accept the element and initiate the measurement.)*

After you accept the elements, the angle between the two elements is displayed in the Status bar/Tool Settings window.

Measure Length

The Measure Length tool measures the total length of an open element or the length of the perimeter of a closed element.

Chapter 9 Measurement and Dimensioning

When you invoke the Measure Length tool, a Tolerance (%) field appears in the Tool Settings window. Tolerance sets the maximum allowable percentage of the distance between the true curve and the approximation for measurement purposes. A low value produces a very accurate measurement but may take a long time to calculate. The default value is sufficient in most cases.

Invoke the Measure Length tool from:

Measure tool box	Select the Measure Length tool (see Figure 9–8).
Key-in window	**measure length** (or **me l**) ENTER

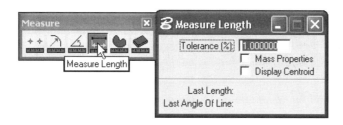

Figure 9–8 Invoking the Measure Length tool from the Measure tool box

MicroStation prompts:

> Measure Length > Identify element *(Identify the element to measure.)*

Once you identify the element, the total length of the element, or element perimeter, is displayed in the Status bar/Tool Settings window.

You can use this tool to measure the cumulative length of several elements by first employing the Element Selection tool to select all the elements you want to include in the measurement. After selecting the elements, select Measure Length from the Measure tool box, and the cumulative length of all selected elements appears in the Status bar/Tool Settings window.

Measure Area

MicroStation provides seven different ways to measure areas. Area options include measuring the area of a closed element; a fence; the intersection, union, or difference of two overlapping closed elements; a group of intersecting elements; or a group of points.

When you select the Measure Area tool, a Tolerance (%) field appears in the Tool Settings window. Tolerance sets the maximum allowable percentage of the distance between the true curve and the approximation for area calculation purposes. A low value produces a very accurate area but may take a long time to calculate. The default value is sufficient in most cases.

387

Measure Area of an Element

The Measure Area tool measures the area of a closed element, such as a circle, ellipse, shape, or block.

Invoke the Measure Area tool from:

Measure tool box	Select the Measure Area tool. In the Tool Settings window, select Element from the Method option menu (see Figure 9–9).
Key-in window	**measure area element** (or **me ar e**) ENTER

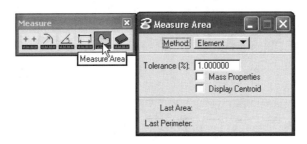

Figure 9-9 Invoking the Measure Area tool and selecting Element method

MicroStation prompts:

Measure Area > Identify element *(Identify the closed element to measure the area.)*

Once you identify the element, the element's area and perimeter length are displayed in the Status bar/Tool Settings window.

You can use this tool to measure the cumulative area of several closed elements by first employing the Element Selection tool to select all the elements you want to include in the area measurement. After selecting the elements, select Measure Area from the Measure tool box, and the cumulative area of all selected elements then appears in the Status bar.

Measure Area of a Fence

This tool measures the area enclosed by a fence.

Place a fence and then invoke the Measure Fence Area tool from:

Measure tool box	Select the Measure Area tool. In the Tool Settings window, select Fence from the Method option menu (see Figure 9–10).
Key-in window	**measure area fence** (or **me ar f**) ENTER

Chapter 9 Measurement and Dimensioning

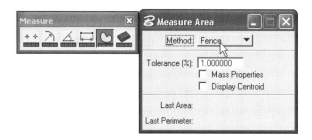

Figure 9-10 Invoking the Measure Area tool and selecting Fence method

MicroStation prompts:

> Measure Fence Area > Accept/Reject Fence Contents *(Click the Data button to accept the fence contents to measure the area, or click the Reject button to disregard the measurement.)*

MicroStation displays the area of the fence in the Status bar/Tool Settings window.

Measure Area Intersection, Union, or Difference

These options measure areas formed by intersecting closed elements. The intersection option allows you to determine the area that is common to two closed elements. The union option allows you to determine the area in such a way that there is no duplication between two closed elements. The difference option allows you to determine the area that is formed from a closed element after removing from it any area that it has in common with the other selected element.

Invoke the Measure Element Union (or Difference, or Intersection) Area tool from:

Measure tool box	Select the Measure Area tool. In the Tool Settings window select Union, Difference, or Intersection from the Method option menu (see Figure 9–11).
Key-in window	**measure area union | difference | intersection** (or **me ar u | d | i**) ENTER

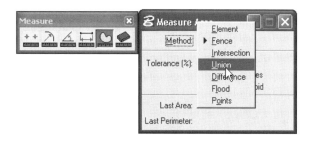

Figure 9-11 Invoking the Measure Area and selecting Union method

389

MicroStation prompts:

> Measure Element Union Area > Identify the element *(Identify the first element.)*
>
> Measure Element Union Area > Accept/Reject (Select next input) *(Identify the second element.)*
>
> Measure Element Union Area > Accept/Reject (Select next input) *(Click the Data button to accept the element, or identify another element.)*
>
> Measure Element Union Area > Identify Additional/Reset to terminate *(Identify additional elements, or click the Reset button to terminate and initiate the measurement.)*

After you select all the elements and place the last data point in space, MicroStation displays an image of only the part of the elements that is included in the type of area you select. Click the Reset Button to cause the elements to reappear, and MicroStation displays the area and perimeter length in the Status bar/Tool Settings window.

Measure Area Flood

This measures the area enclosed by a group of elements. Invoke the Measure Area Enclosing Point tool from:

Measure tool box	Select the Measure Area tool. In the Tool Settings window select Flood from the Method option menu (see Figure 9–12).
Key-in window	**measure area flood** (or **me ar fl**) ENTER

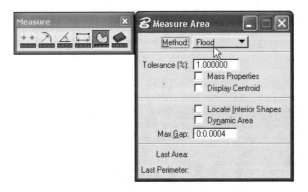

Figure 9-12 Invoking the Measure Area and selecting Flood method

MicroStation prompts:

> Measure Area Enclosing Point > Enter data point inside area *(Click the Data button inside the area enclosed by the elements.)*
>
> Measure Area Enclosing Point > Accept, Initiate Measurement *(Click the Data button to accept and initiate measurement.)*

Chapter 9 Measurement and Dimensioning

After you click the Data button, a small spinner will appear in the Status bar. The spinner spins to indicate that MicroStation is determining the area enclosed by the elements. As MicroStation traces the area, it highlights the elements. When the area has been determined, the spinner stops spinning and the area and perimeter length appear in the Status bar/Tool Settings window.

 Note: The **Max Gap** setting controls the space that is allowed between the elements that define the flood area. If the gap is zero, the elements must be touching.

Measure Area Points

This tool measures the area formed by a set of data points you enter. It assumes the perimeter of the area is formed by straight lines between the data points. An image of the area is displayed as you enter the data pointsInvoke the Measure Area Defined By Points tool from:

Measure tool box	Select the Measure Area tool. In the Tool Settings window select Points from the Method option menu (see Figure 9–13).
Key-in window	**measure area points** (or **me ar p**) ENTER

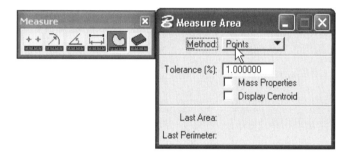

Figure 9-13 Invoking the Measure Area and selecting Points method

MicroStation prompts:

> Measure Area Defined By Points > Enter shape vertex *(Place data points to define the vertices of the area to be measured. When the area is completely defined, click the Reset button to initiate area measurement.)*

As you are entering data points, a closed dynamic image of the area appears on the screen. When you press the Reset button, the area and the perimeter length appear in the Status bar/Tool Settings window.

391

DIMENSIONING

MicroStation's dimensioning features provide an excellent way to add dimensional information to your design, such as lengths, widths, angles, tolerances, and clearances.

Dimensioning of any drawing is generally one of the last steps in manual drawing; however, it does not need to be the last step in your MicroStation drawing. If you place the dimensions and find out later they must be changed because the size of the objects they are related to have changed, MicroStation allows you to stretch or extend the objects and have the dimensions change automatically to the new size. MicroStation provides three basic types of dimensions: linear, angular, and radial dimensioning. Figure 9–14 shows examples of these three basic types of dimensions.

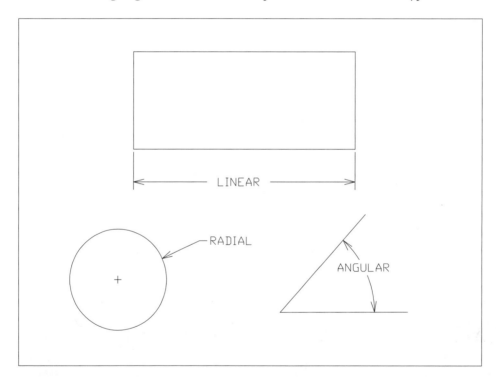

Figure 9-14 Examples of the three basic types of dimensions

All the available dimensioning tools in MicroStation are found in four tool boxes: Linear Dimensions, Angular Dimensions, Radial Dimensions and Misc Dimensions as shown in Figure 9–15.

Chapter 9   Measurement and Dimensioning

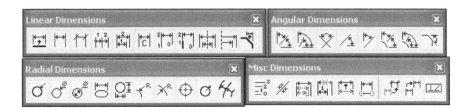

Figure 9-15 Dimension tool boxes

Dimensioning Terminology

The following terms occur commonly in the MicroStation dimensioning procedures.

Dimension Line

This is a line with markers at each end (arrows, dots, tick marks, etc.). The dimensioning text is located along this line; you may place it above the line or in a break in the dimension line. Usually, the dimension line is inside the measured area. If there is insufficient space, MicroStation places the dimensions and draws two short lines outside the measured area with arrows pointing inward.

Extension Lines

The extension lines (also called *witness lines*) are the lines that extend from the object to the dimension line. Extension lines normally are drawn perpendicular to the dimension line. (Several options that are associated with this element will be reviewed later in this chapter.) Also, you can suppress one or both of the extension lines.

Terminators

Terminators are placed at one or both ends of the dimension line, depending on the type of dimension line placed. MicroStation allows you to use arrows, tick marks, or arbitrary symbols of your own choosing for the terminators. You can also adjust the size of the terminator.

Dimension Text

This is a text string that usually indicates the actual measurement. You can accept the default measurement computed automatically by MicroStation, or change it by supplying your own text.

Figure 9–16 shows the different components of a typical dimension.

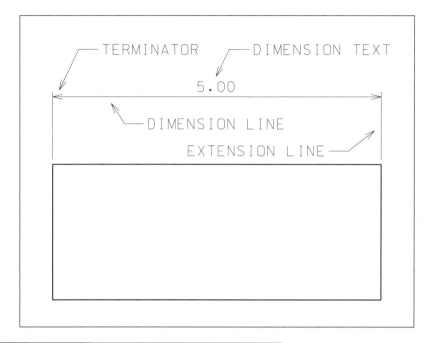

Figure 9-16 Different components of a typical dimension

Leader

The leader line is a line from text to an object on the design, as shown in Figure 9–17. For some dimensioning, the text may not fit next to the object it describes; hence, it is customary to place the text nearby and draw a leader from it to the object.

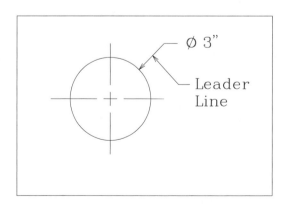

Figure 9-17 Example of placing a leader line

Associative Dimensioning

Dimensions can be placed as either associative dimensions or normal (non-associative) dimensions. Associative dimensioning links dimension elements to the objects being dimensioned. An association point does not have its own coordinates, but is positioned by the coordinates of the point with which it is associated. When you change the size, shape, or position of an element, MicroStation modifies the associated dimension automatically to reflect the change. It also draws the dimension entity at its new location, size, and rotation.

To place associative dimensions, turn ON the **Association** check box in the Tool Settings window when you invoke one of the dimensioning tools, as shown in Figure 9–18. If you place a dimension when the **Association** check box is OFF, the dimension will not associate with the object dimensioned, and if the object is modified, the dimension is not changed.

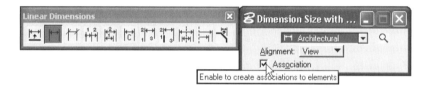

Figure 9-18 Status of the check box for the Association in the Tool Settings window

Placing associative dimensions can significantly reduce the size of a design file that has many dimensions, since a dimension element is usually smaller than its corresponding individual elements.

Alignment Controls

The alignment controls the orientation of linear dimensions. **View**, **Drawing**, **True**, and **Arbitrary** are the available options. The options are selected from the **Alignment** option menu in the Tool Settings window when you invoke one of the linear dimensioning tools.

- The **View** option aligns linear dimensions parallel to the view X or Y axis. This is useful when dimensioning *three dimensional* reference files with dimensions parallel to the viewing plane.

- The **Drawing** option aligns linear dimensions parallel to the design plane X or Y axis.

- The **True** option aligns linear dimensions parallel to the element being dimensioned. The extension lines are constrained to be at right angles to the dimension line.

- The **Arbitrary** option places linear dimensions parallel to the element being dimensioned. The extension lines are not constrained to be at right angles to the dimension line. This is useful when dimensioning elements in *2D* isometric drawings. The **Iso Lock** check box must be set to ON.

Figure 9–19 shows examples of placing linear dimensions with different alignment controls.

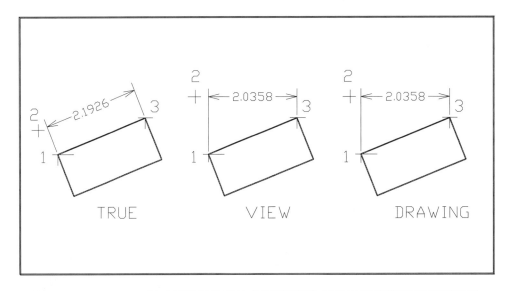

Figure 9-19 Examples of placing linear dimensions with different alignment controls

Dimension Styles

A dimension style is a saved set of dimensioning settings. You can define dimensions styles and apply them to dimension elements during placement. MicroStation allows you to set the active dimension style from the drop-down list box located in the Tool Settings window when you invoke one of the Dimension commands. If the active dimension style is set to (none), the active dimensioning settings are applied. If the dimension style is changed, any dimension with that style also changes to those settings. Dimension styles can be created, customized and saved for easy recall.

Refer to the Dimension Settings section for detailed explanations for defining and modifying dimension styles.

Linear Dimensioning

Linear dimensioning tools allow you to dimension such linear elements as lines and line strings. Following are the tools available to dimension the linear elements.

- Dimension Size with Arrows—This tool allows you to dimension the linear distance between two points (length).

- Dimension Size with Stroke—This is similar to the Dimension Size with Arrow tool, except the terminators are set to strokes instead of arrows.

- Dimension Location—With this tool you can dimension linear distances from an origin (datum), with the dimensions placed in line (chained).

- Dimension Location (Stacked)—This tool also allows you to dimension linear distances from an origin (datum), but with the dimensions stacked.

- Dimension Size Perpendicular to Points—With this tool you can dimension the linear distance between two points. The first two data points entered define the dimension's *Y* axis.

- Dimension Size Perpendicular to Line—This tool dimensions the linear distance perpendicular from an element to another element or point. The dimension's *Y* axis is defined by the element identified.

- Dimension Symmetric—This tool is used to create symmetric dimensions by defining a center point (non-associative).

- Dimension Half—This tool is used to create one-sided dimension by defining a center.

- Dimension Chamfer—This tool is used to place a dimension on a chamfer.

Dimension Size with Arrows

To place linear dimension with arrows, invoke the Dimension Size with Arrows tool from:

Linear Dimensions tool box	Select the Dimension Size with Arrows tool. If necessary, set the Association check box to ON and select an Alignment and Dimension Style in the Tool Settings window (see Figure 9–20).
Key-in window	**dimension size arrow** (or **dim si a**) ENTER

Figure 9–20 Invoking the Dimension Size with Arrow tool from the Linear Dimensions tool box

MicroStation prompts:

Dimension Size with Arrow > Select start of dimension *(Place a data point as shown in Figure 9–21 (point 1) to define the starting point of the dimension.)*

Dimension Size with Arrow > Define length of extension line *(Place a data point as shown in Figure 9–21 (point 2) to define the length of the extension line.)*

Dimension Size with Arrow > Select dimension endpoint *(Place a data point as shown in Figure 9–21 (point 3) to define the end point of the dimension.)*

Dimension Size with Arrow > Select dimension endpoint (*Continue placing data points as shown in Figure 9–21 (points 4 and 5) to continue linear dimensioning in the same direction, or click the Reset button to change the direction, or click the Reset button twice to start all over again.*)

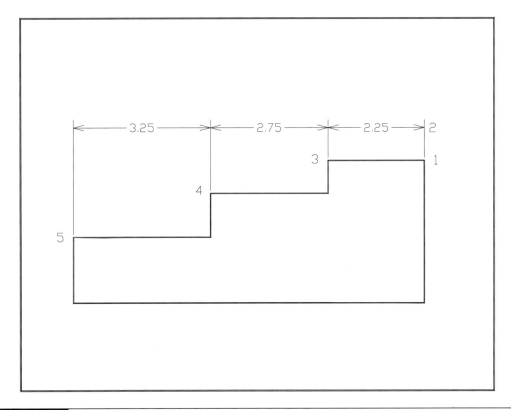

Figure 9-21 Example of placing linear dimensioning with the Dimension Size with Arrows tool

 Note: Use the appropriate Snap Lock when placing the data points for the starting point and end point of the dimension line.

After you place dimensions, the dimension text can be edited using the Edit Text tool in the Text tool box. To edit dimension text, select the Edit Text tool, then select and accept the dimension text to be edited. MicroStation displays the Dimension Text dialog box similar to the one shown in Figure 9–22.

Chapter 9 Measurement and Dimensioning

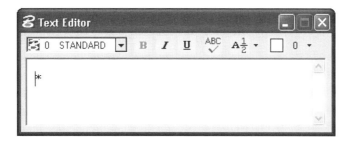

Figure 9-22 Text dialog box

The asterisk (*) in the Primary Text edit field indicates the current default dimension text string.

- If you need to change the default dimension string, delete the asterisk (*) and key-in the new dimension text string.

- If you need to add any prefix and/or suffix text string to the default dimension text string, keep the asterisk (*) and type in the appropriate text string in the Text edit field.

Click the data button to accept the changes or click the Reset button to disregard the changes.

Dimension Size with Stroke

To place linear dimensions with strokes, invoke the Dimension Size Stroke tool from:

Linear Dimensions tool box	Select the Dimension Size Stroke tool. If necessary, set the Association check box to ON and select an Alignment and Dimension Style in the Tool Settings window (see Figure 9–23).
Key-in window	**dimension size stroke** (or **dim si s**) ENTER

Figure 9-23 Invoking the Dimension Size Stroke tool from the Linear Dimensions tool box

399

MicroStation prompts are similar to those for the Dimension Size with Arrows tool. Instead of placing arrows at the end of the dimension line, MicroStation places strokes, as shown in Figure 9–24.

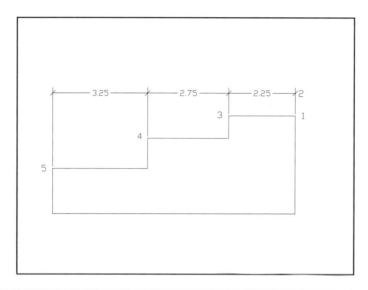

Figure 9-24 Example of placing linear dimensioning with the Dimension Size Stroke tool

Dimension Location

The Dimension Location tool enables you to dimension linear distances from an origin (datum), as shown in Figure 9–25. The dimensions are placed in a line (chain). All dimensions are measured on an element originating from a common surface, centerline, or center plane. The Dimension Location tool is commonly used in mechanical drafting.

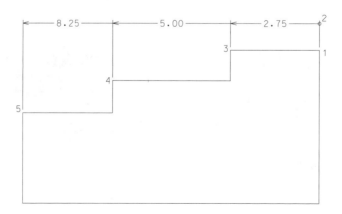

Figure 9-25 Example of placing linear dimensioning from an origin (datum) with the Dimension Location tool

Chapter 9 Measurement and Dimensioning

To place linear dimensions in a chain, invoke the Dimension Location tool from:

Linear Dimensions tool box	Select the Dimension Location tool. If necessary, set the Association check box to ON and select an Alignment and Dimension Style in the Tool Settings window (see Figure 9–26).
Key-in window	**dimension location single** (or **dim lo s**) ENTER

Figure 9–26 Invoking the Dimension Location tool from the Linear Dimensions tool box

MicroStation prompts:

> Dimension Location > Select start of dimension *(Place a data point to define the origin.)*
>
> Dimension Location > Define length of extension line *(Place a data point to define the length of the extension line.)*
>
> Dimension Location > Select dimension endpoint *(Place a data point to define the end point of the dimension. If necessary, press* ENTER *to edit the dimension text.)*
>
> Dimension Location > Select dimension endpoint *(Continue placing data points to continue linear dimensioning in the same direction, or click the Reset button to change the direction, or click the Reset button twice to start all over again.)*

 Note: Use the appropriate Snap Lock when placing the data points for the starting point and end point of the dimension line.

Dimension Location (Stacked)

With the Dimension Location (Stacked) tool you can dimension linear distances from an origin (datum), as shown in Figure 9–27. The dimensions are stacked. All dimensions are measured on an element originating from a common surface, centerline, or center plane. The Dimension Location (Stacked) tool is commonly used in mechanical drafting because all dimensions are independent, even though they are taken from a common datum. If necessary, you can change the stack offset distance; see the later section on Dimension Settings.

401

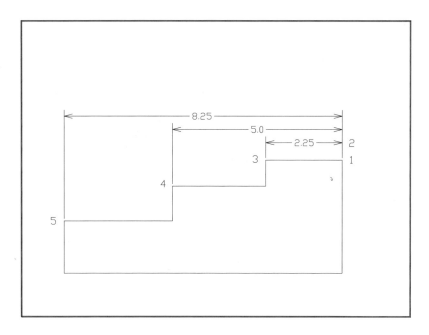

Figure 9-27 Example of placing linear dimensioning (stacked) from an origin (datum) via the Dimension Location (Stacked) tool

To place linear dimensions in a stack, invoke the Dimension Location (Stacked) tool from:

Linear Dimensions tool box	Select the Dimension Location (Stacked) tool. If necessary, set the Association check box to ON and select an Alignment and Dimension Style in the Tool Settings window (see Figure 9–28).
Key-in window	**dimension location stacked** (or **dim lo st**) ENTER

Figure 9-28 Invoking the Dimension Location (Stacked) tool from the Linear Dimensions tool box

MicroStation prompts:

> Dimension Location (Stacked) > Select start of dimension *(Place a data point to define the origin.)*
>
> Dimension Location (Stacked) > Define length of extension line *(Place a data point to define the length of the extension line.)*
>
> Dimension Location (Stacked) > Select dimension endpoint *(Place a data point to define the end point of the dimension line. If necessary, press* ENTER *to edit the dimension text.)*
>
> Dimension Location (Stacked) > Select dimension endpoint *(Continue placing data points to continue linear dimensioning in the same direction, or click the Reset button to change the direction, or click the Reset button twice to start all over again.)*

 Note: Use the appropriate Snap Lock when placing the data points for the starting point and end point of the dimension line.

Custom Linear Dimension

The Custom Linear Dimension tool is used to dimension the linear distance between two points (length) when you desire to use a third type of dimension line terminator and do not want to change the terminator assignments for the Dimension Size with Arrow tool or Dimension Size with Stroke tool. To place Custom linear dimension, invoke the Custom Linear Dimension tool from:

Linear Dimensions tool box	Select the Custom Linear Dimension tool. If necessary, set the Association check box to ON and select an Alignment and Dimension Style in the Tool Setting window (see Figure 9–29).
Key-in window	**dimension linear** (or **dim li**) ENTER

Figure 9-29 Invoking the Custom Linear Dimension tool from the Linear Dimensions tool box

MicroStation prompts are similar to those for the Dimension Size with Arrows tool.

403

Dimension Size Perpendicular to Points

The Dimension Size Perpendicular to Points tool can dimension the linear distance between two points. The first two data points entered define the dimension's *Y* axis, as shown in Figure 9–30.

Figure 9-30 Example of placing linear dimensioning with the Dimension Size Perpendicular to Points tool

To place linear dimensions between two points, invoke the Dimension Size Perpendicular to Points tool from:

Linear Dimensions tool box	Select the Dimension Size Perpendicular to Points tool. If necessary, set the Association check box to ON and select a Dimension Style in the Tool Settings window (see Figure 9–31).
Key-in window	**dimension size perpendicular to points** (or **dim si p p**) ENTER

Figure 9-31 Invoking the Dimension Size Perpendicular to Points tool from the Linear Dimensions tool box

MicroStation prompts:

Dimension Size Perpendicular to Points > Select base of first dimension line *(Place a data point to define the base point of the first dimension line.)*

Dimension Size Perpendicular to Points > Select end of extension line *(Place a data point to define the length of the extension line.)*

Dimension Size Perpendicular to Points > Select dimension endpoint *(Place a data point to define the end point of the dimension line.)*

 Note: Use the appropriate Snap Lock when placing the data points for the starting point and end point of the dimension line.

Dimension Size Perpendicular to Line

The Dimension Size Perpendicular to Line tool dimensions the linear distance perpendicular from an element to another element. The dimension's *Y* axis is defined by the element identified.

To place a linear dimension perpendicular from one element to another, invoke the Dimension Size Perpendicular to Line tool from:

Linear Dimensions tool box	Select the Dimension Size Perpendicular to Line tool. If necessary, set the Association check box to ON and select a Dimension Style in the Tool Settings window (see Figure 9–32).
Key-in window	**dimension size perpendicular to line** (or **dim si p l**) ENTER

Figure 9–32 Invoking the Dimension Size Perpendicular to Line tool from the Linear Dimensions tool box

MicroStation prompts:

Dimension Size Perpendicular to Line > Select base of first dimension line *(Place a data point to define the base point of the first dimension line.)*

Dimension Size Perpendicular to Line > Select end of extension line *(Place a data point to define the length of the extension line.)*

Dimension Size Perpendicular to Line > Select dimension endpoint *(Place a data point to define the end point of the dimension line.)*

 Note: Use the appropriate Snap Lock when placing the data points for the starting point and end point of the dimension line.

Dimension Symmetric

The Dimension Symmetric tool is used to create symmetric dimensions by defining a center point (non-associative). To dimension symmetrically, invoke the Dimension Symmetric tool from:

Linear Dimensions tool box	Select the Dimension Symmetric tool. Select an Alignment and Dimension Style in the Tool Settings window (see Figure 9–33).
Key-in window	**dimension symmetric** (or **dim sy**) ENTER

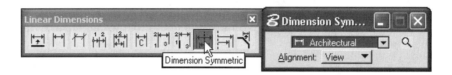

Figure 9–33 Invoking the Dimension Symmetric tool from the Linear Dimensions tool box

MicroStation prompts:

> Dimension Symmetric > Select point on center line *(Place a data point to define the center of the dimension.)*
>
> Dimension Symmetric > Select point to define dimension location *(Place a data point to define the length of the extension line.)*
>
> Dimension Symmetric > Select dimension endpoint *(Place a data point to define the end point of the dimension.)*

 Note: Use the appropriate Snap Lock when placing the data points for the starting point and end point of the dimension line.

Dimension Half

The Dimension Half tool is used to create one-sided dimensions by defining a center. To dimension one-side of an element, invoke the Dimension Half tool from:

Linear Dimensions tool box	Select the Dimension Half tool. Select an Alignment and Dimension Style in the Tool Settings window (see Figure 9–34).
Key-in window	**dimension symmetric** (or **dim h**) ENTER

406

Chapter 9 Measurement and Dimensioning

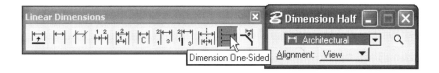

Figure 9-34 Invoking the Dimension Half tool from the Linear Dimensions tool box

MicroStation prompts:

> Dimension Half > Select point on center line *(Place a data point to define the center of the dimension.)*
>
> Dimension Half > Select point to define dimension location *(Place a data point to define the length of the extension line.)*
>
> Dimension Half > Select dimension endpoint *(Place a data point to define the end point of the dimension.)*

 Note: Use the appropriate Snap Lock when placing the data points for the starting point and end point of the dimension line.

Dimension Chamfer

The Dimension Chamfer tool is used to place a dimension on chamfer. To dimension a chamfer, invoke the Dimension Chamfer tool from:

Linear Dimensions tool box	Select the Dimension Chamfer tool. Select an Alignment and Dimension Style in the Tool Settings window (see Figure 9–35).
Key-in window	**dimension chamfer** (or **dim ch**) ENTER

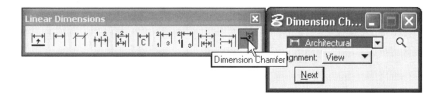

Figure 9-35 Invoking the Dimension Chamfer tool from the Linear Dimensions tool box

407

MicroStation prompts:

> Dimension Chamfer Angle> Select chamfer element *(Identify chamfer element.)*
>
> Dimension Chamfer Angle> Select second element to measure chamfer from *(Identify a second element from which to measure the chamfer.)*
>
> Dimension Chamfer Angle > Define Length of extension line *(Place a data point to define the length of extension line or click **Next** button in the Tool Settings window to cycle through the available tools to dimension the chamfer.)*

Dimension Element

The Dimension Element tool provides a fast way to dimension an element. Simply identify the element and MicroStation selects the type of dimensioning tool it thinks is best. For linear elements it selects one of the linear dimensioning tools; for circles, arcs, and ellipses it selects one of the radial dimensioning tools.

When you invoke the Dimension Element tool, the tool name Dimension Element appears in the Status bar. As soon as you select an element, the name of the dimensioning tool MicroStation intends to use appears in the Status bar. If you don't want that selected dimension tool, press ENTER and MicroStation switches to another dimensioning tool. Keep pressing ENTER until you find the tool you want to use. You can also press the **Next** button in the Tool Settings window to change dimensioning tools.

Invoke the Dimension Element tool from:

Linear Dimensions tool box	Select the Dimension Element tool. If necessary, set the Association check box to ON, and select an Alignment option and Dimension Style (see Figure 9–36).
Key-in window	**dimension element** (or **dim e**) ENTER

Figure 9-36 Invoking the Dimension Element tool from the Linear Dimensions tool box

MicroStation prompts:

> Dimension Element > Select element to dimension *(Identify the element or segment of the block or line string element.)*

Dimension Element > Accept (Press Return to switch tool) *(Place a data point to indicate where you want the dimension placed, or press* ENTER *as many times as necessary to find the dimensioning tool you want to use and place a data point for the location of the dimension element.)*

For most dimension tools, the second data point indicates the length and direction of the extension line. See Figure 9–37 for examples of placing dimensions with the Dimension Element tool.

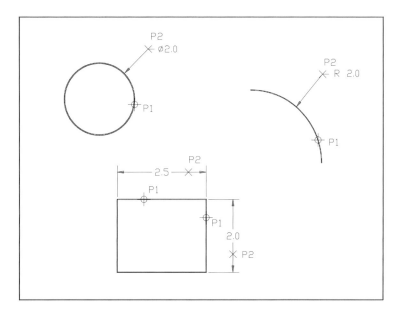

Figure 9–37 Examples of placing dimensioning with the Dimension Element tool

Angular Dimensioning

The angular dimensioning tools create dimensions for the angle between two non-parallel lines, using the conventions that conform to the current dimension variable settings. "Angle" is defined as "a measure of an angle or of the amount of turning necessary to bring one line or plane into coincidence with or parallel to another." Following are the eight different tools available to create angular dimensions.

- Dimension Angle Size—Each dimension (except the first) is computed from the end point of the previous dimension.

- Dimension Angle Location—Each dimension is computed from the dimension origin (datum).

- Dimension Angle Between Lines—To dimension the angle between two lines, two segments of a line string, or two sides of a shape.

- Dimension Angle from *X* axis—To dimension the angle between a line, a side of a shape, or a segment of a line string and the view *X* axis.
- Dimension Angle from *Y* axis—To dimension the angle between a line, a side of a shape, or a segment of a line string and the view *Y* axis.
- Dimension Arc Size—To dimension an arc or circle. Each dimension is computed from the endpoint of the previous dimension, except the first one.
- Dimension Arc Location—To dimension an arc or circle. Each dimension is computed from the dimension origin (datum). The dimensions are stacked.
- Dimension Angle Chamfer—To dimension the angle of a chamfer.

Dimension Angle Size

To place angular dimensioning with the Dimension Angle Size tool, invoke the tool from:

Angular Dimensions tool box	Select the Dimension Angle Size tool. If necessary, set the Association check box to ON and and select a dimension style in the Tool Settings window (see Figure 9–38).
Key-in window	**dimension angle size** (or **dim a s**) ENTER

Figure 9-38 Invoking the Dimension Angle Size tool from the Angular Dimensions tool box

MicroStation prompts:

Dimension Angle Size > Select start of dimension *(Place a data point as shown in Figure 9–39 (point P1) to define the start of the dimension, which is measured counterclockwise from this point.)*

Dimension Angle Size > Define length of extension line *(Place a data point as shown in Figure 9–39 (point P2) to define the length of the extension line.)*

Dimension Angle Size > Enter point on axis *(Place a data point as shown in Figure 9–39 (point P3) to define the vertex of the angle.)*

Chapter 9 Measurement and Dimensioning

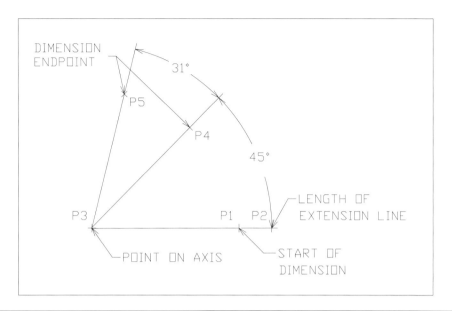

Figure 9-39 Example of placing angular dimensioning via the Dimension Angle Size tool

Dimension Angle Size > Select dimension endpoint *(Place a data point as shown in Figure 9–39 (point P4) to define the end point of the dimension line. If necessary, press* ENTER *to edit the dimension text.)*

Dimension Angle Size > Select dimension endpoint *(Continue placing data points for angular dimensioning, and click the Reset button to complete the tool sequence.)*

Dimension Angle Location

To place angular dimensioning with the Dimension Angle Location tool, invoke the tool from:

Angular Dimensions tool box	Select the Dimension Angle Location tool. If necessary, set the Association check box to ON and select a dimension style in the Tool Settings window (see Figure 9–40).
Key-in window	**dimension angle location** (or **dim a lo**) ENTER

Figure 9-40 Invoking the Dimension Angle Location tool from the Angular Dimensions tool box

MicroStation prompts:

> Dimension Angle Location > Select start of dimension *(Place a data point as shown in Figure 9–41 (point P1) to define the start of the dimension, which is measured counterclockwise from this point.)*
>
> Dimension Angle Location > Define length of extension line *(Place a data point as shown in Figure 9–41 (point P2) to define the length of the extension line.)*
>
> Dimension Angle Location > Enter point on axis *(Place a data point as shown in Figure 9–41 (point P3) to define the vertex of the angle.)*
>
> Dimension Angle Location > Select dimension endpoint *(Place a data point as shown in Figure 9–41 (point P4) to define the end point of the dimension. Press ENTER to edit the dimension text.)*
>
> Dimension Angle Location > Select dimension endpoint *(Continue placing data points for angular dimensioning, and click the Reset button to complete the tool sequence.)*

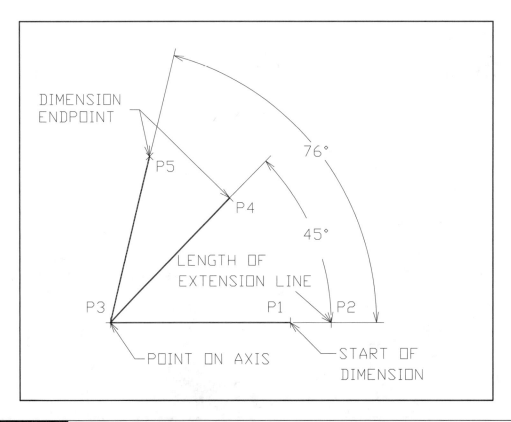

Figure 9-41 Example of placing angular dimensioning via the Dimension Angle Location tool

Chapter 9 Measurement and Dimensioning

Dimension Angle Between Lines

To place angular dimensioning with the Dimension Angle Between Lines tool, invoke the tool from:

Angular Dimensions tool box	Select the Dimension Angle Between Lines tool. If necessary, set the Association check box to ON and select a dimension style in the Tool Settings window (see Figure 9–42).
Key-in window	**dimension angle lines** (or **dim a l**) ENTER

Figure 9–42 Invoking the Dimension Angle Between Lines tool from the Angular Dimensions tool box

MicroStation prompts:

> Dimension Angle Between Lines > Select first line *(Identify the first line or segment, as shown in Figure 9–43 (point P1).)*
>
> Dimension Angle Between Lines > Select second line *(Identify the second line or segment, as shown in Figure 9–43 (point P2).)*
>
> Dimension Angle Between Lines *(Place a data point to define the location of the dimension line, as shown in Figure 9–43 (point P3).)*

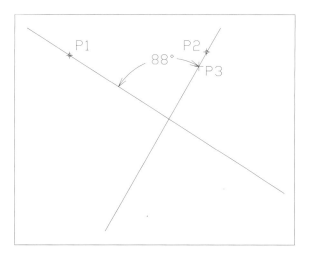

Figure 9–43 Example of placing angular dimensioning with the Dimension Angle Between Lines tool

413

Dimension Angle from X-Axis

To place angular dimensioning with the Dimension Angle from X-axis tool, invoke the tool from:

Angular Dimensions tool box	Select the Dimension Angle from X-axis tool. If necessary, set the Association check box to ON and select a dimension style in the Tool Settings window (see Figure 9–44).
Key-in window	**dimension angle x** (or **dim a x**) ENTER

Figure 9–44 Invoking the Dimension Angle from X-Axis tool from the Angular Dimensions tool box

MicroStation prompts:

> Dimension Angle from X Axis > Identify element *(Identify the element, as shown in Figure 9–45 (point P1).)*
>
> Dimension Angle from X Axis > Accept, define dimension axis *(Place a data point, as shown in Figure 9–45 (point P2), to specify the location and direction of the dimension.)*

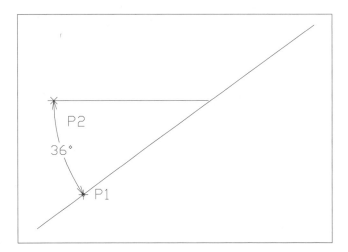

Figure 9–45 Example of placing angular dimensioning with the Dimension Angle from X Axis tool

Chapter 9 Measurement and Dimensioning

Dimension Angle from Y-Axis

To place angular dimensioning with the Dimension Angle from Y-axis tool, invoke the tool from:

Angular Dimensions tool box	Select the Dimension Angle from Y-axis tool. If necessary, set the Association check box to ON and select a dimension style in the Tool Settings window (see Figure 9–46).
Key-in window	**dimension angle y** (or **dim a y**) ENTER

Figure 9–46 Invoking the Dimension Angle from Y-Axis tool from the Angular Dimensions tool box

MicroStation prompts:

> Dimension Angle from Y Axis > Identify element *(Identify the element.)*
>
> Dimension Angle from Y Axis > Accept, define dimension axis *(Place a data point to specify the location and direction of the dimension.)*

Dimension Arc Size

The Dimension Arc Size tool lets you dimension a circle or circular arc. Each dimension is computed from the end point of the previous dimension, except the first one, similar to the Dimension Angle Size tool.

To dimension arcs with the Dimension Arc Size tool, invoke the tool from:

Angular Dimensions tool box	Select the Dimension Arc Size tool. If necessary, set the Association check box to ON and select a dimension style in the Tool Settings window (see Figure 9–47)
Key-in window	**dimension arc size** (or **dim ar s**) ENTER

Figure 9–47 Invoking the Dimension Arc Size tool from the Angular Dimensions tool box

MicroStation prompts:

Dimension Arc Size > Select start of dimension *(Place a data point, as shown in Figure 9–48 (point P1), to define the origin point. The dimension is measured counterclockwise from this point. This point must select an arc, circle, or ellipse.)*

Dimension Arc Size > Define length of extension line *(Place a data point, as shown in Figure 9–48 (point P2), to define the length of the extension line.)*

Dimension Arc Size > Select dimension endpoint *(Place a data point, as shown in Figure 9–48 (point P3), to define the dimension end point. If necessary, press* ENTER *to edit the dimension text.)*

Dimension Arc Size > Select dimension endpoint *(Continue placing data points to continue angular dimensioning, and click the Reset button to terminate the tool sequence.)*

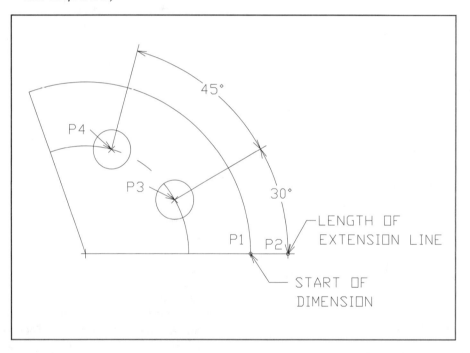

Figure 9-48 Example of placing arc dimensioning with the Dimension Arc Size tool

Dimension Arc Location

The Dimension Arc Location tool enables you to dimension a circle or circular arc. Each dimension is computed from the dimension origin (datum), as shown in Figure 9–49.

Chapter 9 Measurement and Dimensioning

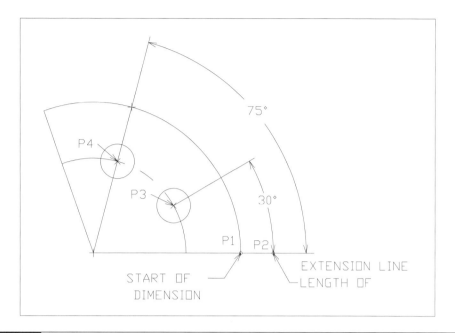

Figure 9–49 Example of placing arc dimensioning via the Dimension Arc Location tool

To dimension arcs with the Dimension Arc Location tool, invoke the tool from:

Angular Dimension tool box	Select the Dimension Arc Location tool. If necessary, set the Association check box to ON and select a dimension style from the Tool Settings window (see Figure 9–50).
Key-in window	**dimension arc location** (or **dim ar l**) ENTER

Figure 9–50 Invoking the Dimension Arc Location tool from the Angular Dimensions tool box

MicroStation prompts:

> Dimension Arc Location > Select start of dimension *(Place a data point to define the origin point. The dimension is measured counterclockwise from this point. This point must select an arc, circle, or ellipse.)*

417

Dimension Arc Location > Define length of extension line *(Place a data point to define the length of the extension line.)*

Dimension Arc Size > Select dimension endpoint *(Place a data point to define the dimension end point. If necessary, press* ENTER *to edit the dimension text.)*

Dimension Arc Size > Select dimension endpoint *(Continue placing data points to continue angular dimensioning, and click the Reset button to terminate the tool sequence.)*

Dimension Angle Chamfer

The Dimension Angle Chamfer is used to dimension the angle of chamfer. To dimension chamfer with Dimension Angle Chamfer tool, invoke the tool from:

Angle Dimension tool box	Select the Dimension Angle Chamfer tool. Select a dimension style from the Tool Settings window (see Figure 9–51).
Key-in window	**dimension angle chamfer** (or dim ang ch) ENTER

Figure 9-51 Invoking Dimension Angle Chamfer tool from the Angular Dimensions tool box

MicroStation prompts:Dimension Chamfer > Select first line *(Identify chamfer line)*

Dimension Chamfer > Select second line *(Identify the base line)*

Dimension Chamfer *(Enter the data position the dimension)*.

Dimension Radial

The Dimension Radial feature provides tools to create dimensions for the radius or diameter of a circle or arc and to place a center mark. Following are the tools available for radial dimensioning:

- Radius—To dimension the radius of a circle or a circular arc.

- Radius Extended—Identical to the Radius tool, except the leader line continues across the center of the circle, with terminators that point outward.

- Diameter—To dimension the diameter of a circle or a circular arc.

- Diameter Extended—Identical to the Diameter tool, except the leader line continues across the center of the circle, with terminators that point outward.
- Center Mark—To place a center mark at the center of a circle or circular arc.
- Dimension Diameter Parallel—To dimension the diameter of a circle using extension lines and a dimension line.
- Dimension Radius/Dimension Note—To place a dimension note on a circle or circular arc.
- Dimension Arc Distance—To dimension the distance between two arcs that have the same center.

Dimension Radius

To place a radial dimension with the Dimension Radius tool, invoke the tool from:

Radial Dimensions tool box	Select the Dimension Radius tool. If necessary, set the Association check box to ON and select a Dimension Style from the Tool Settings window (see Figure 9–52).
Key-in window	**dimension radius** (or **dim radiu**) ENTER

Figure 9–52 Invoking the Dimension Radius tool from the Radial Dimensions tool box

MicroStation prompts:

Dimension Radius > Identify element *(Identify a circle or arc, as shown in Figure 9–53.)*
Dimension Radius > Select dimension endpoint *(Place a data point inside or outside the circle or arc to place the dimension line, as shown in Figure 9–53.)*

 Note: The placement of the dimension text (horizontal or in-line) is set by selecting the Text **Orientation** in the Dimension Settings dialog box.

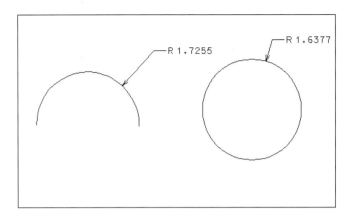

Figure 9-53 Examples of placing radial dimensions with the Dimension Radius tool

Dimension Radius (Extended Leader)

To place radial dimensioning with the Dimension Radius (Extended Leader) tool, invoke the tool from:

Radial Dimensions tool box	Select the Dimension Radius (Extended Leader) tool. If necessary, set the Association check box to ON, and select a dimension style from the Tool Settings window. (see Figure 9–54).
Key-in window	**dimension radius extended** (or **dim radiu e**) ENTER

Figure 9-54 Invoking the Dimension Radial (Extended Leader) tool from the Radial Dimensions tool box

MicroStation prompts:

    Dimension Radius (Extended Leader) > Identify element *(Identify a circle or arc, as shown in Figure 9–55.)*

    Dimension Radius (Extended Leader) > Select dimension endpoint *(Place a data point inside or outside the circle or arc to place the dimension line as shown in Figure 9–55.)*

Chapter 9   Measurement and Dimensioning

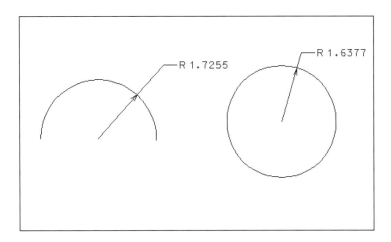

Figure 9-55 Examples of placing radial dimensions with the Dimension Radius (Extended Leader) tool

Dimension Diameter

To place diameter dimensioning with the Dimension Diameter tool, invoke the tool from:

Radial Dimensions tool box	Select the Dimension Diameter tool. If necessary, set the Association check box to ON and select a dimension style from the Tool Settings window. (see Figure 9–56).
Key-in window	**dimension diameter** (or **dim d**) ENTER

Figure 9-56 Invoking the Dimension Diameter tool from the Radial Dimensions tool box

MicroStation prompts:

> Dimension Diameter > Identify element *(Identify a circle or arc, as shown in Figure 9–57.)*
> Dimension Diameter > Select dimension endpoint *(Place a data point inside or outside the circle or arc to place the dimension line, as shown in Figure 9–57.)*

421

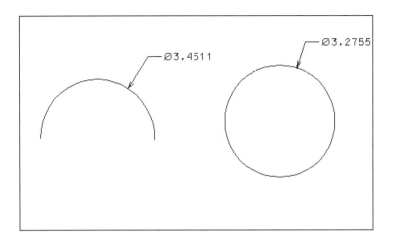

Figure 9-57 Examples of placing diameter dimensions with the Dimension Diameter tool

Dimension Diameter (Extended Leader)

To place diameter dimensioning with the Dimension Diameter (Extended Leader) tool, invoke the tool from:

Radial Dimensions tool box	Select the Dimension Diameter (Extended Leader) tool. If necessary, set the Association check box to ON and select a dimension style from the Tool Settings window. (see Figure 9–58).
Key-in window	**dimension diameter extended** (or **dim d e**) ENTER

Figure 9-58 Invoking the Dimension Diameter (Extended Leader) tool from the Radial Dimensions tool box

MicroStation prompts:

Dimension Diameter (Extended Leader) > Identify element *(Identify a circle or arc, as shown in Figure 9–59.)*

Dimension Diameter (Extended Leader) > Select dimension endpoint *(Place a data point inside or outside the circle or arc to place the dimension line, as shown in Figure 9–59.)*

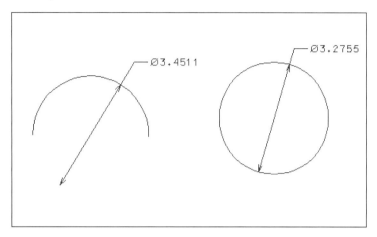

Figure 9-59 Examples of placing diameter dimensions via the Dimension Diameter Extended tool

Place Dimension Diameter Parallel

The Place Dimension Diameter Parallel places a dimension on a circle using extension lines and a dimension line, as shown in Figure 9–60.

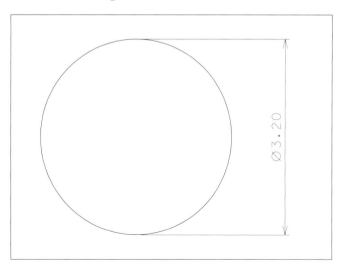

Figure 9-60 Example of placing a Dimension Diameter Parallel on a circle

To place a Dimension Diameter Parallel, invoke the tool from:

Radial Dimensions tool box	Select the Dimension Diameter Parallel tool. If necessary, set the Association check box to ON and select a dimension style from the Tool Settings window (see Figure 9–61).
Key-in window	**dimension diameter parallel** (or **dim d p**) ENTER

Figure 9-61 Invoking the Dimension Diameter Parallel tool from Radial Dimensions tool box

MicroStation prompts:

    Dimension Diameter Parallel > Identify element *(Identify the circle to be dimensioned.)*

    Dimension Diameter Parallel > Select dimension end point *(Drag the dimension line to the desired position and click the Data button to complete the dimension.)*

Place Center Mark

The Place Center Mark tool can place a center mark at the center of a circle or circular arc, as shown in Figure 9–62.

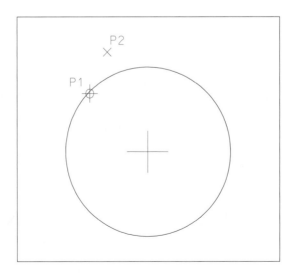

Figure 9-62 Example of placing a center mark by the Place Center Mark tool

To place a center mark with the Place Center Mark tool, invoke the tool from:

Radial Dimensions tool box	Select the Dimension Center tool. If necessary, set Association Lock to ON, select a Dimension Style and key-in the appropriate size of the center mark in the Center Size edit field in the Tool Settings window. (see Figure 9–63).
Key-in window	**dimension center mark** (or **dim c m**) ENTER

Figure 9-63 Invoking the Dimension Center tool from the Radial Dimensions tool box

MicroStation prompts:

    Place Center Mark > Identify element *(Identify a circle or arc to place a center mark on.)*

    Place Center Mark > Accept (next input) *(Click the Accept button, or identify another circle or arc to place the center mark.)*

 Note: If the Center Size text is set to 0.0000, then it applies the text size set in the Text settings box.

Dimension Radius/Diameter Note

The Dimension Radius/Diameter Note tool is used to place a dimension note for a circle or arc indicating either radius or diameter of the selected element. To place a note, invoke the tool from:

Radial Dimensions tool box	Select the Dimension Radius/Diameter Note tool. In the Tool Settings window, select a dimension style and choose either Radius or Diameter information to be included as part of the note (see Figure 9–64).
Key-in window	**dimension radius/diameter note** (or **dim rad/dia no**) ENTER

Figure 9-64 Invoking the Dimension Radius/Diameter Note tool from the Radial Dimensions tool box

MicroStation prompt:

    Dimension Radius > Select circular element to dimension *(Identify circle or arc)*

    Dimension Radius > Define length of extension line *(Enter a data point to define the length of the extension line).*

Dimension Arc Distance

The Dimension Arc Distance tool is used to dimension the distance between two arcs that have the same center. To dimension distance between two arcs, invoke the tool from:

Radial Dimensions tool box	Select the Dimension Arc Distance tool. In the Tool Settings window, select a Dimension Style and View Alignment (see Figure 9–65).
Key-in window	**dimension arc distance** (or **dim ar dist**) ENTER

Figure 9-65 Invoking the Dimension Arc Distance tool from the Radial Dimensions tool box

MicroStation prompt:

    Dimension Arc Distance > Select first arc *(Identify the first circle or arc.)*

    Dimension Arc Distance > Select next arc *(Identify the second circle or arc.)*

    Dimension Arc Distance > Define length of extension line *(Enter a data point to define the length of the extension line.)*

Chapter 9 Measurement and Dimensioning

Miscellaneous Dimensions

The tools in the Misc Dimensions tool box are used to perform dimensioning that is not specific to linear, angular, or radial dimensioning. Following are the tools available in the Misc Dimensions tool box:

- Dimension Ordinate—Labels distances along an axis from a common point of origin.
- Label Line—Labels the length and the direction of a line in a 2D design.
- Insert Dimension Vertex—Adds an extension line to a dimension element.
- Delete Dimension Vertex—Removes an extension line from a dimension element.
- Modify Dimension Location—Moves dimension text or modifies the extension line length of a dimension element.
- Reassociate dimension—Recreates a dimension's association to an element.
- Change Dimension to Active Settings—Changes a dimension to the active dimensioning attributes.
- Match Dimension Settings—Sets the active dimension settings to the dimension attributes of a dimension element.
- Geometric Tolerance—Builds a feature control with geometric tolerance symbols.

Dimension Ordinates

Ordinate dimensioning is common in mechanical designs. It labels distances along an axis from a point of origin on the axis along which the distances are measured. See Figure 9–66 for an example of ordinate dimensioning.

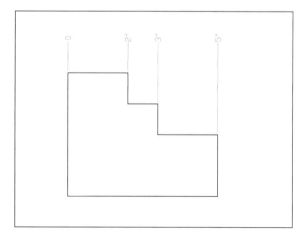

Figure 9-66 Example of ordinate dimensioning

To place ordinate dimensions, invoke the Dimension Ordinates tool from:

Misc Dimensions tool box	Select the Dimension Ordinates tool. If necessary, set the Association check box to ON, and select an Alignment option and Dimension Style in the Tool Settings window (see Figure 9–67).
Key-in window	**dimension ordinate** (or **dim o**) ENTER

Figure 9-67 Invoking the Dimension Ordinates tool from the Misc Dimensions tool box

MicroStation prompts:

> Dimension Ordinates > Select ordinate origin *(Place a data point from which all ordinate labels are to be measured.)*
>
> Dimension Ordinates > Select ordinate direction *(Place a data point to indicate the rotation of the ordinate axis.)*
>
> Dimension Ordinates > Select dimension endpoint *(Place a data point to define the end point of the dimension line. If necessary, press* ENTER *to edit the dimension text.)*
>
> Dimension Ordinates > Select start of dimension *(Place data points at each place where you want an ordinate dimension placed. These points define the base of the extension line. After placing all the points, click the Reset button to terminate the tool sequence.)*

 Note: To prevent the text for a higher ordinate from overlapping the text for a lower ordinate, turn ON the **Stack Dimensions** check box in the Dimension Settings dialog box—Tool Settings category.

Label Line

The Label Line tool places the line length above the line you select and the line rotation angle below the line. You can place a label on a line, line string, block, closed shape, or multi-line.

MicroStation determines which side of the element is the top; angle of rotation depends on how you drew it. The angle of the line or line segment, for example, is measured as a counterclockwise rotation from the first data point to the second. Exercise care in deciding how to draw elements you plan to label with the Label Line tool, or you may not get the angle of rotation you expected. See Figure 9–68 for examples of line labels.

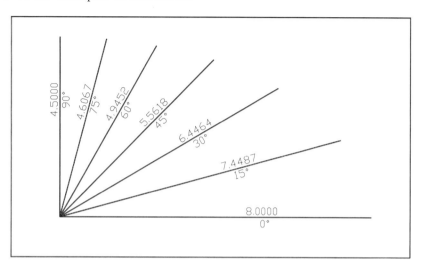

Figure 9-68 Examples of line labels

Invoke the Label Line tool from:

Misc Dimensions tool box	Select the Label Line tool. If necessary, set the Association check box to ON and select a Dimension Style from the Tool Settings window (see Figure 9–69).
Key-in window	**label line** (or **l l**) ENTER

MicroStation prompts:

    Label Line > Identify element *(Identify the element on which you want to place the label.)*

    Label Line > Accept/Reject (Select next input) *(Click the Data button to accept, or click Reset to disregard the dimensioning.)*

Figure 9-69 Invoking the Label Line tool from Misc Dimensions tool box

Insert Dimension Vertex

To add an extension line to a dimension element, invoke Insert Dimension Vertex tool from:

Misc Dimensions tool box	Select the Insert Dimension Vertex tool (see Figure 9–70)
Key-in window	**insert dimvertex** (or **ins dimv**) ENTER

Figure 9–70 Invoking the Insert Dimension Vertex tool from Misc Dimensions tool box

MicroStation prompts:

> Insert Vertex > Identify element *(Identify the dimension line near the desired extension line location.)*
>
> Insert Vertex > Accept/Reject (select next input) *(Enter a data point to position the end of the extension line as shown in Figure 9–71.)*

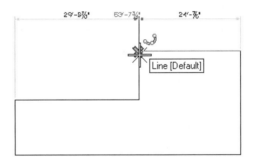

Figure 9–71 An example of inserting dimension vertex

Delete Dimension Vertex

To delete an extension line from a dimension element, invoke the Delete Dimension Vertex tool from:

Misc Dimensions tool box	Select the Remove Dimension tool (see Figure 9–72).
Key-in window	**delete dimvertex** (or **del dimv**) ENTER

Figure 9–72 Invoking the Delete Dimension Vertex tool from Misc Dimensions tool box

Delete Vertex > Identify element (*Identify the extension element.*)

Delete Vertex > Accept/Reject (select next input) (*Click the accept button to accept the deletion.*)

Modify Dimension Location

To move dimension text or modify the extension line length of a dimension element, invoke the Modify Dimension Location tool from:

Misc Dimensions tool box	Select the Modify Dimension tool (see Figure 9–73).
Key-in window	**modify dimension loc** (or **mod dim loc**) ENTER

Figure 9–73 Invoking the Modify Dimension Location tool from Misc Dimensions tool box

Modify element > Identify element (*identify the dimension to modify*)

Modify element > Accept/Reject (*select next input*) (*specify a data point to define the dimension's new position*).

Reassociate Dimension

The reassociate dimension tool is used to recreate a linear or radial dimension's association to an element. You can reassociate dimensions to elements individually, or by using a fence and selection set. The intended elements must appear in the view window for the reassociation of their dimensions to occur. Invoke the Reassociate Dimension tool from:

Misc Dimensions tool box	Select the Reassociate Dimension tool (see Figure 9–74).
Key-in window	**dimension reassociate** (or **dim reas**) ENTER

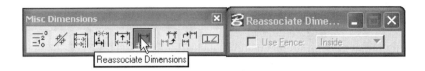

Figure 9-74 Invoking the Reassociate Dimensions tool from Misc Dimensions tool box

Reassociate Dimension > Select dimension (*identify the dimension to reassociate*)
Reassociate Dimension > Accept/Reject (*enter the data point to accept the reassociation*)

 Note: You can reassociate only the dimensions that are dropped with the Drop Association tool.

Change Dimension to Active Settings

The Change Dimension to Active Settings tool is used to change a dimension element to the active dimension attributes. You can change the active dimension attributes through the Dimension Settings dialog box or by changing the active **Dimension Style**. Invoke the Change Dimension to Active Settings tool from:

Misc Dimensions tool box	Select the Change Dimension tool (see Figure 9-75)
Key-in window	**change dimension** (or **cha dim**) ENTER

Figure 9-75 Invoking the Change Dimension tool from Misc Dimensions tool box

Change Dimension to Active Settings > Identify element (*identify the dimension to change active settings*)
Change Dimension to Active Settings > Accept/Reject (*select next input*) (*enter the data point to accept the change*)

Match Dimension Settings

The Match Dimension Settings tool is used to set active dimension settings to the dimension attributes of the selected dimension element. Invoke the Match Dimension Settings tool from:

Misc Dimensions tool box	Select the Match Dimension Attributes tool (see Figure 9–76)
Key-window	**match dimension** (or **mat dim**) ENTER

Figure 9-76 Invoking the Match Dimension Settings tool from Misc Dimensions tool box

> Match Dimension Settings > Identify element (*identify the dimension to change the active dimension settings to the selected dimension element*)

Geometric Tolerance

The Geometric Tolerance settings box (see Figure 9–77) helps you to build feature control frames with geometric tolerance symbols. The feature control frames are used with the Place Note tool and Place Text tool. The **Fonts** drop-down menu available in the settings box allows you to select one of the two fonts, 100—Ansi symbols, and 101—Feature Control Symbols. When a specific font is chosen, the buttons in the settings box reflect the availability of the symbols in the selected font. Whenever you want to add the symbols, click on the buttons as part of placing the text to the Place Note tool and Place Text tool.

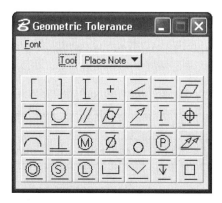

Figure 9-77 Geometric Tolerance settings box

The left bracket ([) and right bracket (]) form the ends of compartments; the vertical line (|) separates compartments.

Invoke the Geometric Tolerance settings box from:

Misc Dimensions tool box	Select the Geometric Tolerance tool (see Figure 9–78).
Key-in window	**mdl load geomtol** ENTER

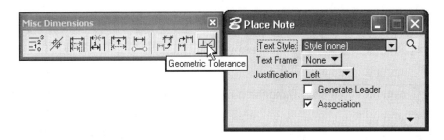

Figure 9-78 Invoking the Geometric Tolerance tool from the Misc Dimensions tool box

Select one of the two tools from the Geometric Tolerance settings box, **Place Note** or **Place Text**. MicroStation sets up the appropriate tool settings in the Tool Settings window. Follow the prompts to place text.

DIMENSION SETTINGS

MicroStation provides dimension settings that allow you to customize dimension tools. The settings are flexible enough to accommodate users in all engineering disciplines. The entire set of dimensioning settings can be saved to a given dimension style name with in the design. You can define dimensions styles and apply them to dimension elements during placement. Dimension styles can be created, customized and saved for easy recall. You can also import dimension styles from another design file to currently working design.

To change the active dimension settings, open the Dimension Settings box from:

Drop-down menu	Element > Dimensions
Key-in window	**dialog dimsettings** (or **di dimset**) ENTER

Chapter 9 Measurement and Dimensioning

MicroStation displays the Dimension Settings box, as shown in Figure 9–79, which serves to control the settings for dimensioning.

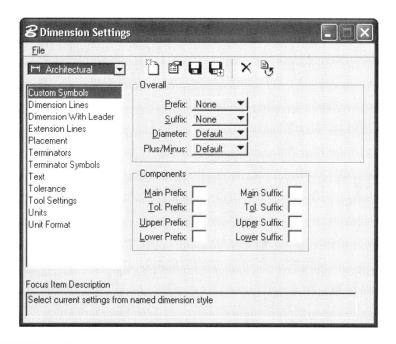

Figure 9-79 Dimension Settings box

The **Dimension Style** drop-down list box displays the active dimension style. If the dimension style is changed, any dimension with that style also changes to those settings. If the active dimension style is set to (none), the active dimensioning settings are applied. To see if a dimension has a style, check the Details tab on the Element Information tool when you identify the dimension element.

To create a new dimension style, first select the settings to reflect your preferences. Click the **Create Dimension Style** icon as shown in Figure 9-80. MicroStation displays the Create New Style dialog box. Type Name and Description for the new dimension style and click the **OK** button. The newly created style name is displayed in the drop-down list box.

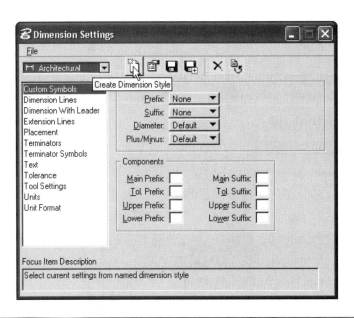

Figure 9-80 Invoking the Create Dimension Style from Dimension Settings box

To modify a dimension style, select the dimension style to modify in the drop-down list box. Make desired changes to the dimension style parameters. Click the **Save Style** icon to save the style. MicroStation displays an Alert box notifying you that this style is in use and changing the settings may change the appearance of existing dimensions.

To delete a dimension style, select the dimension style to delete in the drop-down list box. Click the **Delete Style** icon and the selected dimension style is deleted.

To edit the name and description of the style, select the dimension style in the drop-down list box. Click the **Edit Style Properties** icon. MicroStation displays the Edit Style Properties dialog box with the Name and Description of the selected style. Make necessary changes and click the **OK** button.

To remove overrides made to the dimension settings, click the **Restore Style** icon. The dimension style returns to its original settings.

MicroStation lists the available categories in alphabetical order on the left side of the settings box. Selecting a category causes the appropriate options for the category to be displayed to the right of the category list. The "Focus Item Description" area at the bottom of the window describes the purpose of the option with focus (the one you last clicked on).

Following are the categories and associated options available in the Dimension Settings box.

Custom Symbols

The Custom Symbols category, as shown in Figure 9–79, provides options for inserting symbols (characters from symbol fonts or cells) in the different parts of the dimension text.

Overall Options

The Overall options allow you to place an overall dimension text prefix and suffix, and to replace the default diameter and plus/minus symbols.

There are three options each for placing the Overall **Prefix** and **Suffix**:

- **None**—Do not insert a suffix or prefix.
- **Symbol**—When this option is selected, **Char** and **Font** fields appear in the settings box. Type the number of a font in the **Font** field, and the character to use in the **Char** field. That character will be used as the suffix or prefix.
- **Cell**—When this option is selected, a **Name** field appears in the settings box. Type the name of a cell in the field, it is placed as a shared cell (refer to Chapter 11 for a detailed discussion of cells).

Figure 9–81 shows an example of dual dimensions (English and Metric) with an "X" for the Overall Prefix and a "Y" for the Overall Suffix.

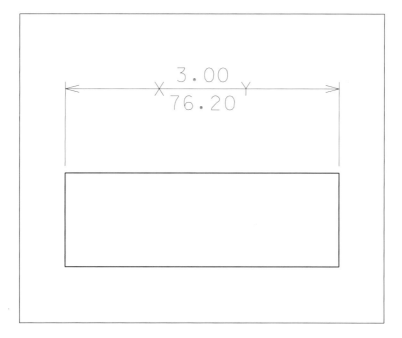

Figure 9–81 Example of Overall Prefix and Suffix in a linear dimension

There are two options each for selecting the Overall **Diameter** and **Plus/Minus**:

- **Default**—Use the default symbol provided by MicroStation.

- **Symbol**—When this option is selected for the Diameter, **Char** and **Font** fields appear in the settings box. Type the number of a font in the **Font** field, and the character to use in the **Char** field. That character will be used as the diameter symbol. When this option is selected for the Plus/Minus, an edit field appears in which you can type the number of a symbol from "0" to "9". (0 is the default symbol.)

Components Options

The Components options allow you to specify characters to use the prefixes and suffixes for parts of the dimension text. These options only allow you to specify the character, but not the font.

- **Main Prefix** and **Main Suffix** are the same as the Overall **Prefix** and **Suffix**.

- **Tol. Prefix** and **Tol. Suffix** place characters before and after the tolerance part of the dimension text, when tolerance is turned on. Figure 9–82 shows an example of placing an "X" before the tolerance part of a dimension and a "Y" after. Tolerance settings are discussed later in this chapter.

- **Upper Prefix** and **Upper Suffix** place characters before and after the upper text of a dual dimension (such as English above the dimension line and Metric below).

- **Lower Prefix** and **Lower Suffix** place characters before and after the lower text of a dual dimension.

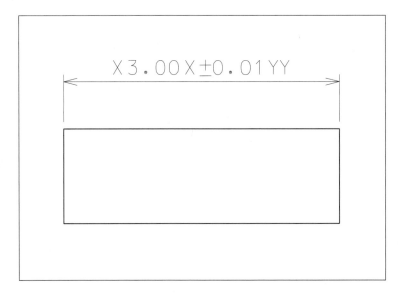

Figure 9-82 Example of Tolerance Prefix and Suffix in a linear dimension

Dimension Lines

The Dimension Lines category, as shown in Figure 9–83, provides options for changing the appearance of dimension lines.

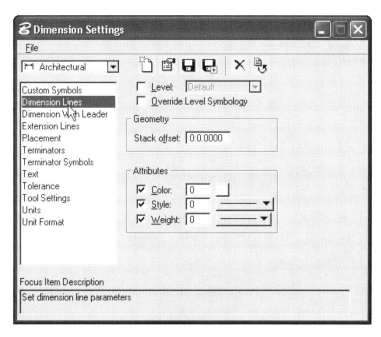

Figure 9-83 Dimension Settings box, Dimension Lines category

Dimension Level

The **Level** check box and drop-down list box allow you to set the level that dimensions are placed on. When the check box is ON, dimensions are placed on the selected level, regardless of what the design's active level is. The complete dimensions are placed on the selected level (not just the dimension lines).

The **Override Level Symbology** check box, when set to ON, shows dimensions using their actual attributes (color, weight, and style), rather than the attributes set by the Level Symbology (refer to Chapter 14 for a detailed description of Level Symbology).

Geometry Option

The **Stack Offset** edit field controls the spacing between dimension lines in stacked dimensions. Enter values in Working Units (MU:SU:PU). When the number is zero, MicroStation sets the space between the stacked dimension lines equal to twice the text height.

Attributes

The Attributes options allow you to set the dimension line **Color**, **Style**, and **Weight**. When an attribute's check box is set to ON, an options menu for the attribute appears to the right of the attribute name. Select the desired attribute value from the options menu, or type number in the provided edit field.

When the check boxes are to set ON, the dimension lines are placed using the attributes defined in this settings box, not the design's active attributes.

Dimension with Leader

The Dimension with Leader category, as shown in Figure 9–84, provides controls that affect the general appearance of dimension leaders.

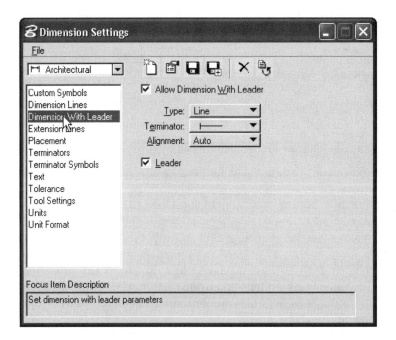

Figure 9–84 Dimension Settings box, Dimension with Leader category

Allow Dimension With Leader

The **Allow Dimension with Leader** check box, when set to ON, allows leader to display with dimensions.

Type

The **Type** option menu allows you to select type of leader line to display from None, Line or Curve.

Terminator

The **Terminator** option menu allows you to select the terminator type from None, Arrow, Slash, empty ball or filled ball.

Alignment

The **Alignment** option menu allows you to select the alignment type from Auto, Left or Right for the placement of the leader.

Leader

The **Leader** check box, when set to ON, adds a horizontal line to the leader.

Extension Lines

The Extension Lines category, as shown in Figure 9–85, provides options for changing the appearance of extension lines.

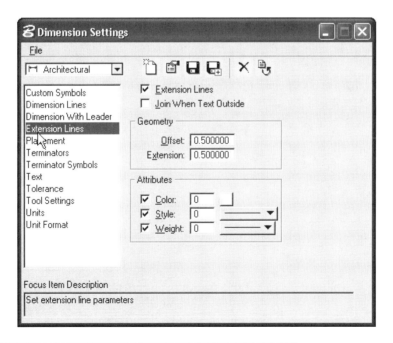

Figure 9-85 Dimension Settings box, Extension Lines category

Extension Line Creation

Two options at the top of the settings box control the creation of extension lines in a dimension:

- If the **Extension Lines** check box is set to ON, the extension lines are drawn with the dimension. If the check box is set to OFF, the dimension is placed in the same way, but the extension lines are not drawn.

- If the **Join When Text Outside** check box is set to ON, the dimension line is drawn between the two extension lines, even when the space between the extension lines is so small that the dimension text must be placed outside of the extension line pair. If the check box is set to OFF, no connecting dimension line will be drawn in tight dimensions. Figure 9–86 shows an example of a tight dimension with check box set to ON and OFF.

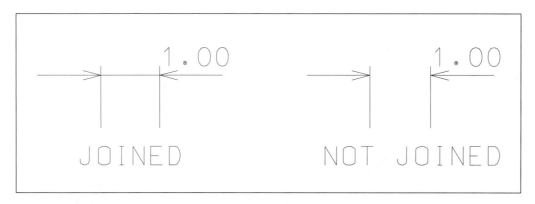

Figure 9-86 Example of a dimension with extension lines joined and not joined

Geometry Options

The **Offset** edit field allows you to set the gap between the object being dimensioned and the start of the extension line. The gap is calculated by multiplying the text height by the value in this field. For example, if the value is 1.0, the offset is equal to the text height.

The **Extension** edit field allows you to set the distance the extension line extends beyond the end dimension line. The extension is calculated by multiplying the text height by the value in this field. For example, if the value is 0.5, the offset is equal to half the text height.

Attributes Options

The Attributes options allow you to set the extension line **Color**, **Style**, and **Weight**. When you turn on an attribute's check box, an options menu for the attribute appears to the right of the attribute name. Select the desired attribute value from the options menu, or type number in the provided edit field.

When the check boxes are ON, the extension lines are placed using the attributes defined in this settings box, not the design's active attributes.

Placement

The Placement category, shown in Figure 9–87, provides options for controlling how dimensions are aligned in relation to the dimensioned element, where the dimension text is placed on the dimension line, and how reference file elements are dimensioned.

The **Alignment** menu provides options that control how the dimension is placed in relation to the dimensioned object:

- **View**—Aligns linear dimensions parallel to the view X or Y axis.
- **Drawing**—Aligns linear dimensions parallel to the design plane X or Y axis.
- **True**—Aligns linear dimensions parallel to the element being dimensioned.
- **Arbitrary**—Aligns linear dimensions parallel to the element being dimensioned. The extension lines are not constrained to be at right angles to the dimension line.

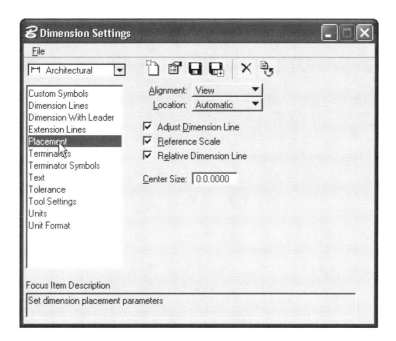

Figure 9–87 Dimension Settings box, Placement category

Figure 9–88 shows examples of placing linear dimensions using different alignment controls.

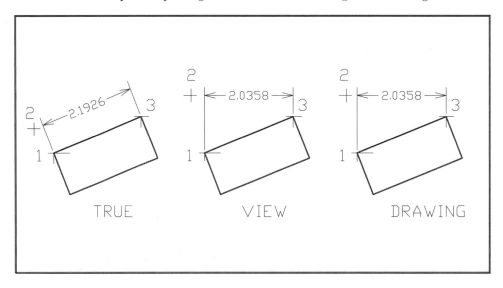

Figure 9–88 Examples of placing linear dimensions via different alignment controls

The **Location** menu provides options for controlling where the dimension text is placed:

- **Automatic**—Automatically places the dimension text according to the justification setting (left, center, or right on the dimension line). Justification is discussed later in this chapter
- **Semi-Auto**—Places the dimension according to the justification setting if the text fits between the extension lines. If not, you must specify the position of the text.
- **Manual**—Specify the dimension text position for all dimensions.

If the **Adjust Dimension Line** check box is set to ON, the dimension line and text are dynamically moved and extension lines are dynamically extended, if necessary, to prevent overlaying existing dimension text. If it is set to OFF, no adjustments are made. Figure 9–89 shows an example of a dynamically adjusted center dimension in a set of three dimensions.

Chapter 9 Measurement and Dimensioning

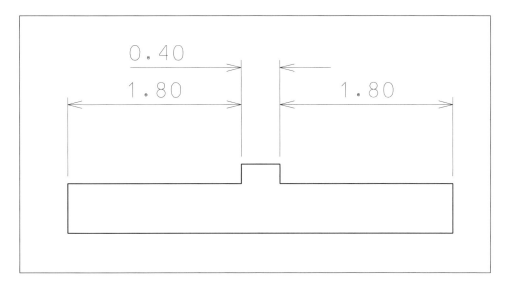

Figure 9-89 Example of dynamically adjusted dimension

If the **Reference File Units** check box is set to:

- ON, the Working Units of the reference file are used when dimensions are placed on elements in the reference file to allow for reference file scaling. (Refer to Chapter 13 for a detailed discussion of Reference Files.)
- OFF, the active design file's Working Units are always used.

The **Relative Dimension Line** check box comes into play when the dimension associated with the element is modified. If the check box is set to:

- ON, the dimension will be moved as necessary to keep the extension line the same length.
- OFF, the dimension stays in the same position and the length of the extension line is varied as necessary to maintain the dimension's relationship to the element.

The **Center Size** edit field allows you to set the default size, in Working Units, for the center mark that is placed from the Dimension Radial tool. The number you enter is half the overall height and width of the center mark. If the field contains a zero, the overall center mark height and width is equal to the text height.

Terminators

The Terminators category, as shown in Figure 9–90, provides options that control the placement of dimension terminators and their appearance after placement.

Figure 9-90 Dimension Settings box, Terminators category

Orientation Options

The terminator orientation options control where the terminators are placed and the style of terminator arrowhead. The default dimension terminator is an arrowhead, but, as we will learn in the Terminator Symbols category, other types of terminators can be used.

The **Terminator** options menu controls the placement of the dimension terminators:

- **Automatic**—If MicroStation decides there is enough room between the extension lines, it will place the terminators and a dimension line between the extension lines. If it decides there is not enough room, it places the terminators and dimension lines outside the extensions (there will not be a dimension line between the extension lines).

- **Inside**—MicroStation always places the terminators and a dimension line between the extension lines.

- **Outside**—MicroStation always places the terminators and dimension lines outside the extension lines (there will not be a dimension line between the extension lines).

- **Reversed**—MicroStation always places the terminators outside the extension lines and places a dimension line both inside and outside the extension lines.

The **Arrowhead** options menu controls the style of arrowhead used for the dimension terminators:

- **Open**—The arrowheads look like angle brackets.
- **Closed**—The arrowheads look like triangles.
- **Filled**—the arrowheads look like triangles and are filled with the outline color.

Figure 9–91 shows examples of each terminator placement position and each style of arrowhead.

Note: To make the filled arrowheads actually appear filled, turn on the Fill View Attribute for the view window you are working in. The View Attributes settings box is opened from the **Settings** drop-down menu.

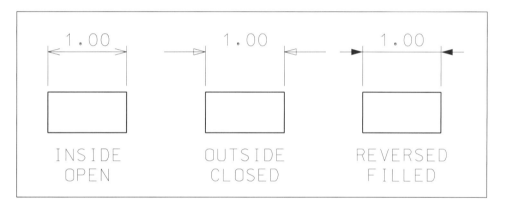

Figure 9-91 Examples of terminator placement positions and styles

Geometry Options

Three Geometry edit fields control the size of arrowhead terminators. Each size is the product of the dimension text height multiplied by the number you enter in the edit field. For example, if the text size is one inch and you enter a 2.0 in the width field, the arrowhead width will be two inches.

- **Width**—Allows you to set the width of the arrowhead from tip to back (parallel to the dimension line).
- **Height**—Allows you to set the height of the arrow head at its widest part (at right angles to the dimension line).
- **Min. Leader**—Allows you to control the minimum length of the dimension line the arrowhead is placed on. For example, if the arrowheads are placed Outside the extension lines, the short pieces of dimension lines on the outside of the extension lines are the minimum leader value in length.

Attributes Options

The Attributes options allow you to set the terminator **Color**, **Style**, and **Weight**. When you set the an attribute's check box ON, an options menu for the attribute appears to the right of the attribute name. Select the desired attribute value from the options menu, or type a number in the provided edit field.

When the check boxes are ON, the terminators are placed using the attributes defined in this settings box, not the design's active attributes.

Terminator Symbols

The Terminator Symbols category, as shown in Figure 9–92, provides controls to specify alternate symbols (characters from symbol fonts or cells) for each of the four types of dimension terminators.

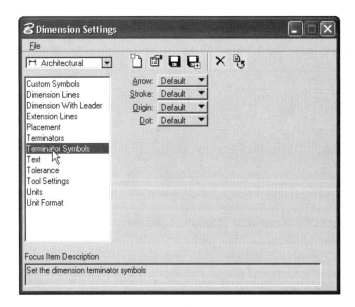

Figure 9-92 Dimension Settings box, Terminator Symbols category

The four types of terminators are:

- **Arrow**—The terminator placed by most of the dimensioning tools.
- **Stroke**—The terminator placed by the Dimension Size Stroke tool.
- **Origin** The start of the dimension symbol placed by the Dimension Location Stacked tool.
- **Dot**—The dimension dot symbol.

Each type of terminator has an options menu that provides the following options:

- **Default**—Use the default symbol provided by MicroStation.
- **Symbol**—When this option is selected, **Char** and **Font** fields appear in the settings box. Type the number of a font in the **Font** field, and the character to use in the **Char** field. That character will be used as the terminator.
- **Cell**—When this option is selected, a **Name** field appears in the settings box. Type the name of a cell in the field, and it is placed as a shared cell (refer to Chapter 11 for a detailed discussion of Cells).

Text

The Text category, as shown in Figure 9–93, provides options for controlling the placement and appearance of dimension text.

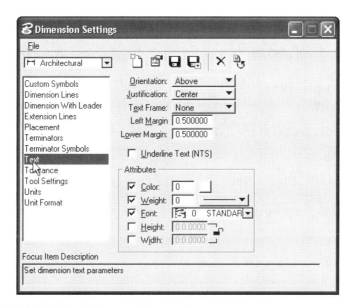

Figure 9–93 Dimension Settings box, Text category

Placement Options

The **Orientation** option menu provides options to control placement of the dimension text relative to the dimension line:

- **In Line**—Places text on the dimension line.
- **Above**—Places text above the dimension line.
- **Horizontal**—Places text horizontally.

Figure 9–94 shows examples of placing linear dimension with different orientation modes.

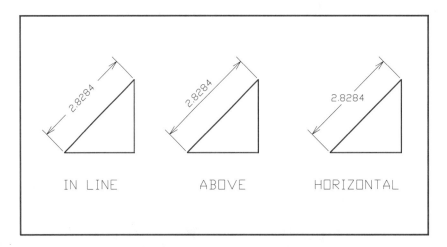

Figure 9–94 Examples of placing linear dimensioning in different Orientation modes

The **Justification** menu provides options to control where on the dimension line the text is placed when the text placement location is set for Automatic or Semi-Auto:

- **Left**—Places the text at the left end of the dimension line.
- **Center**—Places the text at the center of the dimension line.
- **Right**—Places the text at the right end of the dimension line.

Figure 9–95 shows examples of placing linear dimensions with different justification modes.

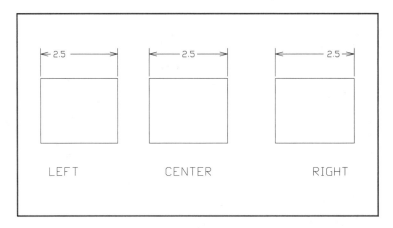

Figure 9–95 Examples of placing linear dimensioning in different Justification modes

The **Text Frame** menu provides options for placing a frame around the dimension text:

- **None**—No frame.
- **Box**—Box frame.
- **Capsule**—Capsule frame.

Figure 9–96 shows examples of placing linear dimensions with different Text frame modes.

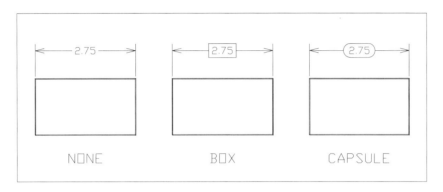

Figure 9–96 Examples of placing linear dimensioning in different Text Frame modes

The **Left Margin** edit field sets the horizontal space in text height units, between the leader line and the dimension text. The gap is the product of the number you enter in the field multiplied by the dimension text height.

The **Lower Margin** edit field sets the vertical space in text height units, between the leader line and the dimension text. The gap is the product of the number you enter in the field multiplied by the dimension text height.

If the **Underline Text (NTS)** check box is set to ON, all dimension text is placed with a line under the characters. If it is set to OFF, no underline is placed.

Attributes Options

The Attributes options allow you to set the dimension text **Color**, **Weight**, **Font**, **Height**, and **Width**. When you set the **Color**, **Weight** or **Font** check box to ON, an options menu for the attribute appears to the right of the attribute name. Select the desired attribute value from the options menu, or type a number in the provided edit field.

When you set the **Height** or **Width** check box to ON, you can enter the height or width, in working units, in the associated edit fields. If you want to force the height and width to always be equal, click on the lock symbol to close it before entering one of the values.

When the check boxes are ON, the terminators are placed using the attributes defined in this settings box, not the design's active attributes.

Tolerance

The Tolerance category, as shown in Figure 9–97, provides options for placing dimension tolerance values in the dimension text.

Tolerance values are added to the dimensions when the **Tolerance Generation** check box is set to ON.

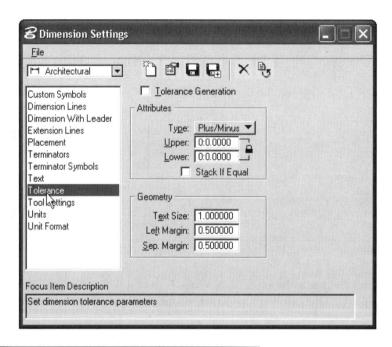

Figure 9-97 Dimension Settings box, Tolerance category

Attributes Options

The **Type** menu provides options for displaying tolerance values as a **Plus/Minus** addition to the dimension value or as a dimension **Limit**.

The **Upper** and **Lower** edit fields allow you to enter the tolerance range maximum and minimum values in working units (MU:SU:MU).

Figure 9–98 shows examples of each tolerance type with the upper and lower tolerance values equal to 0.002.

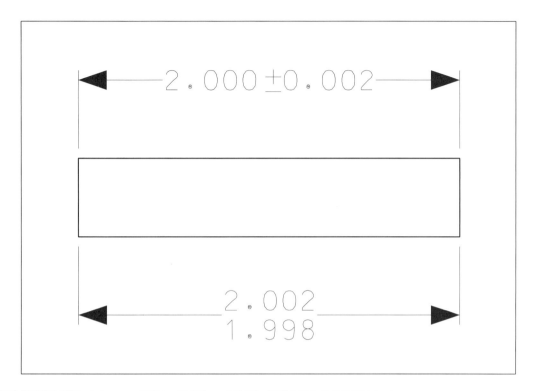

Figure 9-98 Examples of Plus/Minus and Limit tolerance attributes

Geometry Options

The **Geometry** options control the size and position of the tolerance text relative to the dimension text.

- **Text Size**—Sets the tolerance text size, specified as a multiple of the dimension text Height and Width.
- **Left Margin**—Sets the horizontal space, specified as a multiple of text size, between tolerance text and dimension text.
- **Sep. Margin**—Sets the vertical space, specified as a multiple of text size, between tolerance values.

Tool Settings

The Tool Settings category, as shown in Figure 9–99, provides options to customize the settings associated with individual dimensioning tools.

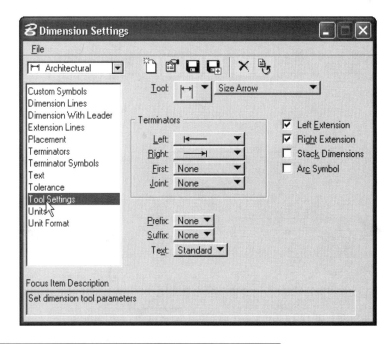

Figure 9–99 Dimension Settings box, Tool Settings category

The two **Tool** menus allow you to select the dimension tool you want to customize. The menu on the left contains pictures of each tool, and the menu on the right contains the names of each tool. Select a tool from either menu.

Each tool has its own set of customization options that appear in the settings box when you select the tool from a Tool menu.

Units

The Units category, as shown in Figure 9–100, consists of controls for settings that affect the type of units used in dimension text. The Primary section contains controls that are used to adjust settings that affect primary (above dimension line) linear dimension text. The Show Secondary Units section contains controls that are used to adjust settings that affect secondary (below the dimension line) linear dimension text.

Chapter 9 Measurement and Dimensioning

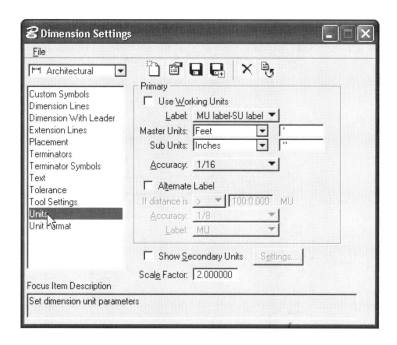

Figure 9-100 Dimension Settings box, Units category

Primary Units

If the **Use Working Units** check box is set to ON, then the working units of the design file are used for the units of dimensioning. The same label as in the Working Units category of the DGN File Settings dialog box is used. If **Use Working Units** is set to OFF, then you can set independently master units and sub units from **Master Units** and **Sub Units** option menus.

The **Label** option menu has seven options for setting the format for the dimension and how it is labeled or not labeled.

The **Accuracy** option menu sets the accuracy with which dimension is displayed, from 0 to 8 decimal places or using scientific notations.

The **Alternate Label** check box allows you to set up alternate dimensions based on the criteria that you define, providing the ability to dimension below or above a certain value. For example, if master units are set to miles, but you want dimensions of less than 1 mile to display in feet, set the If distance in the drop-down menu to less than (<), enter 1 and select MU.

455

Secondary Units

If **Show Secondary Units** check box is set to ON, then secondary dimension text is displayed. Click the **Secondary** button to open Secondary Units dialog box (see Figure 9–101), which sets the measurement system for secondary dimension text. The fields in this dialog box are similar to the Primary section.

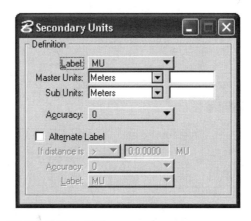

Figure 9-101 Secondary Units dialog box

The **Scale Factor** edit field allows you to enter a dimension scaling value to compensate for a design that was not drawn true size. For example:

- If the design is half its true size, enter a scale factor of 2 to double all dimensions.
- If the design is twice its true size, enter a scale factor of 0.5 to cut all the dimensions in half.

Unit Format

The Unit Format category, shown in Figure 9–102, provides options to change the display format of dimension text.

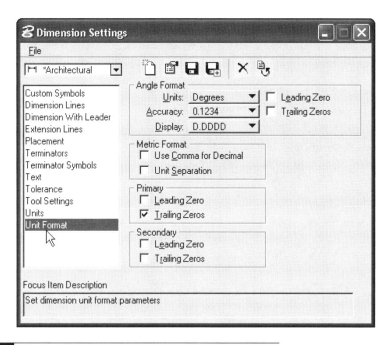

Figure 9-102 Dimension Settings box, Unit Format category

Angle Format Options

The **Angle Format** Options control the dimension units used for the angular dimensions placed by the Dimension Angle Size, Location, Between, From *X*, and From *Y* tools.

- The **Units** menu allows you to place angle dimensions using **Length** in working units, or **Degrees**.

- If the Units are in Degrees, the **Accuracy** menu allows you to set the dimension accuracy from zero to eight decimal places.

- If the Units are in Degrees, the **Display** menu allows you to display the angle as decimal degrees (D.DDDD); degrees, minutes, and seconds (DD°MM'SS"); or Centesimal degrees (C.CCCC).

Note: The Centesimal system measures a right angle by dividing it into 100 equal parts. Units are referred to as "grades," "grads," or "centesimal degrees."

Metric Format

The two **Metric Format** check boxes allow you to set the unit separator symbols to match the usage standards in Europe.

- The **Use Comma for Decimal** check box, when set to ON, causes a switch in the use of the comma and period in numbers (e.g., the comma is used to separate the whole and fractional part of a number).
- The **Unit Separation** check box, when set to ON, causes a space to be placed after the thousands and millions place when dimensions are in the metric format.

Primary and Secondary Options

The **Primary** and **Secondary** check boxes allow you to **Show Leading Zero** and **Show Trailing Zeros** in the dimension text. Figure 9–103 shows an example of a dimension that is set to three decimal places of accuracy, with the zero display check boxes ON and OFF.

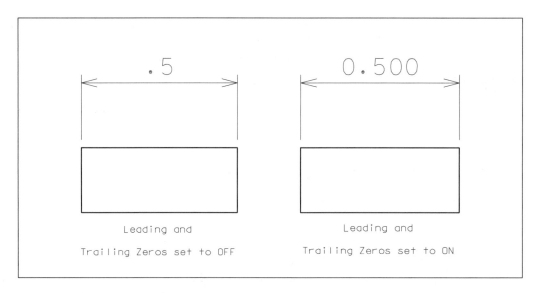

Figure 9-103 Dimension example with leading and trailing zero display ON and OFF

Match Dimension Settings

The steps required to set up dimensioning involve several settings from various categories and are time consuming. After setting everything up, it is all too easy to forget to save it as part of the current design file by using the Save Settings option. If you lose the dimension settings but have placed dimension elements with appropriate settings in your design, the Match Dimension tool can set the current settings by matching them to the settings in effect when the dimensions were placed.

Chapter 9 Measurement and Dimensioning

Invoke the Match Dimension tool from:

Match tool box	Select the Match Dimension Attributes tool (see Figure 9–104).
Key-in window	**Match Dimension** (or **mat d**) ENTER

Figure 9–104 Invoking the Match Dimension Attributes tool from the Match tool box

MicroStation prompts:

> Match Dimension Settings > Identify element *(Identify the dimension element whose dimension settings you want to match.)*
>
> Match Dimension Settings > Accept/Reject (Select next input) *(Click the Data button to set the selected dimension element settings as the current dimension settings, or click the Reject button to reject the settings.)*

Change Dimension

The Change Dimension tool can change a dimension element to the active dimension attributes.

Invoke the Change Dimension tool from:

Misc Dimensions tool box	Select the Change Dimension tool (see Figure 9–105).
Key-in window	**Change Dimension** (or **chan d**) ENTER

Figure 9–105 Invoking the Change Dimension tool from the Dimension tool box

MicroStation prompts:

> Change Dimension to Active Settings > Identify element *(Identify the dimension element to set it to the active dimension attributes.)*
>
> Match Dimension Settings > Accept/Reject (Select next input) *(Click the Data button to update the dimension attributes to the selected element, or click the Reject button to disregard the selection of the element.)*

Open the Exercise Manual PDF file for Chapter 9 on the accompanying CD for project and discipline specific exercises.

REVIEW QUESTIONS

Write your answers in the spaces provided.

1. The dimension line is a _____.

2. The extension lines are _____.

3. The leader line is a _____.

4. Associative dimensioning links _____.

5. To place associative dimensions, the Association Lock must be _____ .

6. What are the three options available for orientation of dimension text relative to the dimension line?
   _____

7. The justification field setting for dimension text does <u>not</u> apply when the _____ placement location mode is selected.

8. Name the two options that are available with dimension text length format.
   _____

9. What are the three types of linear dimensioning available in MicroStation?
   _____

10. The linear dimensioning tools are provided in the _____ tool box.

11. Name the two types of dimensioning included in circular dimensioning.
    _____

12. The Dimension Angle Size tool is used to _____ .

13. The Place Center Mark tool serves to _____ .

14. List the four options available with the Measure Distance tool.
    _____

15. The Measure Radius tool provides information on such elements as:
    _____

Chapter 9 Measurement and Dimensioning

16. List the seven area options available with the Measure Area tool.
    _____
    _____

17. The Measure Area Flood option measures the area enclosed _____.

18. The Geometric Tolerance settings box is used to build _____.

19. Ordinate dimensions are used to label _____.

20. The Label Line tool places the _____ and _____.

chapter 10

Printing

One task has not changed much in the transition from board drafting to CAD, and that is obtaining a hard copy. The term *hard copy* describes a tangible reproduction of a screen image. The hard copy is usually a reproducible medium from which prints are made, and it can take many forms, including slides, videotape, prints, and plots. This chapter describes the most common process for getting a hard copy: printing.

Objectives

After completing this chapter, you will know the following:

- How the printing process works
- What components are involved in the process
- How to create a plot file
- How to create a hard copy
- How to use the Batch Print utility

OVERVIEW OF THE PRINTING PROCESS

In manual drafting, if you need your design to be done in two different scales, you physically have to draw the design for different scales. In CAD, on the other hand, with minor modifications you can print the same design in different scale factors on different-sized paper.

To print a design with MicroStation you carry out a three-step process:

1. Set up the view to be printed or place a fence around the part of the design to be printed.
2. Use the Print dialog box to set the necessary print settings.
3. If the selected printer is the Windows system printer, the printed output is sent to the printer. If the selected printer is not the Window system printer, the Save Print As dialog box opens, to let you save the print file to disk for later submission to the required printer.

The plot file describes all the elements in the print area in a language the printing device can understand, and provides commands to control the printing device. It is separate from your design file, and contains the design as it existed when the plot file was created. If you make changes to the design after creating the plot file and want to print the new design, you must create a new plot file.

MicroStation stores plot files in the directory path contained in the MS_PLTFILES configuration variable. By default, that path is <disk>:Program Files\Bentley\Workspace\projects\examples\generic\out. Replace <disk> with the letter of the disk that contains the MicroStation program. (See Chapter 16 for a detailed explanation of configuration variables.)

Printing devices print the information contained in the plot file on the hard copy page. MicroStation supports many types and models of printing devices. There are electrostatic plotters that provide only shades of gray and more expensive models that print in color. Pen plotters use ink pens contained in a movable rack. A mechanical control mechanism selects pens and moves them across the page under program control.

MicroStation provides a plotter driver file for each supported plotting device. The information contained in this file (combined with the information you supply through the Print settings box) tells MicroStation how to create the plot file and send it to the plotting device.

The plotter driver file specifies the following:

- Printer model
- Number of pens the printer can use
- Resolution and units of distance on the printer
- Pen change criteria
- Name, size, offset, and number for all paper sizes
- Stroking tolerance for arcs and circles
- Border around the print and information about the border comment

- Pen speeds, accelerations, and force, where applicable
- Pen-to-element color or weight mapping
- Spacing between multiple strokes on a weighted line
- Number of strokes generated for each line weight
- Definitions for user-defined line styles (for printing only)
- Method by which prints are generated
- Actions to be taken at print's start and end and on pen changes

Plotter driver files can be edited with any text editor. For more information on the contents of these files, and changes you can make to them, consult the *MicroStation User Guide*.

Note: If you modify a sample plotter driver file, it is a good idea to retain the original file and to save the modified file as a new file with a different name.

PRINTING FROM MICROSTATION

All printing functions can be performed from the Print dialog box. Using the options in this dialog box you can select a printer and adjust various settings that affect printing. Additionally, you can preview the printed output. To Open the Print dialog box from:

Drop-down menu	File > Print
Key-in window	**print** (or **pr**) ENTER

MicroStation displays the Print dialog box, as shown in Figure 10–1.

Figure 10-1 Print dialog box

All options for adjusting printing settings are contained in the menu bar at the top of this dialog box and via the icon bar directly below it. The printer driver, currently selected, displays in the title bar of the dialog box. By default, the printed output is maximized. It will be printed to the largest scale that will fit on the selected paper size. You can see a preview of the printed area, along with the elements in your print. This appears in an expanded portion of the Print dialog box, which you can display or hide as required.

Click the **Show Preview** arrow to the right of the icon bar to display the print preview window.

The Print dialog box expands to display the print preview window as shown in Figure 10–2. The blue rectangle represents the size of the printed output on the selected sheet. To see the part of the drawing that is to be printed, set the **Show Design** check box to ON.

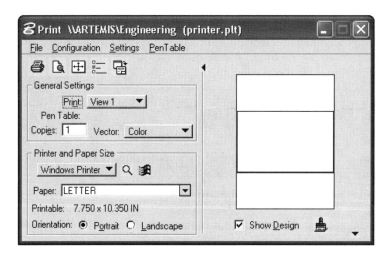

Figure 10-2 The Print dialog box expanded to show the printing details (visually)

In the expanded dialog box, click the **Show Details** arrow at bottom right. The Print dialog box expands to display further printing parameters as shown in Figure 10–3.

Chapter 10 Printing

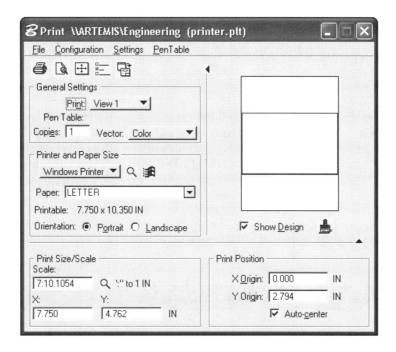

Figure 10-3 The Print dialog box fully expanded to show all settings

Selecting the Area of the Design to Print

By default, MicroStation selects contents of the active view to print. If necessary, you can change the view to print via the **Print** option menu in the **General Settings** section of the Print dialog box. This lets you select any of the open views, or the fence if one is present. If a fence is present in the design file, when the dialog box is opened, the contents of the fence are taken as the default entity to print. Similarly, if you place a fence in the design after the Print dialog box is opened, its contents again become the default entity to print.

Setting the Vector Output Color

MicroStation lets you to print vectors in grayscale or monochrome, using the **Vector** output setting in the Print dialog box. Often, it is advantageous to display the vector information in grayscale or monochrome, rather than the element colors. The **Vector** option menu in the **General Settings** section lets you choose from:

- Monochrome — output is black and white
- Grayscale — design file colors are output as grayscale
- Color — design file colors are used

467

Selecting a Printer

MicroStation lets you work with either of two types of printer drivers: **Windows Printer** or **Bentley Driver**. An option menu in the **Printer and Paper Size** section of the Print dialog box lets you toggle between the two types of printers.

Selecting **Windows Printer** automatically loads the Windows printer driver file (default is "printer.plt"). When you select **Bentley Driver**, by default, the Bentley printer driver file that you last used is loaded. Where required, you can select another Bentley Driver. You can, however, use the configuration variable MS_PLOTDLG_DEF_PLTFILE to define a default printer driver file to be selected each time the Print dialog box is opened. That is, the defined printer driver file will be selected rather than the printer driver file last used.

To select a Bentley printer driver, click the **Select Printer Driver** icon in the **Printer and Paper Size** section of the Print dialog box. The Select Printer Driver File dialog box opens as shown in Figure 10-4.

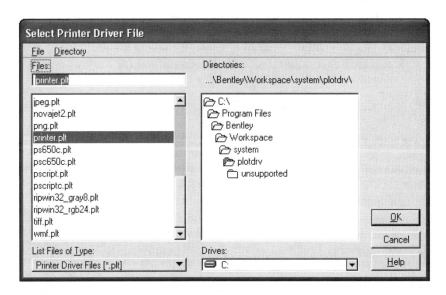

Figure 10-4 Select Printer Driver File dialog box

Choose a printer driver file and Click **OK** button. The Select Printer Driver File dialog box closes, and the name of the selected printer driver file appears in the title bar of the Print dialog box.

All delivered MicroStation printer drivers reference drivers that create print information in industry-recognized formats (such as HPGL/2, HPGL/RTL, ESC/P, TIF, and CGM). If a specific ".plt" file does not exist for the device that you are using, you may be able to use another existing ".plt" file.

Chapter 10 Printing

See Table 10–1 for the list of the supported plotters and corresponding sample plotter driver files supported with MicroStation V8.

Table 10–1 Supported Plotters and Corresponding Sample Plotter Driver Configuration Files

SUPPORTED PLOTTER	PLOTTER DRIVER FILE
CGM	CGM.plt
ESC/P	Epson24.plt
HP-GL/2	hpgl2.plt, hpljet3.plt, hpljet4.plt, hpljet4v.plt, hpdjet.plt, hp650c.plt, drftprop.plt, novajet2.plt
HP-GL/RTL	hpglrtl.plt
PCL	hpljet.plt, hp5xxc.plt, hppcl5.plt
PostScript	epscripc.plt, epscripm.plt, pscript.plt, pscriptc.plt, ps650c.plt, psc650c.plt
Raster File	cals.plt, jpeg.plt, png.plt, tiff.plt, ripwin32_gray8.plt, ripwin32_rgb24.plt

Plotter driver files can be edited with any text editor. For more information on the contents of these files and changes you can make to them, consult the *MicroStation User Guide*.

 Note: If you modify a sample plotter driver file, it is a good idea to retain the original file and to save the modified file as a new file with a different name.

Setting the Printing Parameters

Settings in the Print dialog box let you select the sheet size, set the scale for the print, and position the print on the selected sheet.

The **Printer and Paper Size** section of the Print dialog box lets you select a pre-defined paper size and define the orientation. If you are using a **Bentley Driver**, you can edit the X (width) and Y (height) dimensions of the selected paper size. When you select the **Windows Printer**, these dimensions are not editable. The setting for **Orientation** lets you choose between **Portrait** and **Landscape**.

The **Print Size/Scale** section of the Print dialog box lets you set the print scale. You can define the number of design units (in working units) that equate to printer output unit (printer units). You can key-in this value in the **Scale** edit field, or you can click the **Scale Assistant** icon, and MicroStation displays the Scale Assistant dialog box as shown in Figure 10–5. Define the scale criteria either as Design to Paper or Paper to Design.

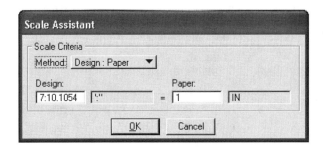

Figure 10-5 Scale Assistant dialog box

As an alternative to setting the scale for the print, you can set the X (width) and Y (height) dimensions for the print. Changing the **Scale**, or either dimension (**X** or **Y**), automatically results in changes to the remaining parameters to maintain the aspect ratio of the print.

If necessary, you can change the printer's units from the **Settings** drop-down menu of the Print dialog box. The settings will remain until they are changed again.

The **Print Position** section of the Print dialog box controls the position of the printable area of the print. You can specify the position of the lower left corner of the print relative to the lower left corner of the page. The **X Origin** value defines the distance horizontally and the **Y Origin** setting defines the distance vertically. To center the print on the page, set the **Auto-center** check box to ON.

The **Preview** section of the Print dialog box displays preview of the print for quickly checking the printing parameters. For more accurate previewing, open the resizable Preview window by selecting **Preview** from **File** drop-down menu.

Setting Print Attributes

The Print Attributes dialog box allows you to change aspects of the printed output's default appearance. Check boxes in the Print Attributes dialog box let you vary the relevant settings for printing purposes. Additionally, you can turn on/off the display of the Fence Boundary and/or the default Print Border for your printer output. Open the Print Attributes dialog box from:

Print dialog box tool bar	Select the Print Attributes tool (see Figure 10–6)
Drop-down menu (Print dialog box)	Settings > Print Attributes

Chapter 10 Printing

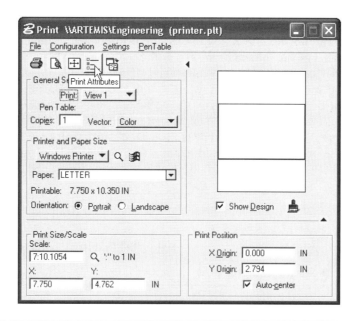

Figure 10-6 Invoking the Print Attributes tool from Print dialog box tool bar

MicroStation displays the Print Attributes dialog box as shown in Figure 10–7.

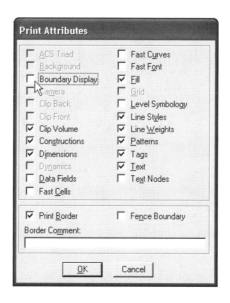

Figure 10-7 Print Attributes dialog box

471

If you set the **Fence Boundary** check box to ON, then the printed output includes the fence shape. If you set the **Print Border** check box to ON, the printed output includes a default border, which can include a label giving information such as the name of the design file and the time and date of the print. By default, the supplied printer drivers have the variables "filenamc" and "time" included in the border record. Additionally, when you turn on **Print Border**, you can add text in the **Border Comment** field. This will appear in the label outside the border and can include configuration variable references, which are expanded in the printed output. For example, if the configuration variable NAME was defined as "Joe Doe" and you keyed-in the description "Name=$(NAME)," the printed output would expand to "Name=Joe Doe" in the printed output.

Creating the Print

To print a design file or to create a plot file, invoke the Print tool from:

Print dialog box tool bar	Select the Print tool (see Figure 10–8).
Drop-down menu (Print dialog box)	File > Print

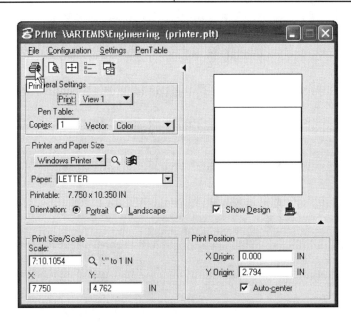

Figure 10-8 Invoking the Print tool from the Print dialog box tool bar

What happens at this stage depends on your system configuration and your selected printer driver. For a standard configuration, with no modifications to printer driver files or configuration variables, the print will either go directly to a printer, or will be saved to disk for later submission to a printer. If **Windows Printer** is selected, then output is sent directly to the Windows system printer. Instead if **Bentley Driver** is selected, the Save Print As dialog box opens as shown in Figure 10–9, to let you specify a name and location for the print to be saved to disk.

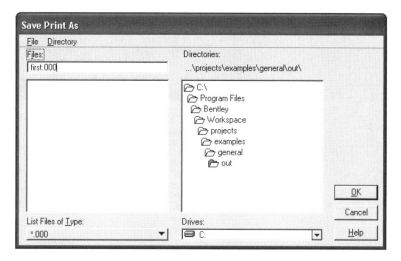

Figure 10-9 Save Print As dialog box

The default plot file name is the same as the design file name, and the .000 extension is added to the file name. If necessary, change the plot file name, then click the **OK** button to create the plot file.

By default, the plot file is saved in the <disk>: Program Files\Bentley\Workspace\projects\examples\generic\out\ directory. Replace <disk> with the letter of the disk that contains the MicroStation program.

If the selected file name already exists, an Alert window opens. Click the **OK** button to overwrite the file, or click the **Cancel** button to return to the Save Print As dialog box and enter a new file name.

SAVING PRINT CONFIGURATION

A print configuration file saves the print information specific to a design file. Print configuration files are a way to streamline repetitive printing tasks. Information saved in a print configuration file includes the following:

- Printing area
- Print option settings
- Fence location
- Displayed levels
- Page size, margin, and scale
- Pen table if attached

Before you create a print configuration file, set the appropriate controls in the Print dialog box. To create a print configuration file, open the Save Print Configuration File As dialog box from:

| Drop-down menu (Print dialog box) | Configuration > Save |

MicroStation displays the Save Print Configuration File As dialog box, as shown in Figure 10–10.

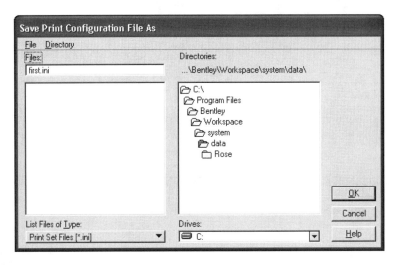

Figure 10-10 Save Print Configuration File As dialog box

Key-in the name of the configuration file in the **Files** edit field and click the **OK** button. MicroStation saves the configuration to the given file name with the .INI extension, and by default it is saved in the <disk>:Program Files\Bentley\Workspace\system\data\ directory.

To open an existing configuration file, invoke the **Open** tool from the **Configuration** drop-down menu in the Print dialog box. MicroStation displays the Select Print Configuration File dialog box. Select the appropriate configuration file and click the **OK** button. MicroStation makes the necessary changes to the print settings.

PEN TABLES

The pen table is a data structure that allows you to modify the appearance of a print without modifying the design file, by performing one or more of the following at print-creation time:

- Changing the appearance of elements
- Determining the printing order of the active design file and its attached reference files
- Specifying text string substitutions

A pen table is stored in a pen table file. The pen table consists of sections that are tested against each element in the design. When a match is found, the output options are applied to the element. The modified element is then converted into print data, which in turn is printed or written to the plot file. At no time are the elements of the design file or its reference files modified.

Creating a Pen Table

To create a pen table, invoke the New tool from:

| Drop-down menu (Print dialog box) | PenTable > New |

MicroStation displays the Create New Pen Table file dialog box. Key-in the pen table file name and click the **OK** button. MicroStation displays the Modify Pen Table settings box, as shown in Figure 10–11.

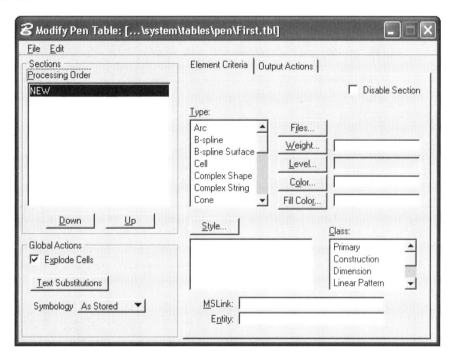

Figure 10–11 Modify Pen Table settings box

By default, MicroStation adds the section called NEW in the **Sections** list box. You can either rename the section or insert a new one and delete the NEW section.

Renaming a Pen Table Section

To rename a section, first select the name of the section in the list box, and then invoke the Rename tool from:

| Drop-down menu (Modify Pen Table settings box) | Edit > Rename Section |

MicroStation displays the Rename Section dialog box. Key-in the new name and click the **OK** button to rename the section.

Inserting a New Pen Table Section

Above an Existing Section

To insert a new section above an existing section, first select the name of the section in the list box, and then invoke the Insert New Section Above tool from:

| Drop-down menu (Modify Pent Table settings box) | Edit > Insert New Section Above |

MicroStation displays the Insert Section dialog box. Key-in the new name and click the **OK** button to insert the new section.

Below an Existing Section

To insert a new section below an existing section, first select the name of the section in the list box, and then invoke the Insert New Section Below tool from:

| Drop-down menu (Modify Pen Table settings box) | Edit > Insert New Section Below |

MicroStation displays the **Insert Section** dialog box. Key-in the new name and click the **OK** button to insert the new section.

Deleting a Pen Table Section

To delete a section, first select the name of the section in the list box, and then invoke the Delete Section tool from:

| Drop-down menu (Modify Pen Table settings box) | Edit > Delete Section |

MicroStation deletes the selected section from the list box.

If necessary, you can change the section's position in the processing order. First select the section in the list box, and then click **Down** or **Up** to change the processing order.

Modifying a Pen Table Section

To modify a pen table section, follow these numbered steps.

STEP 1: Highlight the name of the section to modify from the Sections list box.

STEP 2: *Optional:* Set the check box for **Explode Cells**.

ON — Each cell component element is evaluated independently against the element criteria; the defined output action applies to each element of the cell.

OFF — Each cell header is evaluated against the element criteria; the defined output action applies to each element of the cell.

STEP 3: *Optional:* To substitute text in the design with alternate text for printing, click the **Text Substitutions** button. MicroStation displays the Text Substitutions settings box, as shown in Figure 10–12.

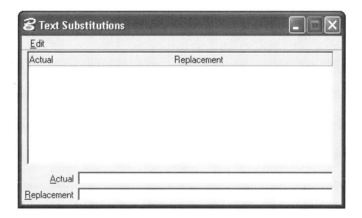

Figure 10-12 Text Substitutions settings box

To insert a text substitution entry, invoke the Insert New tool from:

| Drop-down menu (Text Substitutions settings box) | Edit > Insert New |

An entry labeled "Original" appears in the list box and in the **Actual** edit field. Replace "Original" with the string in the design to be replaced for printing purposes. Type the replacement text string in the **Replacement** edit field and press ENTER.

Note: The defined text string substitutions apply universally to all text elements in the print and only to exact matches of the specified strings.

In addition, you can also replace a text string in a design with a file name, current date, or time.

To replace a text string in a design with a file name, pen table file name, print driver file name, date, or time, select options from the **Edit** drop-down menu (Text Substitutions settings box) as listed in Table 10–2.

Table 10–2 Replacing a Text String with a File Name, Date, or Time

EDIT MENU ITEM	ACTUAL STRING IN THE DESIGN*	REPLACEMENT STRING FOR PRINTING	EFFECT
Insert Short Filename	$FILES$	_FILES_	Replaces the actual text string with the file name of the active design file.
Insert Abbreviated Filename	$FILENAME$	_FILEA_	Replaces the actual text string with the file name of the active design file with the name of the folder.
Insert Long file name	$FILEL$	_FILEL_	Replaces the actual text string with the file name of the active design file name with full path.
Insert Pen Table Short Filename	$PENTBLS$	_PENTBLS_	Replaces the actual text string with the file name of the Attached pen table file.
Insert Pen Table	$PENTBLA$	_PENTBLA_	Replaces the actual text string with the file name of the attached pen table with the name of the folder.
Insert Pen Table Long Filename	$PENTBLL$	_PENTBLL_	Replaces the actual text string with the name of the attached pen table with full path.

Insert Print Driver Short Filename	$PLTDRVS$	_PLTDRVS_	Replaces the actual text string with the file name of the selected printer driver.
Insert Print Driver Abbreviated Filename	$PLTDRVA$	_PLTDRVA_	Replaces the actual text string with the file name of the selected printer driver and name of the folder.
Insert Print Driver Lone Filename	$PLTDRVL$	_PLTDRVL_	Replaces the actual text string with the file name of the Selected printer driver with full path.
Insert Date	$DATE$	_DATE_	Replaces the actual text string with the current date.
Insert Time	$TIME$	_TIME_	Replaces the actual text string with the current time.

* The actual text string is shown with the dollar sign character ($) as the delimiter character just to differentiate it from normal text. It is *not* necessary to have the delimiter character as part of the text string in the design file. You can replace any text string in the design with the file name, date, or time.

To delete a text substitution entry, first highlight the text string substitution, and then invoke the Delete tool from:

| Drop-down menu (Text Substitutions settings box) | Edit > Delete |

The selected text string substitution is deleted from the settings box.

Note: You can see the substitutions to the text string by clicking the **Preview Refresh** icon in the Print Preview settings box.

STEP 4: To set the element criteria, first select the **Element Criteria** tab (shown in Figure 10–13), located in the top right side of the Modify Pen Table settings box.

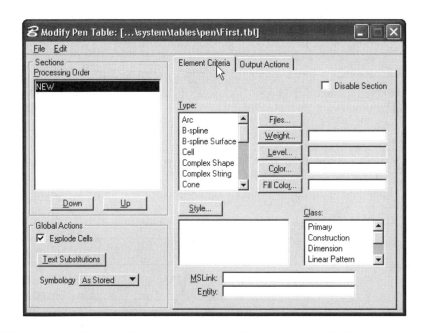

Figure 10-13 Selecting the Element Criteria tab in the Modify Pen settings box

Then select element **Types** from the **Type** list box, and the class or classes of elements to include in the selection criteria. To select multiple items from the list box, hold down CTRL and then select the items. Click the **Files** button to select the file name and reference files if any attached to the active design file to select the order in which the elements will be identified.

You can also select or deselect all the element types, files, and classes by selecting the appropriate tool from the **Edit** drop-down menu in the Modify Pen Table settings box.

In addition, you can set the selection criteria based on weight, level, color, Fill Color, or style by keying-in the appropriate values in the edit fields or by clicking the appropriate button in the Modify Pen Table settings box. MicroStation displays the appropriate dialog boxes. Use the controls in the dialog box to make the selections, and click the **OK** button to close the dialog box.

STEP 5: To set the output actions, first select the **Output Actions** tab (shown in Figure 10-14), located in the top right side of the Modify Pen Table settings box.

Select one of the available options from the **Master Control** option menu. Table 10-3 explains the available Master Control options.

Note: Do not prioritize elements unless it is significant to the print, since prioritized elements require additional processing time and memory.

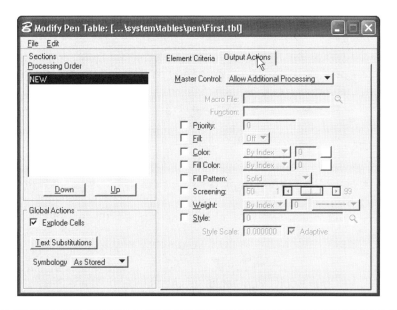

Figure 10-14 Selecting the Output Actions tab in the Modify Pen Table settings box

Table 10-3 Master Control Menu Options

MASTER CONTROL OPTION	EFFECT
Allow Additional Processing (default)	The output actions are applied to the selection criteria, and any loaded MDL applications that you desire to process the element are invoked.
No Additional Processing	The output actions are applied to the selection criteria, and no MDL applications that you desire to process the element are invoked.
Don't Display Element	The elements that satisfy the selection criteria are not printed, and no MDL applications that you desire to process the element are invoked.
Call BASIC Macro Function	The output actions are applied to the selection criteria, and additional processing is performed by the designated function in the designated BASIC macro.

Optional: Set the **Priority** check box to ON to set the priority. Key-in the desired priority value (range: –2147483648 to 2147483647) in the **Priority** edit field. Elements with a lower priority value are printed before elements with a higher priority value. Nonprioritized elements are always printed before all prioritized elements.

Optional: Set appropriate check boxes ON to override the **Fill**, **Color**, **Fill Color**, **Fill Pattern**, **Screening**, **Weight**, or **Style** appropriately for the elements that satisfy the selection criteria. Key-in the values in the edit fields or choose the desired attributes from the pop-up palette. If the element is to be printed with a custom line style, key-in the desired line style scale factor in the **Style Scale** edit field.

STEP 6: Before you print, click the Preview Refresh icon in the Print Preview settings box. MicroStation displays all the elements that are set to print in the Preview box, with appropriate changes as per the settings of the Output Actions. Once you are satisfied with the changes, print the design file.

STEP 7: To save the pen table, invoke the Save tool from:

Drop-down menu (Modify Pen Table settings box)	File > Save

MicroStation saves the modifications made to the existing pen table.

To save the modifications to a different pen table file, invoke the Save As tool from:

Drop-down menu (Modify Pen Table settings box)	File > Save As

MicroStation displays the Create Pen Table File dialog box. Key-in the name of the file to save the settings, and click the **OK** button.

STEP 8: To disable pen table processing, unload the pen table. To unload the pen table, invoke the Unload tool from:

Drop-down menu (Modify Pen Table settings box)	File > Exit/Unload

MicroStation unloads the pen table.

You can also unload the pen table from the **Pen Table** drop-down menu located in the Print settings box.

BATCH PRINTING

MicroStation provides a utility program called Batch Print to print sets of design files. This utility program allows printing of multiple design files. You can compose and re-use job sets that identify design files to be printed and the specifications that describe how they should be printed. You can also print individual files or subsets of the files in large job sets for spot-checking.

To invoke the Batch Print program, select from:

| Drop-down menu | File > Batch Print |

MicroStation displays the Batch Print dialog box, as shown in Figure 10–15.

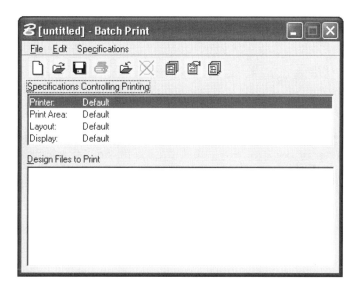

Figure 10-15 Batch Print dialog box

Setting Print Specifications

A print specification is a named group of instructions describing how to perform a certain step in the printing process. The Batch Print utility program provides four specification types: Printer, Print Area, Layout, and Display. The Printer specification type describes the printer, paper size, and post-processing options. The Print Area specification type selects the portions of the design file to print. The Layout specification type places a representation of the given print area on the paper at the specified size and position. And the Display specification allows you to select a pen table and setting the attributes for printing.

Printer Specification

The Printer specification type describes the printer, paper size, and post-processing options. The utility allows you to create a new printer specification, and modify or delete an existing specification. To create a new, modify or delete a printer specification, open the Batch Print Specification Manager from:

| Drop-down menu (Batch Print dialog box) | Specifications > Manage |

MicroStation displays Batch Print Specification Manager as shown in Figure 10–16.

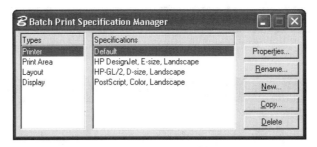

Figure 10-16 Batch Print Specification Manager dialog box

To change the properties of an existing printer specification, select the appropriate specification from the **Specifications** list box, and then click the **Properties** button. MicroStation displays a Printer Specification dialog box similar to Figure 10–17.

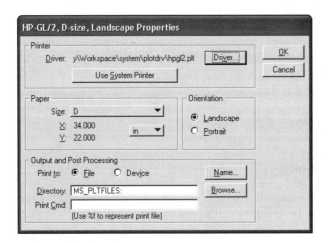

Figure 10-17 Printer Specification dialog box

The Printer Specification dialog box allows you to select a specific printer driver, paper size, orientation, and output and post processing settings. Click the **OK** button to accept the changes and close the dialog box.

Chapter 10 Printing

To create a new printer specification, click the **New** button in the Batch Print Specification Manager. MicroStation displays the New Printer Specification Name dialog box. Key-in the name in the **Name** edit field and click the **OK** button. If necessary, you can change the properties of the selected Printer specifications. MicroStation lists the newly created printer specification in the **Specifications** list box.

To rename a new printer specification, first select the specification you want to rename, and then click the **Rename** button in the Batch Print Specification Manager. MicroStation displays the Rename Printer Specification dialog box. Key-in the new name in the **Name** edit field and click the **OK** button. MicroStation lists the renamed printer specification in the **Specifications** list box.

To create a printer specification from an existing one, first select the specification from which you want to copy, and then click the **Copy** button in the Batch Print Specification Manager. MicroStation displays the New Printer Specification Name dialog box. Key-in the name in the **Name** edit field and click the **OK** button. If necessary, you can change the properties of the newly created Printer specifications. MicroStation lists the newly created printer specification in the **Specifications** list box.

To delete a printer specification, first select the specification you want to delete from the **Specifications** list box, and then click the **Delete** button in the Batch Print Specification Manager. MicroStation deletes the selected printer specification from the **Specifications** list box.

To change the current default selection of the printer specification, invoke the Select tool from:

| Drop-down menu (Batch Print dialog box) | Specifications > Select |

MicroStation displays Select Printer Specification dialog box shown in Figure 10–18.

Figure 10-18 Select Printer Specification dialog box

Select the appropriate printer specification from the list box and click the **OK** button.

If necessary, you can also change the properties of the selected printer specification by double-clicking the name of the specification from the **Specifications** list box in the Batch Print Specifications Manager dialog box. MicroStation displays the Printer Specifications dialog box, similar to Figure 10–17. Make the necessary changes and click the **OK** button to accept the changes and close the dialog box.

485

Print Area

The Print Area specification type allows you to select the portions of the design file to print. The utility allows you to create a new Print Area specification, and modify or delete an existing specification. To create a new, modify or delete a Print Area specification, open the Batch Print Specification Manager from:

Drop-down menu (Batch Print dialog box)	Specifications > Manage

MicroStation displays the Batch Print Specification Manager, shown in Figure 10–16. To change the properties of an existing Print Area specification, first select the **Print Area** from the **Types** list box, and then select the appropriate specification from the **Specifications** list box and click the **Properties** button. MicroStation displays the Print Area Properties dialog box, similar to Figure 10–19.

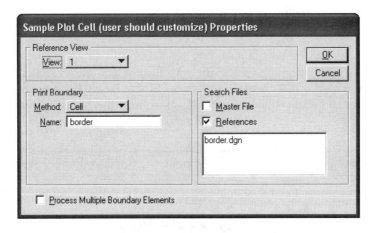

Figure 10-19 Print Area Properties dialog box

The **View** options menu allows you to select the view number or saved view to print.

The **Print Boundary** section of the dialog box allows you to select the boundary-defining elements (similar to placing a fence by snapping to the vertices of a shape). The option menu in the **Print Boundary** section includes three options: **View**, **Shape**, and **Cell**. The **View** selection prints to the extent of the view window. The **Shape** selection prins an area bounded by a particular shape. Specify the attributes of the shape in the appropriate fields in the **Boundary** section of the dialog box. The **Cell** selection prints an area bounded by a cell. Specify the name of the cell in the **Name** edit field in the Boundary section of the dialog box.

The **Master File** and **References** check boxes set the limit for the search for the boundary-defining shape or cell. By default, the utility searches each master file and all of its reference files to find the boundary-defining shape or cell. If necessary, you can restrict the search to specific reference files by typing their logical names or file names in the **References** edit field.

The **Process Multiple Boundary Elements** check box controls whether to generate the prints for each boundary element found in the design file. Set the check box to ON if you want to print for each boundary element found and OFF to generate a print for the first boundary element found only.

Click the **OK** button to accept the changes and close the dialog box.

Similar to Printer specifications, you can also create a new Print Area specification, rename an existing one, or delete one.

To change the current selection of the Print Area specification, invoke the Select tool from:

| Drop-down menu (Batch Print dialog box) | Specifications > Select |

MicroStation displays the Select Print Area Specification dialog box, shown in Figure 10–20.

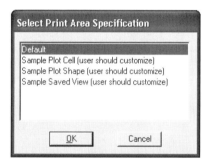

Figure 10–20 Select Print Area Specification dialog box

Select the appropriate Print Area specification from the list box and click the **OK** button.

If necessary, you can also change the properties of the selected Print Area specification by double-clicking the name of the specification from the **Specifications** list box in the Batch Print Specifications dialog box. MicroStation displays the Print Area Properties dialog box, similar to Figure 10–19. Make the necessary changes and click the **OK** button to accept the changes and close the dialog box.

Layout

The Layout specification type describes how the utility program determines the size and position of each print. The utility allows you to create a new Layout specification, and modify or delete an existing specification. To create a new, or modify or delete a Layout specification, open the Batch Print Specification Manager from:

| Drop-down menu (Batch Print dialog box) | Specifications > Manage |

MicroStation displays the Batch Print Specification Manager, as shown in Figure 10–16. To change the properties of an existing Layout specification, first select the Layout from the **Types** list box, and then select the appropriate specification from the **Specifications** list box, and click the **Properties** button. MicroStation displays the Layout Properties dialog box, similar to Figure 10–21.

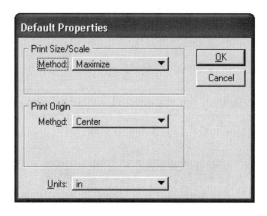

Figure 10-21 Layout Properties dialog box

The **Print Size/Scale Method** menu provides five print size options:

- The **Maximize** selection makes each plot as large as possible, given the paper size and orientation in the job set's printer specification.
- The **Scale** selection allows you to specify a scale factor in terms of master units in the design file to physical units of the output media.
- The **% of Maximum Size** selection allows you to specify an integer value between 10 and 100 percent of its maximum possible size.
- The **X Size** selection allows you to specify an explicit X size (width) for the print.
- The **Y Size** selection allows you to specify an explicit Y size (height) for the print.

The **Plot Origin Method** menu provides two options:

- The **Center** selection centers each print on the output media.
- The **Manual Offset** selection allows you to specify explicit *X* and *Y* offsets for the print. The offsets provided are relative to the media's lower-left margin.

Click the **OK** button to accept the changes and close the dialog box.

Similar to Printer and Print Area specifications, you can create a new Layout specification, rename from an existing one, or delete one.

Chapter 10 Printing

To change the current selection of the Layout specification, invoke the Select tool from:

| Drop-down menu (Batch Print dialog box) | Specifications > Select |

MicroStation displays the Select Layout Specification dialog boxes shown in Figure 10–22.

Figure 10-22 Select Layout Specification dialog box

Select the appropriate Layout specification from the list box and click the **OK** button.

If necessary, you can also change the properties of the selected Layout specification by double-clicking the name of the specification from the **Specifications** list box in the Batch Print Specification Manager dialog box. MicroStation displays the Layout Properties dialog box, similar to Figure 10–21. Make the necessary changes and click the **OK** button to accept the changes and close the dialog box.

Display

The Display specification controls the appearance of printed elements. You can control the printing equivalents of the view attributes, and in addition, you can specify a pen table that will resymbolize the print. To create a new, modify or delete a Display specification, open the BatchPrint Specification Manager from:

| Drop-down menu (Batch Print dialog box) | Specifications > Manage |

MicroStation displays the Batch Print Specification Manager shown in Figure 10–16. To change the properties of an existing Display specification, first select **Display** from the **Types** list box, and then select the appropriate specification from the **Specifications** list box, and click the **Properties** button. MicroStation displays the Display Properties dialog box similar to Figure 10–23

489

Figure 10-23 Display Properties dialog box

The options provided are similar to the View Attributes setting options. Make necessary changes to the display options. In addition, you can also specify the name of the pen table file you want to apply for printing the selected design file. Click the **OK** button to accept the changes and close the dialog box.

Similar to Printer, Print Area, and Layout specifications, you can create a new Display specification, rename an existing one, or delete one.

To change the current selection of the Display specification, invoke the Select tool from:

Drop-down menu (Batch Print dialog box)	Specifications > Select

MicroStation displays the Select Display Specification dialog box shown in Figure 10–24.

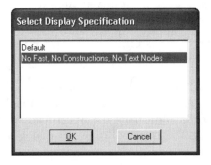

Figure 10-24 Select Display Specification dialog box

Select the appropriate Display specification from the list box and click the **OK** button.

If necessary, you can also change the properties of the selected Display specification by double-clicking the name of the specification from the **Specifications** list box in the Batch Print dialog box. MicroStation displays the Display Properties dialog box, similar to Figure 10–23. Make the necessary changes and click the **OK** button to accept the changes and close the dialog box.

Design Files to Print

To select design files to print, invoke the Add Files tool from:

| Drop-down menu (Batch Print dialog box) | Edit > Add Files |

MicroStation displays the Select Design Files to Add dialog box shown in Figure 10–25.

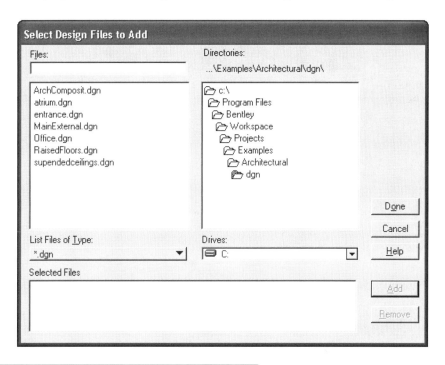

Figure 10-25 Select Design Files to Add dialog box

Select the design files from the appropriate directory and then click the **Add** button. Selected files are added to the **Design Files to Print** list box. Click the **Done** button when the selection is complete. The selected files are printed in the order in which they appear in the list box. If necessary, you can rearrange the files by highlighting their names and invoking one of the modify tools available in the **Edit** drop-down menu.

To save the current selection of the design files set to print, invoke Save tool from:

| Drop-down menu (Batch Print dialog box) | File > Save |

MicroStation displays the Save Job Set File dialog box. Key-in the name of the file in the **Files** edit field to which you want to save the current job set. Click the **OK** button to save the current job set and close the dialog box.

To print the current selection of the design files, invoke the Print tool from:

| Drop-down menu (Batch Print settings box) | File > Print |

MicroStation displays Print Batch dialog box similar to Figure 10–26.

Figure 10-26 Print Batch dialog box

Select one of the two radio buttons available in the **Print Range** section of the dialog box. The **All** option allows MicroStation to print all the design files in the current job set. The **Selection** option prints only the design files that are selected explicitly in the current job set. Click the **OK** button to print the selection and close the dialog box.

 Open the Exercise Manual PDF file for Chapter 10 on the accompanying CD for project and discipline specific exercises.

chapter 11

Cells and Cell Libraries

Objectives

After completing this chapter, you will be able to do the following:

- Create cell libraries
- Attach cell libraries
- Create cells
- Select active cells
- Place cells
- Place line terminators
- Place point elements, characters, and cells
- Maintain cells and cell libraries
- Place and maintain shared cells
- Use and modify cells from the cell selector

CELLS

Cells are like the variously shaped cutouts in a manual drafting template. You draw standard symbols on paper by tracing the outline of the symbol's cutout in the template. In MicroStation you do the same thing by placing a copy of a cell in your design file. Figure 11-1 shows some common uses of cells in various engineering disciplines.

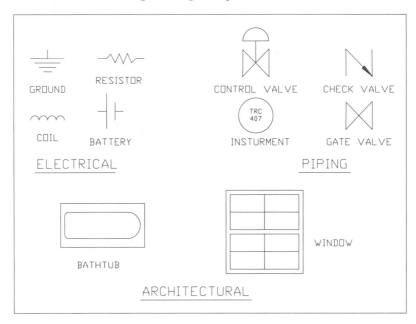

Figure 11-1 Common uses of cells in various engineering disciplines

Even though a cell contains separate elements, the copy you place in a design file acts like a single element when manipulated with tools such as Delete, Rotate, Array, and Mirror. When placing cells, you can change the scale and/or rotation angle of the original object(s). Cells save time by eliminating the need to draw the same thing more than once, and they also promote standardization.

CELL LIBRARIES

If cells are like the holes in a plastic template, cell libraries are the template. A cell library is a file that holds cells. Most engineering companies that use MicroStation have several cell libraries to provide standard symbols for all of their design files. Some companies also create sets of cells stored in libraries that they offer for sale.

You can access any of the cells from the cell library and place copies of them in your current design file. If you create a new cell, it is automatically placed in the library that is attached to your design file.

You can also place a cell that is not in the attached cell library by keying-in the name of the cell. MicroStation searches for the cell in the cell library list specified by the Cell Library List configuration variable (MS_CELLIST). (Refer to Chapter 16 for how to set up the configuration variables).

Although there is no limit to the number of cells you can store in a library, you don't want to end up with a very large, hard-to-manage library. It is advisable to create separate cell libraries for specific disciplines, such as electrical fixtures, plumbing, and HVAC.

Creating a New Cell Library

To start the process of creating a new cell library, invoke the Cell Library box from:

Draw-down menu	Element > Cells (see Figure 11–2).
Key-in window	**dialog cellmaintenance** (or **di ce**) ENTER

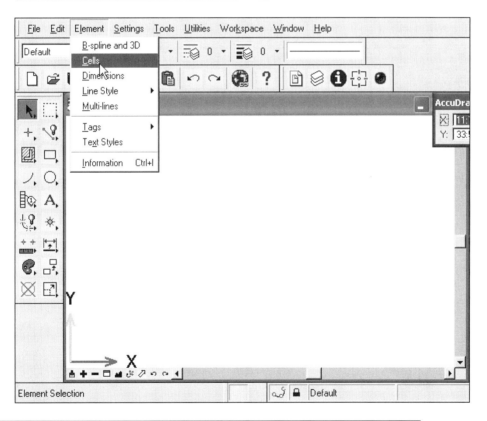

Figure 11–2 Invoking the Cell Library box from the Element drop-down menu.

A settings box appears similar to the one shown in Figure 11-3.

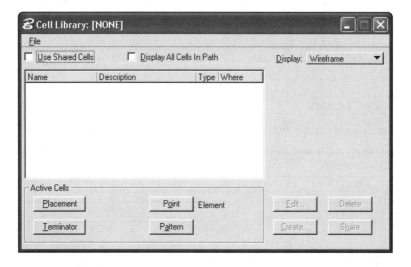

Figure 11-3 Cell Library box

To create a new cell library file, open the Create Cell Library dialog box from:

| Cell Library box | File > New |

MicroStation displays the Create Cell Library dialog box, as shown in Figure 11-4.

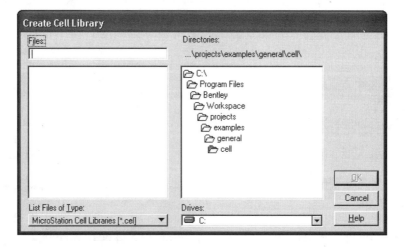

Figure 11-4 Create Cell Library dialog box

Following are the steps to create a new cell library file:

1. In the **Directories** field select the folder where the new cell library is to be placed.
2. Key-in a name for the new library in the **Files** text field. Do *not* type a period or extension. MicroStation appends the ".CEL" extension to the file name.
3. Click the **OK** button to create the new cell library and close the Create Cell Library dialog box.

When you create a new cell library, MicroStation automatically attaches it to your design file, acknowledges the attachment with a message in the status bar, and places the filename in the Cell Library dialog box title bar.

Your cell library is now available for use. You can start creating cells to store in the library.

Attaching an Existing Cell Library

For a cell library to be used in the active design file, it must be attached to the design file. An option in the Cell Library box is used to attach a cell library.

Invoke the Cell Library box from:

Drop-down menu	Element > Cells

To attach a cell library from the Cell Library box, invoke the Attach Cell Library dialog box from:

Cell Library box	File > Attach
Key-in window	**rc=**<name of the library> ENTER

MicroStation displays the Attach Cell Library dialog box.

Following are the steps to attach a cell library:

1. In the **Directories** field select the folder that contains the cell library.
2. In the **Files** area, select the cell library name.
3. Click the **OK** button to attach the cell library to the active design file.

The cell library path and file name are displayed on the Status bar and in the title bar of the Cell Library box, and the name and description of all cells in the library are displayed on the Cell

Library box. Figure 11-5 shows the Cell Library box for the remodel.cel library, one of the libraries provided by MicroStation.

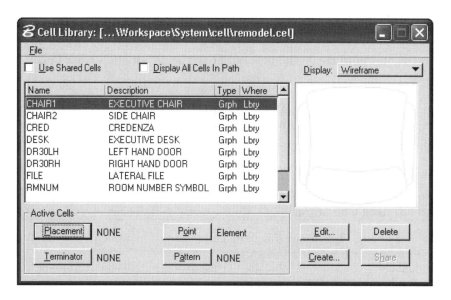

Figure 11-5 Listing of the cell names and descriptions in the Cell Library box

Things to Remember about Cell Library Attachment

- The attachment is permanent as long as MicroStation can find the library file.

- You can attach only one cell library at a time using the Cell Library box. If you attach another library, the first library is detached.

- When the library is attached, all the library's cells are available to you for placement and you can create new cells in the library.

CREATING YOUR OWN CELLS

When you need to place copies of a symbol for which no cell currently exists, you can create your own cell, store it in the attached library, and then use it in any design file.

Before You Start

Here are a few things to consider before starting drawing the elements that will make up your cell.

Cell Elements

A cell is a group of elements that you draw and load into the attached cell library. There is no limit on the size of cells; all element types can be placed in a cell; they can be drawn on any level; and they can be any color, weight, or style.

The original elements do not become a cell. After you create a cell from them, you manipulate or delete the elements.

Cell Rotation

Normally, you draw the object you are going to make into a cell with 0 degrees of rotation. That means it should be upright and facing to the right (see Figure 11–6). Sticking with zero rotation for cells makes it easier to understand what happens to the cell when it is rotated.

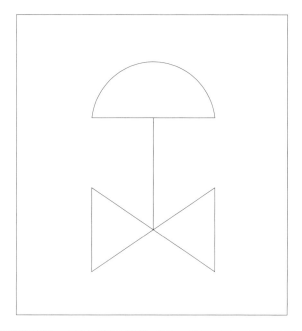

Figure 11-6 Example of a symbol drawn upright and facing to the right

Cell Origin

Each cell has an origin point that determines where the cell is placed in relation to the location of the data point that places it.

The cell is placed such that its origin point is on the data point. For example, if the origin point is in the center of the cell, the cell is centered on the data point. The origin point is defined when the cell is created.

Consider how the cell will be used when deciding where to place the origin when creating the cell. If the cell is to be placed connected to other elements, place the origin point at the connection point. For example, if the cell is a control valve, place the origin where the valve is to be attached to the pipeline, as shown in Figure 11–7.

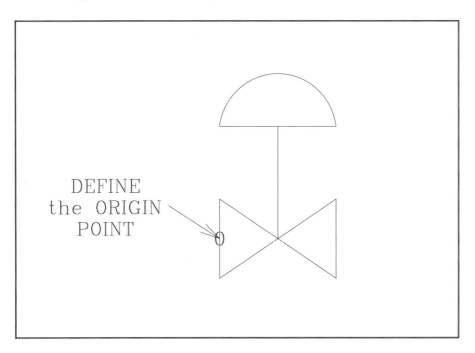

Figure 11-7 Example of defining a cell origin

Steps for Creating Cells

Following are the steps to create new cells.

STEP 1: Select **Element > Cells** to open the Cell Library box.

STEP 2: If you want to put the new cells in an existing cell library, attach the library. Otherwise, create a new cell library to hold the cells.

STEP 3: Draw the elements.

STEP 4: Group the elements by either placing a fence around them or by selecting them with the Element Selection tool to select each element.

Chapter 11 Cells and Cell Libraries

 Note: All of the Fence modes can be used to create a cell from the contents of the fence. Be aware that if the mode is Void or Void-Clip, the cell will be formed from all elements in the design that are *outside* of the fence.

STEP 5: Define the cell's origin point by invoking the Define Cell Origin tool from:

Cells tool box	Select the Define Cell Origin tool (see Figure 11–8).
Key-in window	**define cell origin** (or **de c o**) ENTER

MicroStation prompts:

> Define Cell Origin > Define origin *(Place a data point where the origin is to be located. Snap, if necessary, to place the point precisely.)*

Figure 11-8 Invoking the Define Cell Origin tool from the Cells tool box

 Note: The location of the origin point is indicated by a cross. The cross is not an element, just an indication of the origin location on the view. If you accidentally define the origin point in the wrong place, define another one. The cell will be created using the last origin you define. You can only create a cell after defining an origin point, and the origin point can only be used one time.

STEP 6: Click the **Create** button located in the bottom right corner of the Cell Library box to open the New Cell dialog box, as shown Figure 11–9.

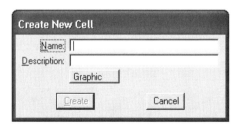

Figure 11-9 Create New Cell dialog box

501

STEP 7: Click the **Name** text field in the Create New Cell dialog box and enter a name for your cell. There is no restriction on the length of the cell name and it can contain any combination of numbers, characters, and symbols (except: & * = < > ? / : " \ |). The name can also contain space characters, except that it cannot start with a space character. Use the name to tell users what the symbol is.

STEP 8: Optionally, click the **Description** text field and enter a description of the cell. There is no restriction on the length of the description and it can contain any combination of numbers, characters, and symbols. Use the description to add notes that will help users understand when and how to use the cell.

STEP 9: Select the appropriate cell type from the drop-down menu located just below the **Description** field. Cells that can be placed in a design can be one of two types—**Graphic** or **Point**. The difference between the two types is described later.

STEP 10: Click the **Create** button to create a new cell from a copy of the selected elements.

 Note: The new cell is placed in the attached library and it appears in the cells list in the Cell Library box, as shown in Figure 11–10. The cell is available for placement in the current design file and in any other design files to which your cell library is attached.

STEP 11: If you no longer need the original elements, delete them from the design file.

STEP 12: Repeat steps 3 though 11 for each cell you need to create.

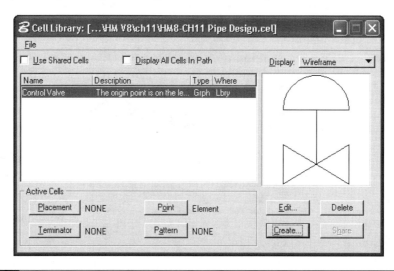

Figure 11-10 Displaying the name and description of the new cell in the Cell Library box

Graphic and Point Cell Types

As just mentioned, two cell types, Graphic and Point, can be placed in your design. The default cell type is Graphic (also called Normal). Here are the differences between Graphic and Point cells.

A Graphic cell, when placed in a design file:

- Keeps the symbology (color, weight, style, and levels) of its elements.
- Remembers the levels on which its elements were drawn. (MicroStation provides two methods for placing the cells—Absolute and Relative modes, discussed in detail later in the chapter.)
- Retains the keypoints of each cell element.

A Point cell, when placed in a design file:

- Takes on the current active color, weight, and style.
- Places all cell elements on the current active level, regardless of what level they were drawn on.
- Has only one keypoint—the cell's origin point. The individual cell elements do not have keypoints.

Selecting the Cell Type

When you create a new cell, you can select either **Graphic** or **Point** from a drop-down menu on the Create New Cell dialog box option. The menu also offers a **Menu** type that is used for customizing the way you use MicroStation and is beyond the scope of this book.

ACTIVE CELLS

To place a copy of a cell in your current design file, select the cell's name in the Cell Library box, click the Active Cells button for the way you want to place the cell, and then select a cell placement tool from the Cells tool box. There are four active cell types that give you the several different ways to use cells in your design file.

- The active cell **Placement** button is used with a group of tools that place cells at a data point.
- The active cell **Terminator** button is used with a tool that places the cell on the end of an element.
- The active cell **Point** button is used with a group of tools that place cells in geometric relations with other elements.
- The active cell **Pattern** button is used with a group of tools that create patterns, such as running bond brick, by placing multiple copies of the cell within a defined area.

When you click one of these buttons, the name of the selected cell is placed next to the button. The Placement, Terminator, and Point tools are described in this chapter. Pattern tools are described in Chapter 12.

Alternate Methods for Selecting the Active Cell

You also can use key-ins to select an active cell for each type of cell placement. In the key-in window, type:

ac=<name>   *and press* ENTER *to select the active placement cell.*
lt=<name>   *and press* ENTER *to select the active terminator cell.*
pt=<name>   *and press* ENTER *to select the active point cell.*
ap=<name>   *and press* ENTER *to select the active pattern cell.*

In each tool, replace <name> with the name of the cell you want to use.

Example: ac=coil ENTER

Several cell placement tools provide a cell name field in the Tool Settings window. You can also select the active cell by typing its name in this field.

Place Cells

To place copies of a cell in your design file, you must select the cell you want to use in the Cell Library box, click the Active Cells **Placement** button to tell MicroStation which cell to use, and then select a placement tool from the Cells tool bar. Three tools are provided to place a cell at data points in your design file:

- Place Active Cell—Places copies of the Active Placement Cell at data points you specify. The cell is placed at the Active Angle and Active Scale.
- Select and Place Active Cell—Lets you select a cell already placed in your design file to be the Active Placement Cell and places copies at the data points you specify.
- Place Active Cell Matrix—Places a rectangular matrix of the active placement cell, with the lower left corner of the matrix at the data point you specify.

Note: MicroStation provides a shortcut for placing individual copies of a cell. When you double-click the cell's name in the Cell Library box, the cell's name is placed next to the Active Cells **Placement** button and the Place Active Cell tool is invoked.

Place Active Cell

The Place Active Cell tool places one copy of the Active Placement Cell at the location you define with a data point. When Place Active Cell is selected, the Tool Settings window provides several options for modifying the way the cell is placed.

The **Active Cell** text field displays the name of the Active Placement Cell, and you can change to a different cell by typing the cell's name in the field. The **Browse Cell(s)** magnifying glass symbol opens the Cell Library box so that you can select another cell from the box. If the settings box is already open when you click the magnifying glass, MicroStation switches focus to the settings box.

Chapter 11 Cells and Cell Libraries

The **Active Angle** field allows you to rotate placements of the Active Placement Cell by the number of degrees you key-in the field. For example, if you key-in 90, the cell is placed flipped over on its left side. Positive angles rotate the cell in a counterclockwise direction and negative angles rotate the cell clockwise. The scroll bar on the right side of the field allows you to select standard rotation angles such as 45, 90, and 180 degrees.

The **X Scale** and **Y Scale** fields allow you to change the size of the cell as it is placed. Numbers greater than zero increase the size of the placed cell and numbers less than zero decrease its size. If the lock next to the two fields is open, you can enter different numbers in each field. If the lock is closed, a number entered in one field is automatically entered in the other field.

If you only see the above settings on the Tool Settings window, click the **Show Cell Placement Options** button on the lower right side of the window to expand the window and reveal more settings fields.

If the cell was created in a design file that has different working units from the active design file, setting the **True Scale** check box ON causes the cell to be "soft scaled" or adjusted to the correct size using the working units of the active design file when it is placed. For example, the Active Placement Cell is a twelve-inch square box that was created in a design whose Master Unit is Inches and the active design file's Master Unit is Feet. If the check box is set to ON, the placed cell is a twelve-inch square box, but if the check box is set to OFF, the placed cell is a twelve-foot square box.

The actions of the **Relative** and **Interactive** check boxes are described later, and the **Flatten** check box and drop-down menu are for 3D designs only.

If the **Association** check box is ON, you can associate the placed cell with an element in the design by snapping to the element and then accepting it. If you associate the cell with an element and move the element, the cell moves with it.

After you make the required settings in the Tool Settings window, invoke the Place Active Cell tool from:

Cells tool box	Select the Place Active Cell tool (see Figure 11–11).
Key-in window	**place cell icon** (or **pl ce i**) ENTER

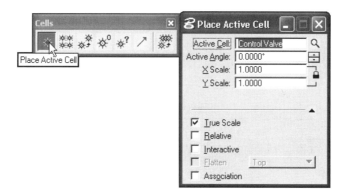

Figure 11–11 Invoking the Place Active Cell tool from the Cells tool box

MicroStation prompts:

> Place Active Cell > Enter cell origin *(Define the origin point at each location where a copy of the cell is to be placed.)*

Each time you place a data point, a copy of the cell is placed such that its origin is on your data point. While this tool is active, the screen cursor drags a dynamic image of the cell. The cell's origin point is at the screen cursor position.

 Note: If you do not have an active cell and invoke the Place Active Cell tool, MicroStation displays the message *No Active Cell* in the Status bar. You must declare an active cell before you can use this tool (discussed earlier in this section).

Place Active Cell using Absolute or Relative Mode

If the **Relative** check box in the Place Active Cell tool's Tool Settings window, is OFF, the tool places cells in Absolute placement mode. If the check box is ON, the tool places cells in Relative placement mode.

In **Absolute** placement mode, the Graphic cell's elements are placed on the same levels as that they were drawn on regardless of the Active Level setting. For example, if a graphic cell contains elements on levels 1 and 3, they are placed in the design file on levels 1 and 3 (see Figure 11–12a).

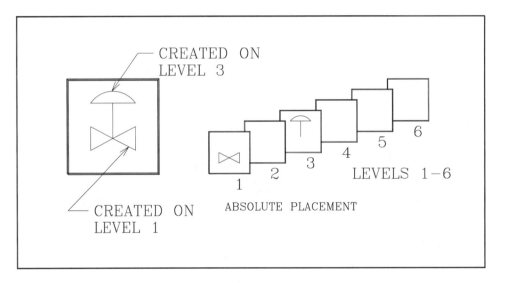

Figure 11-12a Example of placing a cell by Absolute placement mode

In **Relative** placement mode, MicroStation shifts all levels used in the cell such that the lowest level in the cell is on the Active Level in the design. For example, if a Graphic cell has elements on levels 1 and 2 and the active level is 4, the cell elements on level 1 are moved three levels up to level 4 in the design file and the cell elements on level 3 are moved three levels up to level 6 (see Figure 11–12b).

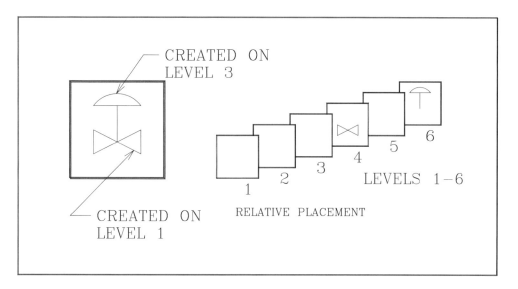

Figure 11-12b Example of placing a cell by Relative placement mode

Place Active Cell using Interactive Placement

Interactive placement is helpful when you need to align your cell with existing elements but you don't know what angle and scale to make it. You define the angle and scale graphically as you place the cell.

To place the active cell interactively, turn the **Interactive** check box ON in the Tool Settings window. When a cell is placed interactively, MicroStation prompts:

Place Active Cell (Interactive) > Enter cell origin *(Define the origin point of the cell.)*

Place Active Cell (Interactive) > Enter Scale or Corner Point *(Either place a data point to scale the cell or key-in the scale factor in the key-in window.)*

Place Active Cell (Interactive) > Enter Rotation by Angle or Point *(Either place a data point to rotate the cell or key-in the rotation angle in the key-in window.)*

Select and Place Cell

Often, when working in an existing design, you will need to place additional copies of a cell that was placed earlier but is no longer the Active Placement Cell. The Select and Place Cell tool allows you to select that cell for placement simply by clicking on a copy of the cell in the design file. The cell you click becomes the active placement cell, and dynamic update shows it at the pointer position. Additional data points place copies of the cell.

The Select and Place Cell tool's Tool Settings window provides the **Active Angle, X Scale**, **Y Scale** fields, the **Browse Cells** magnifying glass, and the **Relative** check box. These fields were described earlier in the discussion of the Place Active Cell tool.

Invoke the Select and Place Cell tool from:

Cells tool box	Select the Select and Place Cell tool (see Figure 11–13).
Key-in window	**select cell icon** (or **se c i**) ENTER

Figure 11–13 Invoking the Select and Place Cell tool from the Cells tool box

MicroStation prompts:

> Select and Place Cell > Identify element *(Select the cell to be copied.)*
>
> Select and Place Cell > Accept/Reject (Select next input) *(Accept the cell by placing a data point at the location where the first copy is to be placed.)*
>
> Select and Place Cell > Enter cell origin *(Place a data point at the location of each additional copy of the cell, or click the Reset button cancel the operation.)*

This tool works only with the copy of the cell in the design, so you can make copies of cells that are not in the currently attached cell library. In fact, the tool works when there is no cell library attached.

Place Active Cell Matrix

Do you need to place two rectangular rows of electronic components in a circuit diagram? The Place Active Cell Matrix tool can do it for you. With this tool, you can quickly place copies of the Active Placement Cell in a rectangular matrix whose parameters you define (see Figure 11–14).

Chapter 11 Cells and Cell Libraries

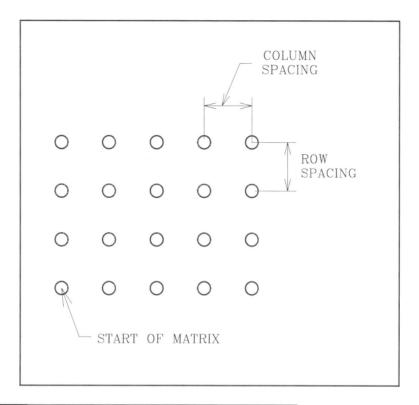

Figure 11-14 Example of placing a cell in a rectangular array

The Place Active Cell Matrix tool's Tool Settings window provides the following fields for defining the size and shape of the matrix.:

- **Active Cell**—The active cell name
- **Browse Cells**—The magnifying glass that opens the Cell Library box
- **Rows**—The number of rows in the matrix
- **Columns**—The number of columns in the matrix
- **Row Spacing**—The space, in Working Units, between the rows
- **Column Spacing**—The space, in Working Units, between the columns

 Note: The row and column spacing is from origin point to origin point, not the distance between the cells.

Invoke the Place Active Cell Matrix tool from:

Cells tool box	Select the Place Active Cell Matrix tool (see Figure 11–15), then set the matrix parameters in the Tool Settings window.
Key-in window	**matrix cell** (or **matr c**) ENTER

Figure 11–15 Invoking the Place Active Cell Matrix tool from the Cells tool box

MicroStation prompts:

Place Active Cell Matrix > Enter lower left corner of matrix *(Place a data point to define the origin location of the cell in the lower left corner.)*

The data point you place to start the matrix designates the position of the origin of the lower left cell in the matrix (see Figure 11–14).

The Place Active Cell Matrix tool places each cell in the matrix at the active angle and active X and Y scales, but these settings are not available in the Tool Settings window. Invoke the Place Active Cell tool or the Select and Place Cell tool to gain access to the **Active Angle**, **X Scale**, and **Y Scale** settings.

Place Active Line Terminator

The Place Active Line Terminator tool places the active terminator cell at the end of the element you select and automatically rotates the cell to match the rotation of the element at the point of connection. See Figure 11–16 for examples of placing line terminators.

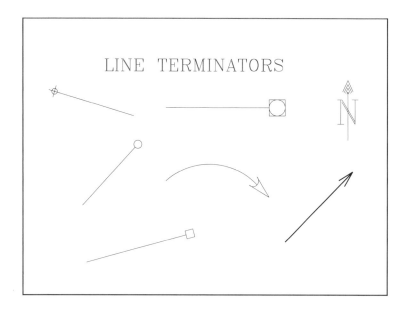

Figure 11-16 Examples of placing line terminators

The tool places the Active Terminator Cell that you select from the Cell Settings box by selecting a cell and then clicking the **Terminator** button.

The Tool Settings window provides fields for modifying the tool's action. The **Terminator** text field displays the name of the Active Terminator Cell and you can type a cell name in the field to change to a different active cell. The **Browse Cells** magnifying glass displays the Cell Library box. The **Scale** field allows you to key-in a scale factor that will be applied to the terminator cell when it is placed. The field sets both the X and Y scale values. A number greater than zero increases the size of the cell and a number less than zero decreases its size.

Invoke the Place Active Line Terminator tool from:

Cells tool box	Select the Place Active Line Terminator tool (see Figure 11–17).
Key-in window	**place terminator** (or **pl t**) ENTER

Figure 11-17 Invoking the Place Active Line Terminator tool from the Cells tool box

MicroStation prompts:

> Place Active Line Terminator > Identify element *(Select the element near the end where the terminator is to be placed.)*
>
> Place Active Line Terminator > Accept/Reject (Select next input) *(Click the Data button again to place the terminator, and, optionally, select another element to place a terminator.)*

 Note: Beginning users of MicroStation often select the element to place a terminator by first snapping to the element with the Tentative button. That is not necessary, because MicroStation finds the end of the element automatically. Just identify the element near the end on which you want to place the terminator.

Point Placement

MicroStation provides six point tools that place a dot, character, or cell in your design file:

- Place Active Point—Places a single point at the data point you specify.
- Construct Points Between Data Points—Places a set of equally spaced points between two data points.
- Project Point Onto Element—Places a point on an element at the point on the element nearest to a data point.
- Construct Point at Intersection—Places a point at the intersection of two elements.
- Construct Points Along Element—Places a set of equally spaced points along an element between two data points on the element.
- Construct Point at @Dist Along Element—Places one point at a keyed-in distance along an element from a data point.

Types of Points

The Tool Settings window for each point tool provides the **Point Type** drop-down menu for selecting the type of point you want to place. The available point types are as follows.

- **Element**—Places dots (0-length lines). If you want to make these dots more noticeable, increase the Active Line Weight before placing them.
- **Character**—Places a text character using the Active Font. Key-in the appropriate character in the **Character** edit field located in the Tool Settings window.
- **Cell**—Places the Active Point Cell as the point. Point cells are placed at the active angle and active scale. You can select the Active Point Cell from the Cell Library box by selecting a cell name and clicking the **Point** button. The Cell Library box can be opened by

clicking the **Browse Cell(s)** magnifying glass. You can also select the point cell by keying-in the cell's name in the **Cell** field on the Tool Settings window of each Point tool.

 Note: Character and cell points are rotated to the Active Angle. Cell points are rotated to the Active Angle and scaled to the Active Scale. Angle and Scale fields are not provided in the Tool Settings window for the Point tools.

Do not confuse using a cell as the active point with point and graphic cells. You can make either a point or a graphic cell the active point cell. The Point tools place graphic cells in Absolute mode only.

Place Active Point

The Place Active Point tool places a single point (element, character, or cell) at the location of the data point you define.

Invoke the Place Active Point tool from:

Points tool box	Select the Place Active Point tool (see Figure 11–18).
Key-in window	**place point** (or **pl po**) ENTER

Figure 11–18 Invoking the Place Active Point tool from the Points tool box

MicroStation prompts:

> Place Active Point > Enter point origin *(Place a data point to place one active point in the design or click Reset to drop the point.)*

Construct Points Between Data Points

The Construct Points Between Data Points tool places a specified number of points (element, character, or cell) between two data points.

In addition to the fields that are in the Tool Settings window for all point tools, the window provides the **Points** field where you can key-in the number of points to place between the two data points. The number you key-in includes the points that will be placed on the two data points.

Invoke the Construct Points Between Data Points tool from:

Points tool box	Select the Construct Points Between Data Points tool (see Figure 11–19) and key-in the number of points in the Points field.
Key-in window	**construct points between** (or **constru po b**) ENTER

Figure 11-19 Invoking the Construct Points Between Data Points tool from the Points tool box

MicroStation prompts:

> Construct Pnts Between Data Points > Enter first point *(Place a data point to define the location of the first point in the series.)*
>
> Construct Pnts Between Data Points > Enter endpoint *(Place a data point to define the location of the last point in the series, or click the Reset button to terminate the tool without placing any points.)*

After you place the first set of points, you can continue placing additional sets. Each additional set uses the last data point of the previous set as its starting point. To start over with a first data point, click the Reset button on your pointing device. See Figure 11–20 for an example of placing 10 points between two data points.

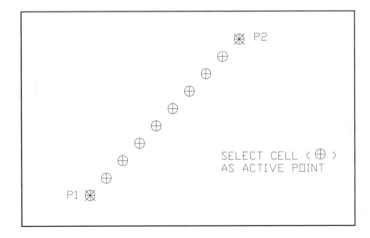

Figure 11-20 Example of placing 10 points between two data points

Project Point Onto Element

The Project Point Onto Element tool places the active point (element, character, or cell) on the selected element at a location projected from the acceptance data point.

Invoke the Project Point Onto Element tool from:

Points tool box	Select the Project Point Onto Element tool (see Figure 11–21).
Key-in window	**construct point project** (or **constru po p**) ENTER

Figure 11–21 Invoking the Project Point Onto Element tool from the Points tool box

MicroStation prompts:

> Construct Active Point Onto Element > Identify element *(Select the element.)*
>
> Construct Active Point Onto Element > Accept/Reject (Select next input) *(Select the point in the design from which to project the point.)*

Figure 11–22 shows examples of projecting the active point onto an element.

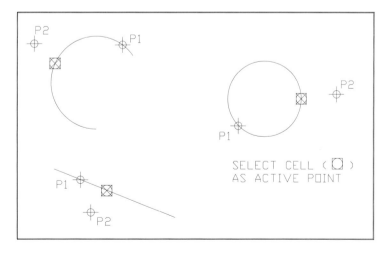

Figure 11–22 Example of placing an active point projected onto an element

Note: The only purpose of the first data point is to identify the element to which the projection will be made. There is no need to use the Tentative button when identifying the element.

Construct Point at Intersection

The Construct Point at Intersection tool places an active point (element, character, or cell) at the intersection of two elements.

Invoke the Construct Point at Intersection tool from:

Points tool box	Select the Construct Point at Intersection tool (see Figure 11–23).
Key-in window	**construct point intersection** (or **constru po i**) ENTER

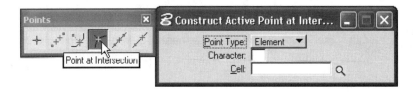

Figure 11–23 Invoking the Construct Point at Intersection tool from the Points tool box

MicroStation prompts:

> Construct Active Point at Intersection > Select element for intersection *(Select one of the elements.)*
>
> Construct Active Point at Intersection > Select element for intersection *(Select the other element.)*
>
> Construct Active Point at Intersection > Accept - Initiate intersection *(Place a data point to place the point.)*

The acceptance data point only initiates placement of the point; it does not identify another element. See Figure 11–24 for examples of placing an active point at an intersection of two elements.

Chapter 11 Cells and Cell Libraries

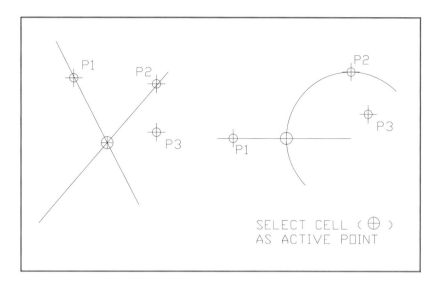

Figure 11-24 Example of placing an active point at an intersection of two elements

 Note: If the two elements intersect more than once (such as a line passing through a circle), identify the two elements close to the intersection where you want the point to be placed. There is no need to use the Tentative button. Just place the Data button close to the intersection.

Construct Points Along Element

The Construct Points Along Element tool places a set of active points (elements, characters, or cells) equally spaced along an element between two data points on the element.

In addition to the fields that are in the Tool Settings window for all point tools, the window provides the **Points** tool, where you can key-in the number of points to place between the two data points. The number you key-in includes the points that will be placed on the two data points.

Invoke the Construct Points Along Element tool from:

Points tool box	Select the Construct Points Along Element tool (see Figure 11–25) key-in the number of points in the Points field.
Key-in window	**construct point along** (or **constru po a**) ENTER

517

Figure 11-25 Invoking the Construct Points Along Element tool from the Points tool box

MicroStation prompts:

Construct Pnts Along Element > Enter first point *(Select the element at the location where the first point is to be placed.)*

Construct Pnts Along Element > Enter endpoint *(Define the location on the element where the last point is to be placed on the element or click the Reset button to terminate the tool without placing any points.)*

You can continue placing points along elements, and, if necessary, you can also change the active point at any time while placing them. See Figure 11-26 for an example of placing 10 points along an element between two data points.

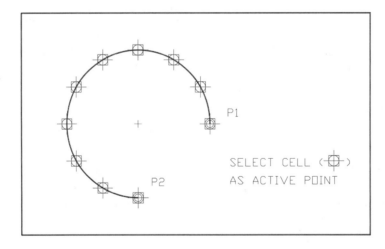

Figure 11-26 Example of placing 10 points along an element between two data points

Point at Distance Along

The Point at Distance Along tool places the active point (element, character, or cell) at a keyed-in distance along an element from the data point that identified the element.

In addition to the fields that are on the Tool Settings window for all point tools, the window provides the **Distance** tool, where you can key-in the distance in working units.

Chapter 11 Cells and Cell Libraries

Invoke the Construct Point at Distance Along tool from:

Points tool box	Select the Point at Distance Along tool (see Figure 11–27) key-in the distance in the Distance field.
Key-in window	**construct point distance** (or **constru po d**) ENTER

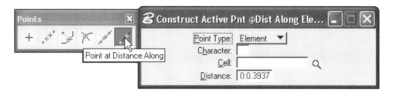

Figure 11-27 Invoking the Point at Distance Along tool from the Points tool box

MicroStation prompts:

> Construct Active Pnt @Dist Along Element > Identify element *(Identify the element.)*
>
> Construct Active Pnt @Dist Along Element > Accept/Reject (Select next input) *(Place a data point to accept the construction or click the Reset button to terminate the tool sequence without placing a point.)*

You can continue selecting elements along which you want to place a point and you can change the distance at any time while using the tool. See Figure 11–28 for an example of placing a point at a specified distance along an element.

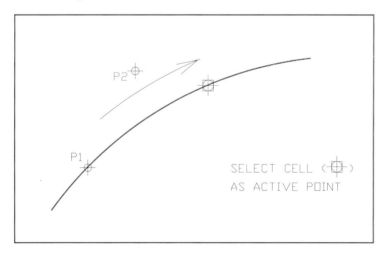

Figure 11-28 Example of placing a point at a specified distance along an element

519

CELL SELECTOR

In addition to the Cell Library box, MicroStation provides the Cell Selector box for previewing and selecting cells. The Cell Selector box has a button for each cell in the attached library. Clicking one of the buttons selects a placement action. In addition to the cell library file attached from the Cell Library box, the Cell Selector box can contain buttons for cells from other cell libraries.

Opening the Cell Selector Box

Open the Cell Selector box from:

| Drop-down menu | Utilities > Cell Selector |

MicroStation displays the Cell Selector box, as shown in Figure 11–29.

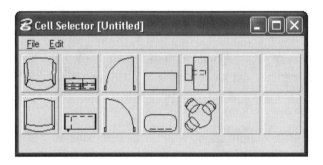

Figure 11-29 Cell Selector box

The contents of the box vary. If no customization has been done to the Cell Selector box, it contains buttons for all cells in the cell library attached to the design file. If no cell library is attached, MicroStation opens the Select Cell Library dialog box so you can select a cell library.

Invoking Tools from the Cell Selector Box

A unique tool action can be assigned to each cell button that is invoked when the button is clicked in the Cell Selector box. The default action is to set the cell assigned to the button as the Active Placement Cell and invoke the Place Active Cell tool.

Customizing the Cell Selector

The **Edit** menu in the Cell Selector box provides options for customizing the action, appearance, and content of the cell buttons.

Chapter 11 Cells and Cell Libraries

Editing A Button's Content

Use the **Button** option to edit the content and appearance of a selected cell button.

Click the button to be edited and invoke the **Button** option from:

| Cell Selector settings box | Edit > Button |

MicroStation displays the Configure Cell Selector Button settings box, as shown in Figure 11–30.

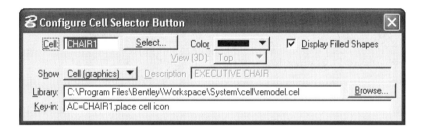

Figure 11-30 Configure Cell Selector Button dialog box

Make the required settings changes and then close the dialog box. Following are descriptions of the fields in the dialog box.

- Click the **Color** button to display a menu of button colors. Click one of the colors to change the button's color. This changes the color of the button, not the color of the cell itself. Later you will learn that you can place the same cell on additional buttons and assign different click actions to each button. Using unique colors for each type of action can help you find the correct button. For example, buttons with the Place Active Cell action can be black and buttons with the Place Terminator action can be green.

- If the **Display Filled Shapes** check box is set to ON, and the cell contains filled elements, the elements are displayed on the button with their fill color. If the check box is set to OFF, no fill colors are displayed on the button. The fill tools are described in Chapter 12.

- The **Show** menu provides options that allow you to control the information displayed on the selected button. Select **Cell (Graphics)** to display a picture of the cell, **Cell Name** to display only the cell's name, **Description** to display the cell's description, or **Cell and Name** to display a picture of the cell on the button and the cell name below the button.

- The cell **Description** text field is available only when the **Description** option is selected in the **Show** menu. It allows you to edit the cell description that appears in the button.

521

- The **Library** text field displays the name of the cell library associated with the selected button. To change to a different library, type the location and file name in the field or click the **Browse** button next to the field to open the Select Cell Library dialog box. If you open the dialog box, select the file location in the **Directories** area and the file name in the **Files** area, and then click the **OK** button to select the library.
- The **Key-in** text field shows the MicroStation key-in that is initiated when you click the cell button. You can enter multiple key-ins by separating them with a semi-colon. For example, "AC=CHAIR1;place cell icon" makes CHAIR1 the Active Placement Cell and activates the Place Active Cell tool.

Inserting a New Button

The **Insert** option inserts a new button immediately after the selected cell button.

Select the button that you want the new button to follow and invoke the Insert tool from:

Cell Selector box	Edit > Insert

MicroStation displays the Define Button dialog box, as shown in Figure 11-31.

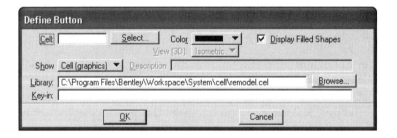

Figure 11-31 Define Button dialog box

The Define Button dialog box contains the same fields as the Configure Cell Selector Button dialog box. Make the required settings for the new button and click the **OK** button to save the settings or click the **Cancel** button to cancel the operation without creating a new button.

Copying and Pasting a Button

The **Copy** option copies the content of a selected cell button and the **Paste** option pastes the copied content into another selected cell button.

To copy the content of a cell button, select the button and invoke the **Copy** option from:

Cell Selector box	Edit > Copy

MicroStation places the contents of the selected button in a temporary "clipboard."

To paste the contents of the clipboard onto another cell button, select the button and invoke the **Paste** option from:

| Cell Selector box | Edit > Paste |

MicroStation places the contents of the clipboard onto the selected cell button.

This procedure is useful when you need to use the same cell for more than one tool action. For example, if you want to use the Arrow cell both for placing copies of the active cell and for placing line terminators, set the action in the first Arrow button the Place Active Cell tool, copy and paste the Arrow button to a second button and set that button's action for the Place Terminator tool.

Cutting a Button

The **Cut** option removes the contents from the selected cell button and places the content onto the temporary clipboard so that you can paste it onto another button.

Select the button whose content is to be cut and invoke the **Cut** option from:

| Cell Selector box | Edit > Cut |

MicroStation immediately cuts the selected button's content and removes the button.

Used with the **Paste** option, this tool is useful for moving a cell to a new location on the Cell Selector box.

Deleting a Button

The **Delete** option deletes the selected cell button from the Cell Selector box.

Click the button to be deleted and invoke the Delete tool from:

| Cell Selector box | Edit > Delete |

MicroStation immediately deletes the configuration of the selected button.

Deleting all Buttons

The **Clear Configuration** option deletes all buttons from the Cell Selector box.

Invoke the Clear Configuration tool from:

| Cell Selector box | Edit > Clear |

MicroStation immediately deletes the content of all cell buttons on the Cell Selector box.

Changing Button and Gap Size

The **Button Size** option allows you to change the size of the button and the gap between buttons. If you want to be able to see more buttons without scrolling the Cell Selector box, make the size and gap smaller. If you are having a hard time reading the content of the buttons, make them larger.

Invoke the Button Size tool from:

Cell Selector box	Edit > Button Size

MicroStation displays the Define Button Size dialog box, as shown in Figure 11–32.

Figure 11-32 The Define Button Size dialog box

The **Button Size (Pixels)** and **Gap Size (Pixels)** text fields set the size in pixels of the buttons and of the gap between the buttons. After keying-in the desired pixel values in the two fields, click the **OK** button to apply the new sizes or click **Cancel** to cancel the operation without changing either size.

 Note: Pixels are the smallest addressable unit on a display screen. When you increase the number of pixels for the button or gap size, the size is increased.

Setting Button Defaults

The **Defaults** option allows you to select default settings for creating new buttons in the Cell Selector box.

Invoke the Define Defaults tool from:

Cell Selector box	Edit > Defaults

MicroStation displays the Define Defaults dialog box, as shown in Figure 11–33.

Chapter 11 Cells and Cell Libraries

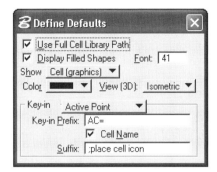

Figure 11-33 Define Defaults dialog box

Make the required settings changes and then close the dialog box. Most of the settings are identical to settings in the Configure Cell Selector Button dialog box that were described earlier in this section. Following are descriptions of the default settings that have not previously been described.

- If the **Use Full Cell Library Path** check box is set to ON, the full path is displayed in the **Library** field of the Define Button dialog box when you insert a new button. If the check box is set to OFF, only the cell library file name is displayed.

- The **Font** text field displays the number of the MicroStation font that is used for the cell name or description on the new button.

- The **View (3D)** drop-down menu sets the default display method for 3D cells and is not used in 2D design files.

- The **Key-in** drop-down menu selects the Active Cell key-in that is placed in the **Key-in** field when the new button is created. The options are **Active Cell** (AC=), **Active Point** (PT=), **Active Pattern** (AP=), **Active Terminator** (LT=), and **User Defined** (you type the key-in in the **Key-in Prefix** field).

- The **Key-in Prefix** contains the first key-in that is placed in the **Key-in** field for new buttons. If one of the active cell options is selected in the **Key-in** menu, the key-in characters for the type of active cell are placed in the **Key-in Prefix** field. If the **User Defined** option is selected, the field is empty. You can add to, change, or replace any text in this field.

- If the **Cell Name** check box is set to ON, the name of the cell assigned to the new button is added to the **Key-in** field for new buttons. This check box is automatically turned on when one of the active cell options is selected in the **Key-in** menu.

- The **Suffix** field contains additional key-in sequences for new buttons. This key-in is placed after the key-in listed in the **Key-in Prefix** field and is separated from it by a semicolon. If you place more than one key-in in this field, separate them with a semi-colon.

525

Creating and Using Cell Selector Files

The **File** options in the Cell Selector box allow you to add cells from other libraries to the Cell Selector box and save the box's configuration to a file that can be opened in other design files.

Adding Buttons from Other Cell Libraries

In the cell selection descriptions thus far, the cells on the buttons have all been from the cell library currently attached to the design file. The Cell Selector's **Load Cell Library** option allows you to place cells from additional cell libraries in the active design file's Cell Selector box. The cell buttons from loaded libraries have the same features as buttons from the attached library. For example, they can be used with the Place Active Cell tools.

Invoke the Load Cell Library tool from:

| Cell Selector box | File > Load Cell Library |

MicroStation displays the Select Cell Library to Load dialog box, as shown in Figure 11-34.

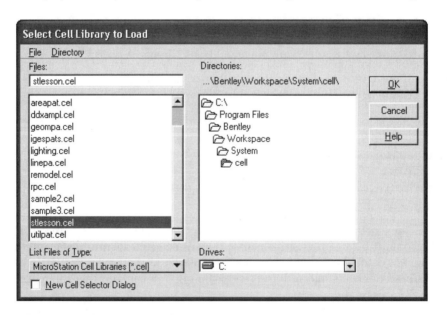

Figure 11-34 The Select Cell Library to Load dialog box

Select the location of the file to be loaded from the **Directories** area and select the file name from the **Files** area. Click the **OK** button to load the selected library or click the **Cancel** button to cancel the operation without loading the library.

Creating a New File

Use the **New** option to create a file that contains the Cell Selector configuration.

Invoke the **New** option from:

| Cell Selector box | File > New |

MicroStation displays the Define Cell Selector File dialog box.

Select the location where the file is to be placed in the **Directories** area and key-in the file name in the **Files** field. Click the **OK** button to create the file or click the **Cancel** button to cancel the operation without creating the file.

After you click the **OK** button, the file location and name are displayed in the Cell Selector box.

Saving the Current Configuration

If you make changes to the configuration after creating a new configuration file, use the **Save** option to save the changes to the file.

Invoke the **Save** option from:

| Cell Selector box | File > Save |

If there is a file location and name in the Cell Selector title bar, MicroStation saves the Cell Selector changes in that file. Otherwise, it invokes the **New** option so you can create a new file to store the configuration.

Saving the Current Configuration to a Different Cell Selector File

Use the **Save As** option to save the current Cell Selector content and configuration to a different file from the one displayed in the title bar.

Invoke the **Save As** option from:

| Cell Selector box | File > Save As |

MicroStation displays the Define Cell Selector File dialog box. Select a location for the file in the **Directories** area and key-in the file name in the **Files** field. Click the **OK** button to create the file or click the **Cancel** button to cancel the operation without creating the file.

Opening a Cell Selector File

Use the **Open** option to replace the current Cell Selector with the Cell Selector stored in a file.

Invoke the **Open** option from:

| Cell Selector box | File > Open |

MicroStation displays the Select Cell Selector File dialog box. Select the file location in the **Directories** area of the box and the file name from the **Files** area. Click **OK** to replace active design file's Cell Selector with the one in the selected file, or click the **Cancel** button to cancel the operation without opening the file.

CELL HOUSEKEEPING

You have seen several tools that allow you to select active cells and place copies of them in your design file. Now let's look at additional tools that work on cells already placed in your design file. The available tools include the following:

- Identify Cell—Displays the name and other related information of the selected cell.
- Replace Cell—Replaces a cell in the design file with another cell from the currently attached cell library.
- Drop Complex Status—Breaks up a cell into its individual elements.
- Drop Fence Contents—Breaks up all cells contained in a fence to their individual elements. The elements lose their identities as cells.
- Fast Cells View—Speeds up view updates by displaying only a box showing the location of all cells in the view, rather than the cell elements.

Identify Cell

The Identify Cell tool displays the name of a selected cell. This tool is useful when you want to use a cell already placed in your design file as a terminator cell or point cell but you don't know the cell's name. The information is displayed in the Status bar.

Invoke the Identify Cell tool from:

Cells tool box	Select the Identify Cell tool (see Figure 11–35).
Key-in window	**identify cell** (or **i c**) ENTER

Chapter 11 Cells and Cell Libraries

Figure 11-35 Invoking the Identify Cell tool from the Cells tool box

MicroStation prompts:

    Identify Cell > Identify element *(Select the cell.)*

    Identify Cell > Accept/Reject (Select next input) *(Place a data point to accept the cell or click the Reset button to cancel the selection.)*

The name of the cell appears in the Status bar after the first data point and again after the second data point.

Replace Cells

The Replace Cells tool updates or replaces cells in the design file. It is useful when the design of a cell is changed and there are copies of the old cell in your design file, and when design changes require different cells (such as a different type of valve).

The Replace Cells Tool Settings Window

The Tool Settings window provides several settings to control the way this tool operates, but not all of them may be visible when the tool is selected. Click the **Show Cell Replacement Options** arrow on the lower right of the window to display all the options, as shown in Figure 11–36.

Figure 11-36 Invoking the Replace Cells tool from the Cells tool box

The **Method** menu allows you to choose between updating and replacing selected cells. The **Update Method** replaces the selected cell in the active design with the cell of the same name in the attached cell library. The action of the **Replace Method** depends on the **Use Active Cell** check box (which is only available when the **Replace Method** is selected). If the check box is set to:

- OFF, you identify the cell in the design that is to be replaced, identify another cell in the design to replace the first cell, and click the Data button a third time to initiate the replacement.

529

■ ON, you identify and accept a cell to have it replaced by the Active Placement Cell. You set the Active Placement Cell by keying-in its name in the **Use Active Cell** text field or by clicking the **Browse Cell(s)** magnifying glass and selecting it from the Cell Library settings box.

The **Mode** menu provides two options that control the way cells are updated or replaced. Select the **Single Method** to update individual cells by selecting and accepting them or select **Global Update** to update or replace all cells with the same name as the selected cell.

If the **Use Fence** check box is ON, and a fence is defined in the design, the tool manipulates all cells within the fence. When this check box is ON, the drop-down menu next to the check box allows you to select the **Fence Mode** to use for the tool operation.

If the **True Scale** check box is ON, and the replacement was created in a design file that used different working units from those in the active design file, the cell is adjusted to its true size using the units of the active file. If the check box is OFF, no size adjustment is made.

The **Replace Tags** and **Replace User Attributes** check boxes control which sets of tags and attributes are associated with the new cell. If the check boxes are ON, the new cell's tags and associations are used. If the check boxes are OFF, the tags and attributes of the old cell are retained.

If the **Relative Levels** check box is ON, the lowest level in the new cell is set equal to the lowest level in the old cell and the other levels in the new cell are adjusted up or down by the same number of levels as the lowest level. If the check box is OFF, no level adjustment is made.

Invoking the Replace Cells tool

Invoke the Replace Cells tool from:

Cells tool box	Select the Replace Cell tool (see Figure 11–36).
Key-in window	**replace cells extended** (or **rep cells e**) ENTER

MicroStation Prompts When a Fence is not Used

If there is no fence defined in the design, or the **Use Fence** check box is OFF, MicroStation's first prompt is:

Replace Cell > Identify Cell *(Identify the cell to be updated or replaced.)*

If the **Update Method** is selected, or the **Replace method** is selected and the **Use Active Cell** check box is turned ON, MicroStation' second prompt is:

Replace Cell > Accept/Reject *(Click the Data button to initiate the replacement action or click the Reset button to cancel the operation.)*

If the **Replace Method** is selected and the **Use Active Cell** check box is OFF, MicroStation's second and third prompts are:

> Replace Cell > Accept/Reject Replacement Cell *(Select the cell that is to replace the first cell you selected.)*

> Replace Cell > Identify Replacement Cell *(Click the Data button to initiate the replacement action or click the Reset button to cancel the operation.)*

Note: If you need to replace a shared cell (discussed later in this chapter), identify one of them, and MicroStation will replace all the instances of the shared cell with the same name.

The replaced cell may shift position in your design file. That happens when the new cell's origin point was not defined in the same relationship to the cell elements as the old cell's origin.

MicroStation Prompts When a Fence is Used

If a fence is defined in the design and the **Use Fence** check box is ON, the fence identifies the cells to be replaced, so MicroStation skips the cell identification prompt.

If the **Update Method** is selected, or the **Replace Method** is selected and the **Use Active Cell** check box is off, MicroStation prompts:

> Replace Cell > Accept/Reject Fence *(Click the Data button anywhere in the design to initiate the replacement action, or click the Reset button to cancel the operation.)*

If the **Replace Method** is selected and the **Use Active Cell** check box is OFF, MicroStation prompts:

> Replace Cell > Accept/Reject Replacement Cell *(Select the cell that is to replace the first cell you selected.)*

> Replace Cell > Identify Replacement Cell *(Click the Data button to initiate the replacement action or click the Reset button to cancel the operation.)*

Drop Complex Status

The cells placed in your design file are "complex shapes" that act like one element when manipulated. If you need to change the shape of a cell in your design file, you must first "drop" the cell to break it into separate elements. A dropped cell loses its identity as a cell and becomes separate, unrelated elements.

To drop a cell, invoke the Drop Element tool from:

Groups tool box	Select the Drop Element tool and turn the Complex check box ON (see Figure 11-37).
Key-in window	**drop element** (or **dr e**) ENTER

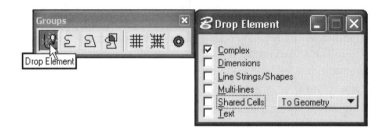

Figure 11-37 Invoking the Drop Element tool from the Groups tool box

MicroStation prompts:

> Drop Element > Identify element *(Identify the cell to be dropped.)*
>
> Drop Element > Accept/Reject (select next input) *(Click the Data button to drop the cell or click the Reset button to cancel the operation.)*

Drop Fence Contents

The Drop Fence Contents tool breaks all the cells enclosed in a fence into separate elements. Before you select this tool, group the cells to be dropped by placing a fence and selecting the appropriate fence mode.

Invoke the Drop Fence Contents tool from:

Fence tool box	Select the Drop Fence Contents tool (see Figure 11-38).
Key-in window	**fence drop complex** (or **f dr c**) ENTER

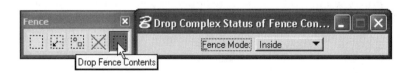

Figure 11-38 Invoking the Drop Fence Contents tool from the Fence tool box

Chapter 11 Cells and Cell Libraries

MicroStation prompts:

> Drop Complex Status of Fence Contents > Accept/Reject fence contents *(Click the Data button to initiate the drop action or click the Reset button to cancel the operation.)*

Fast Cells View

A lot of complex cells in a view may cause the Update View tool to take an unacceptable amount of time to complete refreshing the view. If you see that happening, you can reduce the time by turning the **Fast Cells** check box ON in the View attributes settings box. When the check box is ON for a view, a box is displayed at each cell location, rather than the cell elements.

To display boxes at the locations of cells in the view, do the following:

1. Select **Settings > View Attributes** to open the View Attributes settings box.
2. Turn the **Fast Cells** check box ON.
3. Check the **View Number** menu and, if necessary, change it to the number of the view where you want to display boxes in place of the actual cells.
4. Click the **Apply** button to apply the view attribute settings to the view whose number displayed in the **View** menu or click the **All** button to apply the settings to all open views.
5. If you no longer need the View Attributes settings box, close it.

The view attribute settings stay in effect until you change them or exit from MicroStation. To make the settings permanent, select **File > Save Settings**.

Note: If you print a view that has the **Fast Cells** check box turned ON, only the boxes print.

LIBRARY HOUSEKEEPING

You've seen how to place cells in your design file and take care of them. Now let's look at some housekeeping tools that help you take care of the cells in the attached cell library. The discussion includes explanations of how to do the following:

- Edit a cell's name and description
- Delete a cell from the Cell Library
- Compress the attached Cell Library
- Create a new version of a cell

All these tools affect the cells in the attached library, not the cells in your design file.

533

Edit a Cell's Name and Description

The Edit Cell Information dialog box, which is available from the Cell Library box, allows you to change a cell's name and description. If you need to create a new version of a cell, and you want to keep the old one around, rename it before creating the new version. If the person who designed the cell failed to provide a description, you can provide one to help other users of the cell library figure out what is in it. Following is the step-by-step procedure for renaming a cell and changing the cell description.

1. If the Cell Library box is not already open, select **Element > Cells** to open it.
2. Select the cell you want to edit.
3. Click the **Edit** button to open the Edit Cell Information dialog box (see Figure 11–39).
4. Make the necessary changes to the cell name and description.
5. Click the **Modify** button to make the necessary changes and close the Edit Cell Information dialog box.

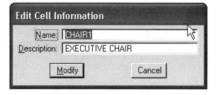

Figure 11-39 Edit Cell Information dialog box

Note: Changing the cell name and description in the library does not affect the cells already in the design file (or any other design file). They keep their old names.

Alternate Method

To rename a cell from the key-in window, type **CR=<old>, <new>**, and press ENTER. Replace <old> with the cell's current name and <new> with the new cell name. The key-in only changes the cell names. You cannot use a key-in to change descriptions.

Delete a Cell from the Library

If a cell becomes obsolete, or if it was not drawn correctly, it can be deleted from the cell library using the **Delete** button on the Cell Library box. When you ask to delete a cell, MicroStation opens an Alert dialog box to ask you if you really want to delete the cell. Following are the step-by-step procedures for deleting a cell from the Cell Library:

1. If the Cell Library box is not already open, select **Element > Cells** to open it.
2. Select the cell you want to delete.

3. Click the **Delete** button.
4. The Alert dialog box opens to ask you if you really want to delete the cell.
5. When the Alert box appears, click the **OK** button to delete it or click the **Cancel** button to cancel the operation.

Note: This procedure deletes a cell from the cell library. It does not delete copies of the cell already placed in the design file (or any other design file).

Alternate Method

To delete a cell from the key-in window, type CD=<name>, and press ENTER. Replace <name> with the cell's name.

When you use the key-in to delete a cell, MicroStation does not open the Alert window to ask you if you really want to delete the cell.

Note: The Undo tool will not undo the deleting of a cell from the attached library.

Compress the Attached Cell Library

Deleting a cell from a cell library does not really delete it. The cell is marked as deleted and is no longer available, but its elements still take up space in the cell library file. To get rid of the no-longer-usable cell elements, compress the cell library with the **Compress** option in the Cell Library box's **File** drop-down menu. Following are the step-by-step procedures to compress the Cell Library:

1. If the Cell Library box is not already open, select **Element > Cells** to open it.
2. From the **File > Compress**.

The attached cell library is compressed, and the Status bar indicates that the Cell Library is compressed.

Create a New Version of a Cell

Occasionally the need will arise to replace a cell with an updated version of that cell. The geometric layout of the object represented by the cell may have changed, or you may have discovered that a mistake was made when the cell was originally drawn. The cell elements cannot be edited in the library. You must create a new cell and place it in the library.

Follow this procedure to replace a cell in the library with a new version of the cell:

1. Place a copy of the cell in a design file that has the same Working Units as the design file in which the cell was created.
2. Drop the cell.

3. Make the required changes to the cell elements in the design.
4. If possible, place the cell origin at the same place as in the old cell.
5. Delete the old cell from the library, or rename it.
6. Create the new cell from the modified elements.
7. If you deleted the old cell, compress the library.
8. If you had already placed copies of the cell in your design files, invoke the Replace Cell tool to update those copies with the new version of the cell.

SHARED CELLS

Thus far, each time you placed a cell, a separate copy of the cell's elements was placed in the design file. That uses up a lot of disk space in a design file that contains many copies of the same cell.

Shared cells can help with the disk space problem. Each time you place a shared cell, the placement refers back to the shared copy rather than placing more elements in your design file. No matter how many copies of a shared cell you place, only one copy is actually in your design file.

When you declare a cell to be shared, MicroStation places a copy of it in your design file. You can have the shared cell's elements stored in your design file even though no copies of the cell have ever been placed in the design. Later, when you use the cell placement tools, the placements refer to the locally stored cell elements rather than to the copy in the library. Shared cells can be placed even when no cell library is attached.

Let's look how you can do the following:

- Turn on the shared cell feature
- Determine which cells are shared
- Declare a cell to be shared
- Place shared cells
- Turn a shared cell into an unshared cell
- Delete the shared cell copy from your design file

Turn on the Shared Cell Feature

The **Use Shared Cells** check box on the Cell Library box (see Figure 11–40) turns the shared cell feature ON or OFF. Each time you click the button, you toggle between using shared cells and not using them.

Chapter 11 Cells and Cell Libraries

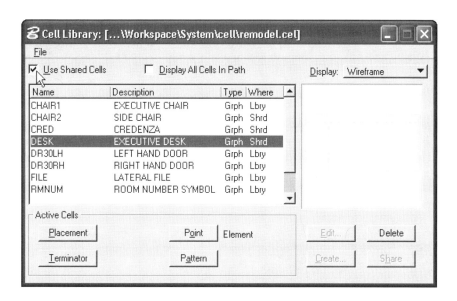

Figure 11-40 Use Shared Cells Check box on the Cell Library box

Determine Which Cells Are Shared

When the shared cells feature is ON, all shared cells in your design file show up in the Cells List in the Cell Library box (see Figure 11–40). Shared cells are indicated by "Shrd" in the **Where** column.

If the shared cell is also in the currently attached cell library, the list shows the shared copy in your design file—not the one in the library.

Each time you select a shared cell in your design file, the Status bar will tell you it is a shared cell.

Declare a Cell to Be Shared

Use the **Share** button located in the bottom right in the Cell Library box to place copies of the shared cells in your design file. Following are the step-by-step procedures:

1. If the Cell Library box is not already open, select **Element > Cells** to open it.
2. Turn the **Use Shared Cells** check box ON.
3. Select the cell you want to be shared.
4. Click the **Share** button.

When you click the **Share** button, "Shrd" appears in the **Where** column of the selected cell, and a copy of the cell is placed in your current design file.

This procedure stores a shared copy of the cell in your design file but does not actually place the cell in the design. This is a handy way to create a seed file for a group of design files that will use the same set of cells. The procedure includes creating a new design file, attaching the cell library, and declaring all required cells shared. Then you can create new design files by copying the seed file. Each copied file contains a copy of the elements that make up the shared cells.

Place Shared Cells

When the **Use Shared Cells** check box is ON, all cell placement tools place shared cells. Placing a cell makes the cell shared, even if you had not previously declared the cell to be shared.

The tools that place the active placement, terminator, and point cells all work the same for shared cells as they do for unshared cells. The Replace Cell tool automatically replaces all the copies of the shared cell when you identify one of them.

Turn a Shared Cell into an Unshared Cell

The Drop Element tool provides a **Shared Cells** check box for turning shared cells either into unshared cells or individual elements in the design. The menu associated with the **Shared Cells** check box controls the way the shared cell is dropped. The **To Geometry** option drops the cells all the way back to individual elements in the design, and the **To Normal Cell** option removes the shared status (each shared cell is replaced with a local copy of the cell).

Invoke the Drop Element tool from:

Groups tool box	Select the Drop Element tool, and, in the Tool Settings window, turn the Shared Cells check box ON, turn the other check boxes OFF, and select the way you want to drop the shared cells (see Figure 11–401).
Key-in window	**drop element** (or **dr e**) ENTER

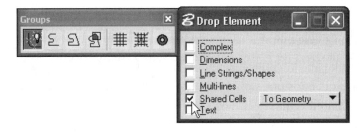

Figure 11-41 The Drop Element tool set to drop a shared cell

MicroStation prompts:

> Drop Element > Identify element *(Select the shared cell to be converted.)*
>
> Drop Element > Accept/Reject (select next input) *(Click the Data button to convert the selected cell or click the Cancel button to cancel the operation without dropping the cell.)*

Delete the Shared Cell Copy from Your Design File

The placements of shared cells can be deleted like any other element, but the actual shared cell copy takes a little more work to delete. It is done in the Cell Library box via the **Delete** button.

Here is the procedure for deleting the shared cell elements from your design file.

1. Either delete all placements of the shared cell or make them into unshared cells (see earlier).
2. If the Cell Library box is not already open, select **Element > Cells** to open it.
3. Turn the **Use Shared Cells** check box ON.
4. Select the shared cell you want to delete and make sure the cell has "Shrd" in the Where column.
5. Click the **Delete** button (located in the bottom right of the Cell Library box).
6. In the Alert box, click the **OK** button to delete the shared cell or click the **Cancel** button to cancel the operation.

This procedure deletes only the copy of the shared cell elements in your design file. It does not delete anything from the attached cell library.

Note: If there are still any placements of the shared cell in your design file when you try to delete the shared cell, an Alert dialog box appears with a message stating that you cannot delete the cell.

Open the Exercise Manual PDF file for Chapter 11 on the accompanying CD for project and discipline specific exercises.

REVIEW QUESTIONS

Write your answers in the spaces provided.

1. Explain briefly the difference between a cell and cell library.
   _____

2. How many cell libraries can you attach at one time to a design file? _____

3. List the steps involved in creating a cell.
   _____
   _____

4. What is the alternate key-in AC= used for? _____

5. What is the file that has an extension of .CEL? _____

6. What does it mean to place a cell Absolute?
   _____

7. What does it mean to place a cell Relative?
   _____

8. Explain briefly the differences between a graphic cell and a point cell.
   _____
   _____

9. How many cells can you store in a library? _____

10. What is the purpose of defining an active cell as a terminator?
    _____

11. What is the purpose of turning ON the Fast Cells View attribute?
    _____

12. Explain briefly the benefits of declaring a cell as a shared cell.
    _____
    _____

chapter 12

Patterning

Patterns are used in drawings for several reasons. Cutaways (cross sections) are hatched to help the viewer differentiate among components of an assembly and to indicate the material of construction. Patterns on surfaces depict material and add to the readability of the drawing. In general, patterns help communicate information about design. Because drawing patterns is a repetitive task, it is an ideal computer-aided drafting application.

Objectives

After completing this chapter, you will be able to do the following:

- Control the display of patterns in view windows
- Place hatching, crosshatching, and area patterns via seven placement methods
- Manipulate patterns
- Fill elements

CONTROLLING THE VIEW OF PATTERNS

Patterning can place so many elements in a design that view updates may start taking an unacceptable time to complete on slow workstations. To overcome that, MicroStation provides the **Patterns** view attribute check box to turn the display of pattern elements ON and OFF.

Turn the **Patterns** check box ON when placing new patterns so the placement results can be seen. When work with patterns is completed, the **Patterns** check box can be turned OFF to speed up view updates. Figure 12–1 shows a crosshatch pattern with the view attribute ON and with it OFF.

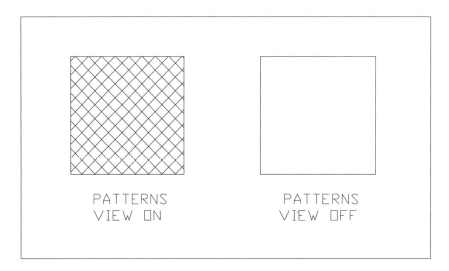

Figure 12-1 Example of turning the Patterns view attribute ON and OFF

To change the **Patterns** view status, invoke the View Attributes settings box from:

Drop-down menu	Select Settings > View Attributes
Key-in window	**dialog viewsettings** (or **di views**) ENTER

MicroStation displays the View Attributes settings box, as shown in Figure 12–2. Turn the **Patterns** check box ON or OFF and select a **View Number** from the top of the settings box. Click the **Apply** button to apply the settings to the selected view or click the **All** button to apply the settings to all open view windows.

 Note: If you place patterns in a view that has the **Patterns** check box turned OFF, the patterns are placed but they do not show up on the screen. To avoid the confusion that that can cause, turn the **Patterns** check box ON before placing patterns.

Chapter 12 Patterning

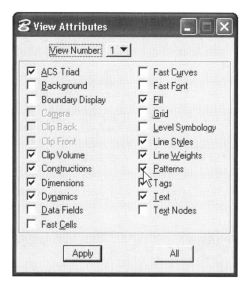

Figure 12-2 View Attributes settings box with the Patterns view attribute set to ON

PATTERNING COMMANDS

MicroStation provides three tools for placing patterns in a design file:

- The Hatch Area tool places a set of parallel lines in the selected pattern area and is invoked from the Patterns tool box, as shown in Figure 12-3a.

Figure 12-3a Hatch Area icon and settings box

- The Crosshatch Area tool places two sets of parallel lines in the selected pattern area and is invoked from the Patterns tool box, as shown in Figure 12-3b.

543

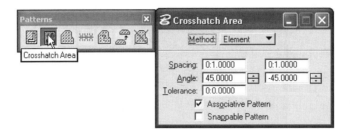

Figure 12-3b Crosshatch Area icon and settings box

- The Pattern Area tool fills the selected pattern area with tiled copies of the active pattern cell and is invoked from the Patterns tool box, as shown in Figure 12–3c.

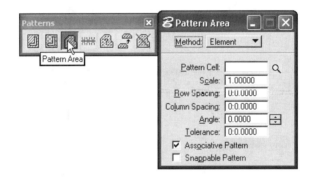

Figure 12-3c Pattern Area icon and settings box

Examples of the patterns placed by each tool are shown in Figure 12–4.

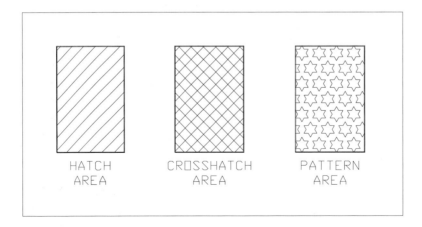

Figure 12-4 Examples of hatching, crosshatching, and patterning

TOOL SETTINGS FOR PATTERNING

The following discussion describes the tool settings for the three patterning tools.

Spacing

The **Spacing** fields set the space in Working Units between patterning elements.

- Hatch Area—One **Spacing** field sets the space between the lines.
- Crosshatch Area—Two **Spacing** fields set the space between each set of lines.
- Pattern Area—Two fields (**Row Spacing** and **Column Spacing**) set the space between the rows and columns of the selected cell.

The **Spacing** fields for the Hatch Area and Crosshatch Area tools cannot be set to zero. If you enter a zero in one of the fields, the field is reset to one Sub Unit. Zeros are allowed in the Pattern Area fields.

Angle

The **Angle** fields set the rotation angle (from the positive X axis direction) for the patterning elements. A positive angle causes counterclockwise rotation.

- Hatch Area—One field sets the angle of the lines.
- Crosshatch Area—Two fields set the angle of each set of lines.
- Pattern Area—One field sets the angle of the rows of cells.

If both crosshatch angles are equal to zero, one set of lines is placed at zero degrees of rotation and the other is placed at 90 degrees of rotation. If the first crosshatch line angle is not equal to zero and the second is equal to zero, the second set is placed at right angles to the first set. For example, if the first set's angle is 45 degrees and the second is zero degrees, the first set is placed at 45 degrees and the second set at 135 degrees (45+90).

Tolerance

The **Tolerance** field sets the variance between the true element curve and the approximation of the curve when a curved element is patterned. The curve is approximated by a series of straight-line segments, and a low tolerance number increases the accuracy of the approximation by reducing the length of each segment. The low tolerance number also means a larger design file and slower pattern placement.

Associative Pattern

The **Associative Pattern** check box, when ON, associates the pattern elements with the element they pattern. Changes to an associated element also affect the pattern. For example, if the element is stretched, the pattern expands to fill the new size. Figure 12–5 shows an example of pattern association.

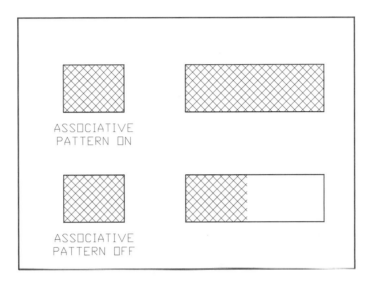

Figure 12-5 Example of the associative pattern setting

Snappable Pattern

The **Snappable Pattern** check box, when ON, allows snapping to, selecting, and manipulating individual elements in the pattern. If the check box is OFF, individual pattern elements cannot snapped to or selected.

 Note: If the **Associative Pattern** and **Snappable Pattern** check boxes are both ON for a pattern, you can snap to individual elements in the pattern, but you cannot select the individual pattern elements.

Pattern Cell

The **Pattern Cell** text field is used by the Pattern Area tool to provide the name of the cell to use for creating the pattern. If no cell name is provided, the Pattern Area tool displays an error message in the Status bar and no pattern is placed. To select a cell for patterning, key-in its name in the field.

An alternate way to select a cell for patterning is to click the **Browse Cell(s)** magnifying glass to open the Cell Library settings box. In this box you can attach the cell library that contains the cell you want to use and select the cell. To select the cell, click its name in the list of cells and click the **Pattern** button.

Scale

The **Scale** text field is used by the Pattern Area tool to scale the cell when it is used to create the pattern. A number less than 1 reduces the size of each copy of the cell; a number greater than 1 increases the size.

 Note: If the cell to be used for the pattern area was created with different Working Unit ratios than those of the design file, the cell may need to be scaled. If you are not sure of the correct scale, make the cell the active placement cell and select the Place Active Cell tool. The dynamic image of the cell shows you what size it is in your design. Adjust the cell placement scale until the dynamic image is the correct size, and then set the Pattern Area tool's **Scale** field to the same value.

Method

The **Method** drop-down menu provides options that control how the patterned area is determined. The specific options are discussed in the next topic.

PATTERNING METHODS

The **Method** drop-down menu provides seven patterning methods for defining the area to be patterned, as shown in Figure 12–6.

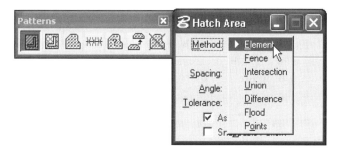

Figure 12-6 Patterns Method drop-down menu

The following list describes the patterning area methods; Figure 12–7 shows examples of each method.

547

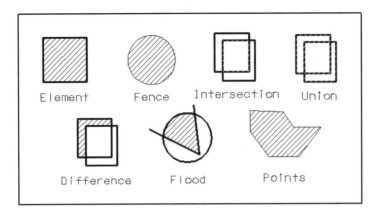

Figure 12-7 Examples of the seven area patterning methods

- **Element**—Patterns a selected closed element, such as a Block or Ellipse, or between components of a multi-line.
- **Fence**—Patterns the area enclosed by a fence.
- **Intersection**—Patterns the area common to two or more overlapping closed elements.
- **Union**—Patterns two or more selected closed elements as if they were one element. The elements do not have to overlap.
- **Difference**—Patterns the part of the first selected closed element that does not overlap the other selected closed elements. If the first element is completely inside the second element, no pattern is placed.
- **Flood**—Patterns the area enclosed by a set of separate elements surrounding the placement data point.
- **Points**—Patterns a temporary closed shape created by a set of data points placed after the tool is invoked.

Chapter 12 Patterning

PLACING PATTERNS

Invoke the pattern placement tools from:

Patterns tool box	Select the required pattern placement tool: • Hatch Area (see Figure 12–3a) • Crosshatch Area (see Figure 12–3b) • Pattern Area (see Figure 12–3c)
Key-in window	• **hatch** (or **h**) ENTER • **crosshatch** (or **cro**) ENTER • **pattern area** (or **pat a**) ENTER

Note: The placement prompts depend on the area definition method selected. The following discussion shows the prompts for each area method and uses the Hatch Area tool as an example. The prompts are the same for Crosshatch Area and Pattern Area.

Element

When the **Element Method** is selected, MicroStation prompts:

> Hatch Area > Identify element *(Select the closed element to be patterned as shown in Figure 12–8.)*
>
> Hatch Area > Accept @pattern intersection point *(Click the Data button again to accept the element and initiate pattern placement.)*

Note: For each patterning method, the acceptance point defines a point through which pattern elements pass—one line for Hatch Area, the intersection of two lines for Crosshatch Area, and the origin point of one of the pattern cells for Pattern Area. If the acceptance point is not within the patterned area, the intersection will be on the edge of the area at the closest extension of the acceptance point.

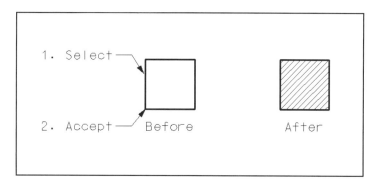

Figure 12–8 Example of patterning a closed element

549

Fence

When the **Fence Method** is selected, MicroStation prompts:

> Hatch Fence > Accept/Reject fence contents *(Place a data point to define an intersection point for the pattern, as shown in Figure 12–9, and initiate patterning.)*

The example in Figure 12–9 is a circular fence. The fence must be placed before it can be patterned. The fence is not part of the pattern and removing it does not affect the pattern.

If the **Associative Pattern** check box is ON, there will be an outline around the pattern in the shape of the fence. If the button is OFF, there will not be an outline.

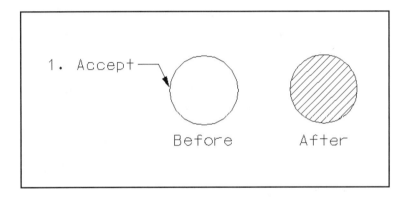

Figure 12-9   Example of patterning a fence

Intersection

When the **Intersection Method** is selected, MicroStation prompts:

> Hatch Element Intersection > Identify element *(Select the first element, as shown in Figure 12–10.)*
>
> Hatch Element Intersection > Accept/Reject (Select next input) *(Select all the other elements that make up the intersection; then click the Data button in space.)*
>
> Hatch Element Intersection > Identify additional/Reset to complete *(Click the Reset button to initiate the patterning.)*

As each element is accepted, the dynamic image changes the appearance of the selected elements to show only the intersection area. The elements are not affected and they reappear after the tool is completed.

Chapter 12 Patterning

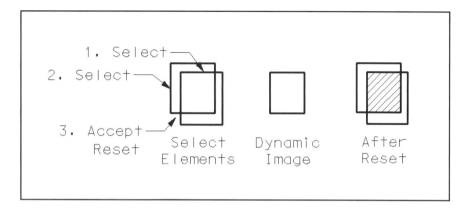

Figure 12-10 Example of intersection patterning

Union

When the **Union Method** is selected, MicroStation prompts:

> Hatch Element Union > Identify element *(Select the first element, as shown in Figure 12–11.)*
>
> Hatch Element Intersection > Accept/Reject (Select next input) *(Select all the other elements that make up the union, then click the Data button in space.)*
>
> Hatch Element Union > Identify additional/Reset to complete *(Click the Reset button to initiate the patterning.)*

As each element is accepted, the dynamic image changes the appearance of the selected elements to show the union area with interior segments removed. The elements are not affected and they reappear after the tool is completed.

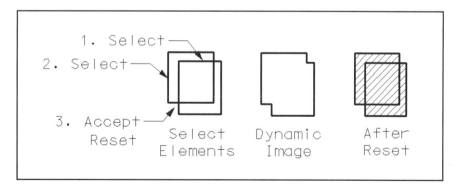

Figure 12-11 Example of union patterning

551

Difference

When the patterning area method is **Difference**, MicroStation prompts:

> Hatch Element Difference > Identify element *(Select the first element, as shown in Figure 12–12.)*
>
> Hatch Element Difference > Accept/Reject (Select next input) *(Select all the other elements that make up the difference; then click the Data button in space.)*
>
> Hatch Element Difference > Identify additional/Reset to complete *(Click the Reset button to initiate the patterning.)*

As each element is accepted, the dynamic image changes the appearance of the selected elements to show only the difference area. The part of the first selected element that does not overlap the other selected elements is patterned. The elements are not affected and they reappear after the tool is completed.

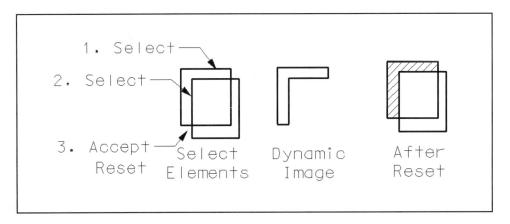

Figure 12-12 Example of difference patterning

Flood

When the **Flood Method** is selected, the Tool Settings window displays additional fields that are unique to this patterning method.

If the **Locate Interior Shapes** check box is ON, the patterning elements do not cover closed elements, such as block elements, inside the area selected for patterning. If the **Locate Text** check box is ON, the patterning elements do not cover text inside the area selected for patterning. Figure 12–13 shows an example of patterning using the **Flood Method** and the two locate check boxes.

Chapter 12 Patterning

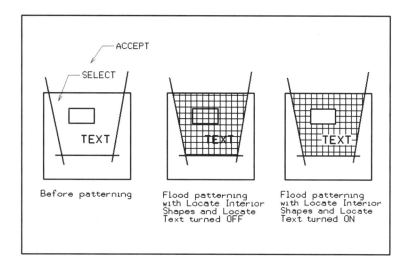

Figure 12-13 Examples of the use of Locate Interior Shapes and Locate Text on flood patterning

If the **Dynamic Area** check box is ON, MicroStation dynamically highlights potential flood areas as you move the pointer around the design plane. If the check box is OFF, you will not see the flood area until you select it with a data point. Another feature provided by the **Dynamic Area** check box is the ability to select several, separate areas for patterning at one time. To select several areas, turn the check box ON, and hold down CTRL while selecting the areas. After all areas are selected, release CTRL and click the Data button to initiate patterning all selected areas, as shown in Figure 12–14.

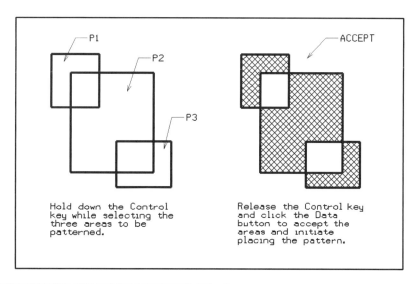

Figure 12-14 Example of using the CTRL key for Flood patterning

553

The **Max Gap** text field allows you to control the maximum space in Working Units between the enclosing elements in Flood patterning mode. If there is a gap in the perimeter of the selected area that is greater than the **Max Gap** setting, the area is not patterned, and an error message appears in the Status bar. If the area is complex, the tool may take a long time to decide if the area is truly enclosed.

When the patterning area method is **Flood**, MicroStation prompts:

> Hatch Area Enclosing Point > Enter data point inside area *(Place a data point inside the area to be patterned, as shown in Figure 12–13.)*
>
> Hatch Area Enclosing Point > Accept @pattern intersection point *(Click the Data button again to initiate patterning.)*

A dynamic image is drawn around the perimeter of the flood area.

Points

When the **Points Method** is selected, MicroStation prompts:

> Hatch Area Defined By Points > Enter shape vertex *(Place the first three vertex points of the shape, as shown in Figure 12–15.)*
>
> Hatch Area Defined By Points > Enter point or Reset to complete *(Continue entering shape vertex points, or click the Reset button to complete the shape and initiate patterning.)*
>
> Hatch Area Defined By Points > Identify additional/Reset to complete *(Click the Reset button to initiate the patterning.)*

After the third point, the dynamic image shows the shape of the pattern area if the Reset button were to be clicked.

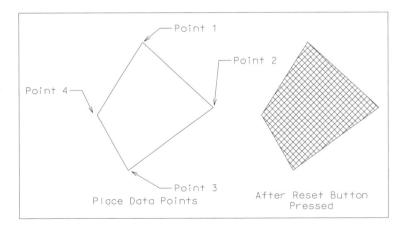

Figure 12-15 Example of patterning an area enclosed by points

USING THE AREA SETTING TO CREATE HOLES

You saw that you can leave holes in patterns when using the **Flood Method** by turning the **Locate Interior Shapes** check box ON. When this check box is ON, enclosed shapes within the flood area are not patterned. You can also create closed shapes, such as blocks, circles, and ellipses with either a **Solid Area** or **Hole Area**. The **Element** and **Fence** methods of the pattern placement tools will not pattern **Hole** area elements when the **Associative Pattern** check box is OFF.

An example of the use of the **Area** setting is in creating elevation views of a building. The walls can be created using **Solid Area** and the windows can be created using **Hole Area**. When the **Element Method** is used with the Pattern Area tool to lay a brick pattern on the walls, the windows will not be patterned, as shown in Figure 12–16.

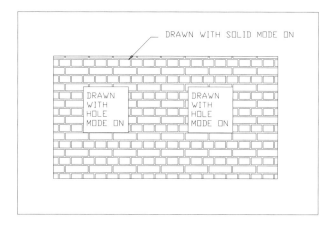

Figure 12-16 Example of the effect of Hole elements within a patterned element

Text placed when **Hole Area** is active will not be patterned, as shown in Figure 12–17.

Figure 12-17 Example of patterning over a text string placed in Hole Area mode

The Tool Settings window for each tool that places closed shapes includes the **Area** menu. An example is shown in Figure 12–18. The **Area** option you select becomes the active setting and all closed elements (and text) that you create after that use the active setting.

Figure 12-18 Example of the Area mode tool setting for a closed element placement tool

 Note: As was stated in the text, only the **Element** and **Fence** patterning methods recognize **Hole** areas, and they only recognize them when the **Associative Pattern** check box is OFF.

Changing an Element's Area

If an element was placed via the wrong **Area** mode, you can change it to the correct area with the Change to Active Area tool. Invoke the tool from:

Change Attributes tool box	Select the Change to Active Area tool and select the required Area (Solid or Hole) in the Tool Settings window (see Figure 12–19).
Key-in window	**change area** (or **chan a**) ENTER

Figure 12-19 Invoking the Change Element to Active Area tool from the Change Attributes tool box

MicroStation prompts:

>Change Element to Active Area > Identify element *(Identify the element.)*
>
>Change Element to Active Area > Accept/Reject (Select next input) *(Click the Data button to initiate the area change or click Reset to reject the selected element. Optionally, also select the next element to be changed.)*

DELETING PATTERNS

A special delete tool for patterns deletes all pattern elements but not the element that contains the pattern. Invoke the Delete Pattern tool from:

Patterns tool box	Select the Delete Pattern tool (see Figure 12–20).
Key-in window	**delete pattern** (or **del pat**) ENTER

Figure 12-20 Invoking the Delete Pattern tool from the Patterns tool box

MicroStation prompts:

Delete Pattern > Identify element *(Identify the pattern to be deleted.)*

Delete Pattern > Accept/Reject (select next input) *(Click the Data button again to accept and delete the pattern or click the Reset button to drop the selected pattern.)*

 Note: Patterns can contain a large number of elements. After deleting patterns, you may want to select **File > Compress Design** to remove the deleted elements from the design file. Remember though, that when you compress the design, you also clear the Undo buffer.

MATCHING PATTERN ATTRIBUTES

If additional patterns need to be placed with the same patterning attribute settings as existing patterns, the Match Pattern Attributes tool provides a fast way to set the attributes. The tool sets the active patterning attributes to match those of a selected pattern.

Invoke the Match Pattern Attributes tool from:

Patterns tool box	Select the Match Pattern Attributes tool (see Figure 12–21).
Key-in window	**match pattern** (or **mat pat**) ENTER

Figure 12-21 Invoking the Match Pattern Attributes tool from the Patterns tool box

MicroStation prompts:

> Match Pattern Attributes > Identify element *(Identify the element containing the pattern to be matched.)*
>
> Match Pattern Attributes > Accept/Reject (Select next input) *(Click the Data button to accept the element or click the Reject button to reject it.)*

When you accept the selected pattern, the active patterning elements are set to match those of the selected pattern and the active settings appear in the Status bar.

 Note: The Match Pattern Attributes tool is also available in the Match tool box that is opened from the Tools drop-down menu.

FILLING AN ELEMENT

The Tool Settings window for tools that place closed elements (such as Circle and Block) include the **Fill Type** and **Fill Color** menus to control what happens to the area enclosed by the element.

The **Fill Type** options are:

- ■ **None**—The element area is transparent and the outline color is set to the **Active Color**.
- ■ **Opaque**—The element area and outline are set to the color selected in the **Fill Color** menu (which is also the **Active Color**).
- ■ **Outlined**—The element area is set to the color selected in the **Fill Color** menu and the element outline is set to the **Active Color**.

The **Fill Color** menu opens a menu that contains all of the colors supported by MicroStation.

- ■ When the **Opaque Fill Type** is selected, the default **Fill Color** is equal to the **Active Color**. Selecting a **Fill Color** also selects the **Active Color**.
- ■ When the **Outlined Fill Type** is selected, the **Fill Color** sets the color of the element area, and the **Active Color** sets the color of the element outline. Changing the **Fill Color** does not change the **Active Color**.

Figure 12–22 shows a typical Tool Settings window with the Fill Type menu open; Figure 12–23 shows examples of opaque and outlined fill types.

Chapter 12 Patterning

Figure 12-22 Fill Type menu in the Tool Settings window

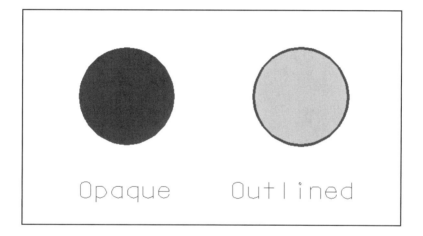

Figure 12-23 Examples of Opaque and Outlined Fill Types

To place filled closed elements, just select the desired placement tool, and then select the desired **Fill Type** and **Fill Color** before placing the element. The Fill settings you select become the active setting, and they are applied to all closed elements you create after that.

Area Fill View Attribute

Before placing filled closed elements, make sure the **Fill** check box is ON in the View Attributes settings box for the view window being used to place the elements. If the check box is OFF, filled elements appear to be transparent, and they print that way.

Turn the **Fill** check box ON, invoke the View Attributes settings box from:

Drop-down menu	Settings > View Attributes
Key-in window	**dialog viewsettings** (or **di views**) ENTER

559

If the **Fill** check box is OFF, turn it ON, as shown in Figure 12–24, and then:

- Click the **Apply** button to apply the view settings to the **View Number** shown at the top of the window, or
- Click the **All** button to apply the view settings to all open view windows.

The applied view attributes stay in effect until they are changed or the file is closed. To make the settings permanent, select **File > Save Settings**.

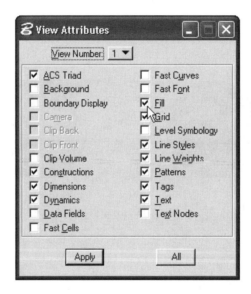

Figure 12-24 The Fill attribute in the View Attributes settings box

Changing the Fill Type of an Existing Element

The Change to Active Fill Type tool changes the fill type of existing elements.

Invoke the tool from:

Change Attributes tool box	Select the Change to Active Fill Type tool and, if necessary, set the desired Fill Type and Fill Color in the Tool Settings window (see Figure 12–25).
Key-in window	**dialog change fill** (or **chan f**) ENTER

Figure 12-25 Invoking the Change Element to Active Fill Type tool from the Change Attributes tool box

MicroStation prompts:

> Change Element to Active Fill Type > Identify element *(Select the element to be changed.)*
>
> Change Element to Active Fill Type > Accept/Reject (Select next input) *(Click the Data button to initiate the Fill Type change or click the Reset button to drop the element.)*

Open the Exercise Manual PDF file for Chapter 12 on the accompanying CD for project and discipline specific exercises.

REVIEW QUESTIONS

Write your answers in the spaces provided.

1. List the three tools MicroStation provides for area patterning.

2. Explain briefly the difference between Hatch Area and Crosshatch Area patterning.

3. List the methods that are available to set the area patterning.

4. Explain with illustrations the differences between the **Intersection**, **Union**, and **Difference** options in area patterning.

5. Explain the difference between the **Element** and **Points** options in area patterning.

6. What is the purpose of providing a second data point in hatching a closed element?

7. The Pattern Area tool places copies of the **Active Pattern** _____ in the area you select.

8. What is the purpose of turning ON the **Associative Pattern** check box?

9. What is the purpose of placing elements in a **Hole Area** mode?

10. Explain briefly the purpose of invoking the Match Pattern Attributes tool.

11. Explain briefly the three options available for area **Fill** placement.

12. Explain the steps involved in switching an existing element between area **Fill** modes.

chapter 13

Attaching References

One of the most powerful timesaving features of MicroStation is its ability to view other models while you are working in your design file. MicroStation lets you attach an unlimited number of models to the active model. This function takes place in the form of references. A reference cannot be modified. You can attach as a reference a model that resides in either the open DGN file or some other DGN file.

Objectives

After completing this chapter, you will be able to do the following:

- Attach a reference
- Move a reference in the design plane
- Scale your view of a reference
- Rotate your view of a reference
- Mirror your view of a reference
- Clip off part of your view of a reference
- Detach a reference
- Reload a reference

OVERVIEW OF REFERENCES

When a model is referenced in your design file, you can view and tentative point snap to all elements in the reference model. The only information about the referenced model that becomes a permanent part of your design file is the name of the referenced model and its directory path. It is sometimes convenient to refer to one part of a model while drawing in another area by attaching the active model to itself. Elements in the reference can be copied into your design file. Elements copied from a reference become part of the current design file. This method is useful if, for example, a model that was drawn previously contains information that can be used in your current design file.

References may be scaled, moved, rotated, and viewed by levels via tools that are specifically programmed to work with references. The reference behaves as one element when you manipulate it with references manipulation tools. The only exception is the regular Copy tool, in which the elements behave as individual elements. All of the manipulations are performed on your view of the referenced, not the actual file. You cannot edit or modify the contents of a referenced. If you need to make any changes to it, you actually have to load that file in MicroStation.

When you attach an external reference, it is permanently attached until you detach it. When you open a design file, MicroStation automatically reloads each reference; thus, you see the current version of each reference when you view your design file.

The reference manipulation tools are available in a tool box and from the References settings box. Open the References tool box from:

Drop-down menu	Tools > Reference Files.
Key-in window	**dialog toolbox reference** (or **di to ref**) ENTER

MicroStation displays the References tool box (see Figure 13–1).

Figure 13-1 The References tool box

Open the References settings box from:

Drop-down menu	File > Reference.
Key-in window	**dialog reference** (or **di ref**) ENTER

MicroStation displays the References settings box. All reference manipulation tools are in the settings box **Tools** drop-down menu (see Figure 13–2).

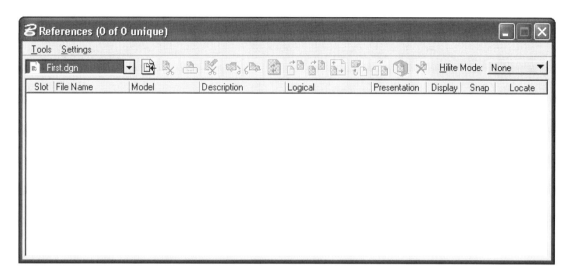

Figure 13-2 The References settings box

EXAMPLES OF USING REFERENCES

Borders and title blocks are an excellent example of design files that are useful as references. The elements that make up the border use considerable space in a file and usually amount to around 40 to 50 KB. If a border and title block are drawn in each design file, it would waste a large amount of space, especially if you multiply 50 KB by 1000 design files. If references are used correctly, they can save a lot of disk space.

Accuracy and efficient drawing time are other important design features that are enhanced through the use of references. As mentioned earlier, when an addition or change is made to a design file that is serving as a reference, all the design files that reference it will reflect the modifications. For example, let's say the name of the company is changed. Just change the company name in the title block design file, and all the models that reference the title block automatically display the new company name the next time they are accessed. (That's much easier than accessing 1000 models to correct one small detail.) References save time and ensure the drawing accuracy required to produce a professional product.

When combined with the networking capability of MicroStation, external references give the project manager powerful new tools for coping with the realities of file management. The project manager can, by combining drawings through the referencing tools, see instantaneously the work of the various departments or designers working on a particular aspect of the project. If necessary, the manager can overlay a drawing where appropriate, track the progress, and maintain document integrity. At the same time, departments need not lose control over individual designs and details.

Let's look at an example of this function in operation. Let's say that you are a supervisor with three designers reporting to you. All three of them are working on a project, and each one is responsible for one-third of the project. As a supervisor you want to know how much progress each designer has made at the end of each day. Instead of calling up each of the three models to see the progress, you can create a dummy design file and attach the three models as references. Every day you can call up the dummy design file, and MicroStation will display the latest versions of the references attached to your design file. This will make your job a lot easier and give you an opportunity to put together all three pieces of the puzzle to see how they fit into the evolving design.

As mentioned earlier, you can attach unlimited references at any time to a design file (earlier versions defaulted to a maximum of 255). In addition to attaching a model as a reference, you can attach a raster image as a reference. Monochrome, continuous-tone (gray-scale), or color images in a variety of supported image formats can be attached.

ATTACHING A REFERENCE

To attach a model as a reference to the active model, invoke the Attach Reference tool from:

References tool box	Select the Attach Reference tool (see Figure 13–3).
References settings box	Tools > Attach
Key-in window	**reference attach** (or **refe a**) ENTER

Figure 13-3 Invoking the Attach Reference tool from the References tool box

MicroStation displays the Attach Reference dialog box. Select the design file that contains the model to attach as a reference and click **OK** button.

The Attach Reference Settings dialog box opens as shown in Figure 13–4.

Chapter 13 Attaching References

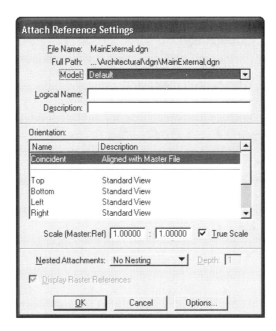

Figure 13-4 Attach Reference Settings dialog box

The Attach Reference Settings dialog box is used to attach a model to the active model. The top part of the dialog box displays the name and path of the design file being attached as a reference. The Model drop-down menu lists the models within the reference. It is used to select the model to be attached to the master model.

Click in the **Logical Name** edit field and type in a name. The logical name is a short identifier or nickname with a 40-character maximum that you can use later to identify the reference file when you need to manipulate it. The logical name is optional, unless you attach the same design file more than once to the current design file.

The next edit field is for a **Description**. The description, which is optional, can serve to describe the purpose of the attachment. If more than two or three reference files are attached, it allows you to identify the purpose of attachment quickly. The description cannot exceed 40 characters.

The **Orientation** list box lists the views of the model being attached. The most common way to attach a reference is to attach it coincidentally, which means that the coordinates of the referenced model's design plane are aligned with those of the active model, without any rotation, scaling, or offset.

If the mode is set to Saved View, you can select one of the available Saved Views from the list box. A saved view allows you to display a clipped portion of the design file. You can attach the saved view at a specified scale factor by entering the appropriate factor in the **Scale (Master:Ref)** edit fields. The Saved View mode is available when the reference contains saved views. (Saved View manipulation was described in Chapter 3.)

The **Scale (Master:Ref)** edit fields allow you to specify the scale factor as a ratio of the Master Units in the active model to the Master Units in the attached model. The **True Scale** check box, when set to ON, aligns the units in the active model one to one with units in the referenced model. For example, if the units in the active model are feet and the units in the referenced model are meters, then meters are converted to feet so the elements are referenced in at the same relative size.

The **Nested Attachments** option menu allows you to select what MicroStation does with references that are attached to the model you reference. If the **No Nesting** option is selected, only the model you select is attached to your design. If the **Live Nesting** option is selected, all models referenced by the model you are attaching will also be attached to your design. If the **Copy Attachments** option is selected, models attached to the attached model copied into the active model.

The **Depth** edit field sets the number of levels of referenced models that are recognized. Models can have their own referenced models, which, in turn, can have more referenced models, and so on. If the **Depth** is set to 0, only the model is attached to the master model; referenced models in the referenced model are ignored.

The **Display Raster References** check box controls whether raster references will be displayed in the master model.

The **Options** button opens the Attachment Settings dialog box as shown in Figure 13–5, used for adjusting the settings of a referenced model.

Figure 13-5 Attachment Settings dialog box

The **Display** check box controls the screen display of the specified reference. If for some reason you don't want to display the attached reference but still want to keep it attached, set the **Display** check box to OFF; MicroStation will not display or plot the specified reference.

Chapter 13 Attaching References

The **Snap** check box controls the ability to tentative point snap to reference elements. If **Snap** is set to OFF, you can see the reference elements, but you cannot tentative snap on them. If **Snap** is set to ON, you can tentative snap to any element in the specified reference.

The **Locate** check box controls the ability to locate reference models and copy them into your master model via the Copy tool. If **Locate** is set to OFF, you cannot identify the elements to use with the Copy tool. If **Locate** is set to ON, you can identify the elements for use with the Copy tool.

The **Clip Back** and **Clip Front** check boxes control the display of the back and front clipping plane displays respectively.

The **Scale Line Styles** check box controls the scaling of custom line styles in the reference. If it is set to ON, custom line style components (for example, dashes) are scaled by the **Scale (Master:Ref)** factor. If it is set to OFF, custom line style components are not scaled.

The **Ignore Attachment When Live Nesting** check box controls whether all models referenced by the model will be attached.

The **Use Lights** check box controls whether lights placed in the reference model are displayed or not.

Click the **OK** button to attach the selected reference to the current model, or click the **Cancel** button to cancel the reference attachment. You can reference the same model more than once. The only restriction is that you must provide a unique logical name for each attachment. Although it is not required, it is advisable to use the **Description** field to explain each attachment. That will save other people time in figuring out why the same model is referenced more than once. For example, the same model might be referenced more than once when different parts of the referenced design serve as details in your design file.

Figure 13–6 shows a design before and after attaching a reference (title block).

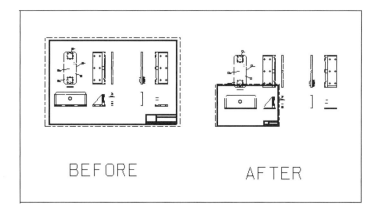

Figure 13-6 Design before and after attaching a reference

Listing the Attached References

To list the references attached to the current model, open the References settings box from:

| Drop-down menu | File > Reference |

MicroStation displays the References settings box, as shown in Figure 13–7. The settings box lists all the references attached to the current design file. In addition to listing the references, the settings box also provides the logical name of the reference, and the status of the **Display**, **Snap**, and **Locate** options of the references.

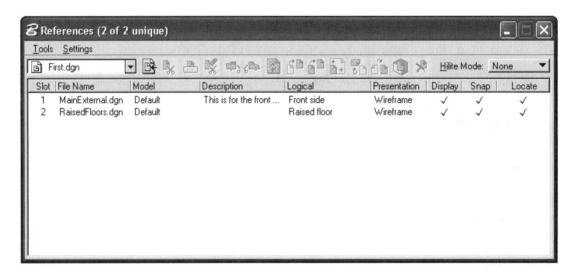

Figure 13–7 References settings box

If necessary, you can also change the attached model, the option to save its full path, its logical name, and its description. Double-click the reference file name in the References settings box, and MicroStation displays the Attachment Settings dialog box, shown in Figure 13–8. Make the required changes in the dialog box, and then click the **OK** button to save the changes, or the **Cancel** button to close the box without making any changes.

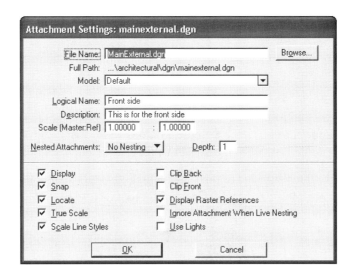

Figure 13-8 Attachment Settings dialog box

REFERENCE MANIPULATIONS

MicroStation provides a set of manipulation tools specifically designed to manipulate references. As mentioned earlier, the reference behaves as one element, and there is no way to drop it. The reference manipulation tools include Move, Copy, Scale, Rotate, Mirror Horizontal, Mirror Vertical, Clip Boundary, Clip Mask, and Clip Mask Delete.

Selecting the Reference

As discussed at the beginning of this chapter, the reference manipulation tools are available in the References tool box and in the **Tools** drop-down menu in the References settings box. The first step in using any of these tools is to select the reference to be manipulated.

- If the References settings box is open, the reference selected in the list field is the one that will be manipulated when you select a manipulation tool from the tool box or settings box.

- If the References settings box is closed, you must use the manipulation tools in the tool box, and each command will start by asking you which reference is to be manipulated. To select a reference, click on an element in the reference, and then click a second time to select it. If you have the Key-in window open, you can also type the reference's logical name to select it.

 Note: The reference manipulation tool discussions that follow assume you have already selected the reference in the References settings box, and they do not show the reference selection prompts. If you are using the tool from the tool box and the settings box is closed, you will be prompted to select the reference to manipulate the tools.

Move Reference

The Move Reference tool allows you to move a reference from one location to another. Before you invoke the tool, select the name of the reference listed in the References settings box.

To move a reference, invoke the Move References tool from:

References tool box	Select the Move References tool (see Figure 13-9).
References settings box	Tools > Move
Key-in window	**references move** (or **ref mo**) ENTER

Figure 13-9 Invoking the Move References tool from the References tool box

MicroStation prompts:

> Move Reference > Enter point to move from *(Place a data point anywhere on the reference file.)*
>
> Move Reference > Enter point to move to *(Place a data point where you want to move the reference file in reference to the first data point and press the reset button to complete command sequence.)*

Figure 13-10 shows a view before and after moving a reference border.

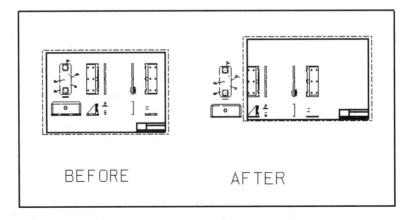

Figure 13-10 The design view before and after moving a reference

Copy Reference Attachment

The Copy Reference tool allows you to make additional attachments of an existing reference file attachment. Everything about the attachment is copied, and a unique logical name is provided for each copy. You attach it one time and then make copies of the attachment for the other details. Before you invoke the tool, select the name of the reference listed in the References settings box.

To copy a reference, invoke the Copy References tool from:

References tool box	Select the Copy References tool (see Figure 13–11).
References settings box	Tools > Copy
Key-in window	**references copy** (or **ref co**) ENTER

Figure 13-11 Invoking the Copy References tool from the References tool box

MicroStation prompts:

> Copy Reference Attachment > Enter point to copy from *(Place a data point anywhere on the reference.)*
>
> Copy Reference Attachment > Enter point to copy to *(Place a data point where you want to copy the reference in reference to the first data point and press the reset button to complete command sequence.)*

Scale Reference

The Scale References tool allows you to enlarge or reduce a reference. Before you invoke the tool, select the name of the reference listed in the References settings box.

To scale a reference, invoke the Scale References tool from:

References tool box	Select the Scale References tool (see Figure 13–12).
References settings box	Tools > Scale
Key-in window	**references scale** (or **ref s**) ENTER

Figure 13-12 Invoking the Scale References tool from the References tool box

In the Tool Settings window, the **Method** option menu sets by which the referenced model is scaled. The **Scale Factor** selection scales the model by a specified factor, the **Absolute Ratio** selection scales the model by a specified ratio, and the **By Points** selection scales the model by points entered.

The **Scale Factor** edit field (available only when the **Scale Factor** option is selected) sets the factor by which the model is scaled. The **Scale (Master:Ref)** edit field (available only when **Absolute Ratio** is selected) sets the active model's Master Units to the referenced model's Master Units.

The **Use References Dialog List** check box, when set to ON, the model(s) selected in the References dialog box is scaled.

When the **Use Fence** check box is set to ON, the fence contents are scaled, and the option menu sets the Fence (Selection) Mode.

MicroStation prompts:

> Scale Reference > Enter point to scale ref file about *(Place a data point about which the file will scale.)*

 Note: The scale factor ratio between the active design file and the reference is *not* cumulative. For instance, if you specify a scale of 3:1 followed by 6:1, the final result will be 6:1. Figure 13-13 shows a view before and after scaling the reference border.

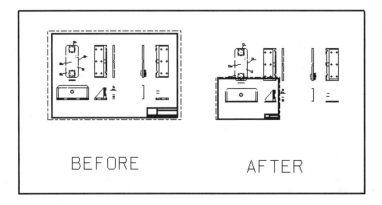

Figure 13-13 The design view before and after scaling a reference

Rotate Reference

The Rotate References tool allows you to rotate a reference to any angle around a pivot point. Before you invoke the tool, select the name of the reference listed in the References settings box.

To rotate a reference, invoke the Rotate Reference tool from:

References tool box	Select the Rotate References tool (see Figure 13–14).
References settings box	Tools > Rotate
Key-in window	**references rotate** (or **ref ro**) ENTER

Figure 13-14 Invoking the Rotate References tool from the References tool box

In the Tool Settings window, the **Method** option menu sets the method by which is model is rotated. The **By Angles** selection sets the rotation angle (2D sets in z-axis only and 3D sets the rotation angle(s) on the x, y, and z axes). The **By Points** selection sets the point about which the model is rotated.

The model(s) selected in the References settings box is scaled when the **Use References Dialog List** check box is set to ON.

The fence contents are scaled when the **Use Fence** check box is set to ON and option menu sets the Fence (Selection) Mode.

MicroStation prompts:

> Rotate Reference > Enter point to rotate reference about *(Place a data point to define a pivot point about which the reference file rotates.)*

For example, an angle of 60 degrees rotates the reference 60 degrees counterclockwise from its current position around the pivot point. If you make a mistake, you can always invoke the Undo tool to undo the last operation.

Mirror Reference

The Mirror Reference tool allows you to mirror a reference about the horizontal (*X axis*) axis or vertical (Y axis). Before you invoke the tool, select the name of the reference listed in the References settings box.

To mirror a reference, invoke the Mirror Reference tool from:

References tool box	Select the Mirror Reference tool (see Figure 13–15).
References settings box	Tools > Mirror Horizontal or Tools > Mirror Vertical
Key-in window	**reference mirror** (or **ref mi**) ENTER

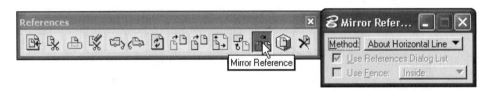

Figure 13-15 Invoking the Mirror Reference tool from the References tool box

In the Tool Settings window, the **Method** option sets the method by which the model is mirrored. The **About Horizontal Line** method selection is used to mirror models about a horizontal axis and the **About Vertical Line** method selection is used to mirror models about a vertical axis.

The model(s) selected in the References settings box is mirrored when the **Use Reference Dialog List** check box is set to ON.

The fence contents are mirrored when the **Use Fence** check box is set to ON and option menu sets the Fence (selection) Mode.

MicroStation prompts:

> Mirror Reference About Horizontal/Vertical > Enter point to mirror about *(Place a data point to define the mirror axis.)*

Clip Reference

The Clip Reference tool lets you display only a desired portion of a reference. You can use a Fence or closed shape to define the clipping boundary. The clipping boundary can have up to 60 vertices, and you can even use a circular fence or a circle as clipping boundary. Nonrectangular clipping boundaries are displayed in a view and plotted only if the **Fast Ref Clipping** View Attribute is set to OFF for the selected view.

Before you invoke the tool, select the reference listed in the References settings box.

To clip a boundary, invoke the Clip Reference tool from:

References tool box	Select the Clip Reference tool (see Figure 13–16).
References settings box	Tools > Clip Boundary
Key-in window	**reference clip boundary** (or **ref c bo**) ENTER

Figure 13-16 Invoking the Clip Reference tool from the References tool box

In the Tool Settings window, the **Method** option menu determines the method by which the reference clip boundary is set. The **Element** method selection sets the clip boundary set by the selected element, and the **Fence** method selection sets the clip boundary set by the contents of a fence.

The model(s) selected in the References settings box is clipped when **Use Reference Dialog Box** check box is set to ON.

MicroStation prompts:

> Set Reference Clip Element > Identify Clipping Element *(Identify the element to clip boundary.)*
>
> Set Reference Clip Element > Accept/Reject Clipping Element *(Click the Accept button to accept the clipping.)*

MicroStation displays only the part of the reference enclosed within the selected element.

Mask Reference

The Mask Reference tool, like the Clip Reference tool, allows you to display only a portion of a reference. Clip Reference displays the part inside a fence, whereas the Mask Reference tool displays the part outside the fence. Place a Fence on the reference to define the desired clipping boundary. The clipping mask can have up to 60 vertices.

Before you invoke the tool, select the reference listed in the References settings box.

To clip a mask, first place a fence, and then invoke the Mask Reference tool from:

References tool box	Select the Mask Reference tool (see Figure 13–17).
References settings box	Tools > Clip Mask
Key-in window	**reference clip mask** (or **ref c m**) ENTER

Figure 13-17 Invoking the Mask Reference tool from the References tool box

MicroStation prompts:

> Set Reference Clip Mask > Accept/Reject Fence Clip Mask *(Click the accept button to accept the clipping.)*

MicroStation displays only the part of the reference outside the fence. If you make a mistake, you can always invoke the Undo tool to undo the last operation.

Delete Clipping Mask(s)

The Delete Clip tool can delete a reference clipping mask. Before you invoke the tool, select the reference listed in the References settings box.

To delete a reference clipping mask, invoke the Delete Clip tool from:

References tool box	Select the Delete Clip tool (see Figure 13–18).
References settings box	Tools > Delete Clip
Key-in window	**reference clip mask delete** (or **ref c m delete**) ENTER

Figure 13-18 Invoking the Delete Clip tool from the References tool box

MicroStation prompts:

> Delete Reference Clip Component > Accept/Reject Delete Clip Mask *(Click the Accept button to delete the clip mask.)*

MicroStation deletes the selected clip mask.

Merging References into Master

MicroStation allows you to merge a DGN file and all attached references into a single output file. Merging ensures that all data attached to a single DGN file (all views, angles of views, rendering settings, and other settings) are stored in a single location. The result of a merging operation is effectively a snapshot of the DGN file.

This capability is especially useful for plotting active files, where many users are frequently changing a file's composition, views, settings, and attachments.

To merge the selected reference(s) from References settings box's list box into the active model, invoke the Merge into Master tool from:

References settings box	Tools > Merge into Master
Key-in window	**mdl load refmerge**

MicroStation prompts:

> Merge References > Select View For Merge *(Click anywhere in the view window to merge the selected references.)*

An alert box asks you to confirm that the selected reference(s) is to be merged into the active model. Click OK button to accept the merge.

Reloading the Reference

MicroStation automatically loads all references attached to a design file only when you first open the design file. To get the latest version of a reference while your design is open, you must use the Reload tool. The Reload tool is helpful, especially in a network environment, for accessing the latest version of the reference while you are working in a design to which it has been attached.

Before you invoke the tool, select the reference listed in the References settings box.

To reload a reference, invoke the Reload Reference tool from:

References tool box	Select the Reload Reference tool (see Figure 13–19).
References settings box	Tools > Reload
Key-in window	**reference reload** (or **ref r**) ENTER

Figure 13-19 Invoking the Reload Reference tool from the References tool box

MicroStation reloads the selected reference. To reload all references attached to the model, invoke the Reload All tool.

Detaching the Reference

When you no longer need a reference, you can detach it from the current model. Once the reference is detached, there is no longer a link between the reference and the current model.

Before you invoke the tool, select the reference listed in the References settings box.

To detach a reference, invoke the Detach Reference tool from:

References tool box	Select the Detach Reference tool (see Figure 13–20).
References settings box	Tools > Detach
Key-in window	**reference detach** (or **ref d**) ENTER

Figure 13-20 Invoking the Detach Reference tool from the References tool box

MicroStation displays an alert box to confirm that the selected reference is to be detached. Click the **OK** button to detach.

To detach all the attached references, invoke the Detach All tool from:

| References settings box | Tools > Detach All |
| Key-in window | **reference detach all** (or **ref d all**) ENTER |

MicroStation displays an alert box to confirm that all the reference are to be detached. Click the **OK** button to detach all the references.

Exchange Reference

MicroStation provides a tool that allows you to make one of the references the active file, so you can make changes to it. When you invoke this tool, the active file is closed and the selected reference is loaded in MicroStation for editing. Before you invoke the tool, select the reference listed in the References settings box.

To exchange a reference for the active design file, invoke the Exchange tool from:

| References settings box | Tools > Exchange |

MicroStation exchanges the files, and the selected reference becomes the active design file.

Reference File Agent Dialog Box

The Reference File Agent dialog box enables you to automatically maintain local copies of remote referenced models. To open Reference File Agent dialog box, invoke the Ref Agent tool from:

| References settings box | Tools > Ref Agent |

MicroStation displays Reference File Agent dialog box similar to Figure 13–21.

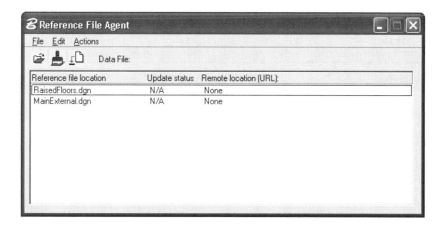

Figure 13-21 Reference File Agent dialog box

The **Open** tool opens a data file from which associated URLs are drawn. The default value (loaded at startup) is the file specified by the environment variable MS_REFAGENTDATA.

The **Refresh** tool recreates the list of referenced models and their associated URLs and sends out new HTTP requests to check whether the remote file is newer than the local copy. Used if the remote file is believed to have been updated since the program was loaded, or if referenced models have been attached or detached and are not yet represented in the list.

The **Edit URL** tool opens the Edit Reference Agent Entry dialog box, where you can associate (or change the association of) a URL with the selected referenced model. If a URL is entered and the **OK** button is selected, an HTTP request is sent to determine the update status of the remote file, and the currently open data file is rewritten to include the new linkage.

The **Clear URL** tool eliminates the association with the remote file by clearing the URL.

The **Out-of-date files** tool downloads the remote files that are known to be newer than their local copies, and attaches them appropriately.

The **Update All files** tool causes the download and reattachment of ALL currently attached referenced models that have associated URLs, regardless of the current update status. Use this to be certain that all local files exactly match those at the remote location.

The **Update Selected files** tool causes only currently selected referenced models to be downloaded and reattached. This allows you to keep a remote copy of a single file, or of a specific set of files, without downloading all new files.

The **Exit** tool closes the Reference File Agent dialog box.

Update Sequence Dialog Box

The Update Sequence dialog box is used to change the display order for all operations involving the updating of the view windows. To open the Update Sequence dialog box, invoke the Update Sequence tool from:

References tool box	Settings > Update Sequence

MicroStation displays the Update Sequence dialog box similar to Figure 13–22.

Chapter 13 Attaching References

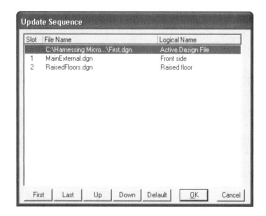

Figure 13-22 Update Sequence dialog box

The Update Sequence dialog box lists the active DGN file and attached references in the order in which they are displayed. To change the position of a file in the sequence, first select the file in the list box (CTRL and SHIFT can be used to select multiple files), and use the navigational buttons provided in the bottom of the dialog box to change position. Click the **OK** button to accept the changes and close the dialog box.

Hilite References

The hilite setting determines if a referenced model is highlighted and/or surrounded by a border when placed in the master model. MicroStation provides three modes of hilite: **Boundaries**, **Hilite** and **Both**. The **Boundaries** mode hilites the border of the selected reference model, and the **Hilite** mode hilites all the elements of selected reference model and **Both** mode hilite elements and border of the selected reference model.

Before you select the **Hilite** method, select the reference(s) listed in the References settings box and choose one of the available modes from the **Hilite** option menu as shown in Figure 13–23.

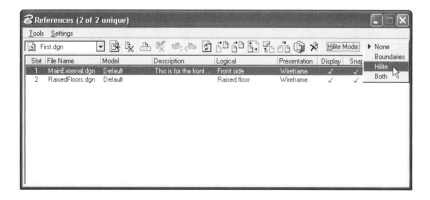

Figure 13-23 Displaying option menu from Hilite Mode in the References settings box.

583

Open the Exercise Manual PDF file for Chapter 13 on the accompanying CD for project and discipline specific exercises.

REVIEW QUESTIONS

Write your answers in the spaces provided.

1. List at least two benefits of using references.

2. By default, how many references can you attach to a design file?

3. What effect does attaching references have on a design file size?

4. List the tools that are specifically provided to manipulate references.

5. Give the steps involved in attaching a reference.

6. Explain the difference between the Reference Clip Boundary tool and the Reference Clip Mask tool.

7. List the steps involved in detaching a reference.

chapter 14

Special Features

MicroStation provides special features that, although used less often than the tools described earlier in this book, provide added power and versatility. This chapter introduces several such features. For more detailed information on each feature, consult the documentation furnished with MicroStation and the online help.

Objectives

After completing this chapter, you will be able to:

- Create and use graphic groups
- Use the Merge utility
- Select groups of elements using selection criteria
- Define a level symbology to help you determine what level elements are on
- Change the highlight and vector cursor colors
- Import and export drawings using the format of other applications
- Manipulate graphic images
- Annotate designs
- Create dimension-driven designs
- Use object linking and embedding

GRAPHIC GROUPS

The Add To Graphic Group tool allows you to group elements so they act as if they were one element when you apply element manipulation tools with the Graphic Group lock turned ON. The Drop From Graphic Group tool allows you to drop elements from a graphic group or drop the entire group.

Note: The elements of a pattern are part of one graphic group, but the elements that contain the pattern are not part of the group.

Adding Elements to a Graphic Group

To add elements to a graphic group, invoke the Add To Graphic Group tool from:

Groups tool box	Select the Add To Graphic Group tool (see Figure 14–1).
Key-in window	**group add** (or **gr a**) ENTER

Figure 14–1 Invoking the Add To Graphic Group tool from the Groups tool box

MicroStation prompts:

> Add to Graphic Group > Identify element *(Identify the element.)*

If the selected element is not already in a graphic group, a new group is started, and MicroStation prompts:

> Add to Graphic Group > Add to new group (Accept/Reject) *(Identify the next element to add to the new group, or click the Reset button to reject the selected element.)*

If the selected element is already in a graphic group, MicroStation prompts:

> Add to Graphic Group > Add to existing group (Accept/Reject) *(Identify the first element to be added to the existing graphic group, or click the Reset button to reject the selected group.)*

If you place the cursor on another element when you accept the currently selected element, MicroStation prompts:

> Add to Graphic Group > Accept/Reject (select next input) *(If more elements are to be added to the new or existing group, select each one. When all elements have been selected, click the Data button in space to accept adding the last selected element.)*

Manipulating Elements in a Graphic Group

The Graphic Group lock controls the effect the element manipulation tools have on elements in a graphic group. If the check box is OFF, the tools manipulate only the selected element. If it is ON, all elements in the graphic group are highlighted and manipulated (even if you cannot see them in the view you are working in). For example, if the **Graphic Group** check box is OFF and you select one of the elements in a graphic group to be deleted, only that element is deleted. If the check box is ON, all elements in the graphic group are deleted.

The **Graphic Group** check box can be turned ON and OFF in any of the Lock settings boxes and menus (see Figure 14–2).

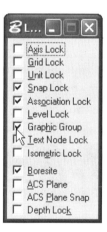

Figure 14-2 The Graphic Group check box in the Lock Toggles box

Note: If the Graphic Group lock is ON, you can determine if the element you select for manipulation is in a group by looking at the element type description in the Status bar. If the element is in a graphic group, the message includes "(GG)."

Copying Elements in a Graphic Group

The Graphic Group lock affects the way the element and fence contents copy tools handle elements in a graphic group.

- If the lock is OFF, the copied elements are not part of any graphic group.
- If the lock is ON, the element copy tool copies the entire graphic group, and the copies form a new graphic group.
- If the lock is ON, only the graphic group elements within the fence (as determined by the Fence Mode) are copied, and the copies become a new, separate graphic group.

Dropping Elements from a Graphic Group

The Drop From Graphic Group tool drops:

- The selected element when the Graphic Group lock is OFF.
- All elements in the group when the Graphic Group lock is ON.

To drop elements from a graphic group, invoke the Drop From Graphic Group tool from:

Groups tool box	Select the Drop From Graphic Group tool (see Figure 14–3).
Key-in window	**group drop** (or **gr d**) ENTER

Figure 14-3 Invoking the Drop From Graphic Group tool from the Groups tool box

MicroStation prompts:

Drop From Graphic Group > Identify element *(Identify the element.)*

Drop From Graphic Group > Accept/Reject (Select next input) *(If more elements are to be dropped, select each one. When all elements have been selected, click the Data button in space to accept dropping the last selected element.)*

Merge Utility

The Merge Utility allows you to copy elements from one or more design files or cell libraries into a design file or cell library. The utility runs from the MicroStation Manager dialog box.

Note: If you have a design file open in MicroStation, select **File > Close** to return to the MicroStation Manager dialog box.

In the MicroStation Manager dialog box, invoke the Merge Utility from:

| Drop-down menu (in MicroStation Manager Window) | File > Merge |

MicroStation displays the Merge dialog box, as shown in Figure 14–4.

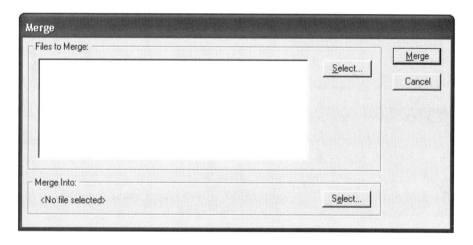

Figure 14-4 The Merge dialog box

To select the file to be merged into the target file, click the top **Select** button (to the right of the **Files to Merge** field). MicroStation opens the Select Files to Merge dialog box, as shown in Figure 14–5.

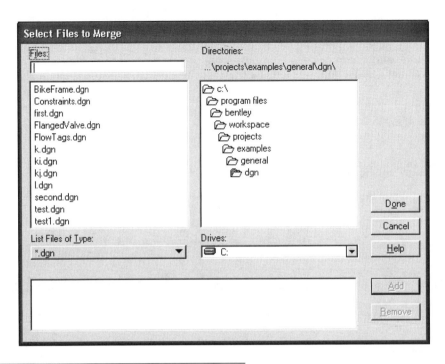

Figure 14-5 The Select Files to Merge dialog box

1. In the Select Files to Merge dialog box, select the location of the file to be merged in the **Directories** field and select the file in the **Files** field.
2. Click the **Add** button (near the bottom of the dialog box) to add the selected file to the list of files to be merged (in the field at the bottom of the dialog box).
3. Repeat the first two steps for each file that is to be merged into the destination file.
4. When all files are selected, click the **Done** button to close the dialog box. The added files are displayed in the Merge dialog box.
5. Click the bottom **Select** button (to the right of the **Merge Into** field) to open the Select Destination File dialog box.
6. In the Select Destination File dialog box, select the location of the file into which the previously selected files will be merged, select the file in the **Files** field, and click the **OK** button.
7. In the Merge dialog box, click the **Merge** button to merge the files and close the Merge dialog box.

After the final step, copies of all elements in the files to be merged are in the destination file.

SELECTION BY ATTRIBUTES

The **Select By Attributes** option in the **Edit** drop-down menu allows you to limit the selection of elements for manipulation to those that meet certain element attributes. For example, if all red ellipses on the Border level need to be changed to the color green, use the Select By Attributes dialog box to select only those elements and apply the color change to the selected elements. Handles appear on the selected elements, just as they did with the selection tool described in Chapter 6. You can delete the selected elements, move them, copy them, change their attributes, and apply several other manipulation tools to them.

Select By Attributes Dialog Box

Open the Select By Attributes dialog box from:

| Drop-down menu | Edit > Select By Attributes |

The dialog box contains several fields, selection menus, and options to open other dialog boxes (see Figure 14–6).

Figure 14–6 Select By Attributes dialog box

The dialog box contains the following parts:

- Selection by **Levels** (upper left of the box).
- Selection by element **Types** (upper right of the box).

- Selection by **Symbology**—color, line style, or line weight (lower left of the box).
- Selection **Mode**, which contains three menus that control the way the selection criteria are applied.
- The **Execute** button at the bottom left of the box that puts into effect the selection criteria currently set in the fields.
- The **Properties** button that opens the Select By Properties dialog box (see Figure 14–7) where you can choose to select elements by property attributes (such as only filled elements) and classes (such as construction elements).
- The **Tags** button that opens the Select By Tags dialog box (see Figure 14–8) where you can limit your selection to elements that contain only certain tags or combinations of tags.

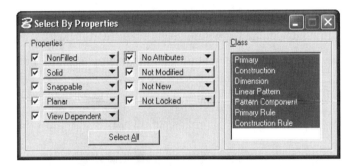

Figure 14-7 Select By Properties dialog box

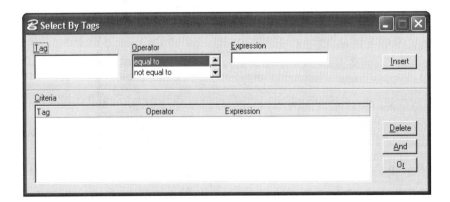

Figure 14-8 Select By Tags dialog box

 Note: Your selection criteria can be based on one item in one of the fields or on a combination of items in one or more fields—for example, all elements on level 10 or only ellipses on level 5 that are filled.

Selection by Level

The **Levels** box lists all level names defined in the design file (See Figure 14–7). If a name is shown with a dark background, it is part of the selection criteria. A light background means the level is not part of the selection criteria.

To select a specific level name and turn off all others, click the Data button on the desired name. To select additional names, hold down CTRL while clicking the Data button on level names. To select a contiguous group of names, drag the screen pointer across them while holding down the Data button. For example, to select only the Border and Roads levels, click the Data button on the Border and CTRL-click the Data button on Roads.

Selection by Type

The **Types** box contains a list of all element types (see Figure 14–7). Type names shown with a dark background are part of the selection criteria. A light background means that type is not part of the selection criteria.

To select a specific element type and turn off all others, click the Data button on the desired type. To select additional types, hold down CTRL while clicking the Data button on type names. To select a contiguous group of types, drag the screen pointer across them while holding down the Data button. For example, to select only ellipses and line strings, click the Data button on the Ellipse type, and then CTRL-click the Data button on the Line String type.

Selection by Symbology

The Symbology area allows you to include element **Color**, **Style**, And **Weight** in the selection criteria (see Figure 14–7). To add a symbology item to the selection criteria, turn the item's check box ON and select a value in the menu to the right of the item's name. For example, turn the **Color** check box ON and select the color **Red** (3) in the menu to limit the selection to only elements that are red.

Controlling the Selection Mode

The three Mode option menus (see Figure 14–7) control the way the selection criteria are applied when you click the **Execute** button.

The first **Mode** option menu has two options:

- **Inclusive** – Select only elements that meet the selection criteria.
- **Exclusive** – Select only elements that do not meet the selection criteria.

For example, if the selection **Type** is set to Ellipse, **Inclusive** mode causes all ellipses to be selected and **Exclusive** mode causes all elements except ellipses to be selected.

The second **Mode** option menu has three options:

- **Selection** – Immediately select (place handles on) all elements that meet the selection criteria when the Execute button is clicked.

- **Location** – Turns on the selection criteria but does not select any elements when the **Execute** button is pressed. Use the Element Selection tool to select elements that meet the criteria in this mode. Elements that do not meet the criteria cannot be selected.

- **Display** – Turns on the selection criteria and makes all elements that do not meet the criteria disappear from the view when the **Execute** button is clicked. No handles are placed on the remaining elements, so the Element Selection tool must be used to select from the elements that still appear in the view.

The third **Mode** option menu has two options:

- **Off** – When the **Execute** button is clicked, turn off the previously set selection criteria so it has no effect on element selection.

- **On** – When the **Execute** button is clicked, turn on the current selection criteria so it can be used.

Note: If you close the Select By Attributes dialog box with a selection criteria in effect, an Alert window appears prompting you to decide what to do with the current selection criteria. Click **OK** to keep the selection criterion in effect or **Cancel** to turn off the criterion.

Select By Properties Dialog Box

Use the Select By Properties dialog box to include additional element attributes in the selection criteria. Open this box by clicking the **Properties** button in the Select By Attributes dialog box.

On the left side of the box, select element **Properties** settings. Each property has an options menu from which you can select the property to be included in the selection criteria. To select a property, turn to the property's check box ON and select a setting from the drop-down menu. For example, the area property options menu allows you to select either **Solid** or **Hole** elements as shown in Figure 14–9.

594

Chapter 14 Special Features

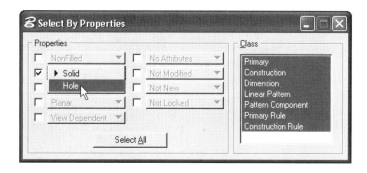

Figure 14-9 Selecting an area property in the Select By Properties dialog box

The command button below the **Properties** area turns all the **Properties** check boxes ON and OFF. The name of the button says what will happen next. If it says **Select All**, it turns ON all the check boxes and if it says **Clear All**, it turns OFF all the check boxes.

On the right side of the box is a field in which you select what class or classes of elements to include in the selection criteria (see Figure 14–9). If the class is shown with a dark background, it is selected. Clicking the Data button on the class name adds it to or removes it from the selection criteria. For example, if you want to select only elements in the **Construction** class, set the word **Construction** in the menu to have a dark background and set all the others to have a light background.

Select By Tags Dialog Box

The Select By Tags dialog box (see Figure 14–10) is used to specify criteria based on tag values. If selection criteria based on tag values are specified, elements that do not have attached tags with the specified tag names are not selected, located, or displayed. To open the Select By Tags dialog box, click the **Tags** button in the Select By Attributes dialog box.

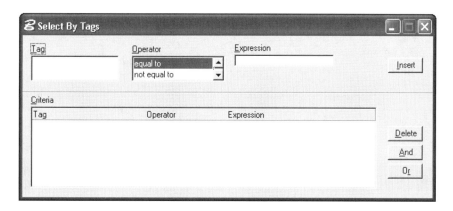

Figure 14-10 Select by Tags dialog box

595

Executing Selection Criteria into Effect

Click the **Execute** button in the Select By Attributes dialog box to select the elements according to the settings. Handles appear (depending on the settings) on the selected elements. After completing the selection criteria, invoke the appropriate element manipulation tool and follow the prompts.

LEVEL SYMBOLOGY

Designers often make extensive use of levels in design files to help organize the various parts of a design. For example, a plot plan may have separate levels for roadways, descriptive text, foundations, utilities, the drawing border, and title block information. An architectural plan might have the walls on one level, the dimensions on another level, electrical information on still another level, and so on. Separating parts of the design by level allows designers to turn on only the part they need to work on and allows them to print parts of the design separately.

Keeping up with what level everything is on can be confusing. Level names help, but additional information can be provided by using level symbology to assign a specific color, style, and weight to each level. You have already used level symbology in some of the chapter exercises when you specified a color, style, and weight for each level and then selected the **ByLevel** option for the Active Color, Active Line Style, and Active Weight.

Setting Symbology for a Level

Level symbology is set in the Level Manager settings box where you define level names. Invoke the Level Manager box from:

Drop-down menu	Settings > Level > Manager

MicroStation displays the Level Manager settings box, as shown in Figure 14–11.

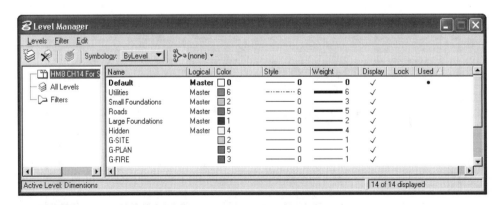

Figure 14-11 The Level Manager settings box

Chapter 14 Special Features

Each row in the large field of the Level Manager settings box describes one level and contains **Color**, **Style**, and **Weight** fields for setting the level symbology. When you click one of the fields, a pop-up dialog box appears that displays the available options. Figure 14–12 shows an example of the **Level Weight** dialog box.

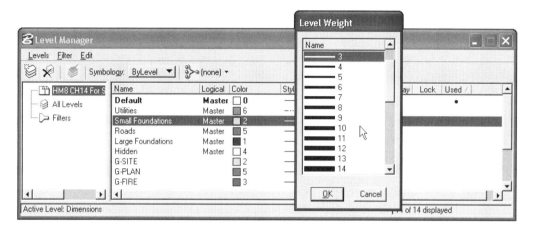

Figure 14-12 The Level Weight dialog box

Select the desired setting from the dialog box and click the **OK** button to apply it to the level from which you opened the dialog box. If you opened the box my mistake, click the **Cancel** button to close it without selecting a value.

Using the Level Symbology in the Design

In the Attributes tool box, each symbology setting (**Active Color**, **Active Line Style**, and **Active Line Weight**) has a **ByLevel** option. Figure 14–13 shows an example of the option in the **Active Line Weight** drop-down menu.

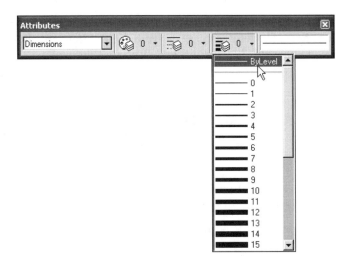

Figure 14-13 The Active Line Weight ByLevel option

If you select the **ByLevel** option for an active symbology, the symbology's value is set to the symbology of the Active Level each time you change the Active Level, and all elements are placed using the level symbology.

CHANGING THE HIGHLIGHT AND POINTER COLOR

The highlight and drawing pointer colors can be changed from the Design File settings window. These colors are used for:

- Highlighting selected elements
- The drawing pointer when a data point is placed and when an element is manipulated
- The locate tolerance circle that appears on the pointer during manipulations

A common case where different colors may be required is when many elements in the design use the same colors as the selection and pointer. In that case, it may be hard to spot selected elements and the pointer can be hard to see.

To change the colors, invoke the Design File dialog box from:

Drop-down menu	Settings > Design File and then select Color from the Category menu (see Figure 14-14).
Key-in window	**mdl load dgnset** (or **md l dgnset**) ENTER

Chapter 14 Special Features

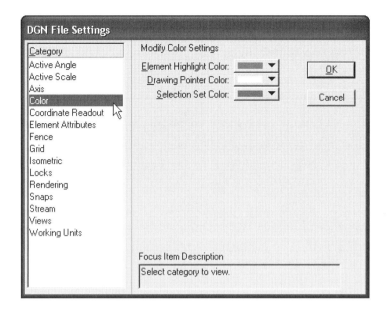

Figure 14-14 Design File dialog box with the Color category selected

In the Design File Settings window:

- Open the **Element Highlight Color** menu and select the desired color for highlighting identified elements.

- Open the **Drawing Pointer Color** menu and select the desired color for the drawing pointer.

- Open the **Selection Set Color** menu and select the desired color for the elements that have been selected with the Element Selection tool when handles are not used.

- Click the **OK** button to make the new color settings active.

- To make the changes permanent in this design file, select **File > Save Settings**.

Note: The color changes apply only to the design file in which they were changed. Each design file has its own highlight and pointer color settings.

Do not set the colors the same as the view window background color. If the color is the same, highlighted elements and the pointer cannot be seen.

599

IMPORTING AND EXPORTING DRAWINGS IN OTHER FORMATS

MicroStation can import files containing graphic information in several file formats and can export drawings to those formats.

- Several formats can be imported from the MicroStation Manager and File Open windows.
- Design files can be exported to several formats from the File Save As window.
- Export and Import options are also provided in the **File** drop-down menu.

The support of other formats allows MicroStation users to share design files with clients and vendors using other CAD and graphic applications.

Supported Formats

Table 14–1 describes file exchange formats that MicroStation can open directly and save and that are also available in the **Import** and **Export** submenus from the drop-down menu File. Table 14–2 describes file exchange formats that are available only from the **Import** and **Export** submenus.

Table 14-1 File Exchange Formats Available for Opening and Saving Files

FORMAT	OPEN	SAVE AS	DESCRIPTION
DGN	Yes	Yes	Native format MicroStation version 7 and version 8 design files.
DWG	Yes	Yes	Native format AutoCAD drawings.
DXF	Yes	Yes	Drawing Interchange Format— developed by Autodesk, Inc., to exchange graphic data among many CAD and graphics applications. DWG and DXF imports are handled identically.
CGM	Yes	Yes	Computer Graphics Metafile Format— an ANSI standard for the exchange of picture data between different graphics applications; device and environment independent.
D	Yes	No	MicroStation TriForma Document Files. TriForma is another Bentley product.
RDL	Yes	Yes	Redline files.

Table 14–2 File Exchange Formats Available in the Import and Export Submenus

FORMAT	IMPORT	EXPORT	DESCRIPTION
IGES	Yes	Yes	A public domain, neutral file format that serves as an international standard for the exchange of data between different CAD/CAM systems.
CGM	Yes	Yes	Computer Graphics Metafile Format— an ANSI standard for the exchange of picture data between different graphics applications; device and environment independent.
Image	Yes	No	Several graphics formats used by text processing and publication graphics packages
Text	Yes	No	ASCII text files (discussed in Chapter 6)
3D	No	Yes	MicroStation's 3-dimensional design file format. If the open design file is 3D, there will be an option to save it as a 2D (2-dimensional) drawing.

Opening a File of Another File Format

The file formats shown in Table 14–1 can be opened directly in MicroStation. The conversion to MicroStation's design file format is done as the file opens.

To open a drawing created in another format from the MicroStation Manager or Open dialog box:

1. Open the **List Files of Type** menu (see Figure 14–15).
2. Select the format of the drawing to be opened.
3. Follow the usual procedures for opening a file.

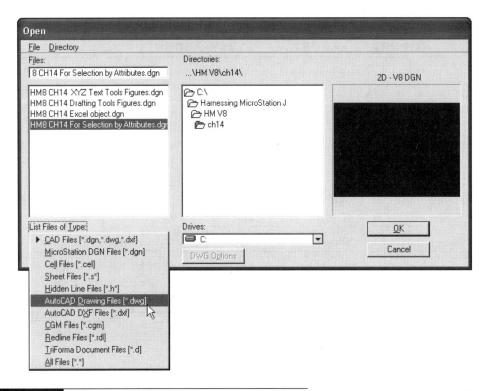

Figure 14-15 Selecting the format of a file to be opened

Saving a Design File Using Another File Format

The file formats shown in Table 14–1 can be saved from the File Save As dialog box. The conversion to the other file format is done as the file is saved. To save a design file in another format:

1. Select **File > Save As**.
2. Open the **List Files of Type** menu (see Figure 14–16).
3. Select the format to be used.
4. Supply a directory path and file name.
5. Click the **OK** button to initiate the conversion and close the dialog box.

The contents of the design file are converted to the other format, and the design file remains open in MicroStation.

Chapter 14 Special Features

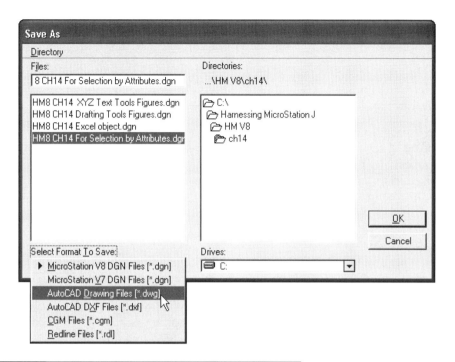

Figure 14-16 Selecting the format to save a design file as

Importing and Exporting Other Formats into an Open Design File

The file formats described in Table 14–2 are available for import and export from an open design file. The **Import** and **Export** options are available from the **File** drop-down menu.

- The **Import** option inserts the imported file contents into the open design file. The elements in the imported file are converted to design file elements. Image files remain as images in the design file and are not an element.

- The action of the **Export** tool varies among the formats. In some cases it opens the Save As dialog box and follows that window with an Export dialog box containing options to control the way the design file is opened. In other cases, the Export dialog box opens directly.

For detailed information on importing and exporting, refer to the technical documentation furnished with MicroStation.

MANIPULATING IMAGES

In MicroStation, the term "image" refers to graphics files that can be inserted in word processing and graphics processing applications. The pictures of MicroStation windows presented in this textbook are examples of such images. MicroStation provides a set of tools for creating, manipulating, and viewing images under the **Utilities** drop-down menu.

For example, a programmer created an automated drawing procedure that involves several custom dialog boxes. A technical writer is creating a training guide for the procedure in a Windows-based word processing package and needs pictures of the dialog boxes. The programmer uses the **Capture** option to capture the dialog boxes as bit-mapped image files that the technical writer can insert in the manual file.

Save

The **Utilities > Image > Save** option opens the Save Image dialog box (see Figure 14–17) from which the contents of one of the eight view windows can be saved as an image file using one of several available image formats. To determine the required image format, refer to the documentation furnished with the application for which the image is being created.

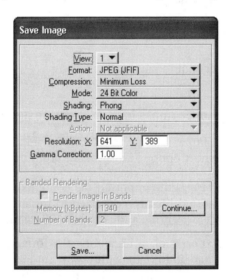

Figure 14–17 Save Image dialog box

Capture

The **Utilities > Image > Capture** option opens the Screen Capture dialog box (see Figure 14–18) from which you can select several methods of capturing all or part of the image on the workstation screen. Each capture method opens a Capture Output dialog box in which you can specify a file name and directory path for the captured screen image.

Chapter 14   Special Features

Figure 14-18   Screen Capture dialog box

The Screen Capture methods include:

- **Capture Screen**—Capture the entire screen.

- **Capture Rectangle**—Capture the contents of a rectangle you define within the MicroStation workspace.

- **Capture View**—Capture the contents of the view you select. Windows on top of the view are also captured.

- **Capture View Window**—Capture the contents and window border of the view you select. Windows on top of the view are also captured.

Convert

The **Utilities > Image > Convert** option opens the Raster Convert dialog box (see Figure 14–19) in which you can select a raster image file and convert it to an Output format file using a file name and path you specify.

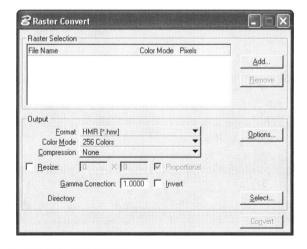

Figure 14-19   Raster Convert dialog box

605

Display

The **Utilities > Image > Display** option opens the Display Image dialog box (see Figure 14–20) from which you can select an image file to view in a separate window (see Figure 14–21).

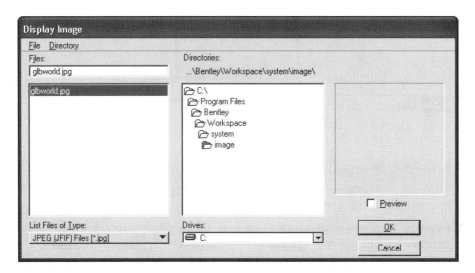

Figure 14-20 Display Image dialog box

Figure 14-21 Example of displaying an image file in the Display Image window

Movies

The **Utilities > Image > Movies** option opens the Movies dialog box (see Figure 14–22) from which an animated sequence file can be selected for viewing in a separate window.

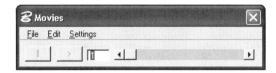

Figure 14-22 Movies dialog box

THE ANNOTATION TOOLS

MicroStation provides Annotation tools that place annotation items in the design such as flags, callout markers, and leaders.

Invoke the Annotation tools from:

Drop-down menu	Tools > Annotation
Key-in window	**dialog toolbox annotation** (or **di to annotati**) ENTER

The **Annotation** submenu contains four options. The first, **Annotation**, opens the Annotation tool frame. The **Annotate**, **Drafting Tools**, and **XYZ Text** options open tool boxes. The three tool boxes can also be opened from the Annotate tool frame. The tool frame and boxes are shown in Figure 14–23.

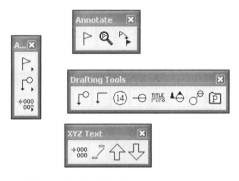

Figure 14-23 The Annotation tool frame and tool boxes

Note: Like the Main tool frame, the Annotation tool frame always displays the last button selected from each tool box in the frame.

The Annotate Tool Box

The Annotate tool box provides tools for placing, editing, and updating design notes that are placed in the design file behind flag symbols. The flag holds the text of the note and provides a visual reminder that there is a note at the location of the flag.

Invoke the Annotate tool box from:

Drop-down menu	Tools > Annotation > Annotate
Annotation tool frame	Click the Annotate icon
Key-in window	**dialog toolbox annotate** (or **di to ann**) ENTER

Place Flag

The Place Flag tool inserts new annotation flags in the design file. Invoke the Place Flag tool from:

Annotate tool box	Select the Place Flag tool (see Figure 14–24).
Key-in window	**place flag** (or **pl fl**) ENTER

Figure 14–24 Invoking the Place Flag tool from the Annotate tool box

The Tool Settings window provides options for setting up the new flag:

- Adjust the size of the placed flag by entering a number in the **Scale** field. Entering a number greater than one increases the size and entering a number less than one decreases the size.

- Select the level on which the flag is placed from the **Level** drop-down menu.

- Place the flag in the **Primary** or **Construction** class using the **Class** drop-down menu. If the **Construction View** attribute is turned OFF, elements in the **Construction** class do not appear in the view.

- Key-in the name of the image you want to use for the flag or click the **Browse** button to open the Select Image File for Flag dialog box. The box opens displaying

MicroStation's default location for images. If you want to search for an image in a different location, find and select that location in the **Directories** field and select the image file name in the **Files** field. See Figure 14–25 for examples of typical flag symbol images and the names of the files containing the images.

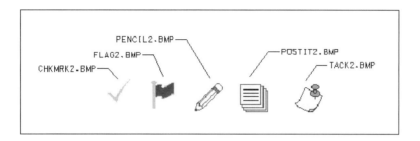

Figure 14-25 Typical flag images and their bit map file names

MicroStation prompts:

Place Flag > Identify location *(Define the point where the flag is to be placed.)*

Defining the flag location opens the Define Flag Information dialog box (see Figure 14–26). Type the annotation text in the field and click **OK** to place the flag and close the dialog box.

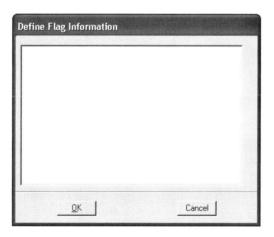

Figure 14-26 Define Flag Information dialog box

 Note: The flag's image appears in the design file but not the text. To view and edit the text, double-click the image to open the Define Flag Information dialog box or click the Show/Edit Flag icon in the Annotate tool box.

Show/Edit Flag Tool

The Show/Edit Flag tool allows you to view and edit the annotation text for a selected flag in the Define Flag Information dialog box. Invoke the Show/Edit Flag tool from:

Annotate tool box	Select the Show/Edit Flag tool (see Figure 14–27).
Key-in window	**show flag** (or **sh f**) ENTER

Figure 14-27 Invoking the Show/Edit Flag tool from the Annotate tool box

MicroStation prompts:

> Show/Edit Flag > Select flag *(Click the Data button on the flag to be viewed or edited.)*
>
> Show/Edit Flag > Accept/Reject *(Click the Data button again to accept the selected flag.)*

Accepting the flag opens the Define Flag Information window, which displays the annotation text and makes it available for editing. After you complete viewing or editing the flag text, click the **OK** button to close the window and save the changes.

Update Flag Tool

The Update Flag tool changes the image of an existing flag to the image currently set in the Place Flag's tool's settings **Image** field. Before invoking the Update Flag command, invoke the Place Flag tool and use the settings box **Browse** button to find and select the correct image.

Invoke the Update Flag command from:

Annotate tool box	Select the Update Flag tool (see Figure 14–28).
Key-in window	**flag update** (or **fl u**) ENTER

Figure 14-28 Invoking the Update Flag tool from the Annotate tool box

Chapter 14 Special Features

MicroStation prompts:

> Update Flag > Select flag *(Select the flag to be updated.)*
>
> Update Flag > Accept/Reject *(Click the Data button again to update the flag symbol.)*

The Drafting Tools

The Drafting Tools tool box provides tools that place callout markers and leaders and a tool to customize the way the elements are placed by the other tools. Following are brief descriptions of each tool in the toolbox.

Place Callout Leader

Figure 14–29 shows the Place Callout Leader tool and examples of callout leaders placed by the tool. Place callout leaders by specifying the terminator location and the center of the callout symbol.

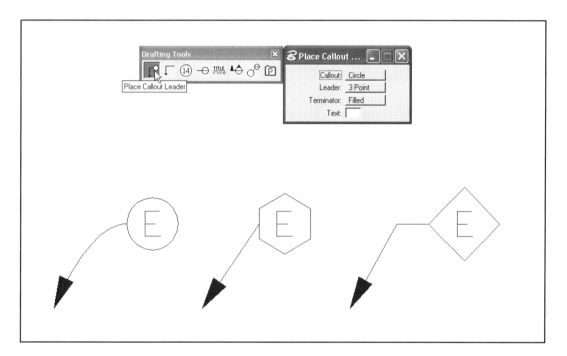

Figure 14-29 The Place Callout Leader tool and examples of callout leaders placed by the tool

The Tool Settings window provides options for setting up the callout leader. Select the callout shape (the figure shows a **Circle**, **Hex 1**, and **Diamond**) in the **Callout** drop-down menu. Select the type of leader line (**Bspline**, **2 Point**, or **3 Point**) in the **Leader** drop-down menu (the figure shows examples of all three). The **Terminator** drop-down menu provides options for selecting the way the interior of the callout is handled. Type the callout text in the **Text** field.

611

Place Leader and Text

Figure 14–30 shows the Place Leader and Text tool and examples of leaders placed by the tool. Place the leaders by specifying the terminator location and the end of the leader.

The Tool Settings window provides options for setting up the leader. Select the type of leader line (**Bspline**, **2 Point**, or **3 Point**) from the **Leader** drop-down menu (the figure shows examples of **Bspline** and **3 Point**). The **Terminator** drop-down menu provides options for selecting the way the interior of the callout should be handled. The field below the drop-down menus provides a place to key-in the leader text.

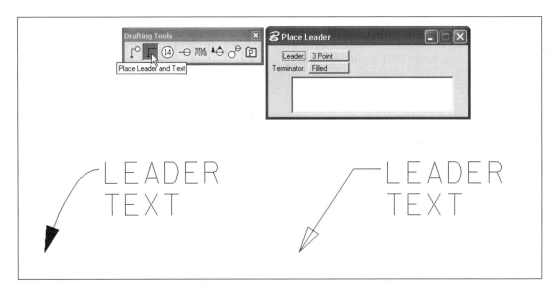

Figure 14–30 The Place Leader and Text tool and examples of leaders placed by the tool

Place Callout Bubble

Figure 14–31 shows the Place Callout Bubble tool and examples of bubbles placed by the tool. Place the bubbles by specifying the center of the bubble.

The Tool Settings window provides options for setting up the callout bubble. Select the callout shape (the figure shows several types) from the **Callout** drop-down menu. Key-in the callout text in the **Text** field.

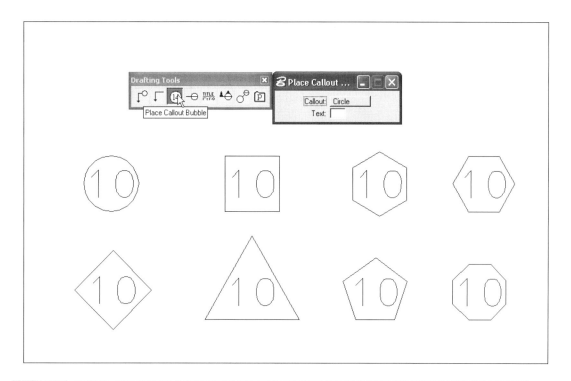

Figure 14-31 The Place Callout Bubble tool and examples of bubbles placed by the tool

Place Section Marker

Figure 14–32 shows the Place Section Marker tool and an example of a marker placed by the tool. Place the markers by specifying the left end of the marker line.

The Tool Settings window provides options for setting up the markers. Key-in the text that is to be placed above the marker line in the **Text** field and the number that is to be placed in the bubble above the line in the **Ref #** field. Key-in the text and number that are to be placed below the line in the **Subtitle** and **Sht #** fields.

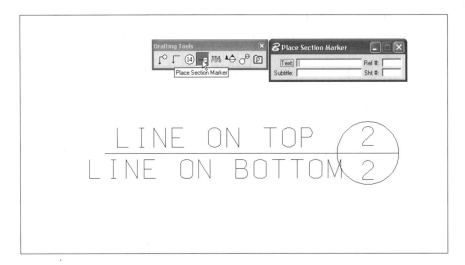

Figure 14-32 The Place Section Marker tool and an example of a marker placed by the tool

Place Title Text

Figure 14-33 shows the Place Title Text tool and an example of titles placed by the tool. Place the title text by specifying the left end of the title line.

The Tool Settings window provides options for setting up the selection markers. Key-in the text that is to be placed above the marker line in the **Text** field and the text that is to be placed below the line in the **Subtitle** field.

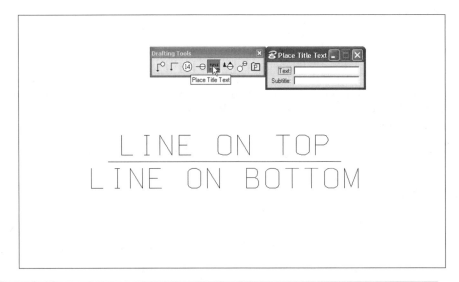

Figure 14-33 The Place Title Text tool and an example of a title placed by the tool

Chapter 14 Special Features

Place Arrow Marker

Figure 14–34 shows the Place Arrow Marker tool and examples of arrow markers placed by the tool. Place the marker by specifying the center of the marker and the direction of the arrow.

The Tool Settings window provides options for setting up the arrow marker. Type the number that is to appear above the line in the **Ref #** field and the number that is to appear below the line in the **Sht #** field.

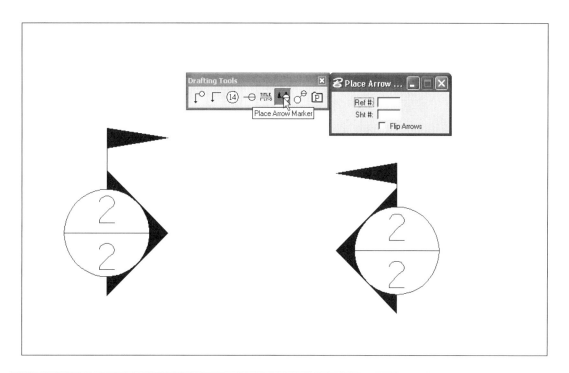

Figure 14-34 The Place Arrow Marker tool and examples of arrow markers placed by the tool

Place Detail Marker

Figure 14–35 shows the Place Detail Marker tool and examples of a detail marker placed by the tool. Place the marker by specifying the center of the detail and edge of the detail circle and the center of the marker.

The Tool Settings window provides options for setting up the detail marker. Type the number that is to appear above the line in the **Ref #** field and the number that is to appear below the line in the **Sht #** field.

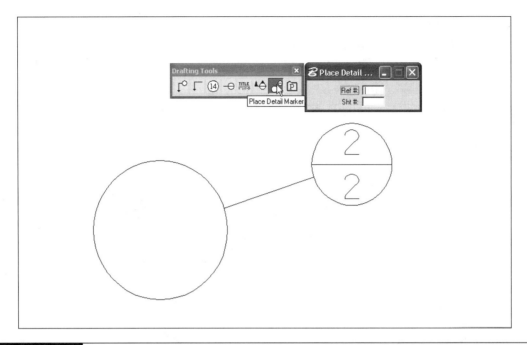

Figure 14-35 The Place Detail Marker tool and an example of a detail marker placed by the tool

Define Properties

Figure 14–36 shows the Drafting Tools Properties settings box that the Define Properties tool opens. The settings box provides settings for customizing the appearance of the elements placed by the Drafting Tools. Options are provided to control the size and attributes of the callout bubbles, leaders, bubbles, and text.

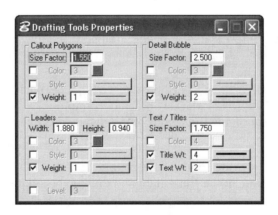

Figure 14-36 The Drafting Tools Properties settings box

XYZ Text Tool Box

The XYZ Text tool box provides tools for placing coordinate labels in the design, exporting coordinate points to an ASCII file, and importing coordinate points from an ASCII file. Following are brief descriptions of each tool in the tool box.

Label Coordinates

Figure 14–37 shows the Label Coordinates tool and examples of placing coordinate labels. You place coordinate labels by placing data points in the design. The Tool Settings window provides options for controlling the placement of the coordinate labels. The options are described below. The text is placed using the active text parameters.

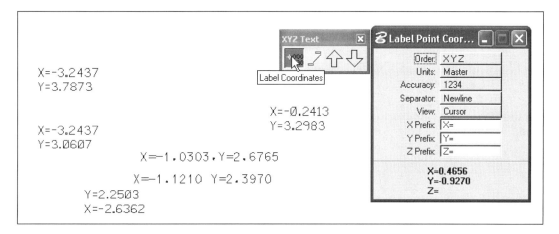

Figure 14-37 The Label Coordinates tool and examples of the labels in a design

The Tool Settings window provides the following options for controlling the placement of the coordinate label text:

- The **Order** menu allows you display the X coordinate in **XYZ** OR **YXZ** order.

- The **Units** menu allows you to display the coordinate values using **Master** Units, **Sub** Units, full **Working** Units, or **UORs** (Units of Resolution).

- The **Accuracy** menu allows you to control the accuracy of the fractional part of the coordinate values.

- The **Separator** menu allows you to have each coordinate value placed on a newline, separated by a **Comma**, or separated by a **Space**.

- The **View** menu is for 3D files only.

- **X**, **Y**, and **Z Prefix** text fields allow you to change the prefix that is placed before each coordinate value.

Note: The coordinates are placed as multi-line text elements. If the **Text Node View Attribute** is ON, the Text Node number and cross are displayed with the coordinate element.

Label Element

Figure 14–38 shows the Label Element tool and examples of placing coordinate labels. The Tool Settings window provides the same coordinate text options as the Coordinate Label tool, and three buttons that control the way elements are selected:

- The **Single** button allows you to place coordinate labels for a single element.
- The **Fence** button allows you to place coordinate labels on all elements grouped by a fence.
- The **All** button allows you to place coordinate labels on all elements in the design. A confirmation box appears asking you to confirm that you really want to put labels on all elements.

Each time you use this tool you must select one of the control buttons. For example; before you can place a coordinate label for one element, you must click the **Single** button. The coordinate text is placed using the active text parameters.

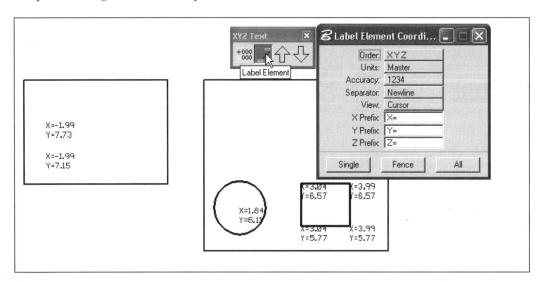

Figure 14-38 The Label Element tool and examples of coordinate labels placed on elements

Note: Coordinate labels are placed at the center of a circle and ellipse, at the center and ends of an arc, and at each vertex of linear elements.

Export Coordinates

Figure 14–39 shows the Export Coordinates tool. This tool allows you to write the coordinates of selected elements to an ASCII file.

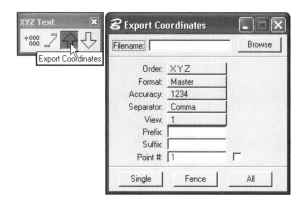

Figure 14-39 The Export Coordinates tool

The Tool Settings window provides options for creating the file to which the coordinates will be exported and how the coordinates will be placed in the file.

- Click the **Browse** button to open the Create Export File dialog box. Select the file location in the **Directories** field and key-in the file name in the **Filename** text field or select an existing file from the **Files** list. Click **OK** to create the file. If there is a file with the same file name, an Alert message box appears asking if you want to overwrite the existing file. Click the **OK** button to overwrite it.

Note: If you are using an existing file, it is not actually overwritten as the Alert message box implies. Later when you place coordinates in the file, another message box will ask if you want to append to the file or overwrite its contents.

- The drop-down menus (**Order**, **Units**, **Accuracy**, **Separator**, and **View**) are the same as described for the Label Coordinates tool.

- The **Prefix** and **Suffix** text fields allow you to key-in text strings that are placed before and after each coordinate in the file.

- If the **Point #** check box is ON, the coordinates are numbered as they are placed in the file and the final coordinate number appears in the **Point #** field. The same numbers are placed in the design at the points where the coordinates were taken.

- The buttons at the bottom of the window allow you to export the coordinates for a **Single** selected element, the elements grouped by a **Fence**, or **All** elements in the design file.

If you are exporting to an existing file, MicroStation displays the Export File Exists dialog box shown in Figure 14–40. If you select **Append existing file**, the coordinates are added to the end of the file. If you select **Overwrite existing file**, the existing information in the file is replaced by the new coordinates. Figure 14–41 shows an example of coordinates placed in an export file.

Figure 14-40 The Export File Exists dialog box

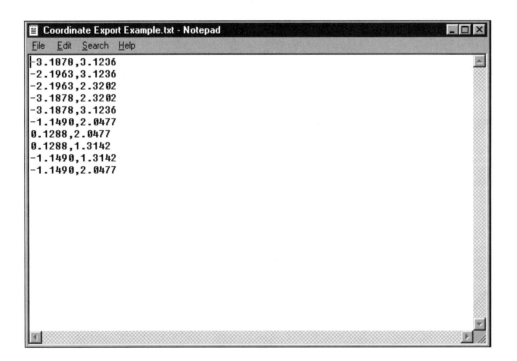

Figure 14-41 Example of exported coordinates

Import Coordinates

Figure 14–42 shows the Import Coordinates tool. This tool allows you to import coordinates from an ASCII file and place them in the active design as a point, text string, or cell.

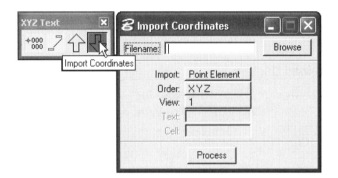

Figure 14-42 The Import Coordinates tool

The associated Tool Settings window provides options to control how the coordinates are imported, and to initiate the importation:

- The **Filename** text field and **Browse** button allow you to either type the name of the file containing the coordinates or browse to it.

- The **Import** menu allows you to choose what is placed at the imported coordinate location. You can choose to place a **Point Element**, **Text** string, or **Cell**.

- The **Order** menu allows you to elect to choose the order in which the coordinates are imported.

- The **View** menu is for 3D designs.

- The **Text** field is enabled only when the **Import** menu is set to **Text**. The field provides a place for you to enter the text string that will be placed at the coordinate locations.

- The **Cell** edit field is enabled only when the **Import** menu is set to **Cell**. The field provides a place for you to enter the name of the cell that will be placed at the coordinate locations. The cell library containing the cell must be attached to the design.

- The **Process** button initiates importing the coordinates from the file whose name is in the **Filename** field.

DIMENSION-DRIVEN DESIGN

Dimension-driven design provides a set of tools to define constraints on the elements that make up a design. These constraints control the size and shape of the design, so the design can be adjusted for changing requirements by simply entering new dimension values.

Dimension-driven cells are cells defined from constrained designs. The dimensions of such cells can be changed as they are placed.

Example of a Constrained Design

We introduce dimension-driven design by describing the steps required to create the model shown in Figure 14–43. Before we start constructing the model, let's look at the constraints on the model and at the effect of changing a constraint.

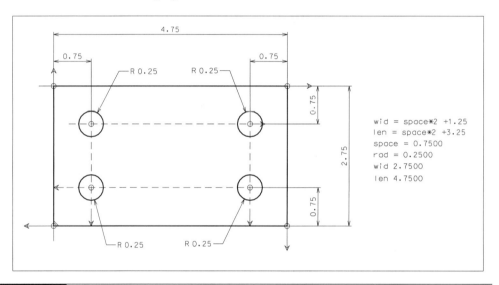

Figure 14-43 Example of a constrained design with the construction elements displayed

The Constraints

The design in the figure contains several "Construction" class elements that graphically represent the constraints:

- The dashed lines with arrows on one end are the construction lines to which the design is attached (some of the construction lines are covered by the lines and circles of the design).
- The equations and variables to the right of the model define the constraints.
- The Construction View attribute controls the display of construction elements. Figure 14–43 shows the view with Construction lines set to ON.
- The following constraints were placed on the design of Figure 14–43:

- The angles of the construction lines are fixed, as indicated by the arrows.
- The construction line intersections are constrained to always be connected, as indicated by the small circles.
- The radius of the circles are set by the "rad = 0.2500" variable.
- The circles are constrained always to be centered on the intersections of the inter-construction lines.
- The space between the center of each circle and the adjacent design edges are set by the "space = 0.7500" variable.
- The overall length of the design is set by the "len = space*2 + 3.25" equation. The equation multiples the circle-center-to-edge space by 2 and adds 3.25 to that total (4.75 = 0.75*2 + 3.25).
- The overall width of the design is set by the "wid = space*2 + 1.25" equation. The equation multiplies the circle-center-to-edge space by 2 and adds 1.25 to that total (2.75 = 0.75*2 + 1.25).

Effect of Changing a Constraint

To illustrate the effect of changing a constraint, we use the Text Edit tool to change the add-on value in the wid equation from 1.25 to 0.75 Master Units ("wid = space*2 + 0.75"), and then we use the Re-solve Constraints tool to solve the design for the new value. Figure 14–44 shows the design change caused by changing the wid add-on:

- We moved the two rows of circles closer together.
- We decreased the overall width of the design from 2.75 to 2.25 master units

The overall size was reduced because the value we changed is part of the constraint on the overall width.

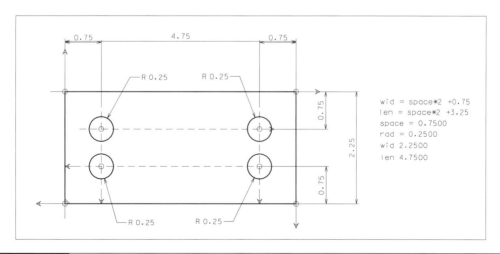

Figure 14-44 Effect of changing the width equation add-on from 2.75 to 2.25 master units

Dimension-Driven Design Terms

Following are common terms in dimension-driven design.

- **Constraint**—Information that controls how a construction is handled within a model. A constraint can be one of the following types:
- **Location**—Fixes the location of a point in the design plane.
- **Geometric**—Controls the position or orientation of two or more elements relative to each other.
- **Dimensional**—Controlled by a dimension.
- **Algebraic**—Controlled by an equation that expresses a relationship among variables.
- **Construction**—An element, such as a line or circle, on which constraints can be placed to control its relation to other constructions in the model.
- **Well-Constrained**—A set of constructions that is completely defined by constraints and has no redundant constraints. It has what is needed to define it and no more.
- **Underconstrained**—A set of constructions that does not have enough constraints to define completely its geometric shape.
- **Redundant**—A constraint applied to a construction that is already well-constrained. It provides no useful information for the construction.
- **Degrees of Freedom**—A number that sums up a dimension driven cell's ambiguity. The cell is underconstrained.
- **Solve**—Constructing the model from the given set of constraints. Each time a constraint is modified or added the model is solved for the new set of constraints. If the model can be solved for the constraints, the model is updated; if not, an error message is displayed in the Status bar.

Dimension-Driven Design Tools

Dimension-driven tool boxes are available in the DD Design submenu. Invoke the submenu from:

Drop-down menu	Select Tools > DD Design
Key-in window	**dialog toolbox dddtools** (or **di to dddt**) ENTER

MicroStation displays the Dimension-Driven tool frame. Figure 14–45 shows the tool frame surrounded by the tool boxes that can be opened from it.

Chapter 14 Special Features

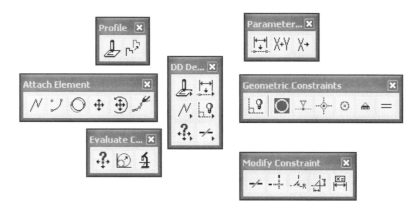

Figure 14-45 Dimension-Driven tool frame and its tool boxes

The tool boxes are:

- **Profile**—Provides tools for drawing a construction profile and for converting elements to a construction profile.

- **Attach Elements**—Provides tools for attaching elements, such as lines and arcs, to the constrained construction elements.

- **Evaluate Constraints**—Provides tools for solving and obtaining information about the construction.

- **Parameter Constraints**—Provides tools for converting dimensions to constraints and for assigning equations and variables.

- **Geometric Constraints**—Provides tools for placing constraints on the construction elements that are to define the model.

- **Modify Constraint**—Provides tools for modifying constraints

Creating a Dimension-Driven Design

A dimension-driven design requires careful planning before starting to define the constraints. Numerous tools are available to constrain the design. A good way to be introduced to the method is to walk through the creation of a design. The following discussion shows one way to create the model shown in Figure 14–46.

Draw Construction Elements

Set the line weight to zero; draw the lines and circles, as shown in Figure 14–46. The elements do not have to be drawn to specific dimensions because the constraints to be added later set the dimensions of the design. Draw the elements in the Primary mode. AccuDraw will help you quickly ensure that the lines are drawn horizontal and vertical.

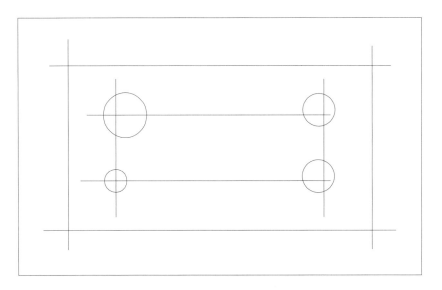

Figure 14-46 Design construction lines and circles

Constrain the angle of the lines

The first constraint we are going to put on the design is to force the lines to stay at the angles at which they were drawn. To constrain the angle of the lines shown in Figure 14–46, invoke the Constrain Elements tool from:

Geometric Constant tool box	Select the Constrain Elements tool. In the Tool Settings window, select the Smart Constrain Elements method (leftmost **Method** button) and turn the **Convert to constructions** check box ON (see Figure 14–47).
Key-in window	**constrain element** (or **con e**) ENTER

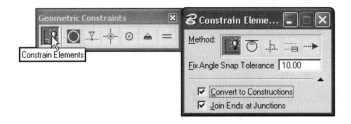

Figure 14-47 Invoking the Constrain Elements tool from the Geometric Constraints tool box

Chapter 14 Special Features

MicroStation prompts:

> Make Smart Constrain Elements > Identify Construction *(Identify one of the lines, and click the Data button again to accept it. The second point does not select another element.)*
>
> Smart Constrain Elements > Identify Second Construction or Same Construction to Fix Angle *(Identify the same line again to fix its angle.)*
>
> Smart Constrain Elements > Accept *(Click the Data button a third time to constrain the angle.)*

An arrow appears at one end of the line, the line's color changes to yellow and line style to dash, and the line is changed to a construction element. This constraint ensures that the lines always remain at their original rotation angle.

Repeat the steps to constrict the angle of the rest of the elements in the design. Figure 14–48 shows the result of constraining the angle of all the lines.

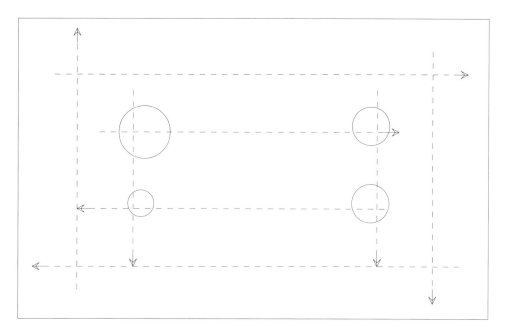

Figure 14-48 Result of constraining the angles of the lines

 Note: If the **Constructions** View Attribute is turned OFF in the view you are working in, the constrained elements disappear from the view.

627

Constrain Point at Intersection

This tool either constrains a point to lie at the intersection of two constructions or forces two constructions to pass through a point. It works with any kind of construction except points. To constrain the line intersections as shown in Figure 14–50, invoke the Constrain Point At tool from:

Geometric Constraints tool box	Select the Constrain Point At tool (see Figure 14–49).
Key-in window	**constrain intersection** (or **con i**) ENTER

Figure 14–49 Invoking the Constrain Point At tool from the Geometric Constraints tool box

MicroStation prompts:

> Constrain Point at Intersection > Identify Construction *(Identify one of the intersecting lines near the point of intersection, then select the other intersecting line.)*
>
> Constrain Point at Intersection > Identify point or Accept+Reset to Create *(Click the Data button in space, then click the Reset button to constrain the point.)*

A small circle appears at the constrained intersection. This constraint ensures that when one line is modified, the constrained lines always pass through the constraining point.

To constrain the remaining intersections, repeat the procedure for each section of the outer and inner set of lines. Figure 14–50 shows the result of constraining the intersections.

Chapter 14 Special Features

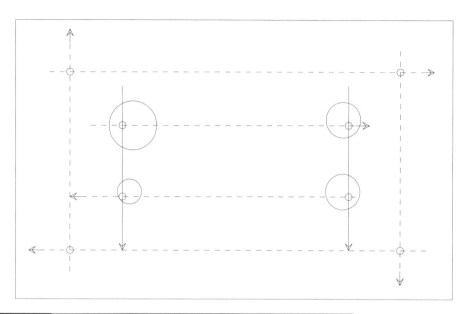

Figure 14-50 Result of constraining the construction intersections

Constrain Two Points to Be Coincident

This tool lets you constrain two points to the same location (coincident), two circles to be concentric (have the same center), or a point to lie at the center of a circle. To constrain the circles to be centered at the intersection points as shown in Figure 14-52, invoke the Constrain Points Coincident tool from:

Geometric Constraints tool box	Select Constrain Points Coincident tool (see Figure 14–51).
Key-in window	**constrain concentric** (or **con c**) ENTER

Figure 14-51 Invoking the Constrain Two Points to Be Coincident tool from the Geometric Constraints tool box

MicroStation prompts:

> Constrain Two Points to Be Coincident > Identify Point (or Ellipse) *(Identify the intersection point the circle is to be centered about.)*
>
> Constrain Two Points to Be Coincident > Identify Next Point (or Ellipse) *(Identify the the circle.)*
>
> Constrain Two Points to Be Coincident > Accept *(Click the Data button in space to complete the constraint.)*

This constraint causes the circles to change to construction elements that are yellow and dashed. The circles are constrained to stay centered over the intersections when the positions of the intersecting lines are modified.

Repeat the procedure to constrain the other three circles to the intersection points. Figure 14–52 shows the result of constraining the circles to the intersection points.

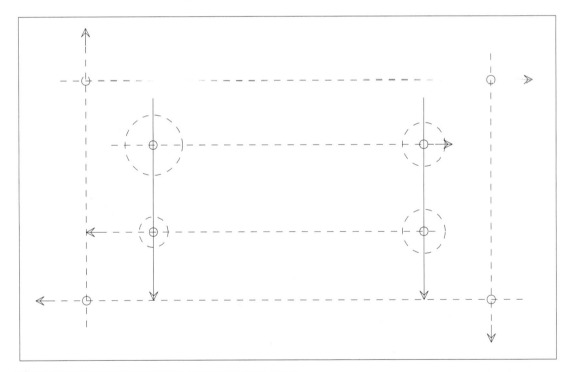

Figure 14-52 Result of constraining the circles

Chapter 14   Special Features

Fix Point at Location

This tool enables you to fix the location of a point (or the center of a circle or ellipse) in the design plane. To attach the lower left intersection of the design to its current location in the design plane, invoke the Fix Point tool from:

Geometric Constraints tool box	Select the Fix Point tool (see Figure 14–53).
Key-in window	**constrain location** (or **con l**) ENTER

Figure 14-53 Invoking the Fix Point tool from the Geometric Constraints tool box

MicroStation prompts:

> Fix Point at Location > Identify Point (or Ellipse) *(Select the lower left intersection point.)*
>
> Fix Point at Location > Accept/Reject *(Click the Data button a second time to accept the point.)*

This tool fixes the design to a location in the design plane, and this point remains fixed when changes are made to the size of the constrained design.

Construct Attached Line-String or Shape

With this tool you can create a line-string or shape with the vertices attached to construction points, circles, or constraints. It is recommended to set a higher value for active line weight so that you can easily distinguish it from the construction elements. To connect the shape as shown in Figure 14–55, invoke the Attach Line-String or Shape tool from:

Attach Element tool box	Select the Attach Line-String Or Shape tool (see Figure 14–54).
Key-in window	**attach lstring** (or **at ls**) ENTER

Figure 14-54 Invoking the Attach Line-String or Shape tool from the Attach Element tool box

MicroStation prompts:

> Construct Attached Line String or Shape > Identify Point or Constraint *(Select one of the constraint points on an intersection of a pair of the outer lines, then select the constraint point at the other end of one of the lines.)*
>
> Construct Attached Line String or Shape > Identify point, or RESET to finish *(Select the other two constraint points on the outer lines, then select the first constraint point again to complete the shape.)*

When the first constraint point is selected again, a closed shape is placed attached to the four points, as shown in Figure 14–55. Changes to the position of the constraint points will change the shape of the attached shape element.

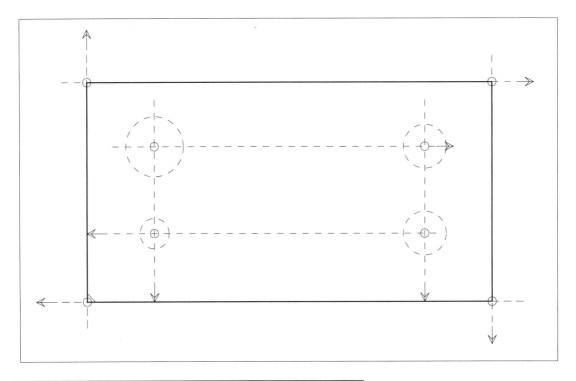

Figure 14-55 Result of drawing the design's outline shape

Chapter 14 Special Features

Construct Attached Ellipse or Circle

This tool enables you to create and attach a circle to a construction circle or point. To attach the design circles to the intersection points as shown in Figure 14–57, invoke the Attach Ellipse tool from:

Attach Element tool box	Select the Attach Ellipse tool (see Figure 14–56).
Key-in window	**attach circle** (or **at c**) ENTER

Figure 14–56 Invoking the Attach Ellipse tool from the Attach Element tool box

MicroStation prompts:

> Construct Attached Ellipse or Circle > Identify Ellipse *(Select one of the construction circles, then click the Data button in space to accept the circle.)*

This attachment causes the design circle to always stay centered on the inner construction line intersections. Repeat the procedure for the other three circles. The result of attaching design circles to each of the construction circles is shown in Figure 14–57.

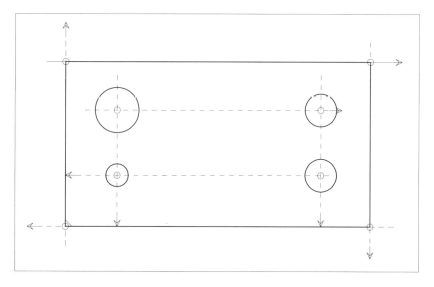

Figure 14–57 Result of drawing the design's circles

Add dimensions on the design as shown in Figure 14–58. Turn the Association check box ON before placing the dimensions. Associating the dimensions with the design elements allows the dimensions to be used as constraints. When you construct the linear dimensions for the circles, start the dimension on the outside rectangle. If you start the dimension at the circles, an error appears.

Also make sure the dimension extension lines are connected to the design elements, not the constructions. To make sure of that, snap to the starting and ending points until the correct element is highlighted, then accept it.

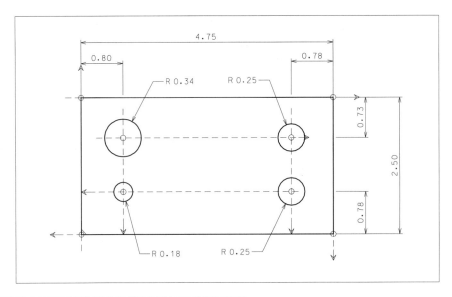

Figure 14-58 Result of dimensioning the design

 Note: Ignore the dimension values now. The dimensions will be constrained and their values set later.

Place the following text strings for the equations and variables by using the Place Text at Origin tool, and place each line of text in separate, one-line text elements. Place the text to the right of the design.

    wid = space*2 + 1.25

    len = space*2 + 3.25

    space = 0.75

    rad = 0.25

    wid

    len

Figure 14–59 shows the design after adding the text.

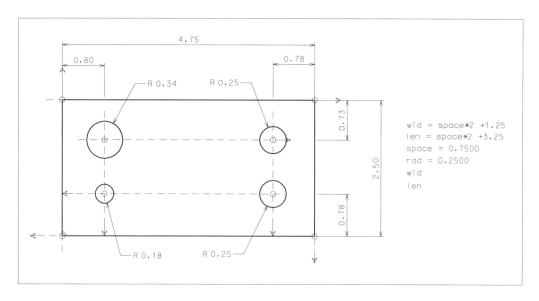

Figure 14-59 Result of adding the text strings to the design

Assign Variable to Dimensional Constraint

With this tool you can assign a constant or variable to a dimensional constraint. The constant or variable then represents the dimension's value in equations. Assign the "rad" variable as the radius of each circle. Invoke the Assign Variable tool from:

Parameter Constraints tool box	Select the Assign Variable tool (see Figure 14–60).
Key-in window	**assign variable** (or **as v**) ENTER

Figure 14-60 Invoking the Assign Variable tool from the Parameter Constraint tool box

MicroStation prompts:

> Assign Variable to Dimensional Constraint > Identify variable *(Select the "rad = 0.25" text string.)*
>
> Assign Variable to Dimensional Constraint > Identify Constraint *(Select the radial dimension of one of the circles.)*
>
> Assign Variable to Dimensional Constraint > Accept *(Click the Data button in space to assign the "rad" variable to the circle's dimension and set the circle's radius to the "rad" value.)*

Repeat the procedure for the radial dimensions of the other three circles.

Similarly, constrain the space between each circle center and the adjacent design edges with the Assign Variable. Assign the "Space" variable to each of the four circle-center-to-edge dimensions.

Assign Equation

This tool assigns an algebraic constraint—an equation that expresses a constraint relationship between variables, numerical constants, and built-in functions and constants—to a model. Create the "wid" and "len" equations by invoking the Assign Equation tool from:

Parameter Constraints tool box	Select the Assign Equation tool (see Figure 14–61).
Key-in window	**assign equation** (or **as e**) ENTER

Figure 14-61 Invoking the Assign Equation tool from the Parameter Constraint tool box

MicroStation prompts:

> Assign equation > Identify equation *(Select the "wid = space*2 + 1.25" text string.)*
>
> Assign equation > Identify variable, or RESET to finish *(Select the "space = 0.75" text string, select the "wid" text string, then click the Data button in space to complete defining the "wid" equation.)*

Similarly, invoke the Assign equation tool again to create the "len" equation by selecting "len = space*2 + 3.25", "space = 0.75", and "len."

Constrain the design's outline dimensions by invoking the Assign Variable tool to assign "len" to the overall horizontal dimension and "wid" to the overall vertical dimension.

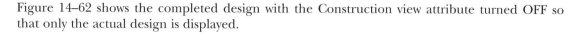

Figure 14–62 shows the completed design with the Construction view attribute turned OFF so that only the actual design is displayed.

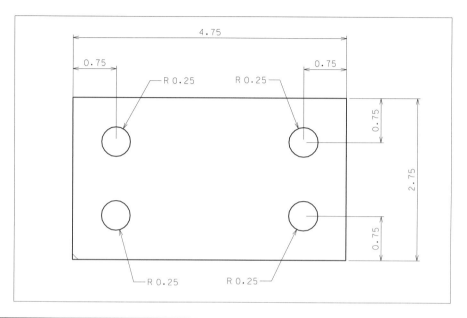

Figure 14-62 Completed design with construction elements turned off

 Note: After the design is completed, it may be necessary to invoke the Modify Element tool to adjust the position of some of the dimension elements and to move the text strings.

Modifying a Dimension-Driven Design

Changing one of the variable values and then re-solving the design modifies dimension-driven designs. Constraint equations and variables can be changed by editing the text strings or by invoking the Modify Value tool. For example, the design we just constructed contains four variables:

- The "wid" and "len" variables are set equal to equations that contain constants (1.25 and 3.25). Editing the equations with the Edit Text tool can change the constants. These two constants control the horizontal and vertical space between the circles.

- The "space" variable is set equal to a constant (0.75) that can be changed either by editing the text string via the Edit Text tool or with the Modify Value tool. Changing this variable changes the position of the circles and, because it appears in the two equations, the overall width and length of the design.

- The "len" variable is set equal to a constant (0.25) that can be changed to control the size of the circles.

Modify Value of Dimension or Variable

This tool can edit the value of a dimensional constraint. To change the variable, invoke the Modify Value tool from:

Modify Constraint tool box	Select the Modify Value tool (see Figure 14–63).
Key-in window	**model edit_dimension** (or **mo e**) ENTER

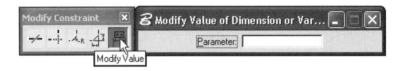

Figure 14–63 Invoking the Modify Value tool in the Modify Constraint tool box

MicroStation prompts:

> Modify Value of Dimension or Variable > Identify element *(Select the dimension or variable to be changed.)*
>
> Modify Value of Dimension or Variable > Accept *(Click the Data button in space to accept the selected element.)*
>
> Modify Value of Dimension or Variable > Enter a value *(Enter the new value in the Settings window edit field, then press* ENTER *to re-solve the design for the new value.)*

Re-solve Constraints

If the Edit Text tool is used to change a variable's value, the design must be "re-solved" to apply the new constraint value. To re-solve the design, invoke the Re-solve Constraints tool from:

Evaluate Constraints tool box	Select the Re-solve Constraints tool (see Figure 14–64).
Key-in window	**update model** (or **up m**) ENTER

Figure 14–64 Invoking the Re-solve Constraints tool from the Evaluate tool box

MicroStation prompts:

> Re-solve Constraints > Identify Element *(Select the variable that has been changed or the text of a dimension that is constrained to the variable, then click the Data button in space to initiate re-solving of the design.)*

Creating a Dimension-Driven Cell

Dimension-driven cells are created from dimension-driven designs when the Construction view attribute is set to ON and all of the construction elements of the design are included in the design. When a dimension-driven cell is placed in a design file, its constraints can be changed.

 Note: If the Construction view attribute is OFF when a cell is created from a dimension-driven design, the cell is not created as a dimension-driven cell.

OBJECT LINKING AND EMBEDDING (OLE)

Object linking and embedding (OLE) is a Microsoft Windows feature that allows a document to contain an object from other application. MicroStation supports OLE and can display information from other documents that also support OLE. For example, a MicroStation design file can contain a Microsoft Excel spreadsheet.

When an object is linked to a MicroStation design file, the object is actually an image of the other document. It displays the current information from the other document and when you double-click the object, it opens the object's application to allow you to edit the object. For example, if you double-click a linked Excel spreadsheet object, the Excel application opens.

When you embed an object, the object is a static snapshot of what was in the other document at the time you embedded it, and there is no link to the original document. Changes made to the original document do not appear in the embedded object.

Linked and embedded objects appear in MicroStation view windows and are printed with the design file.

Embedding MicroStation Elements in Another Document

Elements in a MicroStation design file can be copied into other documents that support graphic formats. To illustrate it, we will describe copying elements from a design file into a Microsoft Word document:

1. Use the Selection Tool to select the elements.
2. Select **Edit > Copy** to copy the selected elements to the Windows clipboard.
3. In the Microsoft Word document, go to the point in the document where the MicroStation elements are to be placed.

4. Select **Edit > Paste Special** to open the Paste Special dialog box as shown in Figure 14–65.
5. In the **AS** field, select the paste method you want to use (such as **Picture**).
6. Click the **OK** button to paste the text.

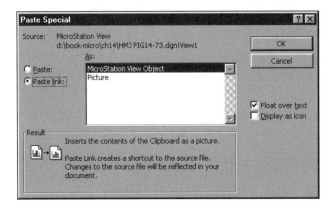

Figure 14-65 Word Paste Special dialog box

Inserting Objects in a MicroStation Design File

MicroStation provides options for inserting objects from other applications. The objects can be static copies of the documents (snapshots) or live links back to the documents that can be updated to display the current content of the documents.

Objects are inserted into MicroStation design file using the Insert Object dialog box. Invoke the dialog box from:

| Drop-down menu | Edit > Insert Object (see Figure 14–66). |

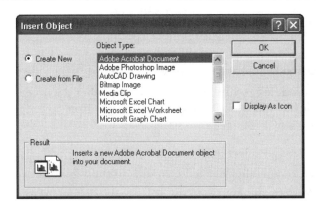

Figure 14-66 The Insert Object dialog box

Chapter 14 Special Features

Options in the Insert Object dialog box determine how the object is inserted in the design file. To create a new object, turn the **Create New** option ON. To attach an existing document, turn the **Create from File** option ON. Select the type of object from the **Object Type** field and click the **OK** button.

If you are creating an object from a file, the options on the dialog box change to a **File** text field, a **Browse** button, and a **Link** check box.

Either type the name and location of the file in the **File** text field, or click the **Browse** button to open the Browse dialog box from when you can select the file's location and name. Turn the **Link** check box ON to create a live link to the file or turn it OFF to insert a static copy of the current contents of the file. Click the **OK** button to create the object, or click the **Cancel** button to cancel the operation.

In the Insert OLE Object Tool Settings window, select the placement **Method** to either place the object by defining **By Corners** or **By Size**. If you select the **By Size Method**, the **Scale** text field provides a place for you to enter a scaling factor. A number greater than one increases the size of the object.

Place a data point to define the location of the object in the design file. If you are placing the object **By Center**, the data point defines the center of the object. If you are placing it **By Corner**, the data point defines the upper, left corner of the object. If the **Link** check box is turned ON, MicroStation inserts the linked object in the design file and opens the objects application so that the object can be edited. If you do not want to edit the object, close the application.

Figure 14–67a shows a Microsoft Excel worksheet in the Excel application, and Figure 14–67b shows the worksheet as an object in a MicroStation design file.

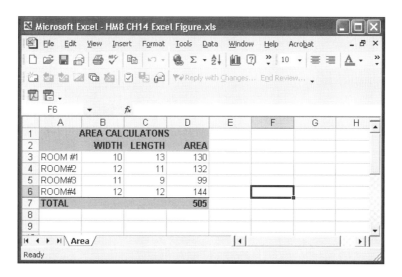

Figure 14-67a An Excel spreadsheet

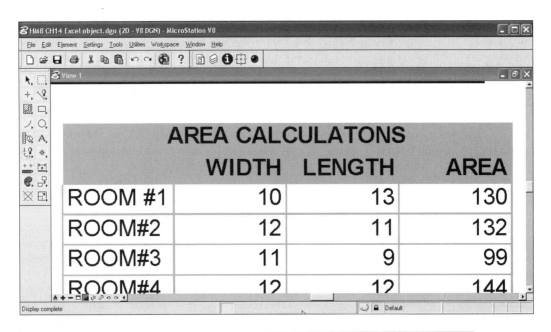

Figure 14-67b The Excel spreadsheet as an object in a MicroStation design file

Updating Objects in a MicroStation Design File

Objects that were inserted with the **Link** check box turned ON are live links to the original document. When you update the links, each one shows the current content of the linked document (provided, of course, that MicroStation can find it). To update all links in the design file, select Update Links from:

| MicroStation menu bar | Select Edit > Update Links |

MicroStation proceeds with updating the live links. While it is updating the links, it displays an Update Links message box that builds a percent complete bar as it completes its work. The message box is closed when all links are updated.

Maintaining Objects in a MicroStation Design File

The Links dialog box provides options for maintaining links to objects. Invoke the dialog box from:

| MicroStation menu bar | Select Edit > Links |

MicroStation displays the Links dialog box, as shown in Figure 14-68.

Chapter 14 Special Features

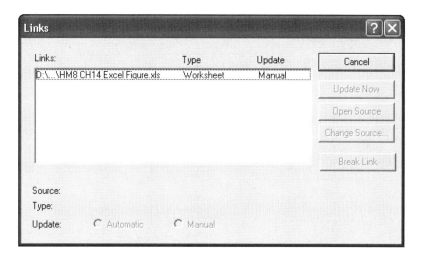

Figure 14-68 The Links dialog box

In the Links dialog box, the **Links** field displays the location and file name of all linked objects. To perform maintenance on a linked object, select the object from the **Links** field and click one of the maintenance buttons:

- Click **Update Now** to update the selected object with the latest information from the object.

- Click **Open Source** to open the object in its native application for editing.

- Click **Change Source** to open the Change Source dialog box from which you can select a document to replace the current object.

- Click **Break Link** to break the link from the object to its native application and turn it into a static copy of the document. When the link is broken, the object cannot be updated from the original document. Before MicroStation breaks the link, it displays an alert message box that asks you to confirm that you really want to break the link. To break the link, click the box's **OK** button.

When you complete performing maintenance on the links, click the **Cancel** button to close the Links dialog box.

REVIEW QUESTIONS

Write your answers in the spaces provided.

1. Explain briefly the purpose of creating a graphic group.

2. List the steps in creating a graphic group.

3. Explain the purpose of the Graphic Group lock setting.

4. Explain briefly the purpose of element selection by element type.

5. Explain briefly the benefits of using level symbology.

6. Describe how to set the symbology for a level.

7. List the steps involved in converting a MicroStation design file to an AutoCAD drawing file.

8. Explain briefly the benefits of dimension-driven design.

chapter 15

Internet Utilities

The Internet is the most important way to convey digital information around the world. You are probably already familiar with the best-known uses of the Internet: e-mail (electronic mail) and the Web (short for "World Wide Web"). E-mail lets users exchange messages and data at very low cost. The Web brings together text, graphics, audio, and movies in an easy-to-use format. Other uses of the Internet include FTP (File Transfer Protocol, for effortless binary-file transfer), Gopher (which presents data in a structured, subdirectory-like format), and Usenet (a collection of more than 100,000 news groups).

MicroStation allows you to interact with the Internet in several ways. Engineering Links are a set of Web-enabling technologies that simplify management of your MicroStation-based engineering projects. In addition, MicroStation includes a Web browser, which enables users to directly access data over the Internet without leaving the application. The browsing tools provide everything the user needs for browsing all of the information on the Web.

Objectives

After completing this chapter, you will be able to do the following:

- Launch the Web browser and open files remotely
- Publish MicroStation data to the Internet
- Link Geometry to Internet URLs

LAUNCHING THE WEB BROWSER AND OPENING FILES REMOTELY

Web browsers are applications designed to provide the user with direct, easy access to the diverse content of the World Wide Web. The original Web browser, created at the NCSA (National Center for Supercomputing Applications at the University of Illinois), is called Mosaic. Revolutionary in its time, Mosaic, and the startling capabilities it introduced to the Internet, played a major role in igniting the exuberant explosion in Web usage, which got underway in the early 1990s. Building on their remarkable success, members of the Mosaic development team went on to create Netscape Navigator. Microsoft's Internet Explorer has functionality similar to Netscape Navigator, and also provides access to the diverse content of the World Wide Web.

Highly intuitive in design and function, Web browsers provide a clean, easily navigable window through which to scan pages on the Web. Customizable to suit the user's aesthetics and interests, browsers typically support a number of secondary functions including sending and receiving e-mail, monitoring newsgroups, or even downloading files directly from Web sites.

MicroStation new browsing tools provide everything users need for browsing all of the information on the Web.

To open the Web browser, invoke the MicroStation Link from:

Drop-down menu	Utilities > Connect Web Browser
E-Links tool box	Select the Connect to Browser tool (see Figure 15–1).

Figure 15-1 Invoking the Connect to Browser tool from the E-Links tool box

MicroStation opens the default Web browser, if not open already, and connects to the Internet. The choice of browsers is defined with the variable MS_USEEXTERNALBROWSER. You can set the variable to open Netscape Navigator or Microsoft Explorer instead of the internal browser, if you prefer. Figure 15–2 shows an example of connecting to the Internet through the Internet Explorer Web browser.

Chapter 15 Internet Utilities

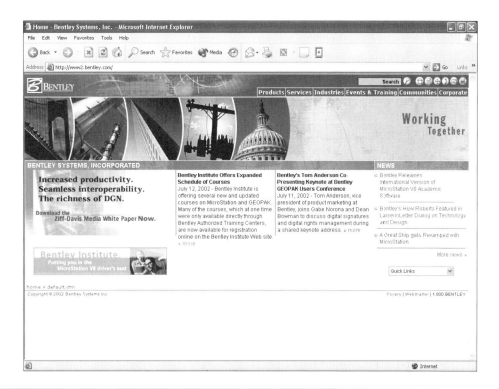

Figure 15-2 Accessing the Internet through the Internet Explorer Web browser

Specify the URL (short for "uniform resource locator") in the **Location** edit field in the browser. The URL is the Web site address, which usually follows the format http://www.bentley.com. The URL system allows you to find any resource (a file) on the Internet. Example resources include a text file, a Web page, a program file, an audio or movie clip—in short, anything that you might also find on your computer. The primary difference is that these resources are located on somebody else's computer. A typical URL looks like the examples in Table 15-1.

Table 15-1 Example URLs

Example	Meaning
http://www.bentley.com	Bentley primary Web site
ftp://ftp.bentley.com	Bentley FTP Server

The "http://" prefix is not required. Most of today's Web browsers automatically add in this routing prefix, which saves you a few keystrokes. URLs can access several different kinds of resources—such as Web sites, e-mail, news groups—but always take on the following format:

   Scheme://netloc

The scheme accesses the specific resource on the Internet, including those listed in Table 15–2.

Table 15–2 URL Prefix Meanings

Scheme Prefix	Meaning
File://	Files on your computer's hard drive or local network
ftp://	File transfer protocol (downloading files)
http://	Hypertext transfer protocol (Web sites)
mailto://	Electronic mail (e-mail)
news://	Usenet news (news groups)
telnet://	Telnet protocol
gopher://	Gopher protocol

The "://" characters indicate a network address. Table 15–3 lists the formats recommended for specifying URL-style file names in the Location: edit field of the Browser.

Table 15–3 Recommend Formats for Specifying URL-Style File Names

Drawing Location	Template URL
Web site	http://servername/pathname/filename
FTP site	ftp://servername/pathname/filename
Local file	file:///drive:/pathname/filename File:///drive|/pathname/filename File://\\localPC\pathname\filename File:////localPC/pathname/filename
Network file	file://localhost/drive:/pathname/filename File://localhost/drive|/pathname/filename

Servername is the name of the server, such as www.bentley.com. The *pathname* is the name of the subdirectory or folder name. The *drive:* is the drive letter, such as C: or D:. A *local file* is a file located on your computer. The *localhost* is the name of the network host computer.

MicroStation allows you to select a URL as a design file location instead of a specific local design file. You can also open files using URLs to open remote settings files, archives, reference files, or cell libraries. Downloaded files from a URL are stored in a directory specified by the configuration variable MS_WEBFILES.

MicroStation allows you to close the Web browser by selecting the same tool that opened the Web browser, whose icon has slightly changed and whose tool tip now says **Disconnect from Browser**. If you had originally opened the browser through the E-Links tool box, selecting **Disconnect from Browser** closes the browser in addition to disconnecting it from MicroStation. If you had opened the browser from outside MicroStation and then connected to it, selecting **Disconnect from Browser** only disconnects MicroStation from the browser without closing the browser.

Remote Opening of Design File

To open a design file located at a remote server by selecting a URL, invoke the Open URL tool from:

| Drop-down menu | File > Open URL |

MicroStation displays a Select Remote Design File dialog box, similar to Figure 15–3.

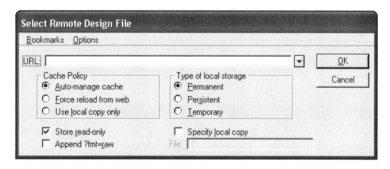

Figure 15-3 Select Remote Design File dialog box

Specify the URL in the edit field. Following are the available options in the dialog box:

Cache Policy

The Cache Policy section of the dialog box allows you to select whether the design files from selected URLs are downloaded or if a previously downloaded local copy is used. Select one of the three following options:

- **Auto-manage cache**—downloads the file from the selected URL only if it is newer than the local version.
- **Force reload from web**—reloads from the Web selection and automatically downloads the file from the selected URL.
- **Use local copy only**—uses the local copy and no Internet request is sent.

Type of Local Storage

The Type of Local Storage section of the dialog box allows you to select how the downloaded file is stored. Select one of the three following options:

- **Permanent**—stores the file that is downloaded and it is not deleted.

- **Persistent**—deletes the file only when it exceeds the size requirements. By default, the total permitted size of local copies of remote files is 25MB.
- **Temporary**—deletes the file at the end of the current session.

Store Read Only

The Store Read Only selection stores downloaded files as read-only. This is a useful reminder that the remote file is not being changed and local modifications could lead to confusion.

Append ?fmt=raw

The Append selection appends the string "?fmt=raw" to the end of the selected URL that is submitted for download. This is required only if the selected site is running ModelServer Publisher.

Specify local copy

If necessary, you can specify a different location in the **File** edit field for downloaded files from selected URLs rather than the default location.

Remote Opening of Settings File

To open a settings file located at a remote server by selecting a URL, first open the Settings Manager from:

| Drop-down menu | Settings > Manage |

MicroStation displays the Select Settings dialog box. Invoke the Open URL tool from:

| Drop-down menu (Select Settings) | File > Open URL |

MicroStation displays Select Remote Settings File dialog box, similar to Figure 15–4.

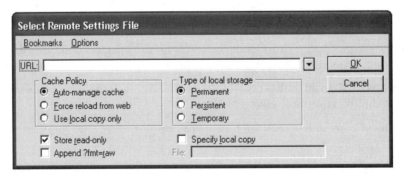

Figure 15-4 Select Remote Settings File dialog box

Chapter 15 Internet Utilities

Specify the URL in the **URL** edit field. The options available in the Select Remote Settings file dialog box are the same explained earlier for Select Remote Design File dialog box.

Remote Opening of Archive File

To open an archive file located at a remote server by selecting a URL, first open the Archive Settings box from:

| Drop-down menu | Utilities > Archive |

MicroStation displays the Archive Settings box. Invoke the Open URL tool from:

| Drop-down menu (Archive Settings) | File > Open URL |

MicroStation displays the Open Remote Archive dialog box, similar to Figure 15–5.

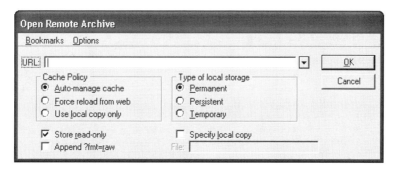

Figure 15-5 Open Remote Archive dialog box

Specify the URL in the **URL** edit field. The options available in the Open Remote Archive dialog box are the same as explained earlier for the Select Remote Design File dialog box.

Remote Opening of Reference File

To open a reference file located at a remote server by selecting a URL, first open the Reference settings box from:

| Drop-down menu | File > Reference |

MicroStation displays the Reference settings box. Invoke the Attach URL tool from:

| Drop-down menu (Reference settings box) | File > Attach URL |

MicroStation displays the Select Remote Design File to Attach: dialog box, similar to Figure 15–6.

651

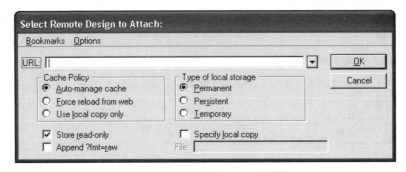

Figure 15-6 Select Remote Design File to Attach dialog box

Specify the URL in the **URL** edit field. The options available in the Select Remote Design File to Attach dialog box are the same as explained earlier for the Select Remote Design File dialog box.

Remote Opening of Cell Library

To open a cell library located at a remote server by selecting a URL, first open the Cell Selector settings box from:

| Drop-down menu | Utilities > Cell Selector |

MicroStation displays the Cell Selector settings box. If you have not attached a cell library to the current design file, then MicroStation displays a Select Cell Library to Load dialog box. Select one of the existing cell libraries to display the Cell Selector settings box. Invoke the Open URL tool from:

| Drop-down menu (Cell Selector Settings) | File > Load Remote Cell Library |

MicroStation displays the Specify URL for Cell Library dialog box, similar to Figure 15-7.

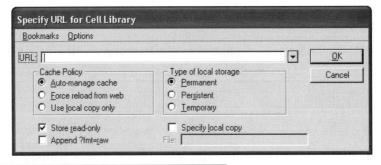

Figure 15-7 Specify URL for Cell Library dialog box

Specify the URL in the **URL** edit field. The options available in the Specify URL for Cell Library dialog box are the same as explained earlier for the Select Remote Design File dialog box.

PUBLISHING MICROSTATION DATA TO THE INTERNET

The HTML (HyperText Markup Language) Author tool allows you to create HTML files that can be viewed via MicroStation Link or any external browser (such as Netscape Navigator or Internet Explorer). HTML files can be created from a Design File Saved View, Design File Snapshot, Cell Library, or BASIC Macros.

Creating HTML File from Design File Saved View

The Design File Saved Views option creates an HTML Web page with the selected view of the design file. To create an HTML file from a design file saved view, first open the HTML Author dialog box from:

| Drop-down menu | Utilities > HTML Author |

MicroStation displays the HTML Author dialog box, similar to Figure 15-8

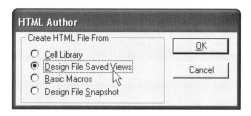

Figure 15-8 HTML Author dialog box

Select the **Design File Saved Views** radio button from the Create HTML File From section of the dialog box and click the **OK** button to close the HTML dialog box. The Select Design File dialog box opens. Select the design file from the appropriate directory and click the **OK** button. MicroStation opens the Design File Walkthrough dialog box, similar to Figure 15-9.

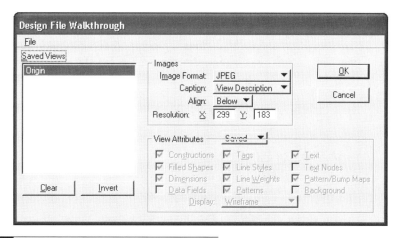

Figure 15-9 Design File Walkthrough dialog box

MicroStation lists all the saved views in the design file in the **Saved Views** list box. You can select any or all of the saved views to be included in the HTML page.

The Images section of the dialog box sets controls for image display. The **Caption** option menu allows you to select whether to display the **View Description**, **View Name**, or **None** as part of the image display on the HTML page. The **Align** option menu controls where the text (selected from the **Caption** option menu) appears with relation to the image on the HTML page. The available options include **Below**, **Left**, **Right**, and **Above**. The **Resolution** edit fields set the resolution of the generated image in pixels.

The View Attributes section of the dialog box lists all the available view attributes and the corresponding settings as saved in the design file. If you need to override the settings, select **Override** from the **View Attributes** option menu and make the necessary changes in the View Attributes settings box.

After making the necessary changes, click the **OK** button to close the Design File Walkthrough dialog box. MicroStation opens the Create HTML File dialog box, similar to Figure 15–10.

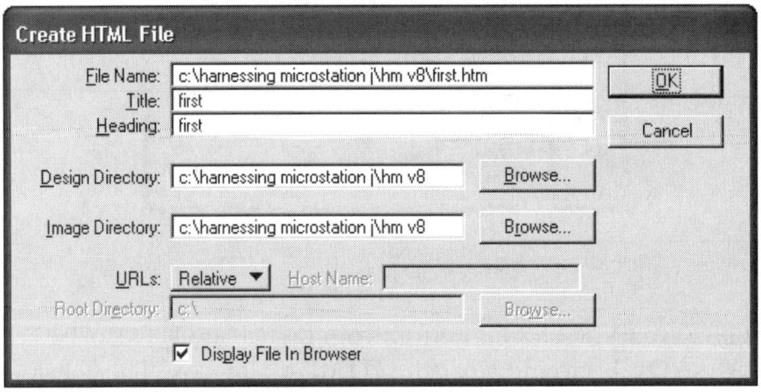

Figure 15-10   Create HTML File dialog box

Specify the file name of the HTML file being created, a descriptive title, and a heading description (appears top of the graphics) in the **File Name**, **Title**, and **Heading** edit fields. Specify the design (location of the design file) and image directories in the **Design Directory** and **Image Directory** fields, respectively. Specify if the URLs being used are absolute or relative in the **URLs** option menu. If it is set to absolute, then specify the host name and root directory in the **Host Name** edit field and **Root Directory** edit fields. Click the **OK** button to create the HTML file from the design file saved view.

Creating HTML File from Design File Snapshot

The Design File Snapshot option creates an HTML Web page with a view-only picture of the selected design file. To create an HTML file from a design file snapshot, first open the HTML Author dialog box from:

| Drop-down menu | Utilities > HTML Author |

MicroStation displays the HTML Author dialog box, similar to Figure 15–8.

Select the **Design File Snapshot** radio button from the Create HTML File From section of the dialog box and click the **OK** button to close the HTML dialog box. The Select Design File dialog box opens. Select the design file from the appropriate directory and click the **OK** button. MicroStation opens the Create HTML File dialog box, similar to Figure 15—11.

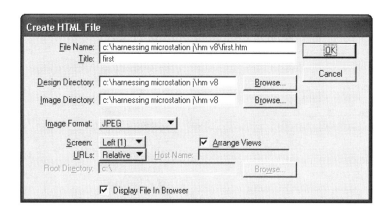

Figure 15-11 Create HTML File dialog box

Specify the file name of the HTML file being created, and the Title in the **File Name** and **Title** edit fields, respectively. Specify the design (location of the design file) and image directories in the **Design Directory** and **Image Directory** fields, respectively. Specify if the URLs being used are absolute or relative in the **URLs** option menu. If it is set to absolute, then specify the host name and root directory in the **Host Name** edit field and **Root Directory** edit field. Click the **OK** button to create the HTML file from the design file snapshot.

Creating HTML File from Cell Library

The Cell Library option creates an HTML Web page from a cell library. To create an HTML file from a cell library, first open the HTML Author dialog box from:

| Drop-down menu | Utilities > HTML Author |

MicroStation displays the HTML Author dialog box, similar to Figure 15–8.

Select the **Cell Library** radio button from the Create HTML File From section of the dialog box and click the **OK** button to close the HTML dialog box. The Select Cell Library to Open dialog box opens. Select the cell library file from the appropriate directory and click the **OK** button. MicroStation opens the HTML Cell Page dialog box, similar to Figure 15–12.

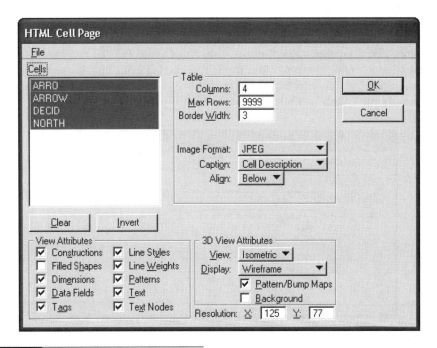

Figure 15-12   HTML Cell Page dialog box

MicroStation lists all the cells in the cell library file in the **Cells** list box. You can select any or all of the cells to be included in the HTML page.

The Tables section of the dialog box allows you to set the settings (such as the number of columns, maximum number of rows, and border width) for the table to be displayed on an HTML page.

The **Caption** option menu allows you to select whether to display the **Cell Description**, **Cell Name**, or **None** as part of the image display on the HTML page. The **Align** option menu controls where the text (selected from Caption option menu) appears in relation to the image on the HTML page. The available options include **Below**, **Left**, **Right**, and **Above**.

The View Attributes section of the dialog box lists all the available view attributes and the corresponding settings as saved in the design file. If necessary, you can make changes in the View Attributes settings.

The 3D View Attributes section of the dialog box sets attributes for rendering generated images including view, display, patterns/bitmaps, and background.

The **Resolution** edit field sets the resolution of the generated image in pixels.

After making the necessary changes, click the **OK** button to close the HTML Cell Page dialog box. MicroStation opens the Create HTML File dialog box, similar to Figure 15–13.

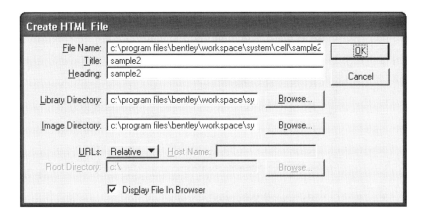

Figure 15–13 Create HTML File dialog box

Specify the file name of the HTML file being created, a descriptive title, and the heading description (appears top of the graphics) in the **File Name**, **Title**, and **Heading** edit fields, respectively. Specify the cell library directory location and image directories in the **Library Directory** and **Image Directory** edit fields, respectively. Specify if the URLs being used are absolute or relative in the **URLs** option menu. If it is set to absolute, then specify the host name and root directory in the **Host Name** edit field and **Root Directory** edit field. Click the **OK** button to create the HTML file from the cell library.

Creating HTML File from Basic Macros

The Basic Macro HTML Page dialog box is used to create an HTML file that references a directory of MicroStation Basic macros. To create an HTML file from a directory of MicroStation Basic macros, first open the HTML Author dialog box from:

| Drop-down menu | Utilities > HTML Author |

MicroStation displays the HTML Author dialog box, similar to Figure 15–8.

Select the **Basic Macros** radio button from the Create HTML File From section of the dialog box and click the **OK** button to close the HTML dialog box. The Select Basic Macro Directory dialog box opens. Select the directory containing basic macro files and click the **OK** button. MicroStation opens the Basic Macro HTML Page dialog box, similar to Figure 15–14.

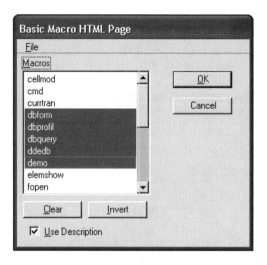

Figure 15-14 Basic Macro HTML Page dialog box

MicroStation lists all the basic macros in the selected directory in the **Macros** list box. You can select any or all of the basic macros to be included in the HTML page. Click the **OK** button to complete the selection.

MicroStation opens the Create HTML File dialog box, similar to Figure 15–15.

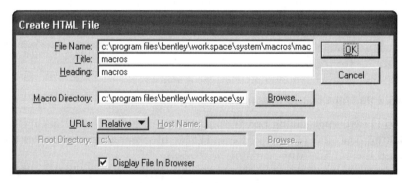

Figure 15-15 Create HTML File dialog box

Specify the file name of the HTML file being created, a descriptive title, and the heading description (appears top of the graphics) in the **File Name**, **Title**, and **Heading** edit fields, respectively. Specify the macro directory location in the **Macro Directory** edit field. Specify if the URLs being used are absolute or relative in the **URLs** option menu. If it is set to absolute, then specify the host name and root directory in the **Host Name** edit field and **Root Directory** edit field. Click the **OK** button to create the HTML file from Basic macros.

Chapter 15 Internet Utilities

LINKING GEOMETRY TO INTERNET URLS

MicroStation allows you to link Internet URLs to geometry. Most browsers highlight a linked entity when the cursor is placed on it and display the link title. Clicking on the linked entity redirects the browser to the linked URL. This feature allows you to connect to component catalogs which can be accessed over the Internet, and, for example, engineers can execute a search over the Web to find a supplier whose components meet their specifications. Having URL links to geometry will also help when someone using the drawing needs more information on that component, since they can just click on the component and it would automatically connect them to the manufacturer's Web page.

Attaching Link to a Geometry

To attach a URL to an element, invoke the Attach Engineering Link tool from:

| E-Links tool box | Select the Attach Engineering Link tool (see Figure 15–16). |

Figure 15-16 Invoking the Attach Engineering Link tool from the E-Links tool box

MicroStation displays the Attach Engineering Link settings box, similar to Figure 15–17.

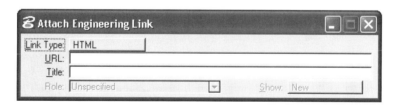

Figure 15-17 Attach Engineering Link settings box

The **Link Type** option menu sets the type of link you are creating: **HTML** or **XML:simple**. If you are creating a **XML:simple** link, you can specify parameters that further define how the link should operate. Specify the URL to attach to an element and the title in the **URL** and **Title** edit boxes. The **Role** option menu specifies what role the object of the link will specify, for example, that of a reference or cell library (enabled only when **Link Type** is set to **XML:simple**). The **Show** option menu determines whether any existing page should be replaced or if a new browser should be opened (enabled only when **Link Type** is set to **XML:simple**).

MicroStation prompts:

> Attach Engineering Link > Identify element *(Identify an element to attach.)*
> Attach Engineering Link > Accept/Reject *(Place a data point to accept the attachment of the link to the selected element.)*

You can attach multiple elements with the same URL.

Displaying Links

To highlight the elements that have engineering links, invoke the Show Engineering Links tool from:

| E-Links tool box | Select the Show Engineering Links tool (see Figure 15–18). |

Figure 15–18 Invoking the Show Engineering Links tool from the E-Links tool box

MicroStation highlights the elements that have engineering links.

Connecting to URLs

To connect to the URL attached to an element, first invoke the Follow Engineering Link tool from:

| E-Links tool box | Select the Follow Engineering Link tool (see Figure 15–19). |

Figure 15–19 Invoking the Follow Engineering Link tool from the E-Links tool box

MicroStation prompts:

> Follow Engineering Link > Identify element *(Identify the element to which the URL is attached.)*
> Attach Engineering Link > Accept/Reject *(Place a data point to accept the access to the URL.)*

Chapter 15 Internet Utilities

MicroStation opens the default Web browser and connects to the URL.

If the Web browser is not able to access the specified URL, then it will display an error message. Make sure proper connection is established to access the Internet.

Edit Engineering Link

To edit the URL attached to an element, first invoke the Edit Engineering Link tool from:

| E-Links tool box | Select the Edit Engineering Link tool (see Figure 15–20). |

Figure 15–20 Invoking the Edit Engineering Link tool from the E-Links tool box

MicroStation prompts:

> Edit Engineering Tags > Identify element *(Identify the element to which the URL is attached.)*

MicroStation displays the Edit Tags [Internet] dialog box similar to Figure 15–21.

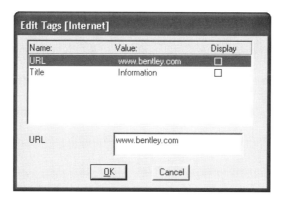

Figure 15–21 Edit Tags [Internet] dialog box

In the list box, select a line item to edit and in the edit text field at the bottom of the dialog box make the necessary changes. If necessary, the appropriate line item **Display** check box can be set to ON, to display the information in the design. Click the **OK** button to accept the changes and close the dialog box.

Delete Engineering Link

To remove an attached engineering link from an element, invoke the Delete Engineering Link tool from:

E-links tool box	Select the Delete Engineering Link tool (see Figure 15–22).

Figure 15–22 Invoking the Delete Engineering Link tool from the E-Links tool box

MicroStation prompts:

> Delete Engineering Link > Identify element *(Identify the element to which the URL is to be deleted.)*
>
> Delete Engineering Link > Accept/Reject *(Place a data point to accept the deletion of the Engineering Link.)*

MicroStation deletes the selected Engineering Link from the element.

chapter 16

Customizing MicroStation

MicroStation is an extremely powerful program off the shelf. But, like many popular engineering and business software programs, it does not automatically do all things for all users. It does, however, permit users to make changes and additions to the core program to suit individual needs and applications. Word processors offer macros that allow you to save a combination of many keystrokes and invoke them at any time with just a few keystrokes. Database management programs have their own library of user functions that can also be combined and saved as a user-named, custom-designed tool. These programs also allow you to create and save standard blank forms for later use, to be filled out as needed. Utilizing these features to make your own copy of a generic program unique and more powerful for your particular application is known as *customizing*.

Customizing MicroStation can include several facets. After completing this chapter, you will be able to do the following:

- Create settings groups and edit existing groups
- Create level groups
- Create custom line styles
- Create and modify workspace components: the project configuration, the user configuration, and the user interface

- Customize the function keys
- Install fonts
- Package design files and their associated resources for sharing on other computers

SETTINGS GROUPS

MicroStation allows you to define a settings group with a user-specified name in three categories: Drawing (default), Scale, and Working Units. Under each settings group of the Drawing category, you can define individual group components. You can set element attributes such as color, weight, line style, level, and class, associating with a primitive tool (such as Place Line, Place Text) as a part of the group component. You can also save the current multi-line definition and active dimension settings. You can define any number of group components for each settings group.

Settings groups and group components can be saved to an external file that, by default, has the extension .STG. This file follows the same concept as the cell library. Once created, the settings file may be attached to any of your design files and activate one of the settings groups and corresponding group component. Selecting a component does the following:

- All element attributes associated with the component are set in the design.
- If a tool key-in is defined for the component, the tool is selected, letting you place an element or elements without invoking the tool from a tool box.

For example, MicroStation provides the **V40 – Dimension Styles** group that sets dimension attributes for several dimensioning standards. Selecting a dimension style from the components list is much faster than manually setting up the many dimension attributes.

Selecting a Settings Group and Corresponding Component

The Select Settings dialog box allows you to select a settings group and corresponding group component from the currently attached settings group file. The settings box also provides an option to attach another settings group file and invoke a settings group and corresponding group component.

To open the Select Settings dialog box, invoke the Manage tool from:

| Drop-down menu | Settings > Manage |

MicroStation displays the Select Settings dialog box, as shown in Figure 16–1.

Figure 16-1 Select Settings dialog box

Chapter 16 Customizing MicroStation

The name of the currently opened settings group file is displayed in the title bar of the settings box. The dialog box is divided into two parts. The **Group** drop-down menu lists the names of the available drawing settings groups. The **Component** drop-down menu lists the name and type of each component in the selected settings **Group**. Each **Component** option includes the name of the component and the settings associated with it.

To invoke one of the components, select the **Group** that contains it and then select the **Component**. If a key-in is defined for the component, the corresponding tool is invoked.

You can dock the Select Settings dialog box to the top or bottom of the Microsoft desktop. The Select Settings dialog box can also be displayed as a large dialog box. To display it as a large dialog box, invoke the Large Dialog tool from:

| Select Settings box | Options > Large Dialog |

MicroStation displays the Select Settings dialog box as shown in Figure 16–2.

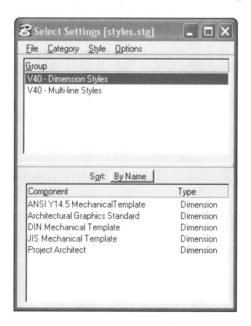

Figure 16-2 Select Settings dialog box (Large Dialog)

The top part lists the names of the available drawing settings groups and the bottom part of the settings box lists the name and type of each component from the selected settings group. The **Category** drop-down menu sets the category for the listing of groups in the **Group** list box. The **Sort** drop-down menu sets the manner in which components are sorted in the **Component** area; **By Name** or **By Type**.

665

Attaching a Settings Group File

To attach a settings group file to a design file, invoke the Open tool from:

Select Settings box	File > Open

MicroStation displays the Open Existing Settings File dialog box. Select the location of the file from the **Directories** area, select the settings group file (extension .STG) from the **Files** area, and click the **OK** button to attach it to the design file. MicroStation lists the available groups and corresponding components in the Select Settings dialog box.

Maintaining a Settings Group

The Edit Settings dialog box allows you to define, modify, and delete settings groups and group components. To open the Edit Settings dialog box, invoke the Edit tool from:

Select Settings dialog box	File > Edit

MicroStation displays the Edit Settings dialog box, as shown in Figure 16–3.

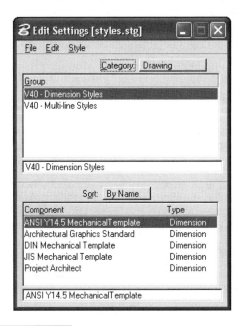

Figure 16-3 Edit Settings dialog box

The name of the currently opened settings group file is displayed in the title bar of the dialog box. By default, the top part of the dialog box lists the names of the available drawing settings groups. The bottom part lists the name and type of each component from the selected settings group.

Creating a Settings Group File

To create a new settings group file, invoke the New tool from:

| Edit Settings dialog box | File > New |

MicroStation displays the Create New Settings File dialog box. Select the location where the new file is to be placed from the **Directories** area, key-in the name of the file in the **Files** text field, and click the **OK** button to create a new settings file.

Creating a Settings Group

To create a new settings group, invoke the Create Group tool from:

| Edit Settings dialog box | Edit > Create > Group |

MicroStation adds a new "Unnamed" group to the **Group** list and places the name in the text box just below the **Group** list. You can change the name from "Unnamed" to any appropriate name in the text field. The maximum number of characters for the group name is 31.

Deleting a Settings Group

To delete a settings group, select the group to be deleted from the **Group** list box and invoke the Delete tool from:

| Edit Settings dialog box | Edit > Delete |

MicroStation opens an Alert message box to confirm the deletion. Click the **OK** button to delete the group and all its components.

Creating a New Component

To create a new component, select the settings group in which you want to create the component and then select one of the seven available component types from:

| Edit Settings dialog box | Edit > Create > component type (see Table 16–1) |

MicroStation adds a new component named "Unnamed" to the Component area and to the text field below the area. You can change the component name in the text field. The maximum number of characters for the group name is 31.

Table 16–1 lists the component types and corresponding tools that can be used with each component type.

Table 16–1 Component Types and Corresponding Tools

COMPONENT TYPE	TOOL
Active Point	Points tool box
Area Pattern	Pattern tool box
Cell	Cells tool box
Dimension	Dimension tool box
Linear	Linear Elements tool box Polygons tool box Arcs tool box Ellipses tool box Curves tool box
Multi-line	The key-in **PLACE MLINE CONSTRAINED** corresponds to the Place Multi-line tool
Text	Text tool box

Modifying a Component

To modify a component, invoke the Modify tool from:

Edit Settings dialog box	Edit > Modify

Select the group to be modified from the **Group** area and select the component to be modified from the **Component** area. Each component type has a unique settings box. In the settings boxes, each setting that can be modified has a check box that must be turned ON before the setting can be modified. At the bottom of each settings box are three buttons:

- Click the **Save** button to save the settings you entered in the settings box fields.

- Click the **Match** button and select an existing element of the same type as the component to match the settings of the selected element, and click the **Save** button to save the settings. For example, if the component type is a Point, selecting a point element in the design file fills in the settings fields with the settings that were in effect when the element was created.

- Click the **Close** button to close the settings box without saving anything.

The following paragraphs describe the settings boxes for each type of component.

Chapter 16 Customizing MicroStation

The Modify Point Component settings box (see Figure 16–4) allows you to modify settings related to Construct Point tools. To set the key-in that will be activated automatically when the component is selected, turn the **Key-in** check box ON and type the appropriate key-in for the Construct Points tool. Set the appropriate element attributes (**Level**, **Color**, **Weight**, and **Line Style**) that will be activated automatically when the component is selected. Select the type of point, **Zero-Length Line**, **Cell**, or **Character**, from the **Type** menu. If you select the **Cell** type, set the appropriate **Cell** settings. If you select the **Character** type, set the appropriate **Character** settings.

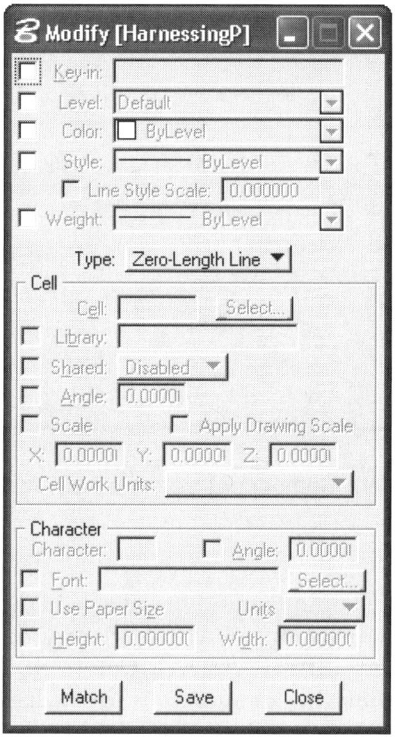

Figure 16-4 Modify Point Component settings box

The Modify Area Pattern Component settings box (see Figure 16–5) allows you to modify settings related to hatching and patterning tools. To set the key-in that will be activated automatically when the component is selected, turn the **Key-in** check box ON and type the appropriate key-in for the hatching or patterning tool. Set the element attributes (**Level**, **Color**, **Weight**, and **Line Style**) that will be activated automatically when the component is selected. Select either **Pattern** or **Hatch** from the **type** menu. If you select the **Pattern** type, set the appropriate **Pattern Settings**. If you select the **Hatch** area, set the appropriate **Hatch Settings**.

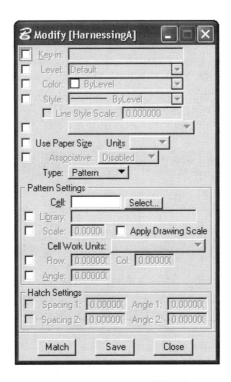

Figure 16-5 Modify Area Pattern Component settings box

The Modify Cell Component settings box (see Figure 16–6) allows you to modify settings related to cell placement tools. To set the key-in that will be activated automatically when the component is selected, turn the **Key-in** check box ON and type the appropriate key-in for the cell placement tool. Set the appropriate element attributes (**Level, Color, Weight**, **Line Style**, and **Angle**) that will be activated automatically when the component is selected. Key-in the name of the cell to be placed in the **Cell** field or click the **Select** button to open the Select Cell dialog box from which you can select the cell you want to use. If a cell library is not attached to your design file, key-in the library location and name in the **Library** field. Select the way you want to use the cell, **Placement** or **Terminator**, from the **Type** menu. If **Placement** is selected, the **Cell** selection becomes the Active Cell when the component is selected in the Select Settings dialog box. If **Terminator** is selected, the **Cell** selection becomes the Active Line Terminator.

Chapter 16 Customizing MicroStation

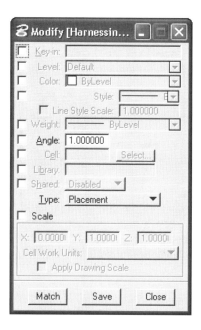

Figure 16-6 Modify Cell Component settings box

 Note: The **Level**, **Color**, **Style**, and **Weight** controls affect placement of a cell using the component only if the specified cell was created as a point cell.

The Modify Dimension Component settings box (see Figure 16–7) allows you to modify settings related to the Dimension component definition. To set the key-in that will be activated automatically when the component is selected, turn the **Key-in** check box ON and type the appropriate key-in for the dimension placement tool. Set the appropriate element attributes (**Level**, **Color**, **Weight**, and **Line Style**) that will be activated automatically when the component is selected. If you want the dimensions associated with the elements they dimension, select **Enabled** from the **Associative** menu. Otherwise, select **Disabled**. Click the **Select** button to open the Select Dimension Definition dialog box from which you can select the type of dimensions to be place with this dimensioning component (such as "ISO Mechanical Template").

671

Figure 16-7 Modify Dimension Component settings box

The Modify Linear Component settings box (see Figure 16–8) allows you to modify settings related to placement of lines, polygons, arcs, circles, ellipses, and curves. To set the key-in that will be activated automatically when the component is selected, turn the **Key-in** check box ON and type the appropriate key-in for the Linear tool. Set the appropriate element attributes (**Level**, **Color**, **Weight**, **Line Style**, **Area**, and **Fill**) that will be activated automatically when the component is selected.

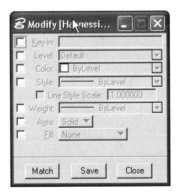

Figure 16-8 Modify Linear Component settings box

The Modify Multi-line Component settings box (see Figure 16–9) allows you to modify settings related to multi-line component definitions. To set the key-in that will be activated automatically when the component is selected, turn the **Key-in** check box ON and type the appropriate key-in for the multi-line placement tool. Set the appropriate element attributes (**Level**, **Color**, **Weight**, and **Line Style**) that will be activated automatically when the component is selected. Click the

Select button to open the Select Multi-line Definition dialog box from which you can select one of several standard multi-line definitions (such as 3-5/8" brick; 2" insul; 8" cmu backup).

Figure 16-9 Modify Multi-line Component settings box

The Modify Text Component settings box (see Figure 16–10) allows you to modify settings related to the placement of text. To set the key-in that will be activated automatically when the component is selected, turn the **Key-in** check box ON and type the appropriate key-in for the place text tool. Set the appropriate element attributes (**Level**, **Color**, **Weight**, and **Line Style**), text attributes (**Font**, **Slant Angle**, **Line Length**, **Fraction**, **Vertical**, and **Underline**), Justification (**Single-line** and **Multi-line**), and Size and Spacing (**Use Paper Size**, **Units**, **Height**, **Width**, **Intercharacter Spacing**, and **Line Spacing**) that will be activated automatically when the component is selected.

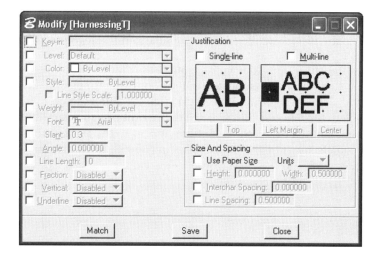

Figure 16-10 Modify Text Component settings box

673

Selecting Categories

In the Select Settings dialog box you can specify the relationship between printing units and design Master Units by selecting one of the available scale settings. Select the relationship from:

Select Settings dialog box	Category > Scale

MicroStation displays the Select Scale dialog box. Select a scale in the dialog box and click the **OK** button to apply it to the design.

You can also set the current Working Units by selecting one of the available Working Units settings. Select the Working Units from:

Select Settings dialog box	Category > Working Units

MicroStation displays the Select Working Units dialog box, which lists a set of standard Working Units. Select a Working Units setup from the dialog box and click the **OK** button to apply it to the design file. An alert box opens for confirming the adjustment of the Working Units settings. Click **OK** to implement the selection or **Cancel** to cancel the selection.

LEVEL FILTERS

In Chapter 3 you were introduced to placing elements on different levels. You learned that you create levels by giving them a name and that there can be an unlimited number of levels. In addition to allowing you to give levels meaningful names, such as "Window Details," MicroStation allows you to create filters to display sets of related levels. When you create a filter, you define filtering criteria to control what levels the filter displays.

For example, if you maintain a seed file that that provides separate levels for each engineering discipline, you can create discipline filters to help users find the levels for their discipline.

Filters are defined in the same Level Manager settings box that you used to create the level names, and a **Filters** icon in the tree view area on the left side of the settings box allows you to view the filters defined for the active design file. When you click the **Filters** icon, the Level Manager settings box displays the defined filters. If there are defined filters, the names of the filters appear below the **Filters** icon and the criteria for each filter appear to the right of the tree area, as shown in Figure 16–11. If you click one of the filter names in the tree area, the names of the levels that meet the filter criteria appear to the right of the tree area.

Chapter 16 Customizing MicroStation

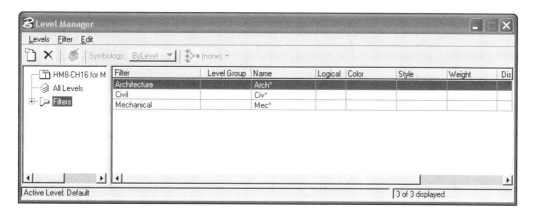

Figure 16-11 Level Filters in the Level Manager settings box

Creating a New Filter

To create a new filter, open the Level Manager settings box, click the **Filters** icon in the tree area of the settings box, and create the new filter from:

| Level Manager settings box | Filter > New |

MicroStation inserts a criteria line for the new filter in the area of the Level Manager settings box to the right of the tree area. The **Object Levels** column in the criteria area lists the filter names and the default name of the new filter ("New Filter") is selected for editing. Type a descriptive name for the new filter and enter the filter criteria in the following columns (see following discussion).

Creating Filter Criteria

To define criteria for the filter, click the filter's name in the tree area. The area to the right of the tree displays a row for entering the filter criteria and all levels that currently match the filter criteria (initially all levels). Filter criteria can be entered in the first row of each column. Examples of typical criteria follow.

The **Name** column allows you to enter text strings that are compared to each level's name. Wild card symbols (* | &) are also provided. Following are examples of typical text strings:

- **obje** — Include only level names that contain the characters "obje" anywhere in the string.
- **obje*** — Include only level names that start with "obje."
- ***obje** — Include only level names that end with "obje."
- **obje*|*ing** — Include only level names that begin with "obje" or end with "ing."

675

- **obje|bord** — Include only level names that contain either the characters "obje" or the characters "bord" anywhere in the name.

- **obje&bord** — Include only level names that contain both "obje" and "bord" anywhere in the name.

- **obje*&*ing** — Include only level names that begin with "obje" and end with "ing."

The **Color**, **Weight**, and **Style** columns allow you to display only levels that match one or more of these element attributes. You enter the criteria by entering the attribute number. If you key-in the number "3" in the **Color** criteria field, only levels that are set to the color Red (3) are displayed by the filter.

 Note: For additional information on entering filter criteria, refer to the MicroStation online help.

CUSTOM LINE STYLES

In addition to the eight standard line styles that were introduced in Chapter 3, MicroStation provides custom line styles and provides tools for creating your own custom line styles. Custom line styles can be selected from the **Active Line Style** menu in the Attributes tool box, where they are listed below the eight standard line styles, as shown in Figure 16–12. When a custom line style is selected, it becomes the Active Line Style and placement tools use it.

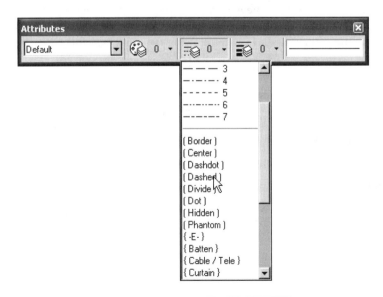

Figure 16-12 Custom line styles in the Active Line Style menu

Chapter 16 Customizing MicroStation

Custom line styles are stored in style library files that have an "RSC" extension. MicroStation provides several style libraries, and more than one library can be attached to your design at the same time. When you open a design file, MicroStation opens the library files that are specified in the MS_SYMBRSRC configuration variable. For example, the following two lines are specified in MS_SYMBRSCR on a workstation that is using one of the Microsoft Windows operating systems:

\Program Files\Bentley\Workspace\System\symb*.rsc

\Program Files\Bentley\Workspace\Standards\symb*.rsc

The * is a wildcard symbol and "*.rsc" tells MicroStation to open all library files found in the two folders.

Note: Configuration variables can be viewed and edited from the Configuration dialog box. Select **Workspace > Configuration** to open the box.

The Line Styles Settings Box

The Line Styles settings box allows you to preview the custom line styles, modify the way the style stroke patterns are placed, and select a style as the Active Line Style.

Invoke the Line Styles settings box from:

Menu bar	Element > Line Style > Custom
Key-in window	**linestyle settings** (or **lines s**) ENTER

MicroStation displays the Line Styles settings box, as shown in Figure 16–13.

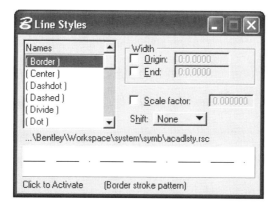

Figure 16-13 Line Styles settings box

The **Names** area on the left side of the setting box lists all available custom line styles. When you select one of the styles by clicking it, a picture of the line style appears in the field near the bottom of the settings box. If you click in the picture of the line style, it is set as the Active Line Style. You can also set a style as the Active Line Style by double-clicking its name in the **Names** area.

When a Active Line Style is a custom line style that contains dash strokes that have width, the **Origin** and **End** fields can be used to change the stroke width. To change one of the widths, turn its check box ON and key-in the width, in Working Units, in the associated text field.

The **Scale factor** sets the scale of all displayable characteristics (dash length and width, point symbol size) of the Active Line Style. To change the scale factor, turn the **Scale factor** check box ON and key-in the scale factor in the associated text field.

The **Shift** menu sets the distance or fraction by which each stroke pattern in the active line style is shifted or adjusted.

The Line Style Editor Settings Box

To create and modify custom line styles, open the Line Style Editor settings box from:

Drop-down menu	Elements > Line Style > Edit
Key-in window	**linestyle edit** (or **lines e**) ENTER

MicroStation opens the Line Style Editor settings box. If no library file was previously opened in the settings box, the box's appearance is as shown in Figure 16–14.

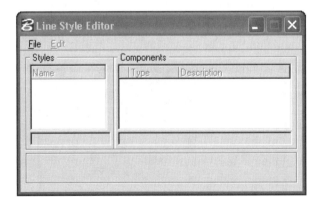

Figure 16-14 The Line Style Editor settings box

If a file was previously opened in the settings box, the size of the settings box varies depending on what type of line style was selected (Stroke Pattern, Point, or Compound). The types are discussed later.

Chapter 16 Customizing MicroStation

The settings box provides two drop-down menus (**File** and **Edit**) in the menu bar. The **Edit** menu is used to create and maintain the line styles in a selected library file. If no file has been opened in the box, only the **File** menu is available for use.

The File Drop-down Menu

The Line Style Editor settings box's **File** drop-down menu provides options for creating and maintaining style library files, as shown in Figure 16–15.

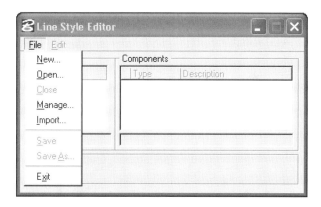

Figure 16–15 The File drop-down menu in the Line Style Editor settings box

The **New** option displays the Create Line Style Library dialog box, from which you can create a new library file. The box opens displaying the first library directory path in the MS_SYMBRSCR configuration variable.

Select the location of the new file in the **Directories** area, key-in the name of the new file in the **Files** text field, and click the **OK** button to create the file. The file location and name appear in the title bar of the Line Style Editor settings box, and the file's only contents, MicroStation's default line styles, are listed in the **Components** area of the box.

The **Open** option displays the Open Line Style Library dialog box, from which you can select an existing library file for editing. Select the file location from the **Directories** area, select the file name from the **Files** area, and click the **OK** button to open the file for editing.

 Note: Only one library file at a time can be open in the Line Style Editor settings box. If there is a file open when you create or open another file, the currently open file is closed.

The **Close** option closes the file currently open in the Line Style Editor settings box, but leaves the settings box open. The file is immediately closed when you select the option. There are no additional prompts.

The **Manage** tool allows you to rename and delete styles from the open library file. If no library file is currently open in the Line Style Editor settings box when you select the option, MicroStation displays the Select Destination Line Style Library dialog box. Navigate to the location of the library file you want to manage in the **Directories** area, select the file in the **Files** area, and click the **OK** button to open the file.

If a library file is already open in the settings box, or after you select and open a file, MicroStation displays the Manage Line Style Definitions dialog box, as shown in Figure 16-16.

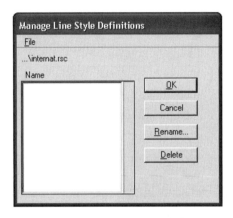

Figure 16-16 The Manage Line Style Definitions dialog box

To rename a line style, select the style's name in the **Name** field and click the **Rename** button. MicroStation opens the Rename Line Style dialog box, in which you can type the new style name. To delete a line style from the file, select the style's name and click the **Delete** button. The style name is removed from the list.

Styles are not actually renamed or deleted until you click the **OK** button to accept the changes and close the settings box. If you click the **Cancel** button, the settings box is closed without making any changes to the library file's styles. If you opened a file from this tool, the file is not loaded in the Line Style Editor settings box.

The **Import** tool allows you to import AutoCAD line styles from an AutoCAD styles file (*.lin). When you select the option, MicroStation displays the Select .lin file to Import From dialog box. After you select the file that contains the styles to be imported, MicroStation opens the Select Linestyles to Import dialog box, from which you can select the styles you want to import.

The **Save** option saves the changes you made to the open library file. The changes do not become part of the design file unless you save them. If you close the box without selecting the **Save** option, all changes are lost.

The **Save As** tool allows you to save the contents of the open library file to another file name. When you select the option, MicroStation opens the Save Line Style Library As dialog box. Select the location were you want to put the new file in the **Directories** area, enter the new file name in the **Files** text field, and click the **OK** button to create the contents of the currently open library file to the new file.

The **Exit** option closes the Line Style Editor settings box. If you previously saved the changes, the setting box closes with no additional prompts. If the box contains unsaved changes, MicroStation displays an Alert box that asks if you want to save the changes to the currently open library file. Click the **Yes** button to save the changes before exiting or click the **No** button to exit without saving the changes.

The Edit Drop-down Menu

The Line Style Editor settings box's **Edit** drop-down menu provides options for creating and maintaining line style names and components in the open library file, as shown in Figure 16–17.

Figure 16-17 The Edit drop-down menu

The **Create** option opens a submenu that provides options for creating new line style names and components. The options are discussed later.

The **Delete** option deletes the selected style name from the **Name** area and the associated style component from the **Components** area. To delete a style, select its name and then select the **Delete** option. The style is immediately deleted. There are no additional prompts.

 Note: Style names that are associated with internal style components cannot be deleted.

The **Duplicate** option creates a copy of the selected style component in the **Components** area. The copy has the same name and content as the original.

The **Line** option links the component selected in the **Components** field to the line style selected in the **Name** field.

The **Snappable** option is a check box that turns on and off the ability to snap to the individual components of the custom line style.

The Create Submenu

When you select **Create** from the **Edit** menu, a submenu opens, as shown in Figure 16–18.

Figure 16-18 The Create submenu

The options in the **Create** submenu create line style components and assign line style names to the components.

The **Name** option creates a new "Unnamed" line style. The new name is automatically selected and is linked to the component selected in the **Components** list box. Replace the default name by keying-in a new one in the text field below the **Names** list.

The **Stroke Pattern** option creates style components that consist of a series of strokes and gaps that are repeated along the length of elements.

Chapter 16 Customizing MicroStation

The **Point** option creates style components that consist of a series of point symbols (such as a shared cell) that are repeated along the length of elements. Point components also contain an association with a stroke pattern component that determines the placement of the point symbol on the element.

The **Compound** option creates style components that are combinations of style components. For example, a stroke pattern might be combined with one of the internal line components.

Creating a Stroke Pattern Line Style Component

When you select the **Edit > Create > Stroke Pattern**, the Line Style Editor settings box expands to show the settings fields for defining a Stroke Pattern, as shown in Figure 16–19.

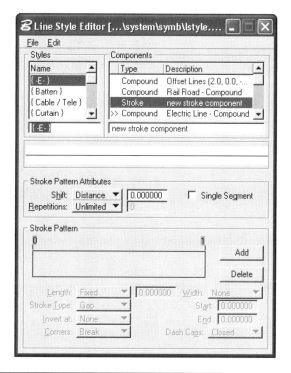

Figure 16-19 Line Style Editor with Stroke Pattern settings

Note: The **Create** menu options create new components, but it is often easier to find and modify a component similar to what you need in one of the library files MicroStation furnishes.

Below the **Name** and **Components** fields is a field that will show an example of the pattern component as you define it. Below the picture are fields for defining the component.

683

The **Stroke Pattern Attributes** area of the settings box contains options that control the way the stroke pattern is placed along elements:

- The **Shift** menu and text field provide options for settings the distance by which the stroke pattern is shifted from the starting point of the element. The shift can be a **Distance** from the beginning, a **Fraction** of the pattern length from the beginning, or the pattern can be **Centered** within the length of the element or element segment.

- The **Repetitions** menu and text field set the number of times the stroke pattern is repeated throughout the length of an element or element segment.

- The **Single Segment** check box controls the truncation of the stroke pattern at the end of each element segment. If it is turned ON, the pattern is shifted to end each segment with a complete stroke.

The **Stroke Pattern** area of the settings box contains options for defining the dashes and gaps that make up the stroke pattern. The pattern is a series of dash and gap strokes, and the options in this area allow you to add, delete, and define the strokes.

- The horizontal bar at the top of the area shows the relative length of each stroke in the pattern. A dark background indicates a dash and a clear background indicates a gap. The small blocks above and below the bar are handles that you can drag to change the length of each stroke. To select a stroke for editing, click its part of the bar. When you click it, the appearance of the stroke changes in the bar to indicate that it is selected.

- The **Add** button adds a new gap stroke to the end of the stroke pattern. After you add it, you can change it from a gap to a dash.

- The **Delete** button removes a selected stroke from the pattern.

The following fields apply to dash and gap strokes:

- The **Length** menu allows you to set a **Fixed** or **Variable** length for the selected stroke. Enter the stroke length in the text field to the right of the menu. The length uses the working units of the design file. This controls how the pattern is shifted or repeated a fixed number of times.

- The **Stroke Type** menu allows you to define the selected stroke as a **Dash** or **Gap**.

- The **Invert at** menu allows you invert the type of the selected stroke when it is used to place an element. The inversion options are **None**, **Origin**, **End**, and **Both**.

- The **Corners** menu controls the behavior of the selected stroke at each element vertex. The stroke can **Break** at the vertex or **Bypass** the vertex. If it bypasses the vertex, the corner may not be sharp.

Chapter 16 Customizing MicroStation

The following options are only available when the selected **Stroke Type** is **Dash**:

- The **Width** menu controls the width of the selected dash stroke. If the **Width** is **None**, the width of the stroke is determined by the Active Line Width. If the **Width** is **Full**, **Left**, or **Right**, the stroke width is controlled by the values entered in the **Start** and **End** fields. The **Left** option applies the width to the left (or above) the center point of the stroke, the **Right** option applies the width to the right (or bottom) of the center point, and the **Both** option applies the width to both sides of the center point.

- The **Start** option applies the **Width** to the starting point of the dash stroke.

- The **End** option applies the **Width** to the ending point of the dash stroke.

- The **Dash Caps** menu sets shape of each end of the selected dash stroke when the ends have a width greater than zero. If the **None** option is selected, the dash stroke is only a line. If any other cap option is selected, the dash stroke is filled.

Figures 16–20a and 16–20b show the stroke settings for a simple stroke pattern that alternates between a dash with width and a gap. The field just above the **Stroke Pattern Attributes** area of the settings box in the figures shows the way the line style will appear when used in the design.

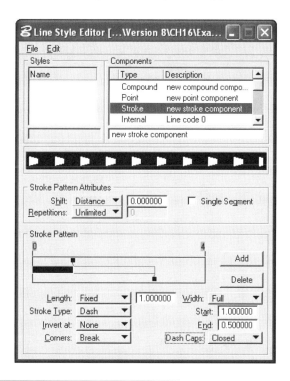

Figure 16–20a Example of a dash stroke definition

685

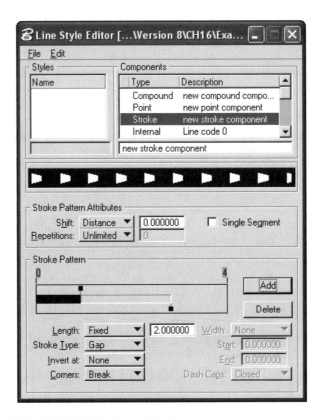

Figure 16-20b Example of a gap stroke definition

Creating a Point Line Style Component

When you select **Edit > Create > Point**, the Line Style Editor settings box expands to show the setting fields for defining a Point component, as shown in Figure 16–21.

Chapter 16 Customizing MicroStation

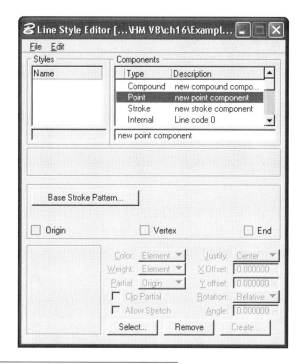

Figure 16-21 Line Style Editor with Point settings

To define a Point component, you select one or more of the Point Symbols defined in the open style library file. If no Point Symbols are defined for the open file, you must create them. The procedure is presented later in this topic.

You can create a point component that only places from one to three Point Symbols on each segment of an element or that places Point Symbols on one or more of the strokes in an associated stroke pattern.

- Below the **Name** and **Components** fields is a field that shows an example of the point component as you define it.
- Below the picture are three check boxes that you can use to place selected Point Symbols at the **Origin**, at each **Vertex**, and **End** of elements.

 Note: The appearance of a slight depression of a check box indicates that it is selected. That is easy to miss.

- Click the **Base Stroke Pattern** button if you want to associated the new Point component with a Stroke Pattern. Selecting this option opens the Base Stroke Pattern dialog box, from which you can select a stroke pattern. When you select a pattern, a bar graph appears below the button that shows the strokes in the pattern.

687

The fields below the **Origin**, **Vertex**, and **End** check boxes are used to assign Point Symbols to positions within the Point component. Before using these commands, you select the position by clicking the **Origin**, **Vertex**, or **End** check boxes, or by clicking one of the strokes in the stroke pattern bar graph. After selecting the position, the following Point Symbol fields become available:

- Click the **Select** button to open the Select Point Symbol dialog box. Select a symbol from the box's **Name** field and click the **OK** button to assign the selected symbol to the element position or pattern stroke you previously selected.

- Click the **Remove** button to remove a Point Symbol from the selected element location or pattern stroke.

- Click the **Create** button to create a new Point Symbol from the contents of a fence in the design file. This button is only available for use when an element position or pattern stroke is selected and a fence is defined in the design file. Here are the steps required to create a new Point Symbol:

    1. Draw the symbol in the design file.

    2. Use the Define Cell Origin tool to define an origin for the Point Symbol.

    3. Place a fence around the elements. The **Fence Mode** setting is ignored by this command and all elements must be inside the fence.

    4. Click the **Create** button to open the Create Point Symbol dialog box.

    5. In the dialog box, key-in the new symbol's name in the **Name** field and click **OK**. The symbol is created but not assigned to the Point component.

- The **Color** menu allows you to choose what color to use for the symbol. Choose **Element** to change the symbol color to the color of the element or **Symbol** to use the color of the symbol.

- The **Weigh** menu allows you to choose what line weight to use for the symbol. Choose **Element** to use the weight of the element or **Symbol** to use the weight of the symbol.

- The **Justify** menu sets the origin point on the **Left**, **Center**, or **Right** end of the selected pattern stroke.

- The **X Offset** and **Y Offset** text fields set the horizontal and vertical offset from the origin point.

- The **Rotation** menu sets the rotation calculation to be **Relative** to the element angle, **Absolute**, or **Adjusted**.

- The **Angle** text field sets the rotation angle value in degrees.

Chapter 16 Customizing MicroStation

Creating a Compound Line Style Component

When you select **Edit > Create > Compound**, the Line Style Editor settings box expands to show the setting fields for defining a Compound component, as shown in Figure 16–22.

Figure 16-22 Line Style Editor with Compound settings

A Compound component consists of Stroke Pattern and Point sub-components. The fields in the settings box allow you to insert and remove sub-components:

- The **Sub-Components** field lists each sub-component. The field shows the **Offset** from the origin, the **Type** of component, and the component **Description**.

- The **Offset** text field sets the distance, in working units measured perpendicular from the work line, by which the selected component is displayed parallel to the work line. If the **Offset** is zero, the selected component is displayed on the work line.

- The **Insert** button opens the Select Component dialog box, from which you can select a sub-component to insert in the compound component selected in the **Components** field.

- The **Remove** button removes the sub-component selected in the **Sub-Components** field from the compound component selected in the **Components** list box.

WORKSPACE

A *workspace* is a customized drafting environment that permits the user to set up MicroStation for specific purposes. You can set up as many workspaces as you need. A workspace consists of "components" and "configuration files" for both the user and the project.

MicroStation comes with a set of workspaces for various disciplines. For example, MicroStation is delivered with a sample "civil" workspace. When the civil workspace is active, the files and tools you need to perform civil engineering design and drafting are available by default.

When the civil workspace is active, tools and tool boxes that are unrelated to that discipline are removed from the interface so that they are out of the way.

Setting Up the Active Workspace

You can select the default workspace from the MicroStation Manager. At the bottom of the MicroStation Manager dialog box (see Figure 16–23), menus are provided that allow you to change the major components of the workspace before opening a design file.

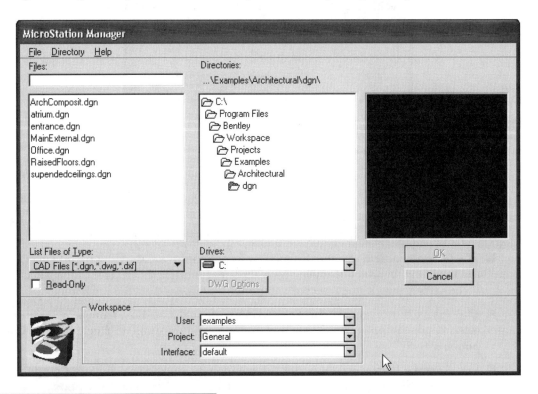

Figure 16–23 MicroStation Manager

Chapter 16   Customizing MicroStation

User Option Menu

The **User** menu in the Workspace section of the dialog box allows you to select the default workspace from the available workspaces. Selecting a workspace from the list reconfigures MicroStation to use that workspace's components. Selecting a workspace also resets the search path to a corresponding folder for loading design files. In addition, MicroStation sets the associated project and user interface.

When MicroStation is started with any workspace as the default workspace, a preference file is created for that workspace, unless one already exists. The settings in the preferences are set either to default settings or to AutoCAD Transition settings, depending on the active user interface component.

To create a new workspace, invoke the **New** option from the **User** menu. MicroStation displays the Create User Configuration File dialog box, similar to Figure 16–24.

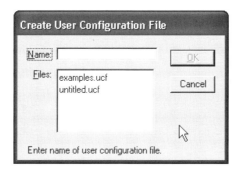

Figure 16-24   Create User Configuration File dialog box

Key-in the name of the new workspace in the **Name** text field and click the **OK** button. MicroStation displays another dialog box, similar to Figure 16–25. Key-in the description (optional) in the **Description** text field. If necessary, change the components by clicking the appropriate **Select** button for **Project** and/or **User Interface**. Click the **OK** button to close the dialog box. MicroStation sets the newly created workspace as the default workspace. A workspace can contain only one project and one interface. These components are attached to a workspace, so, to use two different projects with the same interface or two interfaces for one project, you need to make additional workspaces.

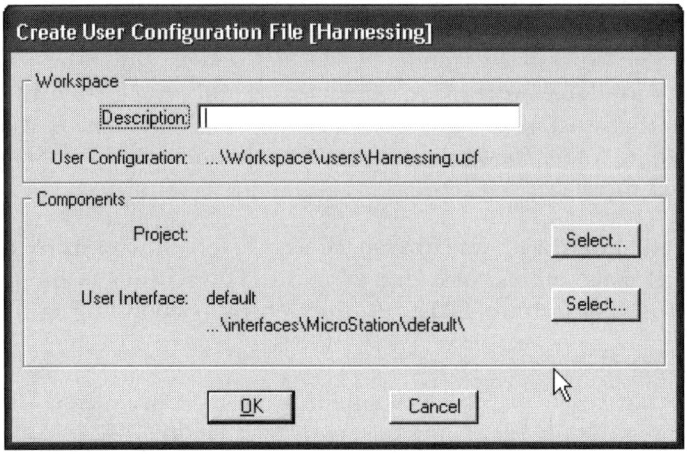

Figure 16-25 Create User Configuration File dialog box

Project Option Menu

The selection of the project sets the location and names of data files associated with a specific design project. If necessary, you can change the selection of the project from the **Project** menu.

You can also create a new Project from the **Project** menu, similar to creating a workspace.

Interface Option Menu

The selection of the interface sets a specific look and feel for MicroStation's tools and general on-screen operation. If necessary, you can change the selection of the interface from the **Interface** option menu. MicroStation comes with discipline-specific interfaces—civil engineering, architecture, mechanical engineering, drafting, and mapping—and it also has interfaces for previous versions (V. 4 and V. 5) and AutoCAD users.

To create a new interface, invoke the **New** option from the **Interface** menu. MicroStation displays the Create User Interface dialog box, similar to Figure 16-26.

Chapter 16 Customizing MicroStation

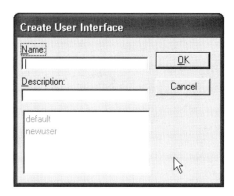

Figure 16-26 Create User Interface dialog box

Key-in the name of the new user interface in the **Name** text field and a description in the **Description** text field (optional). Click the **OK** button to close the dialog box. MicroStation creates an interface directory under the *Bentley\Workspace\interfaces* directory and sets the newly created user interface as the default user interface. The new interface takes the default interface as its starting point. Any changes you make while using this new interface is written only to the new interface.

Setting User Preferences

Preferences are settings that control the way MicroStation operates and the way its tool frames and tool boxes appear on the screen. For example, they affect how MicroStation uses memory on a user's system, how windows are displayed, and how reference files are attached by default. You can change the settings to suit your needs. The user preferences are saved under the same name as the workspace, with the file extension .UCF.

To set the user preferences, invoke the Preferences from:

| Drop-down menu | Workspace > Preferences |

MicroStation displays the Preferences dialog box, similar to Figure 16–27. MicroStation displays the name of the file under which preference settings are saved (if any) as part of the title bar.

693

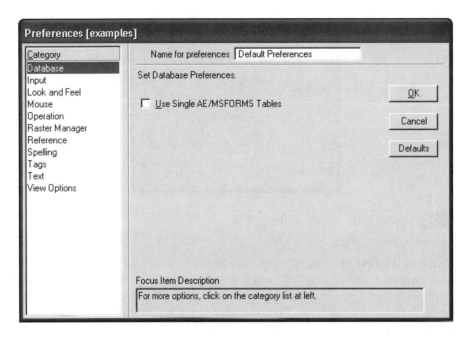

Figure 16-27 Preferences dialog box

The **Name for preferences** text field at the top of the settings box allows you to enter a name for the current set of preferences.

User preferences are divided into areas. The **Category** list box lists all the available categories. Selecting a category causes the appropriate controls to be displayed to the right of the **Category** list. Each category controls a specific aspect of MicroStation's appearance or operation. Table 16–2 describes briefly all the settings available in the Preferences dialog box.

After making the required changes to the preferences settings, click the **OK** button to save the settings and close the Preferences dialog box or click the **Cancel** button to close the dialog box without saving any of the changes. Click the **Default** button to return all settings to the default values.

 Note: Some preferences changes do not go into effect until you close and restart MicroStation.

Table 16-2 Preferences Settings

CATEGORY	PREFERENCE SETTING	DESCRIPTION	DEFAULT
Database	Use Single AE/MSFORMS Tables check box	When ON, single AE and MSFORMS tables are maintained.	OFF
Input	Start in Parse All Mode check box	When ON, the parsing function is enabled as MicroStation starts.	ON
	Disable Drag Operations check box	When ON, MicroStation disregards Data button-up operation when the pointer is in view.	OFF
	Locate Tolerance text field	Sets the size of the search area around the pointer when using manipulation tools.	10
	Locate By Face menu	In 3D designs, allows closed, surface, or solid elements to be identified by a data point on the element's interior as well as its edges.	Rendered Views Only
	Pointer Size menu	Sets the pointer size to Normal (small cross) or Full View (cross the full size of the view window).	Normal
	Pointer Type menu	Sets the alignment of the pointer's crosshairs for Orthogonal or Isometric.	Othogonal
Look and Feel	Single Click menu	When set to Locked, the selected tool remains active after use. When set to Single-click, selected tools are only active for one use.	Locked

Table 16-2 Preferences Settings (continued)

CATEGORY	PREFERENCE SETTING	DESCRIPTION	DEFAULT
Look and Feel (continued)	Default Tool menu	Sets the tool action after the selected tool is used when in Single-click mode. When set to Selection, the Element Selection tool is invoked. When set to None, no tool is selected. When set to Power Selector, a settings box opens that contains options for selecting groups of elements.	Selection
	Highlight menu	Sets the background color of a tool's icon in the tool box when the tool is selected.	Color
	Layout menu	Sets the space between tool icons in the tool boxes to Narrow, Regular, or Wide.	Regular
	Tool size menu	Sets the size of the tool icons in the tool boxes to Small or Large.	Small
	Dialog Font menu	Sets the size, in points, of the text in dialog and settings boxes.	12 pt
	Border font menu	Sets the size, in points, of text in the borders of dialog and settings boxes.	12 pt
	Auto-Focus Tool Settings Window check box	When ON, focus is automatically set to the Tool Settings window when a tool is selected.	ON
	Borderless Icons check box	When turned ON, borders only appear around the icon of the selected tool in tool boxes. When OFF, the icon borders are always visible.	ON

Table 16-2 Preferences Settings (continued)

CATEGORY	PREFERENCE SETTING	DESCRIPTION	DEFAULT
Look and Feel (continued)	Colorize Only Highlighted Tools check box	When turned ON, all tool icons, except the icon of the selected tool, are shown in gray-scale.	OFF
	Tool colors button	When clicked, opens the Tool Colors dialog box in which tool color schemes can be selected.	Default
Mouse	Wheel menu	If your workstation mouse has a wheel, the wheel action can be set to Zoom In/Out, Pan Left/Right, Pan Up/Down, Pan Radial, Pan with Zoom, or None.	Pan Left/Right
	CTRL + Wheel menu	Provides the same set of mouse wheel options as the Wheel menu.	Pan Up/Down
	SHIFT + Wheel menu	Provides the same set of mouse wheel options as the Wheel menu.	Pan Left/Right
	Pan Ratio text field	Sets the pan ratio increment for each turn of the mouse wheel. The number must be one or higher, and the higher the number the greater the amount of panning with each turn.	2.000
	Zoom Ratio text field	Sets the zoom ration increment for each turn of the mouse wheel. The number must be one or higher.	2.000
	Radial Pan Angle text field	Controls the incremental angle of the panning.	45.00

Table 16-2 Preferences Settings (continued)

CATEGORY	PREFERENCE SETTING	DESCRIPTION	DEFAULT
Operation	Open two Application Windows	When ON, MicroStation opens two application windows.	OFF
	Disable Edit Handles check box	When ON, selected elements are shown in the highlight color rather than with handles.	OFF
	Save Settings on Exit check box	When ON, design savings are automatically saved when the design file is closed.	OFF
	Immediately Save Design Changes check box	When ON, changes to the design file are immediately saved. When OFF, changes are only saved when File > Save is selected (If you exit without saving the file, you will be prompted to save).	ON
	Compress File On Exit check box	When ON, deleted files are automatically removed from the design file when it is closed. When OFF, the file is compressed only when File > Compress Design is selected.	OFF
	Enter into Untitled Design check box	When ON, when you start MicroStation, a new design file is automatically created.	OFF
	Display Broken Associations with Different Symbology check box	When ON, elements that lose associativity by changing their weight and style are automatically identified.	ON
	Reset Aborts Fence Operations check box	When ON, clicking the Reset button during a fence operation halts the operation.	ON

Table 16-2 Preferences Settings (continued)

CATEGORY	PREFERENCE SETTING	DESCRIPTION	DEFAULT
Operation (continued)	Level Lock Applies for Fence Operations check box	When OFF, fence contents manipulations ignore the Level Lock setting.	ON
	Optimized Fence Clipping check box	When ON, closed shapes, solids and surfaces are maintained when clipping.	ON
	Display Active Level in All Views check box	When ON, the active level displays across all views.	ON
	Resource Cache text box	Sets the amount of memory, in KB, reserved for resources read from MicroStation resource files and application resource files.	1024
	Font Cache text box	Sets the maximum size, in KB, of the section of memory reserved for data used to display text elements.	256
Raster Manager	Display Border Around Selected Raster check box	When ON, the selected raster image highlights in the view(s).	ON
	File Georeference Has Priority when Raster Is Loaded check box	When ON, gives priority for file location information to georeference over attachment.	OFF
	Save Location Info in Sister File if Required checkbox	When ON, saves location information — only for formats that do not support georeference — in a sister file (HGR or ESRI formats).	OFF
	Open Raster Files Read-Only checkbox	When ON, opens raster files as read-only.	ON

Table 16-2 Preferences Settings (continued)

CATEGORY	PREFERENCE SETTING	DESCRIPTION	DEFAULT
Raster Manager (continued)	Update MS_RFDIR Automatically for Raster Attachments check box	When ON, automatically appends the path of attached raster images to the MS_RFDIR configuration variable. The raster attachments are searched for in the directories named by the configuration variable. When OFF, the paths are appended to the variable and the images is searched for only in the paths already listed in the MS_RFDIR variable.	OFF
	Use Unit Definition Geokey if Present [override PCs unit] check box	When ON, overrides Projected Coordinate System (PCS) unit with Unit Definition Geokey for Geotiff raster file.	ON
	Geotiff Default Unit: 1 Unit = text field and menu	If Use Unit Definition Geokey if Present (override PCS unit) is on, defines each Geotiff measurement unit.	1 Unit = 1 Meters
	WorldFile Default Unit 1 Unit = text field and menu	Defines each measurement unit used with the WorldFile (TFW) sister file.	1 Unit = 1 Meters
	Recent Files List contains text field	Specifies the number of filenames, from 1–10, listed in the Raster Manager File menu.	4
Reference	Locate On When Attached check box	When ON, the capability to identify elements in a particular reference is turned on when the reference is attached.	ON
	Snap On When Attached check box	When ON, the capability to snap to elements in a particular reference is turned on when the reference is attached.	ON

Chapter 16 Customizing MicroStation

Table 16-2 Preferences Settings (continued)

CATEGORY	PREFERENCE SETTING	DESCRIPTION	DEFAULT
Reference (continued)	Use Color Table check box	When OFF, MicroStation ignores any color table attached to a reference for display purposes.	ON
	Remap Colors on Copy check box	When ON, uses colors with the closest color match from the master file's color table when copying elements.	ON
	Cache When Display Off check box	When OFF, memory caching of references that are not displayed is disabled.	OFF
	Reload When Changing Files check box	When OFF, cached references are kept, when possible, in memory when one DGN file is closed and another is opened.	OFF
	Save Settings to save Changes	When OFF, the results of reference manipulations are immediately permanent.	OFF
	Ignore Update Sequence check box	When ON, the Update Sequence menu item is disabled in the Attachment Settings dialog box's Settings menu.	OFF
	Allow Editing of Self References check box	When ON, you can modify self-attached reference elements, and the changes are incrementally displayed.	ON
	Default Nesting menu	Allows you to ignore (No Nesting), continuously update (Live Nesting), or copy nested attachments.	No Nesting

Table 16-2 Preferences Settings (continued)

CATEGORY	PREFERENCE SETTING	DESCRIPTION	DEFAULT
Reference (continued)	Nest Depth text field	Sets the number of levels of nested attachments that are included when attaching a reference.	1
	Copy Levels During Copy menu	Controls how levels in references are copied to the active model. Copy only levels that don't exist in the active model (If Not Found), copy levels if the active model settings are different from the attachment's settings (If Overrides Exist), or copy all levels from a reference into the active model (Always).	If Not Found
Spelling	Case Sensitive check box	Treats words with different letter-case patterns as different words.	ON
	Suggest split words check box	Flags words that appear to be better suited as two words.	ON
	Report double words check box	Flags words that appear twice in the same row.	ON
	Ignore domain names check box	Ignores words that appear to be Internet domain names.	OFF
	Ignore mixed case check box	Ignores words with an unusual mixture of upper and lowercase letters. An example of a "usual" mixture might be words like "MicroStation." An "unusual" mixture might be something like "MiCroSTAtioN."	ON
	Ignore words with numbers check box	Ignores words that contain a mixture of letters, digits, or other symbols.	ON

Table 16-2 Preferences Settings (continued)

CATEGORY	PREFERENCE SETTING	DESCRIPTION	DEFAULT
Spelling (continued)	Ignore words in UPPERCASE check box	Ignore words consisting of all uppercase letters.	OFF
	Ignore capitalized words check box	Ignores words that begin with an uppercase letter.	OFF
	Language menu	Sets the supported language that Spell Checker will use.	American English
Tags	Prompt on Duplicate Tag Sets check box	When ON, if a tag set exists in the DGN file with the same name as a tag set in a cell library that contains a cell being placed, an alert box is displayed to confirm the replacement of the DGN file version of the tag set with the cell library version.	ON
	Use Design File Tag Sets by Default check box	When ON, tag sets in the DGN file cannot be replaced by tag sets of the same names in cell libraries from which cells are placed.	OFF
	Place Tags in Same Graphic Group check box	When ON, if a tag set is attached to an element, all tags in the set become members of the same graphic group.	ON
Text	Display Text with Line Styles check box	When OFF, text in traditional MicroStation fonts is displayed with the standard Solid line style.	OFF
	Fit Text by Inserting Space check box	When OFF, MicroStation places fitted text by enlarging or shrinking the characters of text so that they fit between two data points.	OFF

Table 16-2 Preferences Settings (continued)

CATEGORY	PREFERENCE SETTING	DESCRIPTION	DEFAULT
Text (continued)	Fixed-Width Character Spacing check box	When OFF, the spacing between characters is measured from the end of one character to the beginning of the next character.	OFF
	Preserve Text Nodes check box	When ON, any text that was placed as a text node will remain a text node, even if edited down to one line.	OFF
	Justify Enter Data Fields Like IGDS check box	When OFF, the odd space in a center-justified enter data field containing an odd number of extra blank spaces is positioned at the beginning of the enter data field.	OFF
	ED Character text field	Sets the text character that denotes each character in an enter data field.	_(underbar)
	Smallest Text text field	Sets the size threshold, in pixels, above which text is drawn.	4
	Underline Spacing % text field	Sets the distance, as a percentage of the text height, between the baseline and underlining.	20
	Degree Display Char text field	Sets the ASCII character used to display the degree symbol (°).	176
	Text Editor Style menu	Sets the type of text editing interface: Word Processor, Dialog Box, WYSIWYG, or Key-in.	Word Processor

Table 16–2 Preferences Settings (continued)

CATEGORY	PREFERENCE SETTING	DESCRIPTION	DEFAULT
View Options	Scroll Bars on View Windows check box	When ON, view windows are displayed with borders, including scroll bars and view control bars.	On
	Black Background -> White check box	When ON, the view background color, if set to black, is displayed in white.	OFF
	Tile Counter Clockwise check box	When OFF, four views are tiled in the same manner as in IGDS.	OFF
	Preserve Aspect Ratio of Views	When ON, MicroStation attempts to open views based on the aspect ratio of views from the last saved version. It only applies if you have changed the size of MicroStation's application area since saving a file.	OFF
	Update Frequency (secs) text field	Sets the frequency (in seconds) of the display update when rendering.	0.5
	Gamma Correction text field	Affects the brightness of rendered images.	1.00
	Max Grid Pts/View text field	Sets the maximum number of displayable grid points in a view, counted horizontally.	90
	Max Grid Refs/View text field	Sets the maximum number of displayable grid references (crosses) in a view, counted horizontally.	40
	Line Weights button	Sets the display width (in pixels) for each of the 32 line weights.	1:1

Working with Configuration Variables

MicroStation provides configuration variables that control the way MicroStation works. For example, when you displayed the Open Line Style Library dialog box and selected the line style library file you wanted to edit, the files names that appeared in the list were in locations specified by the MS_SYMBRSCR variable.

You can create, edit, and delete configuration variables from the Configuration dialog box. Invoke the dialog box from:

| Drop-down menu | Workspace > Configuration |

MicroStation displays the Configuration dialog box similar to the Figure 16-28.

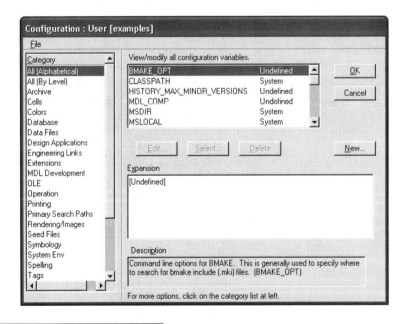

Figure 16-28 Configuration dialog box

The Configuration Dialog Box

The **Category** field contains options for displaying related groups of configuration variables. For example, all variables related to cells are displayed when you click the **Cell** category.

When you select a category, MicroStation lists the variables in the selected category in the variables field at the top, center of the dialog box. The name of the field changes depending on what category is selected. The two **All** categories display the name of each variable. Most of the other options display a short, descriptive name for each variable.

When you select one of the variables in the variables field, the variable is expanded in the **Expansion** field. For example, if you select **Symbology** from the **Category** field and **Symbology Resources** (the descriptive name for MS_SYMBRSCR), the **Expansion** field displays the directory paths that are searched for line style library files when you display the Open Line Style Library dialog box.

Control buttons on the dialog box allow you to **Edit** a variable's action, **Select** specific items to add to a variable's expansion, **Delete** variables, and create **New** variables. If a control button is not available, that maintenance action is not allowed on the variable.

 Note: The content of the Configuration dialog box is different from that described above for the **Design Applications Category**.

Editing Configuration Variables

To modify a configuration variable's action, click the **Edit** button, which opens the Edit Configuration Variable dialog box, as shown in Figure 16–29.

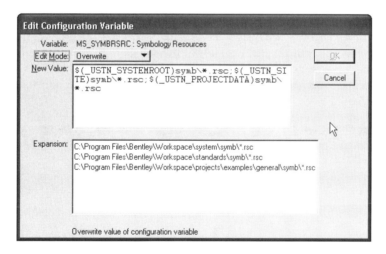

Figure 16-29 Edit Configuration Variable dialog box

The variable's name and descriptive name (if there is one) is displayed at the top of the dialog box. Below the names is the **Edit Mode** drop-down menu from which you can elect to **Overwrite** the current variable's action, **Append** new actions after the existing action, or **Prepend** new actions before the existing action. The **New Value** field provides a place for you to edit the variable's action, and the **Expansion** field shows the results of the variable's action.

The action for most, but not all, variables is a directory path definition that includes other variables and specific folder and file names. For example, the Symbology Resources action shown in Figure 16–29, finds all line style library files in two directory paths. The paths are formed from the contents of two other configuration variables (_USTN_SYSTEMROOT and _USTN_SYSTEMROOT), a specific folder (symb), and a specific type of file (*.rsc). The asterisk (*) is a wildcard that tells MicroStation to include all file names with the rsc extension.

When **Overwrite** is selected, the variable's current action appears in the **New Value** field and you can edit or replace the action. If **Append** or **Prepend** is selected, the **New Value** field is empty and you can only add new information. To view the action again after appending or prepending additional actions, select **Edit Mode > Overwrite**. When you complete editing the variable, click the **OK** button to save your changes and close the dialog box or click the **Cancel** button to close the dialog box without saving the changes.

Note: Describing all possible variable action sequences is beyond the scope of this book. For more information on editing and creating configuration variables, consult the MicroStation online help.

Selecting Configuration Variable Items

When you click the **Select** button, a dialog box opens and allows you to select a specific item to add to a configuration variable's expansion. The dialog box layout varies depending on what type of variable is selected. For example if the action of the selected variable is to find files, the Select File List dialog box appears, as shown in Figure 16–30. The dialog box allows you to add a specific file to the variable's expansion. Select the file's location in the **Directories** field and the file name in the **Files** field.

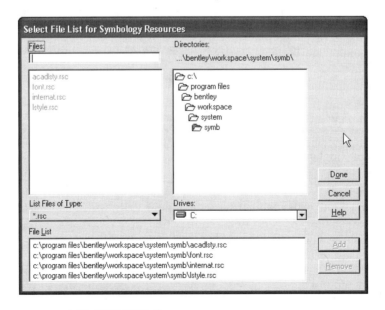

Figure 16-30 Select File List dialog box

Deleting Configuration Variables

To delete a configuration variable, select it and click the **Delete** button. The selected variable is immediately deleted but remains in the variables list with a different color background until you click the **OK** button to close the Configuration dialog box and save the changes.

Note: Only user-defined configuration variables can be deleted.

Creating New Configuration Variables

To create a new configuration variable click the **New** button, which opens the New Configuration Variable dialog box, as shown in Figure 16–31.

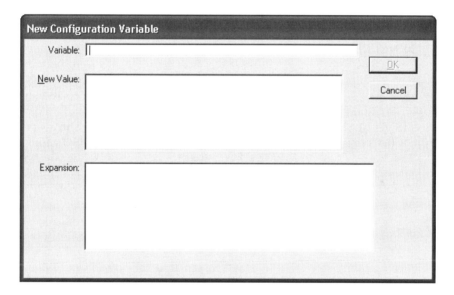

Figure 16–31 New Configuration Variable dialog box

Key-in a name for the new variable in the **Variable** field and key-in the variable's action in the **New Value** field. The results of your action are shown in the Expansion field. If the results are correct, click the **OK** button to save the new variable and close the dialog box.

Table 16-3 Configuration Variables by Category

Category	Variable	Short Name	Description
Archive	MS_ARCHIVE	"Archive Search Path"	Search path for archive files.
	MS_ARCHIVECLASS	"Archive Class Search Path"	Search path for archive class files.
	MS_KEYPAIRLIST	"KeyPair File List"	Digital signature KeyPair file list.
Cells	MS_CELL	"Cell Library Directories"	Search path(s) for cell libraries.
	MS_CELLLIST	"Cell Library List"	List of cell libraries to be searched for cells not found in the current library.
	MS_CELLSELECTORDIR	"Cell Selector Directory"	Directory for Cell Selector button configuration (.csf) files.
	MS_CELLSELECTOR	"Cell Selector File"	Default Cell Selector button configuration file.
	MS_CELLOUT	"Output Cell Libraries"	Default directory for newly created cell libraries.
	MS_MENU	"Menu Cells"	Cell library file containing menu cells.
Colors	MS_DEFCTBL	"Default Color Table"	Default color table if DGN file has none.
	MS_RMENCTBL	"Right Menu Color Table"	Default menu colors (dialog boxes, borders, etc.) for right screen — specifies a color table (.tbl) file.

Table 16–3 Configuration Variables by Category (continued)

Category	Variable	Short Name	Description
Colors (continued)	MS_LMENCTBL	"Left Menu Color Table"	Default menu colors (dialog boxes, borders, etc.) for left screen — specifies a color table (.tbl) file.
Database	MS_DBASE	"Database Files"	Search path(s) for database files.
	MS_SERVER	"Server Loader"	MDL application to load the database interface software.
	MS_DBEXT	"Database Server"	The database interface "server" application.
	MS_LINKTYPE	"Database Linkages"	User data linkage types recognized by the database interface software. See MS_LINKTYPE.
Data Files	MS_CUSTOMUNITDEF	"Unit Definitions for Upgrading"	Directory containing unit definition file used when upgrading pre-V8 files to V8 DGN files.
	MS_HTMLDGNDIR	"HTML Template Directory"	Directory containing template DGN files used to generate cell images for HTML documents.
	MS_SETTINGS	"Settings Resource"	Current settings resource file.
	MS_SETTINGSOUTDIR	"Settings Output Directory"	Directory for newly created settings files.

Table 16-3 Configuration Variables by Category (continued)

Category	Variable	Short Name	Description
Data Files (continued)	MS_SETTINGSDIR	"Settings Directory"	Directory containing settings resource files.
Design Applications	MS_DGNAPPS	N/A	List of MDL applications to load automatically when a DGN file is opened.
Engineering Links	MS_WEBLIB_HISTORY	"Weblib history"	History file for the Weblib shared library.
	MS_BOOKMARKS_IMAGE	"Image bookmarks"	Bookmark file for remote images.
	MS_BOOKMARKS_RSC	"Resource bookmarks"	Bookmark file for remote resource files.
	MS_BOOKMARKS_ARCHIVE	"Archive bookmarks"	Bookmark file for remote archives.
	MS_BOOKMARKS_CELL	"Cell library bookmarks"	Bookmark file for remote cell libraries.
	MS_BOOKMARKS_DGN	"DGN bookmarks"	Bookmark file for remote DGN files.
	MS_REFAGENTDATA	"RefAgent data file"	File containing URL information for the Reference Agent.
	MS_WEBKIOSKMODE	"Kiosk mode"	Set to 1 to turn on kiosk mode.
	MS_WEBTYPESFILE	"Special filetype handling"	Specifies a file that contains special handling instructions for various file types.
	MS_WEBFILES_DIR	"Weblib local storage"	Directory in which copies of remote files are stored.

Table 16-3 Configuration Variables by Category (continued)

Category	Variable	Short Name	Description
Engineering Links (continued)	MS_WEBDOWNLOADDIR	"Download directory"	Sets the directory in which WWW downloads are stored.
	MS_BROWSERMAKE-CHILDWINDOW	"Reparent browser"	When set to 1, the current browser is reparented inside of the MicroStation main window, allowing the viewer to always be seen.
	MS_USEEXTERNAL-BROWSER	"External browser name"	Sets an external browser.
Extensions	MS_CUSTOMUNITDEF		Sets custom unit definitions.
	MS_PLTDATE		Date that last print was generated.
	MS_PLTDGNDIR		Directory containing the last DGN file printed.
	MS_PLTDGNFILE_ABBREV		Last DGN file printed (shortened version).
	MS_PLTDGNFILE_LONG		Last DGN file printed (full path).
	MS_PLTDGNFILE_SHORT		Last DGN file printed (without path).
	MS_PLTDRVFILE_ABBREV		Print driver file used during last print (shortened version).
	MS_PLTDRVFILE_LONG		Print driver file used during last print (full path).

Table 16-3 Configuration Variables by Category (continued)

Category	Variable	Short Name	Description
Extensions (continued)	MS_PLTDRVFILE_SHORT		Print driver file used during last print (without path).
	MS_PLTOUT		Name of last generated print output file.
	MS_PLTPENTBL_ABBREV		Pen table used during last print (shortened version).
	MS_PLTPENTBL_LONG		Pen table used during last print (full path).
	MS_PLTPENTBL_SHORT		Pen table used during last print (without path).
	MS_PLTSCALE		Scale used for last print (as a character string). For example, 70.0739 m:cm / IN.
	MS_PLTSHEETNAME		Name of sheet size used. For example, ISO A3.
	MS_PLTSHEETSIZE		Size of sheet printed. For example, 11x17 (in).
	MS_PLTSTATE		Set to "PREVIEW" or "REALPLOT". Used to distinguish a paper print from a preview (so previews aren't logged to the accounting log).

Table 16-3 Configuration Variables by Category (continued)

Category	Variable	Short Name	Description
Extensions (continued)	MS_PLTTIME		Time that last print file was generated.
MDL Development	MS_RDE_SYSINC	"Runtime resource compiler"	Text to be inserted at the beginning of the command line by the resource compiler DLM. Generally used to specify the location of include files.
	MS_DBGSOURCE	"MDL Source"	Location of source code for MDL applications (used by MDL debugger).
	MS_MDLTRACE	"MDL Trace"	If set, additional debugging print statements are provided when debugging MDL applications.
	MS_ASSERT_MESSAGE	"Assertion Failure Message"	Format of assertion messages. The default is "Assertion Failure %s %s(%d)".
	MS_ASSERT_HANDLING	"Assertion Failure Handling"	Specifies handling of assertion failure: 0 (default) = display error message; 1 = log error message; 2 = stop in debugger
	MS_DEBUGFAULT	"Debug Fault"	If set, automatically invoke the debugger when a fault is detected while an MDL application is active.

Table 16-3 Configuration Variables by Category (continued)

Category	Variable	Short Name	Description
MDL Development (continued)	MS_DEBUG	"Time Out"	If set to an integer with bit 1 on, do not time out.
	MS_TRAP	"Exception Handling"	Exception handling flag — "ALL" (default), "MDL," or "NONE."
OLE	MS_OLESERVE_EMBED_REFFILES * *variable is one word*	"Embed References"	If set, references are embedded along with the master DGN file.
Operation	MS_DGNAUTOSAVE	Design File Auto-Save	Determines the frequency, in seconds, of the auto-save timer.
	MS_AUTORESTORESTATUSBAR * *variable is one word*	"Auto-restore Status Bar"	If set to 1, the status bar is restored to its default appearance whenever the pointer is moved within.
	MS_FKEYMNU	"Function Key Menu"	Open function key menu file.
	MS_ACCUDRAWKEYS	"AccuDraw Shortcuts"	Text file listing AccuDraw shortcut keys.
	MS_GRAPHICSACCELERATOR * *variable is one word*	"Graphics Accelerator"	Sets the name of the graphics accelerator application.
	MS_SAVEMENU	"Attached Menus"	File containing information about attached menus.

Chapter 16 Customizing MicroStation

Table 16-3 Configuration Variables by Category (continued)

Category	Variable	Short Name	Description
Operation (continued)	MS_APPMEN	"Application Menus"	Location of application and sidebar menus.
	MS_WORKSPACEOPTS	"Workspace Options"	Sets how MicroStation Manager displays workspace options: 0 or not set (default) = workspace remains same; 1 = workspace options visible but disabled; 2 = workspace options hidden, MM resized.
	MS_ USECOMMANDWINDOW *variable is one word*	"Use Command Window"	If set, 0 sets the status bar interface, and 1 sets the Command Window interface. If not set, the Style setting in MicroStation Manager determines the interface style.
	MS_FILEHISTORY	"File History"	If set, MicroStation saves the last four files and directories for each file type.
	MS_READONLY	"Read Only"	If set, the active DGN file is read-only.
	MS_WORKMODE	"Work Mode"	Will gain full functional capacity creating elements. Must restart to take effect of changed value.

717

Table 16–3 Configuration Variables by Category (continued)

Category	Variable	Short Name	Description
Operation (continued)	MS_DISALLOWFULL-REFPATH	"Disallow Full Ref Path"	If set, you cannot save a reference attachment with a full path specification.
Printing	MS_BATCHPLT_SPECS	"Batch Print Specifications"	Name of the file that contains the batch printing specifications.
	MS_PLTFILES	"Output Directory"	Directory for printing output files (plotfiles).
	MS_PLOTINI	"Plot Configuration Files"	Directory for plot configuration files.
	MS_PLTR	"Printer Driver File"	Name of printer driver file.
	MS_PENTABLE	"Pen Tables"	Search path for pen table files.
Primary Search Paths	MS_DEF	"Design Files"	Search path(s) for DGN files.
	MS_RFDIR	"References"	Search path(s) for references.
	MS_MDLAPPS	"Visible MDL Applications"	Search path(s) for MDL applications displayed in the MDL dialog box.
	MS_MDL	"MDL Applications"	Search path(s) for MDL applications or external programs loaded by MDL applications.
	MS_MACRO	"Macros"	Search path for macros.

Table 16-3 Configuration Variables by Category (continued)

Category	Variable	Short Name	Description
Primary Search Paths (continued)	MS_RSRCPATH	"Resource Files"	Search path(s) for resource files loaded by MDL applications.
	MS_LIBRARY_PATH	"Library Path"	Search path for dynamic link libraries.
	MS_LEVEL_LIB_DIR	"Level Library Path"	Directory containing level library data files.
Rendering/ Images	MS_MTBL	"Material Tables"	Search path(s) for material tables.
	MS_MATERIAL	"Material Palettes"	Search path(s) for material palettes.
	MS_PATTERN	"Pattern Maps"	Search path(s) for pattern maps.
	MS_BUMP	"Bump Maps"	Search path(s) for bump maps.
	MS_IMAGE	"Images"	Search path(s) for images.
	MS_IMAGEOUT	"Image Output"	Directory in which created image files are stored.
	MS_SHADOWMAP	"Shadow Maps"	Directory where shadow maps will be read from and written to.
	MS_LIGHTING	"Lighting Files"	Directory where IES lighting data will be stored.
	MS_RENDERLOG	"Rendering statistics log"	File in which rendering statistics are logged.

Table 16-3 Configuration Variables by Category (continued)

Category	Variable	Short Name	Description
Rendering/ Images (continued)	MS_PTDIR	"Particle Tracing Work Dir"	Working directory for temporary Particle Tracing files.
Seed Files	MS_SEEDFILES	"Seed File Location"	Search path(s) for all seed files.
	MS_DESIGNSEED	"Default Design File Seed"	Default seed file.
	MS_TRANSEED	"Default Translation Seed"	Default seed file for DWG, CGM, and IGES translations.
	MS_SHEETSEED	"Drawing Sheet Seed File"	Seed file used when creating drawing sheets.
	MS_USERPREFSEED	"User Preference Seed"	Name of seed file used to create user preference resource file.
Symbology	MS_SYMBRSRC	"Symbology Resources"	List of symbology resource files — last one in list has highest priority.
System Environment	MS_CONFIG	"Main Configuration File"	Main MicroStation configuration file — sets up all configuration variables.
	RSC_COMP	"Resource Compiler Command"	Text string to be inserted at the beginning of the command line by the resource compiler (used to specify where to search for include files).

Table 16–3 Configuration Variables by Category (continued)

Category	Variable	Short Name	Description
System Environment (continued)	MDL_COMP	"MDL Compiler Command Line"	Text string to be inserted at the beginning of the command line by the MDL compiler (used to specify where to search for include files).
	BMAKE_OPT	"BMAKE Options"	Command line options for BMAKE. Used to search for bmake include (.mki) files.
	MS_DEBUGMDLHEAP	"Extended Malloc"	If set (to the base name of an MDL application or "ALL), use extended malloc for debugging.
Tags	MS_TAGOUTPUT	"Tag Output"	Output directory for tag data.
	MS_TAGREPORTS	"Tag Reports"	Output directory for tag reports.
	MS_TAGTEMPLATES	"Tag Templates"	Directory containing tag report templates.
Temp and Backup Files	MS_BACKUP	"Backup Files"	Default directory for backup files.
	MS_TMP	"Temporary Files"	Directory for temporary files created and deleted by MicroStation.
	MS_SCR	"Scratch Files"	Directory for scratch files created by MicroStation.

Table 16–3 Configuration Variables by Category (continued)

Category	Variable	Short Name	Description
DWG/DXF Search Path	MS_DWGDATA	"DWG Data Directory"	Local directory to store DWG settings file.
	MS_ACIS	"ACIS Directory"	Directory containing the ACIS subsystem.
	MS_DWGPATFILE	"AutoCAD Patterns File"	Filename containing AutoCAD Pattern Definitions.
	MS_BLOCKLIST	"AutoCAD Block List"	List of DWG/DXF files to be searched for AutoCAD blocks.
	MS_DWGFONTPATH	"AutoCAD SHX Font Directory"	Path(s) that contain AutoCAD SHX fonts.
	See MS_DWGFONTPATH	"AutoCAD Program Directory"	Directory containing AutoCAD. This is used to find DWG support files such as fonts.
Translation — CGM	MS_CGMIN	"CGM Input Directory"	Input directory for CGM translations.
	MS_CGMOUT	"CGM Output Directory"	Output directory for CGM translations.
	MS_CGMLOG	"CGM Log Files"	Output directory for CGM log files.
	MS_CGMTABLES	"CGM Configuration Tables"	Directory containing the CGM translation tables.
	MS_CGMINSET	"CGMIN Settings File"	Settings file for the CGMIN application.
	MS_CGMOUTSET	"CGMOUT Settings File"	Settings file for the CGMOUT application.

Table 16-3 Configuration Variables by Category (continued)

Category	Variable	Short Name	Description
Translation — IGES	MS_IGESIN	"IGES Input Directory"	Input directory for IGES translations.
	MS_IGESOUT	"IGES Output Directory"	Output directory for IGES translations.
	MS_IGESLOG	"IGES Log Files"	Output directory for IGES log files.
	MS_IGESINSET	"IGESIN Settings File"	Settings file for IGES import.
	MS_IGESOUTSET	"IGESOUT Settings File"	Settings file for IGES export.
Translation — Step	MS_STEPOUT	"Step Output Directory." directory."	Output directory for directory.
	MS_STEPLOG	"Step Log Files."	Output directory for Step log files.
	MS_ROSEDB	"Rose database directory."	Rose database directory.
	MS_APP	"Apps from "TSK" statements"	Search path(s) of applications started from "TSK" statements.
Visual Basic for Applications	MS_VBAAUTOLOADPROJECTS * variable is one word	"Names of standard projects"	Names of the projects that are opened when the VBA dialog box is opened.
	MS_VBASEARCHDIRECTORIES * variable is one word	"Directories to search for VBA projects"	Directories that are searched when opening an existing VBA project.
	MS_VBANEWPROJECTDIRECTORY * variable is one word	"Directory for new projects"	Directory that is used whena new project is created.

Table 16-4 Uncategorized Configuration Variables

CLASSPATH	Path to Java/JMDL classes.
HISTORY_MAX_MINOR_VERSIONS	The number of minor versions for the Limit History command to preserve in the history file.
MSDIR	MicroStation root installation directory.
MSLOCAL	Specifies the base directory path for where the required writable portions (that is, local for a network installation) of MicroStation are installed.
MS_BASICEXT_LOAD	List of MDL applications that implement extensions to the BASIC language (MS_BASIC_LOAD).
MS_CMDTABLE_LOAD	List of MDL applications that will have their key-in tables auto-loaded.
MS_CMDWINDRSC	Command Window resource file. Default is used if undefined.
MS_CODESET	MDL application for handling multi-byte character sets.
MS_DATA	Directory for data files created or used by MicroStation.
MS_DEFCHARTRAN	Default character translation table.
MS_DEFAULTSHEETRGB	Allows you to specify a different background color, which is used when you create the first sheet model. Expects an RGB triplet separated by commas.
MS_DGNOUT	Directory containing DGN files created as a result of "on the fly" translation from other file formats.
MS_DWGSEED	Specifies a DWG seed file.
MS_GUIHAND	Identifies auxiliary handlers.

Table 16-4 Uncategorized Configuration Variables (continued)

MS_HELPLOAD_APPLICATION	Determines which application help is based on. Current options are HTMLHelp (default), or StaticWeb. StaticWeb refers to static HTML pages on a website.
MS_HELPLOAD_SERVER	Specifies the root location of the help content. Defaults to $(USTN_DOCUMENTATIONROOT), but can be set to a file server or URL. This variable does not determine if you are using uncompiled HTMLHelp content. If you set this variable, the value of MS_HELPLOAD_SERVER should be the location containing the product directory (i.e., MicroStation)
MS_HELPPATH	Path to help files.
MS_INITAPPS	List of initial start-up MDL applications.
MS_OLDUSERLICENSE	File that contains old user license information. Required for when installing a product upgrade.
MS_OPENDESIGNFILEFILTER	Initial file filter for the MicroStation Manager and Open dialog boxes.
MS_OPENREFFILEFILTER	Initial file filter for the Attach Reference File dialog box.
MS_REFCOLORTABLE	If defined and set to 1, MicroStation always uses the reference's color table. If defined and set to any value other than 1, MicroStation never uses the reference's color table. If not defined, MicroStation uses the the user preference to determine whether or not to use the reference's color table.
MS_RIGHTLOGICKB	If set to 1, keyboard input is right to left.
MS_RSRC	Main MicroStation resource file. Typically set to "ustation.rsc."
MS_SELECTSERVICESHOME	URL of Bentley SELECTservices homepage.
MS_SPELLINGDICTIONARYPATH	Specifies the directory to search for core dictionaries.

Table 16-4 Uncategorized Configuration Variables (continued)

Variable	Description
MS_SPELLINGLANGUAGE	Specifies the language. Valid languages are: AmericanEnglish, BritishEnglish, Brazilian (Brazilian Portuguese), Danish, Dutch, Finnish, French, German, Italian, Norwegian, Spanish.
MS_SPELLINGUSERDICTIONARY	Specifies the user dictionary.
MS_THUMBNAIL	Stores a thumbnail image in each DGN file that can be seen in the MicroStation file open dialog boxes, as well as Windows Explorer. Can be used to control the behavior of thumbnail generation.
MS_USERPREF	User preference resource file.
MS_USERPREF_APPS	Set by application configuration files to add user preferences to standard dialogs.
MS_VBAV8TOV7DIALOG	Specifies the name of a VBA macro that opens a dialog to collect information to be used in translating a DGN file from V8 to V7 format.
MS_VBAV8TOV7FILTER	Specifies the name of a VBA macro that sets up a translation filter.
PZIP_OUT	Specifies default directory for creation of Packager files.

Customizing the User Interface

MicroStation allows you to customize any or all of the parts of the active workspace user interface:

- Tool frames and tool boxes
- View border tools
- Drop-down menus

Customization of MicroStation's user interface is stored in user interface modification files. When you create a new user interface, MicroStation creates a folder under *<disk>:\Bentley\Workspace\interfaces\MicroStation* with the given name of the new user interface. By default, MicroStation places the default user interface file (ustn.r01) in the new folder. If part of the interface is modified, the file name of the user interface modification file is "ustn" with the file name extension "m" followed by a number from 01 to 99. Each time an interface modification file is saved with the same file name, the number in its suffix is incremented. Therefore, the files might be named "ustn.m01," "ustn.m02," and so on. If the tool box or dialog box modified is provided by an MDL (MicroStation Development Language) application, the file name of the user interface modification file is the file name of the MDL application.

Customizing Tool Boxes

Follow these steps to modify or create a tool box. Open the Customize settings box from:

| Drop-down menu | Workspace > Customize |

MicroStation displays a Customize settings box similar to Figure 16–32.

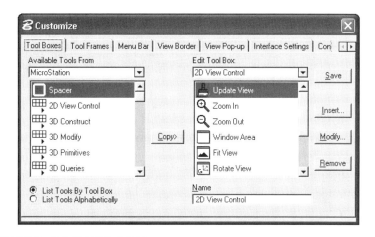

Figure 16-32 Customize settings box

Follow these steps to modify or create a new tool box:

STEP 1: Click the **Tool Boxes** tab on the Customize settings box, and MicroStation lists the tool frame options, as shown in Figure 16–32.

STEP 2: Select an interface from the **Available Tools From** menu. MicroStation lists the available tool boxes for the selected interface in the field below the menu.

STEP 3: Select the tool box you need to customize from the **Edit Tool Box** drop-down menu,

or

Select **Create Tool Box** from the **Edit Tool Box** menu to create a new tool box. MicroStation displays the Create Tool Box dialog box, as shown in Figure 16–33. Key-in the name of the new tool box in the **Name** field and click the **OK** button. MicroStation displays the name of the new tool box in the **Name** field at the bottom of the Customize settings box. When you copy tools into the new tool box, the tool names are displayed in the large field above the **Name** field.

Figure 16-33 Create Tool Box dialog box

STEP 4: From the available tool boxes listed on the **Available Tools From** field (on the left side of the settings box), double-click the name of an interface that contains tools you want to put in your new tool box. MicroStation expands the selected interface to display its tools.

STEP 5: Select the tool you want to insert in your new tool box and click the **Copy** button to put a copy of it in your tool box on the **Edit Tool Box** field. You can also drag a tool into your tool box.

STEP 6: Rearrange the tools by dragging and dropping them in the **Edit Tool Box** field.

STEP 7: Remove a tool from your tool box in the **Edit Tool Box** field by selecting the tool and clicking the **Remove** button. MicroStation removes the selected tool from the tool box.

STEP 8: To insert a new tool, click the **Insert** button, and MicroStation displays the Insert Tool dialog box, as shown in Figure 16–34.

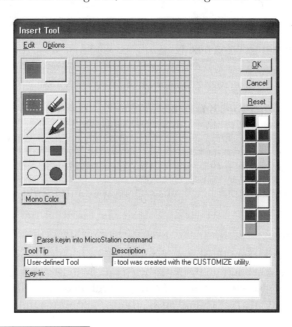

Figure 16-34 Insert Tool dialog box

Chapter 16 Customizing MicroStation

Use the drawing tools to create an icon for the new tool. Key-in the tool tip text and description for the new tool in the **Tool Tip** and **Description** text fields. Key-in the action strings to be associated with the tool in the **Key-in** text field. If a multiple key-in action string is specified, separate the key-ins with a semi-colon (;). For example, the following key-in sets the Active Color to red, the Active Line Weight to 3, and invokes the Place Line tool:

CO=RED;WT=3;PLACE LINE

Click the **OK** button to accept the changes and close the dialog box.

STEP 9: To modify a tool, select the tool from the **Edit Tool Box** field and click the **Modify** button. MicroStation displays the Insert Tool dialog box with the selected tool loaded in the box's fields, as shown in Figure 16–35.

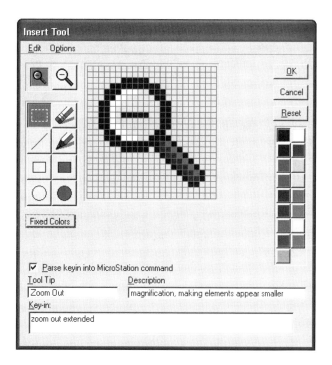

Figure 16-35 Insert Tool dialog box when called from the Modify button

If necessary, use the drawing tools to modify the icon. Make the necessary modifications to the tool tip text, description, and key-ins. Click the **OK** button to accept the changes and close the dialog box.

STEP 10: Click the **Save** button to save the tool box changes. MicroStation displays the customized version of the tool box(es).

STEP 11: If no further changes have to be made, close the Customize settings box.

Customizing Tool Frames

The Customize settings box also can be used to customize tool frames. Open the Customize settings box from:

| Drop-down menu | Workspace > Customize |

Follow these steps to modify or create a new tool frame:

STEP 1: Click the **Tool Frames** tab on the Customize settings box, and MicroStation lists the tool frame options, as shown in Figure 16–36.

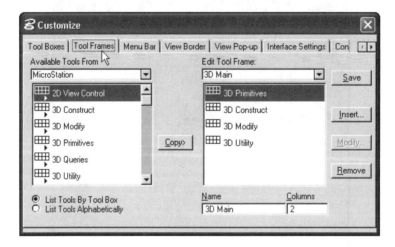

Figure 16–36 Customize settings box with the Tool Frames tab selected

STEP 2: Select the name of the interface that contains the toolboxes you want to add to your frame from the **Available Tools From** menu. MicroStation lists the available tool boxes in the field below the menu.

STEP 3: Select one of the available tool frames to customize from the **Edit Tool Frame** menu,

or

Select **Create Tool Frame** from the **Edit Tool Frame** menu to create a new tool frame. MicroStation displays the Create Tool Frame dialog box, as shown in Figure 16–37. Key-in the name of the new tool frame in the **Name** field and click the **OK** button. MicroStation displays the name of the new tool frame in the **Name** field at the bottom of the Customize dialog box. When you copy tool boxes into the new frame, the tool box names are displayed in the large field above the **Name** field.

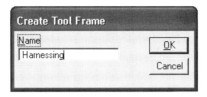

Figure 16-37 Create Tool Frame dialog box

STEP 4: If you want your tool frame to have more than one column, key-in the number of columns in the **Columns** field to the right of the **Name** field.

STEP 5: From the available tool boxes listed on the **Available Tools From** field (on the left side of the settings box), select the tool box you want to insert into the tool frame and click the **Copy** button to add it to the selected tool frame. You can also drag and drop tool boxes from the left side of the dialog box to the right side to add them to the tool frame.

STEP 6: Rearrange the tool boxes by dragging and dropping in the **Edit Tool Frame** list box.

STEP 7: Remove a selected tool box from the tool frame by clicking the **Remove** button. MicroStation removes the selected tool box from the tool frame.

STEP 8: Click the **Save** button to save the tool frame changes. MicroStation displays the customized version of the tool frame(s).

STEP 9: If no further changes need to be made, close the Customize settings box.

Customizing the Menu Bar

The Customize settings box allows you to customize the MicroStation's menu bar by adding, deleting, and creating drop-down menus, and by changing the options in the menus. Open the Customize settings box from:

Drop-down menu	Workspace > Customize

Follow these steps to modify a drop-down menu:

STEP 1: Click the **Menu Bar** tab, and MicroStation lists the available options, as shown in Figure 16–38.

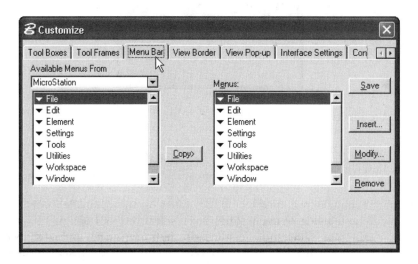

Figure 16-38 Customize settings box with the Menu Bar tab selected

 Note: The left side of the dialog box lists all the menus that can be added to the MicroStation menu bar. Each interface contains its own set of menus. The right side of the menu bar lists all menus on the MicroStation menu bar and is where you create, add, delete, and modify menus.

STEP 2: From the **Available Menus From** field, select the interface that contains the menus you want to add to the MicroStation menu bar. MicroStation displays the menus in the field below the menu.

STEP 3: To change the name of a menu currently on the MicroStation menu bar, select the menu's name in the **Menus** field and click the **Modify** button. MicroStation displays the Modify Menu dialog box, as shown in Figure 16–39.

Chapter 16 Customizing MicroStation

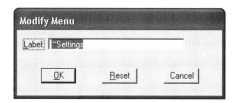

Figure 16-39 Modify Menu dialog box

Key-in the new menu name in the **Label** field and click the **OK** button. The new name appears in the **Menus** field of the Customize dialog box, but it does not appear in the menu bar until you click the **Save** button.

Notes: Spaces are not allowed in menu names. If you key-in a space as part of the menu name, MicroStation removes it when you click the **OK** button.

In the MicroStation menu bar, each menu name has an underlined shortcut character. You can open the menu without using the mouse by tapping the shortcut character while holding down the ALT. To indicate which character in the menu name is the shortcut, key-in the tilde (~) character immediately before the character. There can only be one shortcut key in a menu name and each shortcut key should only be used one time in the menu bar.

STEP 4: To make changes to the items in one of the existing menus on the MicroStation menu bar, double-click the menu name in the **Menus** field to expand the menu and display its items below the menu name, as shown in Figure 16–40.

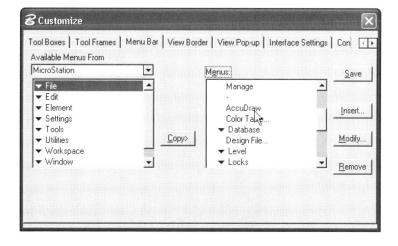

Figure 16-40 An expanded menu in the Customize settings box

733

STEP 5 Select the menu item you want to modify and click the **Modify** button. MicroStation displays the Modify Menu Item dialog box, as shown in Figure 16–41.

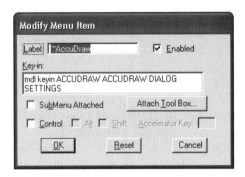

Figure 16-41 Modify Menu Item dialog box

- To change the item's name, key-in the new name in the **Label** text field.

Notes: Spaces are allowed in menu item names.

Each item in a menu has an underlined shortcut key. If you tap a shortcut key when a menu is expanded on the MicroStation menu bar, the option that contains the shortcut key is activated. To indicate which character in the item name is the shortcut, key-in the tilde (~) character immediately before the character. There can only be one shortcut key in an item name and each shortcut key should only be used one time within the menu.

- Turn the **Enabled** check box ON to enable users to select the item or OFF to prevent users from selecting the item.

- Make any required changes to the action that the menu item initiates in the **Key-in** field.

- If the item is to be a submenu that displays a list of additional items, turn the **Submenu Attached** check box ON. To add items to the submenu, double-click its name and use the same procedures described here to insert and modify submenu items.

Note: When you turn the **Submenu Attached** check box ON, the action in the **Key-in** field is ignored and the **Attach Tool Box** button is not available for use.

- To open a tool box from the menu item, click the **Attach Tool Box** button. MicroStation displays the Select Tool Box dialog box, as shown in Figure 16–42. Select a tool box from the list and click the **OK** button to attach the Tool Box to the item.

Chapter 16 Customizing MicroStation

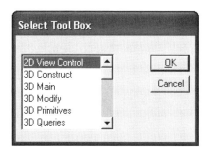

Figure 16-42 Select Tool Box dialog box

- If you want to create an accelerator key sequence that can select the item directly from the keyboard without first selecting the menu, turn the **Control** check box ON. To use an accelerator key sequence, the user types a character key while holding down one or two control keys. To set the control keys, turn one or both of the **Alt** and **Shift** check boxes ON. Key-in the accelerator key in the **Accelerator Key** field.

- Click the **OK** button to close the Modify Menu Item dialog box and save your changes. If you want to reset the item to its default settings before saving it, click the **Reset** button. If you want to close the dialog box without making any changes, click the **Cancel** button.

STEP 6: To create a new menu on the MicroStation menu bar, select the name of the existing menu before which you want to insert the new menu and click the **Insert** button. MicroStation displays an Insert Menu dialog box, as shown in Figure 16-43.

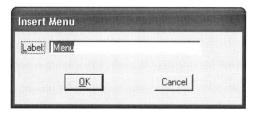

Figure 16-43 Insert Menu dialog box

Key-in the name of the new menu name in the **Label** text field and click the **OK** button to close the dialog box.

STEP 7: To create a menu item, double-click the menu name to expand the menu, select the menu item before which you want to insert the new item, and click the **Insert** button. MicroStation displays the Insert Menu Item dialog box, as shown in Figure 16–44.

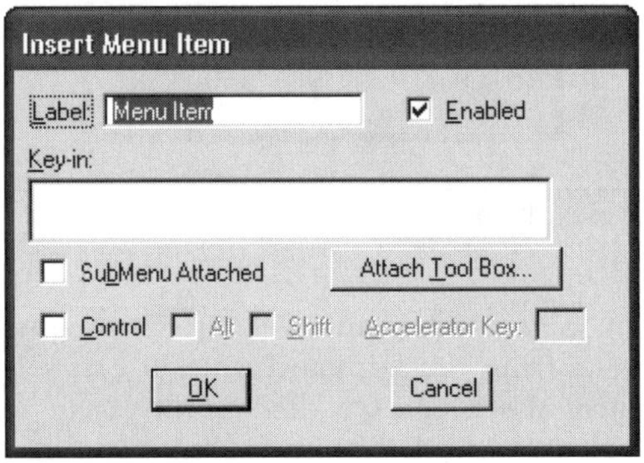

Figure 16-44 Insert Menu Item dialog box

Use the appropriate controls in the Insert Menu Item dialog box to set the menu item and click the **OK** button to save the new item and close the dialog box.

STEP 8: To add an existing menu to the MicroStation menu bar, select the menu in the **Menus** field that the added menu will immediately precede, select the menu to be added in the **Available Menus From** field, and click the **Copy** button to add the menu to the **Menus** field. You can also drag and drop a menu from the **Available Menus From** field to the **Menus** field.

STEP 9: To add an existing menu item to a menu on the MicroStation menu bar, double-click the menu in the **Menus** field to which the new item is to be added, click the item that the added item will immediately precede, double-click the menu in the **Available menus From** field that contains the item to be added, click the item to be added, and click the **Copy** button to add the item to the expanded menu in the **Menus** field. You can also drag and drop an item from the **Available Menus From** field to the **Menus** field.

STEP 10: To rearrange the menus or items in a menu, drag and drop them in the **Menus** field.

STEP 11: To remove the menu or menu item from the **Menus** field, select the menu or menu item and click the **Remove** button. MicroStation removes the selected menu or menu item from the tool box.

STEP 12: To save the menu changes, click the **Save** button. MicroStation displays the customized version in the menu bar.

STEP 13: If no further changes have to be made, close the Customize settings box.

Customizing View Border Tools

The Customize settings box can add and remove tools in the view menu bar on the border of view windows. Open the Customize settings box from:

| Drop-down menu | Workspace > Customize |

Follow these steps to modify view border tools:

STEP 1: Click the **View Border** tab, and MicroStation lists the available options, as shown in Figure 16–45.

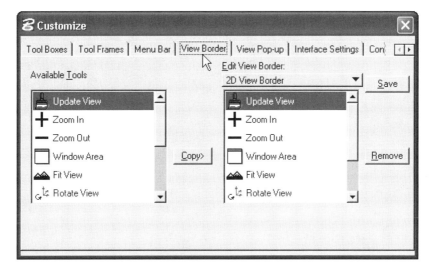

Figure 16-45 Customize settings box with View Border tab selected

STEP 2: Select the view border you want to customize (**2D View Border** or **3D View Border**) from the **Edit View Border** menu.

STEP 3: To add a tool to the selected view border, click the tool in the **Edit View Border** field that you want to be immediately to the right of the added tool, click the tool to be added in the **Available Tools** field, and click the **Copy** button. A copy of the tool selected in the **Available Tools** field is placed just above the selected tool in the **Edit View Border** field. You can also drag and drop tool from the **Available Tools** field to the **Edit View Border** field.

STEP 4: To change the order of the tools in the View Border, drag and drop them in the **Edit View Border** field.

STEP 5: To remove a tool from the View Border, select the tool in the **Edit View Border** field and click the **Remove** button. MicroStation removes the selected tool from the View Border.

STEP 6: To save the view border changes, click the **Save** button. MicroStation displays the customized version of the view border.

STEP 7: If no further changes have to be made, close the Customize settings box.

FUNCTION KEYS

Personal computer keyboards have a set of function keys at the top of the keyboard. MicroStation has a utility that assigns actions to function keys F1 through F12. Once the assignment is made, all you have to do is tap the function key and the action is activated.

There are only 12 functions on the keyboard, but MicroStation allows you to make up to 96 function key assignments by combining the function keys with SHIFT, ALT, and/or CTRL. For instance, you can assign F1 with the following combinations:

    F1

    SHIFT + F1

    ALT + F1

    CTRL + F1

    SHIFT + ALT + F1

    SHIFT + CTRL + F1

    ALT + CTRL + F1

    SHIFT + ALT + CTRL + F1

Function key assignments are stored in an ASCII file, and by default the extension for the function key file is .MNU. This file follows the same concept as the cell library file. Once created, the function key file may be attached to any of your design files to activate the function keys. The default function key menu file is called "funckey.mnu." It is stored in the */Bentley/Worskpace/interfaces/fkeys/* path.

Creating and Modifying Function Key Definitions

To create and modify function key definitions, open the Function Keys dialog box from:

Drop-down menu	Workspace > Function Keys

MicroStation displays the Function Keys dialog box, as shown in Figure 16–46.

Chapter 16 Customizing MicroStation

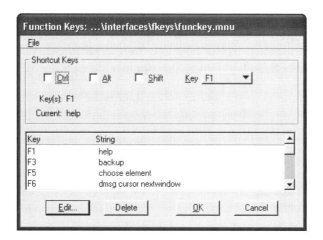

Figure 16-46 Function Keys dialog box

The title bar identifies the file name of the function key menu and the **Key/String** field lists currently defined function key combinations and their definitions.

To select a key sequence to create or edit, turn ON one or more of the **Ctrl**, **Alt**, and **Shift** check boxes, and select a function key from the **Key** drop-down menu. You can also select an existing key sequence by clicking it in the **Key/String** field.

Editing Function Key Definitions

To assign an action to a new key sequence and to change the action of an existing key sequence, click the **Edit** button. MicroStation opens the Edit Key Definition dialog box, as shown in Figure 16–47.

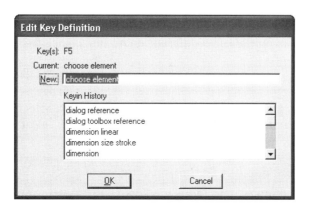

Figure 16-47 Edit Key Definition dialog box

The selected function key sequence is shown at the top of the dialog box, and the action initiated by the sequence is displayed in the **New** text field. If no action has been assigned to the key sequence, **undefined** is displayed in the **New** text field. The **Keyin History** field displays commands that have previously been keyed-in to the **Key-in** settings box.

To change the action initiated by the key sequence, either key-in the action in the **New** field or select an action from the **Keyin History** field. When you click the **OK** button, the new action is displayed in the **Key/String** field of the Function Keys dialog box, but it is not implemented until you save the dialog box.

You can define multiple actions for a key sequence by separating the action strings with semicolons. For example, the following key-in sets the Active Color to red, the Active Line Weight to 2, and invokes the place line tool:

    co=red;wt=2;place line

Deleting Function Key Definitions

To delete a function key definition, select the definition from the list box (or use the check boxes and drop-down menu) and click the **Delete** button. MicroStation immediately removes the selected function key definition from the **Key/String** field in the Function Keys dialog box, but it does not actually delete it until you save the dialog box.

Saving the Function Key Definitions

To save the function key definitions, invoke the Save tool from:

Drop-down menu (Function Keys dialog box)	File > Save

MicroStation saves the current status of the function key definitions to the function key menu file.

Saving the Function Key Definitions to a Different File

To save the function key definitions to a different menu file, invoke the Save As tool from:

Drop-down menu (Function Keys dialog box)	File > Save As

MicroStation opens the Save Function Key Menu As dialog box. Select the location where you want to put the file in the **Directories** field, key-in the name of the menu file in the **Files** text field, and click the **OK** button.

Opening a Different Function Key Definitions File

To open a different function key definitions file, invoke the Open tool from:

Drop-down menu (Function Keys dialog box)	File > Open

MicroStation opens the Select Function Key Menu dialog box. Select the location of the file in the **Directories** field, select the file name of the menu file in the **Files** area, and click the **OK** button. MicroStation opens the selected file and displays its function key definitions in the **Key/String** field of the Function Keys dialog box.

INSTALLING FONTS

Each time a design file is displayed, a font resource file that contains font definitions is opened. MicroStation, by default, uses a resource file called *FONT.RSC* allows you to add new fonts to the font resource file via the utility called Font Installer.

The Font Installer dialog box helps you import fonts from different sources into the MicroStation font library, and you can rename and renumber fonts.

To open the Font Installer dialog box, invoke Install Fonts from:

| Drop-down menu | Utilities > Install Fonts |

MicroStation displays the Font Installer dialog box, as shown in Figure 16–48.

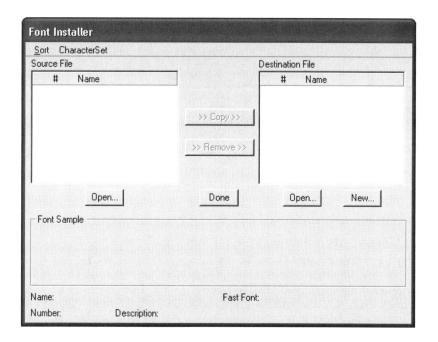

Figure 16-48 Font Installer dialog box

Selecting the Source Font Files

To select the source fonts, click the **Open** button located below the **Source File** field. MicroStation displays the Open Source Font Files dialog box, as shown in Figure 16–49.

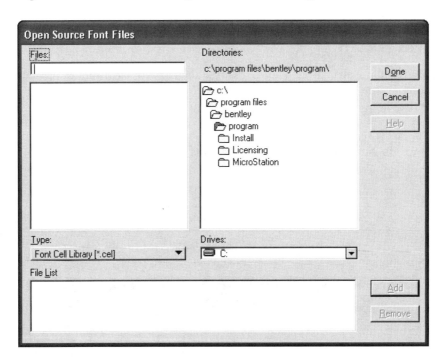

Figure 16-49 Open Source Font Files dialog box

The **Type** drop-down menu provides options to set the file type to one of the following:

> **Font Cell Library**—MicroStation font cell library, a standard cell library that contains cells that define the characters and symbols in a traditional MicroStation font and the font's attributes (*.cel)
>
> **Font Library**—MicroStation Version 5 font library (*.rsc)
>
> **uSTN V4/IGDS Fontlib**—Version 4.1 or earlier (or IGDS) font library

Select the location of the font file you want to open from the **Directories** field, select the font file from the **Files** field, and click the **Add** button to open the selected file. Before clicking the **Add** button, you can select multiple files as the source font files. To remove selected files, click the **Remove** button. When you complete selecting files to open, click the **Done** button to open the files and close the dialog box. The **Source File** field in the Font Installer dialog box lists all the fonts contained in the opened source files.

Selecting the Destination Fonts File

To select the destination fonts file in which you want to put fonts from the source files, click the **Open** button located beneath the **Destination File** list box. MicroStation opens the Open Font Library dialog box, similar to Figure 16–50. Select the location of the font library file from the **Directories** field, the font library file name from the **Files** field, and click the **OK** button to close the dialog box. The **Destination File** field in the Font Installer dialog box lists the fonts in the selected library.

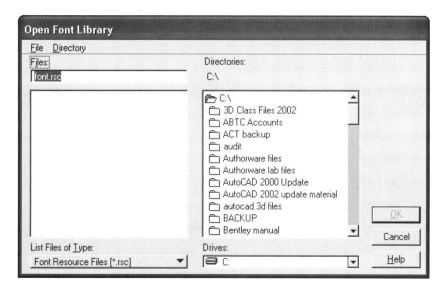

Figure 16-50 Open Font Library dialog box

Importing Fonts

To import a font from the source to the destination file, select the font in the **Source File** field and click the **Copy** button. MicroStation copies the selected font into the **Destination File** field. To remove a font in the **Destination File** field, select the font to be removed from the list and click the **Remove** button. MicroStation removes the selected font from the **Destination File** field.

You can change a font's name, number, and description. Select the font in the **Destination File** field, and MicroStation displays the font's information at the bottom of the dialog box. Make the necessary changes in the appropriate edit fields.

Click the **Done** button to close the Font Installer dialog box.

PACKAGING THE MICROSTATION ENVIRONMENT

The Packaging utility allows you to package all the design files and other associated files in a workspace so that you can share them with users on other computers and/or in other companies, or create a backup copy. The files and operating environment are saved and can be reproduced on the other computer so that that computer has the same work environment as your computer.

Invoke the Packager from:

| Drop-down menu | Utilities > Packager |

MicroStation opens the Create Package wizard, as shown in Figure 16–51.

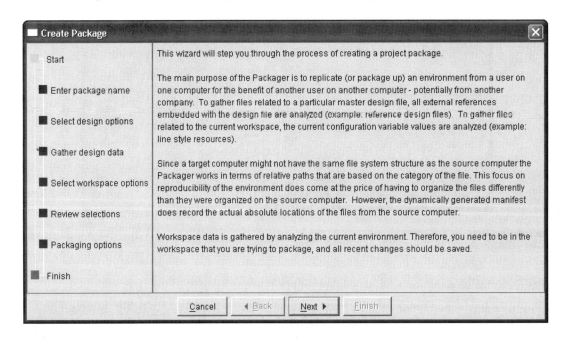

Figure 16-51 The Create Package wizard

The wizard guides you through the process of creating a package that contains the folders and files that make up your MicroStation work environment. Control buttons on the wizard take you through the packaging steps and instructions at each step tell you what to do.

The files are compressed using a standard compressing format and placed in a "zip" file that has the PZIP extension. MicroStation supplies an extractor tool that can uncompress and extract the files on the target computer. Other commercial extraction applications can also uncompress and extract the files.

Chapter 16 Customizing MicroStation

To extract the files, display the contents of the directory that contains the PZIP file and double-click the file name. MicroStation opens the Extractor dialog box, as shown in Figure 16–52.

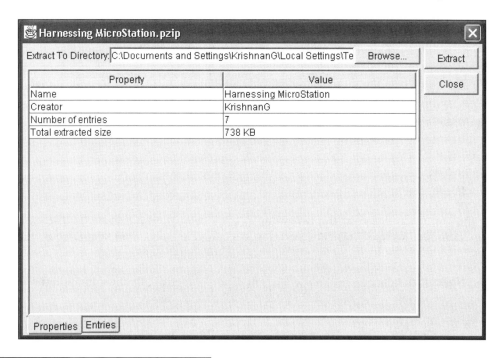

Figure 16-52 The Extractor dialog box

The name of the dialog box will be the name of the package file you selected, and the large field in the center of the box will display either the package properties or package entries (the files). To switch between the two views, click the **Properties** tab or the **Entries** tab at the bottom of the dialog box.

When you display the entries, each file is displayed on one row of the list. Check boxes on each entry line indicate whether or not the file will be extracted when you click the **Extract** button. You can turn the check boxes ON and OFF by clicking them.

Click the **Browse** button to open the Select Directory dialog box and use the **Look in** drop-down menu to find and select the folder where you want to place the extracted copies of the folders and files in the package file. Click the **Select Directory** button to complete the selection process, and then click the **Extract** button to extract the contents of the package file to the selected folder.

When all folders and files are extracted, click the **Close** button to close the extractor dialog box.

SCRIPTS AND MACROS

MicroStation has two application software tools that automate often-used command sequences. Depending on the complexity of the operation to be performed, you can choose to use either scripts or macros.

Scripts

A script is the simplest software application in which you create an ASCII file containing the key-in sequence of the MicroStation commands. For example, the following script sets the Active Color, Active Line Weight, and Active Level:

>   active color red
>   active weight 6
>   active level Borders

To load and run a script, key-in the following in the Key-in window: **@<script_file>** and press ENTER (for example, **@attachRef**). MicroStation executes the command sequence in the order that the script is set.

Note: If the script is not in the current directory, script_file must include the full path to the script.

Macros

Macros are Microsoft Visual Basic for Applications (VBA) programs that automate often-used sequences of operations. Many MicroStation-specific extensions have been added to the VBA language to customize it for the MicroStation environment. Macros select tools and view controls, send key-ins, manipulate dialog boxes, modify elements, and so forth.

For detailed information about creating macros, refer to the MicroStation VBA Guide that comes with the MicroStation software.

To load and run a macro, invoke the Macro tool from:

Drop-down menu	Utilities > Macro > Macros
Key-in window	**Macro** <macro_name> ENTER

MicroStation displays the Macros dialog box. Select a macro from the **Macro name** field and click the **Run** button to execute the macro. Macros can provide a variety of operations. Some will prompt you for input, some may only display messages in the status bar as during execution, and others may complete their operations without any prompts or messages.

chapter 17

3D Design and Rendering

WHAT IS 3D?

In *two dimensional* drawings, you work with two axes, *X* and *Y*. In *three dimensional* drawings, in addition to the *X* and *Y* axes, you work on the *Z* axis, as shown in Figure 17–1. Plan views, sections, and elevations represent only two dimensions. Isometric, perspective, and axonometric drawings, on the other hand, represent all three dimensions. For example, to create three views of a cube, the cube is simply drawn as a square with thickness. This is referred to as *extruded 2D.* Only objects that are extrudable can be drawn by this method. Any other views are achieved by simply rotating the viewpoint of the object, just as if you were physically holding the cube. You can also get an isometric or perspective view by simply changing the viewpoint.

Drawing objects in 3D provides three major advantages:

- An object can be drawn once and then can be viewed and plotted from any angle
- A 3D object holds mathematical information that can be used in engineering analysis, such as finite-element analysis and computer numerical control (CNC) machinery
- Shading can be added for visualization

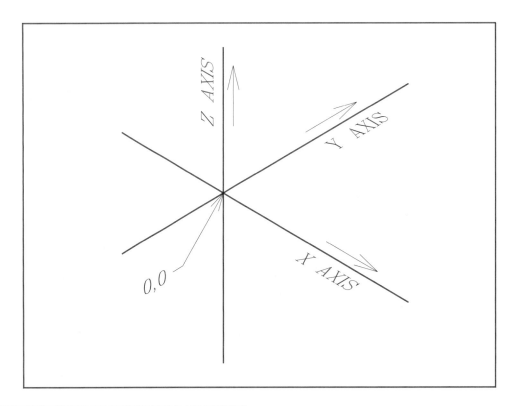

Figure 17-1 X, Y, and Z axes for 3D Design

MicroStation provides two types of 3D modeling: Surface and Solid. Surface modeling defines the edges of a 3D object in addition to surfaces, whereas solid models are the unambiguous and informationally complete representation of the shape of a physical object. Fundamentally, solid modeling differs from surface modeling in two ways:

- The information is more complete in the solid model.
- The method of construction of the model itself is inherently straightforward.

Using MicroStation's SmartSolids and SmartSurfaces tools you can quickly construct complex 3D models. With primitive solids or surfaces, you can add finishing touches, such as fillets and chamfers, and you can create a hollow solid with defined wall thickness with the ShellSolid tool.

This chapter provides an overview of the tools and specific commands available for 3D design.

Chapter 17 3D Design and Rendering

CREATING A 3D DESIGN FILE

The procedure for creating a new 3D design file is similar to that for creating a new 2D design file, except you have to use a seed file that is designed specifically for 3D. To create a new 3D design file, invoke the New tool from:

Drop-down menu	File > New
Key-in window	**create drawing** (or **cr d**) ENTER

MicroStation opens the New dialog box. Click the **Select** button, and MicroStation displays a list of seed files available, as shown in Figure 17–2. Select one of the 3D seed files from the **Files** list box and click the **OK** button. Key-in the name for your new 3D design file in the **Name** edit field and click the **OK** button. MicroStation highlights the name of the file you just created in the **Files** list box of the Create Design dialog box by default. To open the new design file, click the **OK** button and your screen will look similar to the one shown in Figure 17–3. By default, MicroStation displays View 2 with Isometric orientation as shown in Figure 17–3. As part of the title of the view window, MicroStation displays the name of the view being displayed.

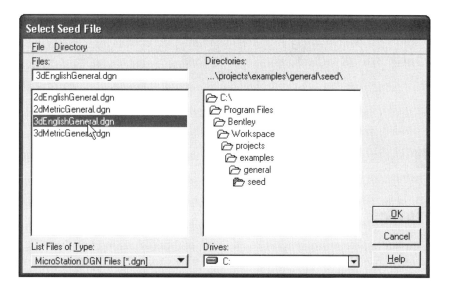

Figure 17-2 Select Seed File dialog box

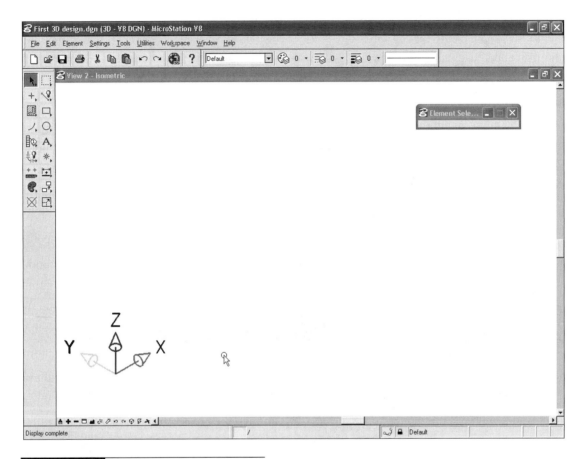

Figure 17-3 MicroStation screen display

VIEW ROTATION

There are seven standard view orientations defined in MicroStation: top, bottom, front, back, right, left, and isometric. You can use one of the four tools available in MicroStation to display the standard view orientation in any of the view windows.

Rotate View

The Rotate View tool displays one of the standard view orientations. In addition, it can rotate the view dynamically. Invoke the Rotate View tool from:

| View Control bar | Select the Rotate View tool (see Figure 17–4). |

Chapter 17 3D Design and Rendering

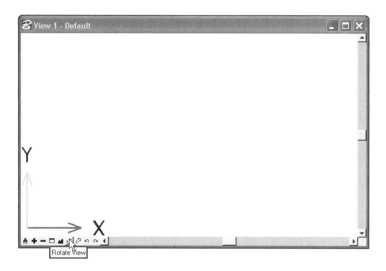

Figure 17-4 Invoking the Rotate View tool from the View Control bar

Select one of the ten available options from the **Method** option menu in the Tool Settings window. To position the view to one of the standard view orientations, first select the standard view orientation (Top, Bottom, Front, Back, Right, Left, Isometric [Top, Front, or Left] or Isometric [Top, Front, or Right]). Then place a data point in the view window where you want to display the view orientation. Selection of the Dynamic option allows you to position the view at any angle around a data point. The three points allow you to rotate the view by defining three data points (origin, direction of the X axis, and a point defining the Y axis).

Changing View Rotation from the View Rotation Settings Box

You can also rotate the view by specifying the angle of rotation in the View Rotation settings box. Open the View Rotation settings box from:

3D View Control tool box	Select the Change View Rotation tool (see Figure 17-5).
Key-in window	**dialog view rotation** (or **di viewro**) ENTER

Figure 17-5 Invoking the Change View Rotation tool from the 3D View Control tool box

MicroStation displays the View Rotation settings box, similar to Figure 17-6.

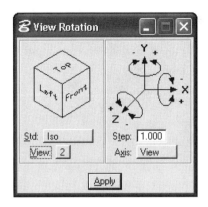

Figure 17-6 View Rotation settings box

The view to be manipulated is selected in the **View** option menu. Key-in the rotation increment in degrees in the **Step** edit field. Click the "+" control to rotate the view in the positive direction by the Step amount around the specified Axis. Click the "–" to rotate the view in the negative direction by the Step amount around the specified Axis. If you want to reposition the view to one of the standard view orientations, select the view orientation from the **Std** option menu. Click the **Apply** button to rotate the selected view to the specified rotation.

Rotating View by Key-in

You can also rotate the view via one of the three key-in tools from:

Key-in window	**vi =** <name of the view> ENTER **rotate view absolute=**<xx,yy,zz> ENTER or **rotate view relative=**<xx,yy,zz> ENTER

In the **VI** key-in, you can specify either one of the standard view rotations (top, bottom, front, back, right, left, or iso) or the name of the saved view.

In the Rotate View Absolute key-in, xx, yy, and zz are the rotations, in degrees, about the view X, Y, and Z axes (by default, 0 for each).

In the Rotate View Relative key-in, xx, yy, and zz are the relative, counter-clockwise rotations, in degrees, about the view X, Y, and Z axes. This key-in follows what is commonly called the "right-hand-rule." For example, if you key-in a positive X rotation and point your right thumb in the view's positive X direction, then the way your fingers curl is the direction of the rotation.

DESIGN CUBE

Whenever you start a new 2D design, you get a design plane—the electronic equivalent of a sheet of paper on a drafting table. The 2D design plane is a large, flat plane covered with an invisible matrix grid along the *X* and *Y* axes. In 3D design, you use that same *XY* plane plus a third-dimension *Z* axis. The *Z* axis is the depth in the direction perpendicular to the *XY* plane. The volume defined by *X*, *Y*, and *Z* is called the *design cube*. Similar to the design plane, the design cube is covered with an invisible matrix grid along each of the *X*, *Y*, and *Z* axes. The global origin (0,0,0) is at the very center of the design cube (see Figure 17–7).

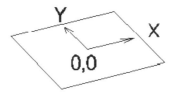

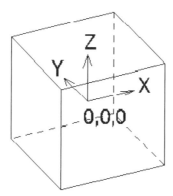

Figure 17-7 Design cube

DISPLAY DEPTH

Display depth enables you to display a portion of your design rather than the entire design. The ability to look at only a portion of the depth comes in handy—especially if the design is complicated. Display depth settings define the front and back clipping planes for elements displayed in a view and is set for each view. Elements not contained in the display depth do not show up on the screen. If you need to work with an element outside the display depth, you must change the display depth to include the element.

Setting Display Depth

To set the display depth, invoke the Set Display Depth tool from:

3D View Control tool box	Select the Set Display Depth tool (see Figure 17–8).
Key-in window	**depth display** (or **dep d**) ENTER

Figure 17-8 Invoking the Set Display Depth tool from the 3D View Control tool box

MicroStation prompts:

> Set Display Depth > Select view for display depth *(Place a data point in a view where you want to set the display depth.)*
>
> Set Display Depth > Define front clipping plane *(Place a data point in any view where you can identify the front clipping plane.)*
>
> Set Display Depth > Define back clipping plane *(Place a data point in any view where you can identify the back clipping plane.)*

You can also set the display depth by keying-in the distances in MU:SU:PU along the view Z axis by absolute or relative coordinates. To set the display depth by absolute coordinates, key-in:

Key-in window	**dp =** <front,back> ENTER

The <front,back> are the distances in working units along the view Z axis from the global origin to the desired front and back clipping planes. MicroStation prompts:

> Set Display Depth > Select view *(Identify the view with a data point to set the display depth.)*

To set the display depth by relative coordinates, key-in:

Key-in window	**dd =** <front,back> ENTER

The <front,back> are the distances in working units, and they add the keyed-in values to the current display depth settings. MicroStation prompts:

Chapter 17 3D Design and Rendering

Set Display Depth > Select view *(Identify the view with a data point to set the display depth.)*

To determine the current setting for display depth, invoke the Show Display Depth tool from:

3D View Control tool box	Select the Show Display Depth tool (see Figure 17–9).
Key-in window	**dp = $** (or **dd = $**) ENTER

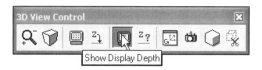

Figure 17-9 Invoking the Show Display Depth Tool from the 3D View Control tool box

MicroStation prompts:

Show Display Depth > Select view *(Place a data point anywhere in the view window.)*

MicroStation displays the current setting of the display depth in the Status bar.

 Note: When you are setting up the display depth in all views, you will notice dashed lines indicating the viewing parameters of the selected view. Both the display volume of the view and the active depth plane are dynamically displayed, with different-style dashed lines.

In Figure 17–10, the display depth is set in such a way that only the square box is displayed but not the circles.

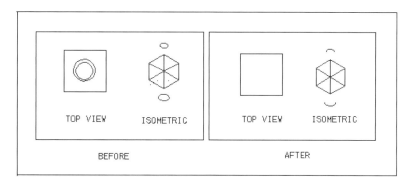

Figure 17-10 Example of setting up the display depth

755

Fitting Display Depth to Design File Elements

A fast way to display the entire design is to invoke the Fit View tool and select the appropriate view. In 2D design the Fit tool adjusts the view window to include all elements in the design file. Similarly, in 3D design the Fit View tool adjusts both the view window and the display depth to include all the elements in the design file. The Fit View tool automatically resets the display depth to the required amount to display the entire design file.

ACTIVE DEPTH

MicroStation has a feature that allows you to place an element in front of or behind the *XY* plane (front and back views), to the left or right of the *YZ* plane (right and left views), and above or below the *XY* plane (top and bottom views). This can be done by setting up the *active depth*. The active depth is a plane, parallel to the screen in each view, where elements will be placed by default. Each view has its own active depth plane, which you can change at any time.

Elements are placed at the active depth if you don't either tentatively snap to an existing element or use a precision input. In the top and bottom views (*XY* plane), the depth value is along its *Z* axis. In the front and back views (*XZ* plane), the depth value is along its *Y* axis. And in the case of right and left views (*YZ* plane), the depth value is along its *X* axis.

Setting Active Depth

To set the active depth, invoke the Set Active Depth tool from:

3D View Control tool box	Select the Set Active Depth tool (see Figure 17–11).
Key-in window	**depth active** (or **dep a**) ENTER

Figure 17–11 Invoking the Set Active Depth tool from the 3D View Control tool box

MicroStation prompts:

> Set Active Depth > Select view *(Place a data point in a view where you want to set the active depth.)*
>
> Set Active Depth > Enter active depth point *(Place a data point in a different view where you can identify the location for setting up the active depth.)*

Chapter 17 3D Design and Rendering

 Note: When you are setting up the active depth in all views, you will notice dashed lines indicating the viewing parameters of the selected view. Both the display volume of the view and the active depth plane are dynamically displayed, with different-style dashed lines.

You can also set the active depth by keying-in the distances in MU:SU:PU along the view Z axis by absolute or relative coordinates. To set the active depth by absolute coordinates, key-in:

| Key-in window | **az** = <depth> ENTER |

The <depth> is the distance in working units along the view Z axis from the global origin to the desired active depth. MicroStation prompts:

>   Set Active Depth > Select view *(Identify the view with a data point to set the active depth.)*

To set the active depth by relative coordinates, key-in:

| Key-in window | **dz** = <depth> ENTER |

The <depth> is the distance in working units, and it adds the keyed-in values to the current active depth setting. MicroStation prompts:

>   Set Active Depth > Select view *(Identify the view with a data point to set the active depth.)*

To determine the current setting for active depth, invoke the Show Active Depth tool from:

3D View Control tool box	Select the Show Active Depth tool (see Figure 17–12).
Key-in window	**az** = **$** (or **dz** = **$**) ENTER

Figure 17–12 Invoking the Show Active Depth tool in the 3D View Control tool box

MicroStation prompts:

>   Show Active Depth > Select view *(Place a data point anywhere in the view window.)*

MicroStation displays the current setting of the active depth in the Status bar.

757

Note: Before you place elements, make sure you are working at the appropriate active depth and display depth.

If you set the active depth outside the range of the display depth, then MicroStation displays the following message in the error field:

Active depth set to display depth

The active depth is set to the value closest to the display depth.

Let's say the current display depth is set to 100,450 and you set the active depth to 525. Since MicroStation sets the active depth to the closest value, in this case the active depth is set to 450. Make sure that MicroStation sets the value for the active depth to the value intended. If necessary, change the display depth and then set the active depth.

Boresite Lock

The Boresite lock controls the manipulation of the elements at different depths. If the Boresite lock is set to ON, you can identify or snap to elements at any depth in the view; elements being moved or copied will remain at their original depths. If it is set to OFF, you can identify only those elements at, or very near, the active depth of a view. You can turn ON or OFF for the Boresite lock from the Lock Toggles settings box, as shown in Figure 17–13.

Figure 17-13 Lock Toggles settings box

Note: Tentative points override the Boresite lock. You can tentatively snap to elements at any depth regardless of the Boresite lock setting.

PRECISION INPUTS

When MicroStation prompts for the location of a point, in addition to providing the data point with your pointing device, you can use precision input tools that allow you to place data points

Chapter 17 3D Design and Rendering

precisely. Similar to 2D placement tools, 3D tools also allow you to key-in by coordinates. MicroStation provides two types of coordinate systems for 3D design: the drawing coordinate system and the view coordinate system.

Drawing Coordinate System

The drawing coordinate system is the model coordinate system fixed relative to the design cube, as shown in Figure 17–14. For example, in the TOP view, X is to the right, Y is up, and Z is out of the screen (right-hand rule). In the RIGHT view, Y is to the right, Z is up, and X is out of the screen, and so on. Following are the two key-ins available for the drawing coordinate system:

XY=<X,Y,Z>

DL=<delta_x,delta_y,delta_z>

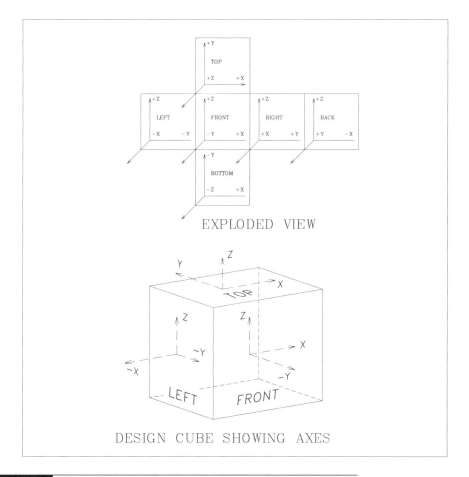

Figure 17–14 Design cube showing the drawing coordinate system

The XY= key-in places a data point measured from the global origin of the drawing coordinate system. The <X,Y,Z> are the X, Y, and Z values of the coordinates. The view being used at the time has no effect on them. The DL= places a data point to a distance along the drawing axes from a previous data (relative) or tentative point. The <delta_x,delta_y,delta_z> are the relative coordinates in the X, Y, and Z axes relative to the cube previous data point or tentative point.

View Coordinate System

The view coordinate system inputs data relative to the screen, where X is to the right, Y is up, and Z comes directly out from the screen in all views, as shown in Figure 17–15. The view coordinate system is view-dependent, that is, depends on the orientation of the view for their direction. Following are the two key-ins available for the view coordinate system:

DX=<delta_x,delta_y,delta_z>

DI=<distance,direction>

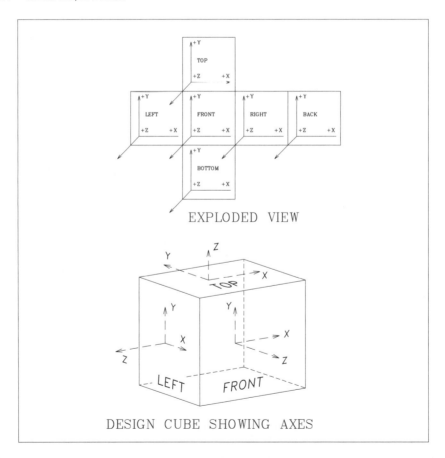

Figure 17-15 Design cube showing the view coordinate system

The DX= key-in places a data point to a distance from the previous data or tentative point (relative) in the same view where the previous point was defined. The <delta_x,delta_y,delta_z> are the relative coordinates in the *X*, *Y*, and *Z* axes relative to the previous data point or tentative point. The DI= key-in places a data point a certain distance and direction from a previous data or tentative point (relative polar) in the same view where the previous point was defined. The <distance,direction> are specified in relation to the last specified position or point. The distance is specified in current working units, and the direction is specified as an angle, in degrees, relative to the *X* axis.

Note: With key-in precision inputs, MicroStation assumes that the view you want to use is the one you last worked in—that is, the view in which the last tentative or data point was placed. The easiest way to make a view current is to place a tentative point and then press the Reset button. Updating a view is also another way to tell MicroStation that the selected view is the view last worked in.

AUXILIARY COORDINATE SYSTEMS (ACS)

MicroStation provides a set of tools to define an infinite number of user-defined coordinate systems called *auxiliary coordinate systems*. An auxiliary coordinate system allows the user to change the location and orientation of the *X*, *Y*, and *Z* axes to reduce the calculations needed to create 3D objects. You can redefine the origin in your drawing, and establish positive *X* and *Y* axes. New users think of a coordinate system simply as the direction of positive *X* and positive *Y*. But once the directions *X* and *Y* are defined, the direction of *Z* will be defined as well. Thus, the user only has to be concerned with *X* and *Y*. For example, if a sloped roof of a house is drawn in detail using the drawing coordinate system, each end point of each element on the inclined roof plane must be calculated. On the other hand, if the auxiliary coordinate system is set to the same plane as the roof, each object can be drawn as if it were in the plan view. You can define any number of auxiliary coordinate systems, assigning each a user-determined name. But, at any given time, only one auxiliary coordinate system is current with the default system.

MicroStation provides a visual reminder of how the ACS axes are oriented and where the current ACS origin is located. The *X*, *Y*, and *Z* axis directions are displayed using arrows labeled appropriately. The display of the ACS axes is controlled by turning ON/OFF the ACS Triad in the View Attributes settings box, as shown in Figure 17–16.

MicroStation provides you with three types of coordinate systems for defining an ACS: rectangular, cylindrical, and spherical coordinate systems.

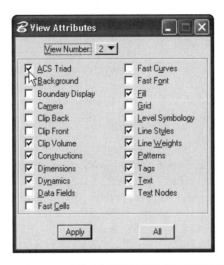

Figure 17-16 View Attributes settings box showing the ACS Triad set to ON

Rectangular Coordinate System

The rectangular coordinate system is the same one that is available for the design cube and is also the default type to define an ACS.

Cylindrical Coordinate System

The cylindrical coordinate system is another 3D variant of the polar format. It describes a point by its distance from the origin, its angle in the *XY* plane from the *X* axis, and its *Z* value. For example, to specify a point at a distance of 4.5 units from the origin, at an angle of 35 degrees relative to the *X* axis (in the *XY* plane), and with a *Z* coordinate of 7.5 units, you would enter: **4.5,35,7.5**.

Spherical Coordinate System

The spherical coordinate system is another 3D variant of the polar format. It describes a point by its distance from the current origin, its angle in the *XY* plane, and its angle up from the *XY* plane. For example, to specify a point at a distance of 7 units from the origin, at an angle of 60 degrees from the *X* axis (in the *XY* plane), and at an angle 45 degrees up from the *XY* plane, you would enter: **7,60,45**.

Precision Input Key-in

Similar to the key-ins available for drawing and view coordinates, MicroStation provides key-ins to input the coordinates in reference to the auxiliary coordinate system. Following are the two key-ins available for the auxiliary coordinate system:

AX=<X,Y,Z>
AD=<delta_x,delta_y,delta_z>

Chapter 17 3D Design and Rendering

The AX= key-in places a data point measured from the ACS origin and is equivalent to the key-in XY=. The <X,Y,Z> are the *X, Y,* and *Z* values of the coordinates. The AD= places a data point to a distance along the drawing axes from a previous data (relative) or tentative point and is equivalent to the key-in DL=. The <delta_x,delta_y,delta_z> are the relative coordinates in the *X, Y,* and *Z* axes relative to the previous data point or tentative point.

Defining an ACS

MicroStation provides three different tools to define an ACS. The tools are available in the ACS tool box. Before you select one of the three tools, select the coordinate system you wish to use with the new ACS from the **Type** option menu in the Tool Settings window. In addition, you can control the ON/OFF check box for two locks in the Tool Settings window. When the **ACS Plane Lock** is set to ON, each data point is forced to lie on the active ACS's *XY* plane (Z=0). When the **ACS Plane Snap** Lock is set to ON, each tentative point is forced to lie on the active ACS's *XY* plane (Z=0).

Defining an ACS by Aligning with an Element

This option lets you define an ACS by identifying an element where the *XY* plane of the ACS is parallel to the plane of the selected planar element. The origin of the ACS is at the point of identification of the element. Upon definition, the ACS becomes the active ACS.

To define an ACS aligned with an element, invoke the Define ACS (aligned with element) tool from:

ACS tool box	Select the Define ACS (Aligned with Element) tool (see Figure 17–17).
Key-in window	**define acs element** (or **de a e**) ENTER

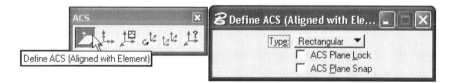

Figure 17-17 Invoking the Define ACS (Aligned with Element) tool from the ACS tool box

MicroStation prompts:

> Define ACS (Aligned with Element) > Identify element *(Identify the element with which to align the ACS and define the ACS origin.)*
>
> Define ACS (Aligned with Element) > Accept/Reject (Select next input) *(Place a data point to accept the element for defining an ACS or click the Reject button to cancel the operation.)*

Defining the ACS by Points

This option is the easiest and most used option for controlling the orientation of the ACS. It allows the user to place three data points to define the origin and the directions of the positive X and Y axes. The origin point acts as a base for the ACS rotation, and when a point is selected to define the direction of the positive X axis, the direction of the Y axis is limited because it is always perpendicular to the X axis. When the X and Y axes are defined, the Z axis is automatically placed perpendicular to the XY plane. Upon definition, the ACS becomes the active ACS.

To define an ACS by Points, invoke the Define ACS (By Points) tool from:

ACS tool box	Select the Define ACS (By Points) tool (see Figure 17–18).
Key-in window	**define acs points** (or **de a p**) ENTER

Figure 17–18 Invoking the Define ACS (By Points) tool from the ACS tool box

MicroStation prompts:

> Define ACS (By Points) > Enter first point @x axis origin *(Place a data point to define the origin.)*
>
> Define ACS (By Points) > Enter second point on X axis *(Place a data point to define the direction of the positive X axis, which extends from the origin through this point.)*
>
> Define ACS (By Points) > Enter point to define Y axis *(Place a data point to define the direction of the positive Y axis.)*

Defining the ACS by Aligning with a View

In this option, the ACS takes the orientation of the selected view. That is, the ACS axes align exactly with those of the view selected. Upon definition, the ACS becomes the active ACS.

To define an ACS by aligning with a view, invoke the Define ACS (Aligned with View) tool from:

ACS tool box	Select the Define ACS (Aligned with View) tool (see Figure 17–19).
Key-in window	**define acs view** (or **de a v**) ENTER

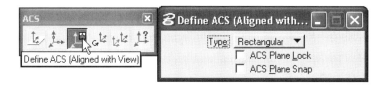

Figure 17-19 Invoking the Define ACS (Aligned with View) tool from the ACS tool box

MicroStation prompts:

> Define ACS (Aligned with View) > Select source view *(Place a data point to select the view with which the ACS is to be aligned and define the ACS origin.)*

Rotating the Active ACS

The Rotate Active ACS tool rotates the Active ACS. The origin of the ACS is not moved. To rotate the active ACS, invoke the Rotate Active ACS tool from:

| ACS tool box | Select the Rotate Active ACS tool (see Figure 17–20). |

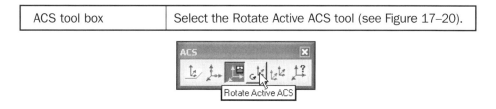

Figure 17-20 Invoking the Rotate Active ACS Tool from the ACS tool box

MicroStation displays the Rotate Active ACS dialog box, as shown in Figure 17–21.

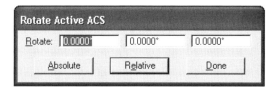

Figure 17-21 Rotate Active ACS dialog box

Key-in the rotation angles, in degrees, from left to right, for the X, Y, and Z axes. Click the **Absolute** button to rotate the ACS in relation to the unrotated (top) orientation. Click the **Relative** button to rotate the ACS in relation to the current orientation. When you are finished, click the **Done** button to close the Rotate Active ACS dialog box.

Moving the Active ACS

The Move ACS tool allows you to move the origin of the Active ACS, leaving the directions of the X, Y, and Z axes unchanged. To move the ACS, invoke the Move ACS tool from:

ACS tool box	Select the Move ACS tool (see Figure 17–22).
Key-in window	**move acs** (or **mov a**) ENTER

Figure 17-22 Invoking the Move ACS tool from the ACS tool box

MicroStation prompts:

> Move ACS > Define origin *(Place a data point to define the new origin.)*

Selecting the Active ACS

The Select ACS tool allows you to identify an ACS for attachment as the active ACS from the saved ACS in each view. To select the ACS, invoke the Select ACS tool from:

ACS tool box	Select the Select ACS tool (see Figure 17–23).

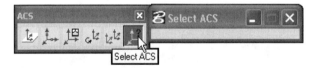

Figure 17-23 Invoking the Select ACS tool from the ACS tool box

MicroStation prompts:

> Select ACS > Select auxiliary system @ origin *(Identify the ACS origin from the coordinate triad displayed.)*

Saving an ACS

You can define any number of ACSs in a design file. Of these, only one can be active at any time. Whenever you define an ACS, you can save it for future use. The Auxiliary Coordinate Systems settings box is used to name, save, attach, or delete an ACS.

Open the Auxiliary Coordinate Systems settings box from:

Drop-down menu	Utilities > Auxiliary Coordinates

MicroStation displays an Auxiliary Coordinate Systems settings box, similar to Figure 17–24.

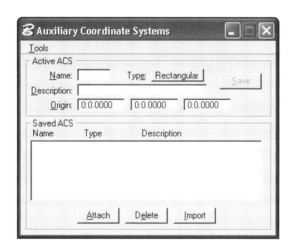

Figure 17-24 Auxiliary Coordinate Systems settings box

Key-in the name for the active ACS in the **Name** edit field. The name is limited to six characters. Select the coordinate system you wish to save with the active ACS from the **Type** option menu. Key-in the description (optional) of the active ACS in the **Description** edit field. The description is limited to 27 characters. If necessary, you can change the origin of the active ACS by keying-in the coordinates in the **Origin** edit field. Click the **Save** button to save the active ACS for future attachment. MicroStation will display the ACS name, type, and description in the Saved ACS list box.

To attach an ACS as an active ACS, select the name of the ACS from the **Saved ACS** list box and click the **Attach** button. To delete an ACS, first highlight the ACS in the Saved ACS list box, and then click the **Delete** button.

Note: All the tools available in the ACS tool box are also available in the Tools drop-down menu of the Auxiliary Coordinate Systems settings box.

3D PRIMITIVES

MicroStation provides a set of tools to place simple 3D elements that can become the basic building blocks that make up the model. The primitive tools include slab, sphere, cylinder, cone, torus, and wedge.

Two important settings need to be taken into consideration before the primitives are created that control the way in which solids and surfaces are created and displayed on the screen. The settings include: display method and number of rule lines that represent a surface with a full 360 degree curvature and selection of solids and surfaces.

Display Method and Surface Rule Lines

MicroStation provides two options: wireframe and surfaces, for on-screen display of the solids and surfaces. By default, it is set to wireframe, which is the more efficient mode for working with solids and surfaces in a design session. The surfaces display mode should be used only where the design is to be rendered with an earlier version of MicroStation.

Surface rule lines provide a visual indication of a surface's curvature. By default, it is set to 4—a full cylindrical solid is displayed with 4 surface rules lines. If necessary, you can increase the setting to display with additional surface rule lines.

To change the display method and set the surface rule lines, open the Change SmartSolid Settings box from:

| 3D Utility tool box | Select the Change SmartSolid Display tool (see Figure 17-25) |

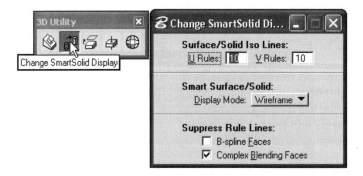

Figure 17-25 Invoking the Change SmartSolid Display tool from the 3D Utility tool box

The **Surface/Solid Iso Lines** section of the Tool Settings window sets the number of rule lines that represent a full 360° of curvature of curved surfaces for SmartSolids and SmartSurfaces. The **Smart Surface/Solid** section lets you set the Display Mode for SmartSurfaces and SmartSolids. The **Suppress Rule Lines** section lets you suppress, or turn off, the display of rule lines for particular faces on SmartSolids and SmartSurfaces.

Selection of Solids and Surfaces

MicroStation provides three options in the selection of solids and surfaces. By default, surfaces and solids may be identified with a data point anywhere on their surface, not necessarily on an edge line or surface rule line. If necessary, you can select the option that will allow you to select solids and surfaces with a data point on an edge or surface rule line. To change the selection mode, open the Preferences dialog box from:

Drop-down menu	Workspace > Preferences

MicroStation displays the Preferences dialog box Select Input category from the **Category** list box. The **Locate By Faces**: option menu provides the following three selection modes in the selection of solids and surfaces:

- **Never**—Solids and surfaces can only be identified with a data point on an edge or surface rule line.
- **Rendered Views Only**—Solids and surfaces rendered with any of the rendering options may be identified with a data point anywhere on their surface.
- **Always**—Solids and surfaces whether rendered or not may be identified with a data point anywhere on their surface.

Place Slab

The Place Slab tool places a volume of projection with a rectangular cross section. To place a slab, invoke the Place Slab tool from:

3D Primitives tool box	Select the Place Slab tool (see Figure 17–26).
Key-in window	**place slab** (or **pl sl**) ENTER

Figure 17-26 Invoking the Place Slab tool from the 3D Primitives tool box

Select the type of surface from the **Type** option menu in the Tool Settings window. The surface (not capped) option is considered to be open at the base and top, whereas the solid (capped) option is considered to enclose a volume completely.

From the **Axis:** option menu in the Tool Settings window, select the direction in which the height is projected relative to the view or design file axes. If set to Screen X, Screen Y, or Screen Z, the height is projected with the selected screen (view) axis. If set to Drawing X, Drawing Y, or Drawing Z, the height is projected with the selected design file axis.

If necessary, turn ON the check boxes for **Orthogonal**, **Length**, **Width**, and **Height** in the Tool Settings window. If **Orthogonal** is set to ON, the edges are placed orthogonally. If you turn on the constraints for **Length**, **Width**, and **Height**, make sure to key-in appropriate values in the edit fields.

MicroStation prompts:

> Place Slab > Enter start point *(Place a data point or key-in coordinates to define the origin.)*
>
> Place Slab > Define Length *(Place a data point or key-in coordinates to define the length and rotation angle. If the Length constraint is set to ON, this data point defines the rotation angle.)*
>
> Place Slab > Define Width *(Place a data point or key-in coordinates to define the width. If the Width constraint is set to ON, this data point accepts the width.)*
>
> Place Slab > Define Height *(Place a data point or key-in coordinates to define the height. If the Height constraint is set to ON, this data point provides the direction.)*

Note: To place a volume of projection with a nonrectangular cross section, use the Extrude tool in the 3D Construct tool box.

Place Sphere

The Place Sphere tool can place a sphere, in which all surface points are equidistant from the center. To place a sphere, invoke the Place Sphere tool from:

3D Primitives tool box	Select the Place Sphere tool (see Figure 17–27).
Key-in window	**place sphere** (or **pl sp**) ENTER

Figure 17–27 Invoking the Place Sphere tool from the 3D Primitives tool box

Select the type of surface from the **Type** option menu in the Tool Settings window.

From the **Axis** option menu in the Tool Settings window, select the direction of the sphere's axis relative to the view or design file axes. If set to Screen X, Screen Y, or Screen Z, the sphere's axis is set with the selected screen (view) axis. If set to Drawing X, Drawing Y, or Drawing Z, the sphere's axis is set with the selected design file axis.

If necessary, turn ON the check box for **Radius** constraint, and key-in the radius in the **Radius** edit field.

MicroStation prompts:

> Place Sphere > Enter center point *(Place a data point or key-in coordinates to define the sphere's center.)*
>
> Place Sphere > Define radius *(Place a data point or key-in coordinates to define the radius. If Radius is set to ON, then the data point accepts the sphere.)*

 Note: To place a volume of revolution with a noncircular cross section, use the Extrude tool in the 3D Construct tool box.

Place Cylinder

The Place Cylinder tool places a cylinder of equal radius on each end and similar to an extruded circle. To place a cylinder, invoke the Place Cylinder tool from:

3D Primitives tool box	Select the Place Cylinder tool (see Figure 17–28).
Key-in window	**place cylinder** (or **pl cy**) ENTER

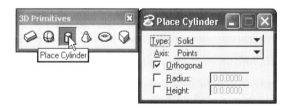

Figure 17-28 Invoking the Place Cylinder tool from the 3D Primitives tool box

Select the type of surface from the **Type** option menu in the Tool Settings window.

From the **Axis** option menu in the Tool Settings window, select the direction of the cylinder's axis or its height relative to the view or design file axes. If set to Screen X, Screen Y, or Screen Z, the direction of the cylinder's axis or height is set with the selected screen (view) axis. If set to Drawing X, Drawing Y, or Drawing Z, the direction of the cylinder's axis or height is set with the selected design file axis.

If necessary, turn ON the check boxes for **Orthogonal**, **Radius**, and **Height** in the Tool Settings window. If **Orthogonal** is set to ON, the cylinder is a right cylinder. If you turn on the constraints for **Radius** and **Height**, make sure to key-in appropriate values in the edit fields.

MicroStation prompts:

> Place Cylinder > Enter center point *(Place a data point or key-in coordinates to define the center of the base.)*
>
> Place Cylinder > Define radius *(Place a data point or key-in coordinates to define the radius. If Radius is set to ON, then the data point accepts the base.)*
>
> Place Cylinder > Define height *(Place a data point or key-in coordinates to define the height. If Height is set to ON, then the data point accepts the cylinder.)*

Place Cone

The Place Cone tool places a cone of unequal radius on each end. To place a cone, invoke the Place Cone tool from:

3D Primitives tool box	Select the Place Cone tool (see Figure 17-29).
Key-in window	**place cone** (or **pl con**) ENTER

Figure 17-29 Invoking the Place Cone tool from the 3D Primitives tool box

Select the type of surface from the **Type** option menu in the Tool Settings window.

From the **Axis:** option menu in the Tool Settings window, select the direction of the cone's axis or its height relative to the view or design file axes. If set to Screen X, Screen Y, or Screen Z, the direction of the cone's axis or height is set with the selected screen (view) axis. If set to Drawing X, Drawing Y, or Drawing Z, the direction of the cone's axis or height is set with the selected design file axis.

If necessary, turn ON the check boxes for **Orthogonal**, **Top Radius**, **Base Radius**, and **Height** in the Tool Settings window. If **Orthogonal** is set to ON, the cone is a right cone. If you turn ON the constraints for **Top Radius**, **Base Radius**, and **Height**, make sure to key-in appropriate values in the edit fields.

Chapter 17 3D Design and Rendering

MicroStation prompts:

> Place Cone > Enter center point *(Place a data point or key-in coordinates to define the center of the base.)*
>
> Place Cone > Define radius *(Place a data point or key-in coordinates to define the base radius. If Base Radius is set to ON, then the data point accepts the base.)*
>
> Place Cone > Define height *(Place a data point or key-in coordinates to define the height and top's center. If Height is set to ON, then the data point defines the top's center; if Orthogonal is set to ON, then the data point defines the direction of the height only.)*
>
> Place Cone > Define top radius *(Place a data point or key-in coordinates to define the top radius. If Top Radius is set to ON, then the data point accepts the cone.)*

Place Torus

The Place Torus tool creates a solid or surface with a donut-like shape. To place a torus, invoke the Place Torus tool from:

3D Primitives tool box	Select the Place Torus tool (see Figure 17–30).
Key-in window	**place torus** (or **pl to**) ENTER

Figure 17-30 Invoking the Place Torus tool from the 3D Primitives tool box

Select the type of surface from the **Type** option menu in the Tool Settings window.

Select the direction of the axis of revolution relative to the view or design file axes from the **Axis** option menu in the Tool Settings window. If set to Screen X, Screen Y, or Screen Z, the axis or revolution is set with the selected screen (view) axis. If set to Drawing X, Drawing Y, or Drawing Z, the axis of revolution is set with the selected design file axis.

If necessary, turn ON the check boxes for **Primary Radius**, **Secondary Radius**, and **Angle** in the Tool Settings window. If you turn ON the constraints for **Primary Radius**, **Secondary Radius**, and **Angle**, make sure to key-in appropriate values in the edit fields.

MicroStation prompts:

> Place Torus > Enter start point *(Place a data point or key-in coordinates to define the start point.)*
>
> Place Torus > Define center point *(Place a data point or key-in coordinates to define the center point, primary radius, and start angle. If Primary Radius is set to ON, then the data point defines the center and the start angle.)*
>
> Place Torus > Define angle and secondary radius *(Place a data point or key-in coordinates to define the secondary radius and the sweep angle. If Secondary Radius is set to ON, then the data point defines the sweep angle; if Angle is set to ON, then the data point defines the secondary radius; and if both Secondary Radius and Angle are set to ON, then the data point defines the direction of the sweep angle rotation.)*

Place Wedge

The Place Wedge tool creates a wedge—a volume of revolution with a rectangular cross section. To place a wedge, invoke the Place Wedge tool from:

3D Primitives tool box	Select the Place Wedge tool (see Figure 17–31).
Key-in window	**place wedge** (or **pl w**) ENTER

Figure 17-31 Invoking the Place Wedge tool from the 3D Primitives tool box

Select the type of surface from the **Type** option menu in the Tool Settings window.

Select the direction of the axis of revolution relative to the view or design file axes from the **Axis** option menu in the Tool Settings window. If set to Screen X, Screen Y, or Screen Z, the axis of revolution is set with the selected screen (view) axis. If set to Drawing X, Drawing Y, or Drawing Z, the axis of revolution is set with the selected design file axis.

If necessary, turn ON the check boxes for **Radius**, **Angle**, and **Height** in the Tool Settings window. If you turn ON the constraints for **Radius**, **Angle**, and **Height**, make sure to key-in appropriate values in the edit fields.

Chapter 17 3D Design and Rendering

MicroStation prompts:

> Place Wedge > Enter start point *(Place a data point or key-in coordinates to define the start point.)*
>
> Place Wedge > Define center point *(Place a data point or key-in coordinates to define the center point and the start angle. If Radius is set to ON, then the data point defines the start angle.)*
>
> Place Wedge > Define angle *(Place a data point or key-in coordinates to define the sweep angle. If Angle is set to ON, then the data point defines the direction of the rotation.)*
>
> Place Wedge > Define Height *(Place a data point or key-in coordinates to define the height. If Height is set to ON, then the data point defines whether the wedge is projected up or down from the start plane.)*

CHANGING THE STATUS—SOLID OR SURFACE

The Convert 3D tool can change the status of an element from surface to solid, or vice versa. To change the status, invoke the Convert 3D tool from:

Modify Surfaces tool box	Select the Convert 3D tool (see Figure 17–32).
Key-in window	**change surface cap** (or **chan su c**) ENTER

Figure 17–32 Invoking the Convert 3D tool from the Modify Surfaces tool box

Select the type of surface whose status you wish to change from the **Convert To** option menu in the Tool Settings window.

MicroStation prompts:

> Convert 3D > Identify solid or surface *(Identify the element whose status is to change.)*
>
> Convert 3D > Accept/Reject *(Place a data point to accept the change in status, or click the Reject button to reject the operation.)*

775

USING ACCUDRAW IN 3D

AccuDraw 3D provides the ability to work in a pictorial view rather than the standard orthogonal views. AccuDraw automatically constrains data points to its drawing plane regardless of its orientation to the view.

Open the AccuDraw window from:

Primary Tools tool box	Select the Toggle AccuDraw tool (see Figure 17–33).
Key-in window	**accudraw activate** (or **acc a**) ENTER

Figure 17–33 Invoking the Toggle AccuDraw tool from the Primary Tools tool box

The AccuDraw window opens, either as a floating window or docked at the top of the MicroStation workspace.

In 3D, when using rectangular coordinates, the AccuDraw window has an additional field for the Z axis. For polar coordinates in 3D, the AccuDraw window has the same two fields as in 2D.

By using AccuDraw keyboard shortcuts you can rotate the drawing plane axes, making it convenient to draw in an isometric view. For example, it is easy with AccuDraw to place a nonplanar complex chain or complex shape in an isometric view in any direction without reverting to an orthogonal view.

AccuDraw's ability to adhere to the standard view axes while manipulating your drawing in a pictorial view is so important that it maintains the current orientation from tool to tool.

By default, AccuDraw orients the drawing plane to the view axes, similar to working with 2D design. You can return AccuDraw to this orientation any time the focus is in the AccuDraw window by pressing the **V** key.

To rotate the drawing plane axes to align with the standard top view, focus in the AccuDraw window and press the **T** key. AccuDraw dynamically rotates the compass to indicate the orientation of the drawing plane.

To rotate the drawing plane axes to align with the standard front view, focus in the AccuDraw window and press the **F** key. AccuDraw dynamically rotates the compass to indicate the orientation of the drawing plane.

To rotate the drawing plane axes to align with the standard side (left or right) view, focus in the AccuDraw window and press the **S** key. AccuDraw dynamically rotates the compass to indicate the orientation of the drawing plane.

To rotate the drawing plane axes 90 degrees about an individual axis, focus in the AccuDraw window and press letters **R** and **X** to rotate 90 degrees about the *X* axis, **R** and **Y** to rotate 90 degrees about the *Y* axis, and **R** and **Z** to rotate 90 degrees about the *Z* axis.

To rotate the drawing plane axes interactively, focus in the AccuDraw window and press letters **R** and **A**. Place data points to locate the *X* axis origin, the direction of the *X* axis, and the direction of the *Y* axis.

PROJECTED SURFACES

The Extrude tool creates a unique 3D object from 2D elements. Line, line string, arc, ellipse, text, multi-line, complex chain, complex shape, and B-spline curve are the elements that can be projected to a defined distance. Surfaces formed between the original boundary element and its projection are indicated by straight lines connecting the keypoints.

To project a boundary element, invoke the Extrude tool from:

3D Construct tool box	Select the Extrude tool (see Figure 17–34).
Key-in window	**construct surface projection** (or **constru su p**) ENTER

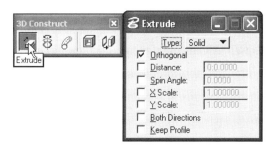

Figure 17–34 Invoking the Extrude tool from the 3D Construct tool box

Select appropriate options in the Tool Settings window. Select the type of surface from the **Type** option menu. When the **Orthogonal** check box is set to ON, the profile element is extruded orthogonally. The **Distance** check box, when set to ON, sets the distance in working units the element is extruded. The **Spin Angle** check box, when set to ON, sets the spin angle. The **X Scale** and **Y Scale** check boxes, when set to ON, sets the scale factor in the x-direction and y-direction respectively. When the **Both Directions** check box is set to ON, the profile element is extruded in both directions. When the **Keep Profile** check box is set to ON, the original profile element is kept in the design.

MicroStation prompts:

> Extrude > Identify profile *(Identify the boundary element.)*
>
> Extrude > Define distance *(Place a data point to define the height. If Distance is set to ON, then the data point provides the direction.)*

EXTRUDE ALONG A PATH

The Extrude Along Path tool creates a tubular surface or solid extrusion along a path. Line, line string, arc, ellipse, text, multi-line, complex chain, complex shape, and B-spline curve are the elements that can be projected along a path. Straight lines connecting the keypoints indicate surfaces formed between the original boundary element and its projection.

To project a boundary element along a path, invoke the Extrude Along Path tool from:

3D Construct tool box	Select the Extrude Along Path tool (see Figure 17–35).
Key-in window	**construct extrude along** (or **constru e a**) ENTER

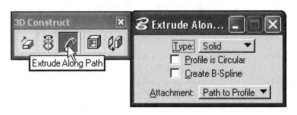

Figure 17–35 Invoking the Extrude Along Path tool from the 3D Construct tool box

Select the type of surface from the **Type** option menu in the Tool Settings window.

When the **Profile is Circular** check box is set to OFF, the surface/solid is constructed by extruding one element (the profile) along another element (the path). Orientation of the profile changes continually to follow the orientation of the path. If it is set to ON, the Tool Settings window expands to display **Inside** and **Outside** radius settings for a tube with circular cross-section to be generated.

When the **Create B-Spline** check box is set ON, a B-Spline surface or solid is created.

The **Attachment** option (available only when **Profile is Circular** check box is set to OFF) sets the attachment mode for the profile to the path.

MicroStation prompts:

> Extrude Along Path > Identify path *(Identify the boundary element.)*
>
> Extrude Along Path > Identify profile or snap to profile at attachment point *(Identify profile or snap to profile at attachment point)*
>
> Extrude Along Path > Accept to create *(Click the Accept button to create the profile.)*

Chapter 17 3D Design and Rendering

SURFACE OF REVOLUTION

The Construct Revolution tool is used to create a unique 3D surface or solid of revolution that is generated by rotating a boundary element about an axis of revolution. Line, line string, arc, ellipse, shape, complex chain, complex shape, and B-spline curve are the elements that can be used in creating a 3D surface or solid. Surfaces created by the boundary element as it is rotated are indicated by arcs connecting the keypoints.

To create a 3D surface or solid of revolution, invoke the Construct Revolution tool:

3D Construct tool box	Select the Construct Revolution tool (see Figure 17–36).
Key-in window	**construct surface revolution** (or **constru su rev**) ENTER

Figure 17-36 Invoking the Construct Revolution tool from the 3D Construct tool box

Select the type of surface from the **Type** option menu in the Tool Settings window.

Select the direction of the axis of revolution relative to the view or design file axes from the **Axis** option menu in the Tool Settings window. If set to Screen X, Screen Y, or Screen Z, the axis of revolution is set with the selected screen (view) axis. If set to Drawing X, Drawing Y, or Drawing Z, the axis of revolution is set with the selected design file axis.

If necessary, turn ON the check boxes for **Angle** and **Keep Profile** in the Tool Settings window. If you turn ON the constraint for **Angle**, make sure to key-in the appropriate value in the edit field.

MicroStation prompts:

> Construct Revolution > Identify profile *(Identify the boundary element.)*
>
> Construct Revolution > Define axis of revolution *(Place a data point or key-in coordinates. If Axis is set to Points, this data point defines one point on the axis of revolution and subsequently MicroStation prompts you for a second data point. If not, this data point defines the axis of revolution.)*
>
> Construct Revolution > Accept, continue surface/reset to finish *(Place additional data points to continue, and/or press the Reset button to terminate the sequence.)*

SHELL SOLID

The Shell Solid tool creates a hollowed out solid for one or more selected faces of a defined thickness.

To create a hollowed out solid, invoke the Shell Solid tool from:

3D Construct tool box	Select the Shell Solid tool (see Figure 17–37).
Key-in window	**construct shell** (or **constru s**) ENTER

Figure 17-37 Invoking the Shell Solid tool from the 3D Construct tool box

Specify the shell thickness in the **Shell Thickness** edit field in the Tool Settings window. Set the **Shell Outward** check box to OFF to create a hollowed out solid for one or more selected faces. If set to ON, the material is added to the outside and the original solid defines the inside of the walls.

MicroStation prompts:

Shell Solid > Identify target solid *(Identify the target solid.)*

Shell Solid > Identify face to open *(Move the screen pointer over the solid, the face nearest the pointer highlights and data point selects the highlighted face)*

Shell Solid > Accept/Reject (select next face) *(Select additional faces, click the Reset button to deselect an incorrect face; to accept the selection of faces, click the Accept button to complete the selection)*

THICKEN TO SOLID

The Thicken To Solid tool is used to add thickness to an existing surface to create a solid. You can specify the thickness by keying-in a distance or graphically.

To add thickness to an existing surface to create a solid, invoke the Thicken To Solid tool from:

3D Construct tool box	Select the Thicken To Solid tool (see Figure 17–38).
Key-in window	**construct thicken** (or **constru th**) ENTER

Figure 17-38 Invoking the Thicken To Solid tool from the 3D Construct tool box

Chapter 17 3D Design and Rendering

If you need to add the thickness to both sides of the selected surface, set the **Add To Both Sides** check box to ON. To increase the thickness to a specific value, key-in the values in the **Thickness** edit field and turn ON the check box. To keep the original profile element, turn ON the check box for **Keep Original**.

MicroStation prompts:

> Thicken to Solid > Identify surface *(Identify the surface.)*
>
> Thicken to Solid > Define thickness *(An arrow is displayed, move the pointer to the side you want to increase the thickness and click the Accept button)*

PLACING 2D ELEMENTS

Any 2D elements (such as blocks and circles) that you place with data points without snapping to existing elements will be placed at the active depth of the view. Also, they will be parallel to the screen. Elements that require fewer than three data points to define (such as blocks, circles with radius, circles with diameter/center, and polygons) take their orientation from the view being used. The points determine only their dimensions, not their orientation. Elements that require three or more data points to describe (shapes, circles by edge, ellipses, and rotated blocks) also provide their planar orientation. Once the first three points have been specified, any further points will fall on the same plane.

CREATING COMPOSITE SOLIDS

MicroStation provides three tools that can create a new composite solid by combining two solids by Boolean operations. There are three basic Boolean operations that can be performed in MicroStation:

- Union
- Intersection
- Difference

Union Operation

The union is the process of creating a new composite solid from two solids. The union operation joins the original solids in such a way that there is no duplication of volume. Therefore, the total resulting volume can be equal to or less than the sum of the volumes in the original solids. To create a composite solid with the union operation, invoke the Construct Union tool from:

3D Modify tool box	Select the Construct Union tool (see Figure 17–39).
Key-in window	**construct union** (or **constru u**) ENTER

Figure 17-39 Invoking the Construct Union tool from the 3D Modify tool box

The **Keep Originals** options menu determines whether or not the original solids are retained after constructing the solid. The **None** selection does not retain any of the originals, **All** selection retains all the original solids, **First** selection retains first solid identified, and **Last** selection retains last solid identified.

MicroStation prompts:

> Construct Union > Identify first solid *(Identify the first solid element for union.)*
> Construct Union > Identify next solid *(Identify the second solid element for union)*
> Construct Union > Identify next solid, or data point to finish *(Identify the third element for union or click the data point to accept the union of two selected solids.)*

See Figure 17–40 for an example of creating a composite solid by joining two cylinders with the Construct Union tool.

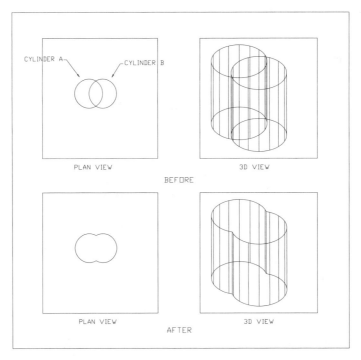

Figure 17-40 Creating a composite solid by joining two cylinders via the Construct Union tool

MicroStation also provides a tool to create a new composite solid from two surfaces. To combine two surfaces by union, invoke the Boolean Surface Union tool (available only by key-in) and select the surfaces to make it into a composite solid. The parts of the solids left are determined by their surface normal orientations. The surface normals can be changed by the Change Surface Normal tool.

Intersection Operation

The intersection is the process of forming a composite solid from only the volume that is common to two solids. To create a composite solid with the intersection operation, invoke the Construct Intersection tool from:

3D Modify tool box	Select the Construct Intersection tool (see Figure 17–41).
Key-in window	**construct intersection** (or **constru i**) ENTER

Figure 17–41 Invoking the Construct Intersection tool from the 3D Modify tool box

The **Keep Originals** options menu determines whether or not the original solids are retained after constructing the solid. The **None** selection does not retain any of the originals, **All** selection retains all the original solids, **First** selection retains first solid identified and **Last** selection retains last solid identified.

MicroStation prompts:

> Construct Intersection > Identify first solid *(Identify the first element for intersection.)*
>
> Construct Intersection > Identify next solid *(Identify the second element for intersection.)*
>
> Construct Intersection > Identify next solid, or data point to finish *(Identify the next element or click the data button to complete the selection)*

See Figure 17–42 for an example of creating a composite solid by joining two cylinders via the Construct Intersection tool.

MicroStation also provides a tool to create a new composite solid from two surfaces. To combine two surfaces by intersection, invoke the Boolean Surface Intersection tool (available only by key-in) and select the surfaces to make it to a composite solid. The parts of the solids left are determined by their surface normal orientations. The surface normals can be changed by the Change Surface Normal tool.

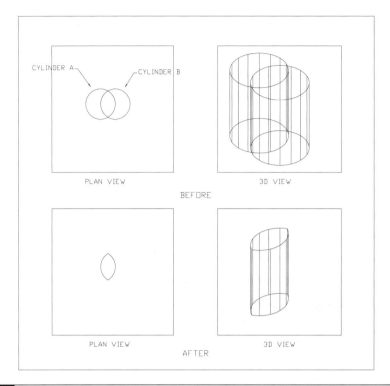

Figure 17-42 Creating a composite solid by intersecting two cylinders with the Construct Intersection tool

Difference Operation

The difference operation is the process of forming a composite solid by starting with a solid and removing from it any volume it has in common with a second object. If the entire volume of the second solid is contained in the first solid, then what is left is the first solid minus the volume of the second solid. However, if only part of the volume of the second solid is contained within the first solid, then only the part that is duplicated in the two solids is subtracted. To create a composite solid with the difference operation, invoke the Construct Difference tool from:

3D Modify tool box	Select the Construct Difference tool (see Figure 17–43).
Key-in window	**construct difference** (or **constru d**) ENTER

Figure 17-43 Invoking the Construct Difference tool from the 3D Modify tool box.

Chapter 17 3D Design and Rendering

The **Keep Originals** options menu determines whether or not the original solids are retained after constructing the solid. The **None** selection does not retain any of the originals, **All** selection retains all the original solids, **First** selection retains first solid identified and **Last** selection retains last solid identified.

MicroStation prompts:

> Construct Difference > Identify solid to subtract from *(Identify the first element for the difference operation.)*
>
> Construct Difference > Identify next solid or surface to subtract *(Identify the second solid or surface to subtract.)*
>
> Construct Difference > Identify next solid or surface to subtract, or data point to finish *(Identify the next element to subtract, or click the data button to complete the selection.)*

See Figure 17–44 for an example of creating a composite solid by joining two cylinders via the Construct Difference Between Surfaces tool.

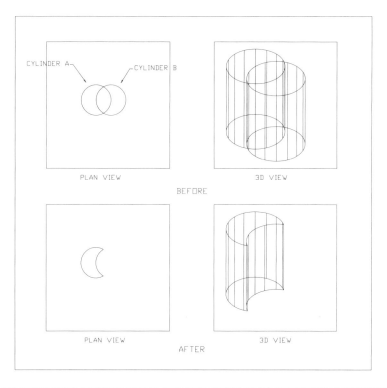

Figure 17-44 Creating a composite solid by subtracting Cylinder B from Cylinder A using the Construct Difference Between Surfaces tool

MicroStation also provides a tool to create a new composite solid from two surfaces. To combine two surfaces by the difference operation, invoke the Boolean Surface Difference tool (available only by key-in) and select the surfaces to make it into a composite solid. The parts of the solids left are determined by their surface normal orientations. The surface normals can be changed by Change Surface Normal tool.

CHANGE NORMAL

The Change Normal Direction tool can change the surface normal direction for a surface. This is useful to control the way the elements are treated while performing the Boolean operations.

To change the surface normal of an element, invoke the Change Normal Direction tool from:

Modify Surfaces tool box	Select the Change Normal Direction tool (see Figure 17–45).
Key-in window	**change surface normal** (or **chan su n**) ENTER

Figure 17-45 Invoking the Change Normal Direction tool from the Modify Surfaces tool box

MicroStation prompts:

> Change Surface Normal > Identify element *(Identify the element; surface normals are displayed.)*
>
> Change Surface Normal > Reverse normals, or RESET *(Place a data point to accept the change in the normal direction.)*

MODIFY SOLID

The Modify Solid tool lets you relocate a face of a solid outward (positive) or inward (negative), relative to the center of the solid. To modify a solid, invoke the Modify Solid tool from:

3D Modify tool box	Select the Modify Solid tool (see Figure 17–46).
Key-in window	**stretch faces** (or **str f**) ENTER

Figure 17-46 Invoking the Modify Solid tool from the 3D Modify tool box

Chapter 17 3D Design and Rendering

To modify the selected solid face by key-in distance, set the **Distance** check box to ON and key-in the distance in the **Distance** edit field in the Tool Settings window.

MicroStation prompts:

> Modify Solid > Identify target solid *(Identify the solid to modify.)*
> Modify Solid > Select face to modify *(Select the face to modify.)*
> Modify Solid > Define distance *(Using the arrow as the guide, move the pointer to define the distance dynamically when the Distance check box is set to OFF and direction.)*

REMOVE FACES

The Remove Faces tool lets you remove an existing face or a feature from a solid and then close the opening. It can also remove faces that are associated with a cut, a solid that has been added to or subtracted from the original, a shell solid, a fillet, or a chamfer. To remove a face or a feature, invoke the Remove Faces tool from:

3D Modify tool box	Select the Remove Faces tool (see Figure 17–47).
Key-in window	**remove faces** (or **rem f**) ENTER

Figure 17–47 Invoking the Remove Faces tool from the 3D Modify tool box

Select one of the two available methods to remove faces from the **Method** option menu in the Tool Settings window. The **Faces** selection allows you to remove one or more faces from a selected solid feature. The **Logical Groups** selection allows you to remove faces that are associated with a cut, a solid that has been added to or subtracted from the original, a shell solid, a fillet, or a chamfer. When the **Add Smooth Faces** check box is set to ON, any tangentially continuous faces are included with the selected face. If set to OFF, only the selected face is considered.

MicroStation prompts:

> Remove Faces and Heal > Identify target solid *(Identify the solid to modify.)*
> Remove Faces and Heal > Identify first face to remove *(Select the face to remove.)*
> Remove Faces and Heal > Accept/Reject (select next face) *(Click the Accept button to remove the selected face, if any select additional faces to continue the selection or click the Reject button to cancel the operation.)*

CUT SOLID

The Cut Solid tool lets you split a solid into two or more segments using a cutting profile. Cutting profiles may be open or closed elements. The open element profile must extend to the edge of the solid. To cut a solid, invoke the Cut Solid tool from:

3D Modify tool box	Select the Cut Solid tool (see Figure 17–48).
Key-in window	**construct cut** (or **constru cut**) ENTER

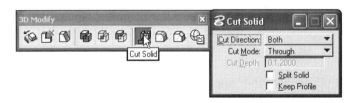

Figure 17-48 Invoking the Cut Solid tool from the 3D Modify tool box

The **Cut Direction** option menu in the Tool Settings window sets the direction of the cut relative to the cutting profile's Surface Normal. Available selections include:

- **Both**—Selects in both directions from the profile's plane.
- **Forward**—Selects from the forward direction from the profile's plane.
- **Back**—Selects from the backward direction from the profile's plane.

The **Cut Mode** option menu in the Tool Settings window sets the limits of the cut. Available selections include:

- **Through**—Cuts through all faces of the solid.
- **Define Depth**—Cuts into the solid a defined distance. Key-in the distance in the Cut Depth edit field.

The **Split Solid** check box controls whether the material is removed or not when it is split into segments. When it is set to ON, no material is removed from the solid, and it is split into two or more segments.

The **Keep Profile** check box controls whether the cutting profile remains in the design. When it is set to ON, the original cutting profile remains in the design.

MicroStation prompts:

Cut Solid > Identify target solid *(Identify the solid.)*
Cut Solid > Identify cutting profile *(Identify the cutting profile.)*
Cut Solid > Accept/Reject *(Click the Accept button to complete the cut or click the Reject button to cancel the operation.)*

CONSTRUCT FILLET

The Fillet Edges tool is used to fillet or round for one or more edges of a solid, projected surface, or a surface of revolution. To construct a fillet, invoke the Fillet Edges tool from:

3D Modify tool box	Select the Fillet Edges tool (see Figure 17–49).
Key-in window	**fillet edges** (or **fill e**) ENTER

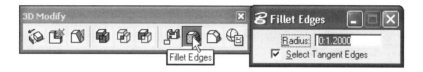

Figure 17-49 Invoking the Fillet Edges tool from the 3D Modify tool box

Key-in the radius for the fillet in the **Radius** edit field in the Tool Settings window. When the **Select Tangent Edges** check box is set to ON, the edges that are tangentially continuous are selected and rounded in one operation.

MicroStation prompts:

> Fillet Edges > Identify Edge to Fillet *(Identify the solid.)*
>
> Fillet Edges > Accept/Reject (select next edge) *(Click the Accept button to accept the fillet for the selected edge, and if necessary, select additional edges to fillet or click the Reject button to cancel the operation.)*

The Fillet Surfaces tool helps you construct a 3D fillet between two surfaces. The fillet is placed by sweeping an arc with a specified radius along the common intersecting curve. The fillet is created in the area pointed to by the surface normals of both surfaces.

To fillet between surfaces, invoke the Construct Fillet Between Surfaces tool from:

Fillet Surfaces tool box	Select the Fillet Surfaces tool (see Figure 17–50).
Key-in window	**fillet surface** (or **fill su**) ENTER

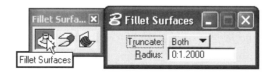

Figure 17-50 Invoking the Fillet Surfaces tool from the Fillet Surfaces tool box

Key-in the radius of the fillet to be drawn in the **Radius** edit field.

The **Truncate** option menu sets which surface(s) are truncated at the point of tangency with the fillet.

MicroStation prompts:

>Fillet Surfaces > Identify first surface *(Identify the first surface.)*
>
>Fillet Surfaces > Identify second surface *(Identify the second surface.)*
>
>Fillet Surfaces > Accept/Reject *(Click the Data button to accept the fillet, or click the Reject button to cancel the operation.)*

CONSTRUCT CHAMFER

The Chamfer Edges tool is used to chamfer one or more edges of a solid, projected surface, or a surface of revolution. To construct a chamfer, invoke the Chamfer Edges tool from:

3D Modify tool box	Select the Chamfer Edges tool (see Figure 17–51).
Key-in window	**chamfer edges** (or **ch e**) ENTER

Figure 17-51 Invoking the Chamfer Edges tool from the 3D Modify tool box

Key-in chamfer distances in the **Distance 1** and **Distance 2** edit fields. The **Select Tangent Edges** check box, when set to ON, makes edges that are tangentially continuous and are selected and rounded in one operation. The **Flip Direction** check box, when set to ON, reverses the direction of the chamfer and sets the values that the faces are trimmed.

MicroStation prompts:

>Chamfer Edges > Identify Edge to Chamfer *(Identify the solid.)*
>
>Fillet Edges > Accept/Reject (select next edge) *(Click the Accept button to accept the chamfer for the selected edge, if necessary, select additional edges to chamfer or click the Reject button to cancel the operation.)*

The Construct Chamfer Between Surfaces tool enables you to construct a 3D chamfer between two surfaces by a specified length along the common intersection curve. The chamfer is created in the area pointed to by the surface normals of both surfaces.

To chamfer between surfaces, invoke the Construct Chamfer Between Surfaces tool:

Key-in window	**chamfer surface** (or **cham s**) ENTER

The **Truncate** option menu sets which surface(s) are truncated at the point of tangency with the chamfer.

Key-in the Chamfer length in the **Chamfer Length** edit field.

The **Tolerance** check box sets the override for the system tolerance.

MicroStation prompts:

> Construct Chamfer Between Surfaces > Identify first surface *(Identify the first surface.)*
>
> Construct Chamfer Between Surfaces > Accept/Reject *(Click the Data button to accept the first surface selection.)*
>
> Construct Chamfer Between Surfaces > Identify second surface *(Identify the second surface.)*
>
> Construct Chamfer Between Surfaces > Accept/Reject *(Click the Data button to accept the second surface selection.)*
>
> Construct Fillet Between Surfaces > Accept/Reject *(Click the Data button to accept the chamfer, or click the Reject button to cancel the operation.)*

PLACING TEXT

MicroStation provides two options to place text in a 3D design: (1) placing text (view-dependent) in such a way that it appears planar to the screen in the view in which the data point is placed but rotated in the other views, or (2) placing text (view-independent) in such a way that it appears planar to the screen in all views.

To place text (view-dependent), click the Place Text icon in the Text tool box, select **By Origin** from the **Method** option menu, and follow the prompts. To place text (view-independent), click the Place Text icon in the Text tool box, select **View Ind** from the **Method** option menu, and follow the prompts. Text parameters are set up in the same way as in the 2D design.

FENCE MANIPULATIONS

Fences are used in a 3D design in much the same way as in a 2D design (see Chapter 6). The difference is that a 3D fence defines a volume. The volume is defined by the fenced area and the display depth of the view in which the fence is placed. The Fence lock options work the same way as in a 2D design.

CELL CREATION AND PLACEMENT

The procedure for creating and placing cells in 3D design is the same as in 2D design (see Chapter 11). Before you create a 3D cell, make sure the display depth is set to include all the elements to be used in the cell and the origin is defined at an appropriate active depth. If a normal cell was created in the top view and then placed in the front view, it will appear as it did in the top view and rotated in other views. In other words, the normal cell is placed as view-dependent, whereas a point cell when placed will appear planar to the screen in all views. A point cell is placed view-independent.

You can attach a 2D cell library to a 3D design file, but the cells will have no depth and will be placed at the active depth of the view in which you are working.

DIMENSIONING

The procedure for dimensioning setup and placement in 3D design is similar to that for 2D design (see Chapter 9). The main difference is that you have to consider on which plane you want the dimensioning to be located. Before you place dimensions in a 3D design, make sure the appropriate option is selected from the **Alignment** option menu in the Linear Dimension tool box. The view measurement axis measures the projection of the element along the view's horizontal or vertical axis. The true measurement axis measures the actual distance between two points, not the projected distance. And the drawing measurement axis measures the projection of an element along the design cube coordinate system's axis.

RENDERING

Shading, or rendering, can turn your 3D model into a realistic, eye-catching image. MicroStation's rendering options give you complete control over the appearance of your final images. You can add lights and control lighting in your design and also define the reflective qualities of individual surfaces in your design, making objects appear dull or shiny. You can create the rendered image of your 3D model entirely within MicroStation. This section provides an overview of the various options available for rendering. Refer to the *MicroStation Reference Guide* for a more detailed description of various options.

Setting Up Cameras

In establishing a viewing position in MicroStation, the assumption you must make is that you are, as it were, looking through a camera to see the image. By default, MicroStation places the camera at a right angle to a view's *XY* plane. If necessary, you can move or reposition the camera to view the model from a different viewing angle.

Chapter 17 3D Design and Rendering

To enable or disable the default camera setting and make changes to the camera setup, invoke the Camera Settings tool from:

| 3D View Control tool box | Select the Camera Settings tool (see Figure 17–52). |

Figure 17-52 Invoking the Camera Settings tool from the 3D View Control tool box

Select one of the available options from the **Camera Settings** option menu in the Tool Settings window.

- **Turn On**—Turns on the camera in a view or views.
- **Turn Off**—Turns off the camera in a view or views.
- **Set Up**—Turns on the camera in a view and sets the camera target and position. The target is the focal point (center) of a camera view. The position is the design cube location from which the model is viewed with the camera. Objects beyond the camera target appear smaller; objects in front of the camera target appear larger and may be outside of the viewing pyramid.
- **Move**—Moves the camera position.
- **Target**—Moves the target.

Select one of the available options from the **Image Plane Orientation** option menu in the Tool Settings window:

- **Perpendicular**—Perpendicular to the camera direction
- **Parallel to X axis**—Parallel to the view X axis; analogous to a bellows camera
- **Parallel to Y axis**—Parallel to the view Y axis; analogous to a bellows camera
- **Parallel to Z axis**—Parallel to the view Z axis; all vertical lines (along the axis) appear parallel

Set the lens angle in degrees and the lens focal length in millimeters in the **Angle** and **Focal Length** edit fields, respectively.

Select one of the available options from the **Standard Lens** option menu if you wish to use the standard lens type commonly used by photographers. MicroStation sets the appropriate lens angle and focal length.

MicroStation prompts depend on the options selected in the Tool Settings window.

Placement of Light Sources

The lighting setup is equally as important as setting the camera angle for producing a high-quality rendered image. MicroStation allows you four types of lighting:

- Ambient lighting
- Flashbulb lighting
- Solar lighting
- Source lighting, including point, spot, and distant

Ambient lighting is a uniform light that surrounds your model. *Flashbulb lighting* is a localized, intense light that appears to emanate from the camera position. *Solar lighting* is sunlight. By defining your location on the earth in latitude, longitude, day, month, and time, you can simulate lighting for most exterior architectural projects. Ambient, flashbulb, and solar lighting are set in the Global settings box invoked from the Rendering submenu of the drop-down Settings menu.

Source lighting is achieved by placing light sources in the form of cells. The MicroStation program comes with three light source cells (point lights—PNTLT, spot lights—SPOTLT, and distant lights—DISTLT) provided in the LIGHTING.CEL cell library.

Point light can be thought of as a ball of light. It radiates beams of light in all directions. Such a light also has more natural characteristics. Its brilliance may be diminished as an object moves away from the source of light. An object that is near a point light will appear brighter; an object that is farther away will appear darker.

Spot lights are very much like the kind of spotlight you might be accustomed to seeing at a theater or auditorium. Spot lights produce a cone of light toward a target that you specify.

Distant light gives off a fairly straight beam of light that radiates in one direction, and its brilliance remains constant so that an object close to the light will receive as much light as a distant object.

Before you place the light cells, adjust the settings in the Source Lighting settings box invoked from the Rendering submenu of the drop-down Settings menu. MicroStation provides various options under the Tool drop-down menu, in the Settings box.

Rendering Methods

MicroStation provides seven different tools to render a view. Depending on the needs and availability of the hardware, you can choose one of the seven tools to render the model.

Chapter 17　3D Design and Rendering

To render a view, invoke the Render tool from:

| 3D View Control tool box | Select the Render tool (see Figure 17–53). |

Figure 17-53 Invoking the Render tool from the 3D View Control tool box

Select the type of area or element to be rendered from the **Target** option menu in the Tool Settings window. The available options include View, Fence (contents), and Element.

Select one of the available options (Wiremesh, Hidden Line, Filled Hidden Line, Constant, Smooth, or Phong) from the **Render Mode** option menu in the Tool Settings window.

- **Wiremesh**—Similar to the default wireframe display, all elements are transparent and do not obscure other elements.

- **Hidden Line**—Displays only the element parts that would actually be visible.

- **Filled Hidden Line**—Identical to a Hidden Line option display except that the polygons are filled with the element color.

- **Constant**—Displays each element as one or more polygons filled with a single (constant) color. The color is computed once for each polygon, from the element color, material characteristics, and lighting configurations.

- **Smooth**—Displays the appearance of curved surfaces more realistically than in constant shaded models because polygon color is computed at polygon boundaries and color is blended across polygon interiors.

- **Phong**—Displays the image after re-computing the color of each pixel. Phong shading is useful for producing high-quality images when speed is not critical.

The **Shading Type** option menu sets the rendering method.

- **Anti-alias**—Displays the image with reduced jagged edges that are particularly noticeable on low-resolution displays. The additional time required for anti-aliasing is especially worthwhile when saving images for presentation, publication, or animated sequences.

- **Stereo**—Renders a view with a stereo effect that is visible when seen through 3D (red/blue) glasses. Stereo Phong shading takes twice as long as Phong because two images—one each from the perspective of the right and left eyes—are rendered and combined into one color-coded image.

DRAWING COMPOSITION

One of MicroStation's useful features is the ability to compose multiple views (standard and saved) on a drawing sheet. This will allow you to plot multiple views on one sheet of paper—what-you-see-is-what-you-get (WYSIWYG). Drawing Composition is a process by which you attach multiple views. The views are attached as references. An attached view in a sheet file can be any standard (top, bottom, right, left, front, back, or isometric), fitted view, or any saved view of a model file. Standard views can be clipped or set to display only certain levels. A view of the model file can be attached in any position at any scale. MicroStation provides a tool that allows you to group a set of views. A group of attached views can be moved, scaled, or detached as one. If necessary, you can remove or add a view to a group. The tools in the Reference attachments (for a detailed explanation about References, refer to Chapter 13) simplify the process of creating sheet views.

 Open the Exercise Manual PDF file for Chapter 17 on the accompanying CD for project and discipline specific exercises.

appendix A

MicroStation Tool Boxes

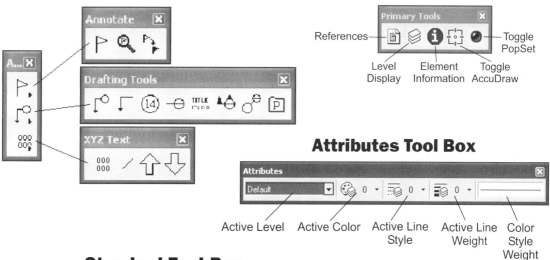

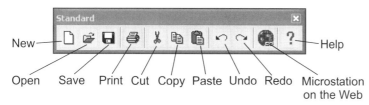

Main Tool Box

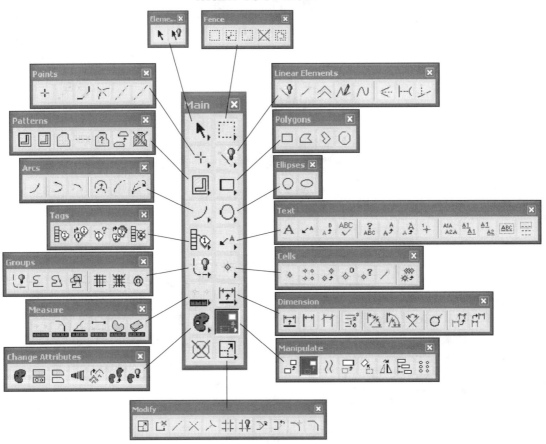

DD Design Tool Box

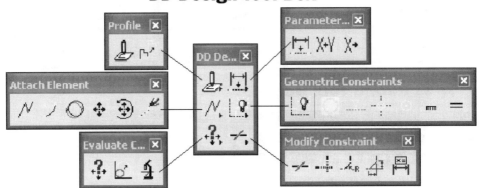

appendix B

Key-in Commands

 Note: MicroStation is not case-sensitive. That is, key-ins can be uppercase or lowercase, or even a mixture of the two, and MicroStation will still understand the key-ins in the same way.

TOOL NAME	KEY-IN
Add to Graphic Group	GROUP ADD
Attach Active Entity	ATTACH AE
Attach Active Entity to Fence Contents	FENCE ATTACH
Attach Displayable Attributes	ATTACH DA
Attach Reference File	REFERENCE ATTACH (RF=)
Automatic Create Complex Chain	CREATE CHAIN AUTOMATIC
Automatic Create Complex Shape	CREATE SHAPE AUTOMATIC
Automatic Fill in Enter Data Fields	EDIT AUTO
B-spline Polygon Display On/Off	MDL LOAD SPLINES; CHANGE BSPLINE POLYGON
Chamfer	CHAMFER

TOOL NAME	KEY-IN
Change B-spline Surface to Active U-Order	MDL LOAD SPLINES; CHANGE BSPLINE UORDER
Change B-spline Surface to Active U-Rules	MDL LOAD SPLINES; CHANGE BSPLINE URULES
Change B-spline Surface to Active V-Order	MDL LOAD SPLINES; CHANGE BSPLINE VORDER
Change B-spline Surface to Active V-Rules	MDL LOAD SPLINES; CHANGE BSPLINE VRULES
Change B-spline to Active Order	MDL LOAD SPLINES; CHANGE BSPLINE ORDER
Change Element to Active Class	CHANGE CLASS
Change Element to Active Color	CHANGE COLOR
Change Element to Active Level	CHANGE LEVEL
Change Element to Active Line Style	CHANGE STYLE
Change Element to Active Line Weight	CHANGE WEIGHT
Change Element to Active Symbol	CHANGE SYMBOLOGY
Change Fence Contents to Active Color	FENCE CHANGE COLOR
Change Fence Contents to Active Level	FENCE CHANGE LEVEL
Change Fence Contents to Active Style	FENCE CHANGE STYLE
Change Fence Contents to Active Symbology	FENCE CHANGE SYMBOLOGY
Change Fence Contents to Active Weight	FENCE CHANGE WEIGHT
Change Fill	CHANGE FILL
Change Text to Active Attributes	MODIFY TEXT
Circular Fillet (No Truncation)	FILLET NOMODIFY
Circular Fillet and Truncate Both	FILLET MODIFY
Circular Fillet and Truncate Single	FILLET SINGLE
Closed Cross Joint	MDL LOAD CUTTER; JOIN CROSS CLOSED
Closed Tee Joint	MDL LOAD CUTTER; JOIN TEE CLOSED

Appendix B Key-in Commands

TOOL NAME	KEY-IN
Complete Cycle Linear Pattern	PATTERN LINE SCALE
Construct Active Point at Distance Along an Element	CONSTRUCT POINT DISTANCE
Construct Active Point at Intersection	CONSTRUCT POINT INTERSECTION
Construct Active Points Between Data Points	CONSTRUCT POINT BETWEEN
Construct Angle Bisector	CONSTRUCT BISECTOR ANGLE
Construct Arc Tangent to Three Elements	CONSTRUCT TANGENT ARC 3
Construct B-spline Curve by Least Squares	MDL LOAD SPLINES; CONSTRUCT BSPLINE CURVE LEAST SQUARE
Construct B-spline Curve by Points	MDL LOAD SPLINES; CONSTRUCT BSPLINE CURVE POINTS
Construct B-spline Curve by Poles	MDL LOAD SPLINES; CONSTRUCT BSPLINE CURVE POLES
Construct B-spline Surface by Cross-Section	MDL LOAD SPINES; CONSTRUCT BSPLINE SURFACE CROSS
Construct B-spline Surface by Edges	MDL LOAD SPLINES; CONSTRUCT BSPLINE SURFACE EDGE
Construct B-spline Surface by Least Squares	MDL LOAD SPLINES; CONSTRUCT BSPLINE SURFACE LEAST SQUARE
Construct B-spline Surface by Points	MDL LOAD SPLINES; CONSTRUCT BSPLINE SURFACE POINTS
Construct B-spline Surface by Poles	MDL LOAD SPLINES; CONSTRUCT BSPLINE SURFACE POLES
Construct B-spline Surface by Skin	MDL LOAD SPLINES; CONSTRUCT BSPLINE SURFACE SKIN
Construct B-spline Surface by Tube	MDL LOAD SPLINES; CONSTRUCT BSPLINE SURFACE TUBE
Construct B-spline Surface of Projection	MDL LOAD SPLINES; CONSTRUCT BSPLINE SURFACE PROJECTION
Construct B-spline Surface of Revolution	MDL LOAD SPLINES; CONSTRUCT BSPLINE SURFACE REVOLUTION
Construct Circle Tangent to Element	CONSTRUCT TANGENT CIRCLE 1

TOOL NAME	KEY-IN
Construct Circle Tangent to Three Elements	CONSTRUCT TANGENT CIRCLE 3
Construct Line at Active Angle from Point (key-in)	CONSTRUCT LINE AA 4
Construct Line at Active Angle from Point	CONSTRUCT LINE AA 3
Construct Line at Active Angle to Point (key-in)	CONSTRUCT LINE AA 2
Construct Line at Active Angle to Point	CONSTRUCT LINE AA 1
Construct Line Bisector	CONSTRUCT BISECTOR LINE
Construct Line Tangent to Two Elements	CONSTRUCT TANGENT BETWEEN
Construct Minimum Distance Line	CONSTRUCT LINE MINIMUM
Construct Perpendicular from Element	CONSTRUCT PERPENDICULAR FROM
Construct Perpendicular to Element	CONSTRUCT PERPENDICULAR TO
Construct Points Along Element	CONSTRUCT POINT ALONG
Construct Surface/Solid of Projection	SURFACE PROJECTION
Construct Surface/Solid of Revolution	SURFACE REVOLUTION
Construct Tangent Arc by Keyed-in Radius	CONSTRUCT TANGENT ARC 1
Construct Tangent from Element	CONSTRUCT TANGENT FROM
Construct Tangent to Circular Element and Perpendicular to Linear Element	CONSTRUCT TANGENT PERPENDICULAR
Construct Tangent to Element	CONSTRUCT TANGENT TO
Convert Element to B-spline (Copy)	MDL LOAD SPLINES; CONSTRUCT BSPLINE CONVERT COPY
Convert Element to B-spline Original	MDL LOAD SPLINES; CONSTRUCT BSPLINE CONVERT ORIGINAL
Copy Fence Content	FENCE COPY
Copy Parallel by Distance	COPY PARALLEL DISTANCE
Copy Parallel by Key-in	COPY PARALLEL KEYIN
Corner Joint	MDL LOAD CUTTER; JOIN CORNER
Create Complex Chain	CREATE CHAIN MANUAL
Create Complex Shape	CREATE SHAPE MANUAL

Appendix B Key-in Commands

TOOL NAME	KEY-IN
Crosshatch Element Area	CROSSHATCH
Cut All Component Lines	MDL LOAD CUTTER; CUT ALL
Cut Single Component Line	MDL LOAD CUTTER; CUT SINGLE
Define ACS (Aligned with Element)	DEFINE ACS ELEMENT
Define ACS (Aligned with View)	DEFINE ACS VIEW
Define ACS (By Points)	DEFINE ACS POINTS
Define Active Entity Graphically	DEFINE AE
Define Cell Origin	DEFINE CELL ORIGIN
Define Reference File Back Clipping Plane	REFERENCE CLIP BACK
Define Reference File Clipping Boundary	REFERENCE CLIP BOUNDARY
Define Reference Clipping Mask	REFERENCE CLIP MASK
Define Reference File Front Clipping Plane	REFERENCE CLIP FRONT
Define True North	DEFINE NORTH
Delete Element	DELETE ELEMENT
Delete Fence Contents	FENCE DELETE
Delete Part of Element	DELETE PARTIAL
Delete Vertex	DELETE VERTEX
Detach Database Linkage	DETACH
Detach Database Linkage from Fence Contents	FENCE DETACH
Detach Reference File	REFERENCE DETACH
Dimension Angle Between Lines	DIMENSION ANGLE LINES
Dimension Angle from X-Axis	DIMENSION ANGLE X
Dimension Angle from Y-Axis	DIMENSION ANGLE Y
Dimension Angle Location	DIMENSION ANGLE LOCATION
Dimension Angle Size	DIMENSION ANGLE SIZE
Dimension Arc Location	DIMENSION ARC LOCATION
Dimension Arc Size	DIMENSION ARC SIZE

TOOL NAME	KEY-IN
Dimension Diameter	DIMENSION DIAMETER
Dimension Diameter (Extended Leader)	DIMENSION DIAMETER EXTENDED
Dimension Diameter Parallel	DIMENSION DIAMETER PARALLEL
Dimension Diameter Perpendicular	DIMENSION DIAMETER PERPENDICULAR
Dimension Diameter	DIMENSION DIAMETER
Dimension Element	DIMENSION ELEMENT
Dimension Location	DIMENSION LOCATION SINGLE
Dimension Location (Stacked)	DIMENSION LOCATION STACKED
Dimension Ordinates	DIMENSION ORDINATE
Dimension Radius	DIMENSION RADIUS
Dimension Radius (Extended Leader)	DIMENSION RADIUS EXTENDED
Dimension Size (Custom)	DIMENSION LINEAR
Dimension Size with Arrow	DIMENSION SIZE ARROW
Dimension Size with Strokes	DIMENSION SIZE STROKE
Display Attributes of Text Element	IDENTIFY TEXT
Drop Association	DROP ASSOCIATION
Drop Complex Status	DROP COMPLEX
Drop Complex Status of Fence Contents	FENCE DROP
Drop Dimension	DROP DIMENSION
Drop from Graphic Group	GROUP DROP
Drop Line String/Shape Status	DROP STRING
Drop Text	DROP TEXT
Edit Text	EDIT TEXT
Element Selection	CHOOSE ELEMENT
Extend 2 Elements to Intersection	EXTEND ELEMENT 2
Extend Element to Intersection	EXTEND ELEMENT INTERSECTION
Extend Line	EXTEND LINE DISTANCE

Appendix B Key-in Commands

TOOL NAME	KEY-IN
Extend Line By Key-in	EXTEND LINE KEYIN
Extract Bspline Surface Boundary	MDL LOAD SPLINES; EXTRACT BSPLINE SURFACE BOUNDARY
Fence Stretch	FENCE STRETCH
Fill in Single Enter Data Field	EDIT SINGLE
Freeze Element	FREEZE
Freeze Elements in Fence	FENCE FREEZE
Generate Report Table	FENCE REPORT
Global Origin	ACTIVE ORIGIN (GO=)
Group Holes	GROUP HOLES
Hatch Element Area	HATCH
Horizontal Parabola (No Truncation)	PLACE PARABOLA HORIZONTAL NOMODIFY
Horizontal Parabola and Truncate Both	PLACE PARABOLA HORIZONTAL MODIFY
Identify Cell	IDENTIFY CELL
Impose Bspline Surface Boundary	MDL LOAD SPLINES; IMPOSE BSPLINE SURFACE BOUNDARY
Insert Vertex	INSERT VERTEX
Label Line	LABEL LINE
Load Displayable Attributes	LOAD DA
Load Displayable Attributes to Fence Contents	FENCE LOAD
Match Pattern Attributes	ACTIVE PATTERN MATCH
Match Text Attributes	ACTIVE TEXT
Measure Angle Between Lines	MEASURE ANGLE
Measure Area	MEASURE AREA
Measure Area of Element	MEASURE AREA ELEMENT
Measure Distance Along Element	MEASURE DISTANCE ALONG
Measure Distance Between Points	MEASURE DISTANCE POINTS

TOOL NAME	KEY-IN
Measure Minimum Distance Between Elements	MEASURE DISTANCE MINIMUM
Measure Perpendicular Distance From Element	MEASURE DISTANCE PERPENDICULAR
Measure Radius	MEASURE RADIUS
Merged Cross Joint	MDL LOAD CUTTER; JOIN CROSS MERGE
Merged Tee Joint	MDL LOAD CUTTER; JOIN TEE MERGE
Mirror Element About Horizontal (Copy)	MIRROR COPY HORIZONTAL
Mirror Element About Horizontal (Original)	MIRROR ORIGINAL HORIZONTAL
Mirror Element About Line Copy	MIRROR COPY LINE
Mirror Element About Line (Ordinal)	MIRROR ORIGINAL LINE
Mirror Element About Vertical (Copy)	MIRROR COPY VERTICAL
Mirror Element About Vertical (Original)	MIRROR ORIGINAL VERTICAL
Mirror Fence Contents About Horizontal (Copy)	FENCE MIRROR COPY HORIZONTAL
Mirror Fence Contents About Horizontal (Original)	FENCE MIRROR ORIGINAL HORIZONTAL
Mirror Fence Contents About Line (Copy)	FENCE MIRROR COPY LINE
Mirror Fence Contents About Line (Original)	FENCE MIRROR ORIGINAL LINE
Mirror Fence Contents About Vertical (Copy)	FENCE MIRROR COPY VERTICAL
Mirror Fence Contents About Vertical (Original)	FENCE MIRROR ORIGINAL VERTICAL
Mirror Reference File About Horizontal	REFERENCE MIRROR HORIZONTAL
Mirror Fence About Vertical	REFERENCE MIRROR VERTICAL
Modify Arc Angle	MODIFY ARC ANGLE
Modify Arc Axis	MODIFY ARC AXIS
Modify Arc Radius	MODIFY ARC RADIUS
Modify Element	MODIFY ELEMENT
Modify Fence	MODIFY FENCE
Move ACS	MOVE ACS

Appendix B Key-in Commands

TOOL NAME	KEY-IN
Move Element	MOVE ELEMENT
Move Fence Block/Shape	MOVE FENCE
Move Fence Contents	FENCE MOVE
Move Reference File	REFERENCE MOVE
Multi-Cycle Segment Linear Pattern	PATTERN LINE MULTIPLE
Open Cross Joint	MDL LOAD CUTTER; JOIN CROSS OPEN
Open Tee Joint	MDL LOAD CUTTER; JOIN TEE OPEN
Pattern Element Area	PATTERN AREA ELEMENT
Pattern Fence Area	PATTERN AREA FENCE
Place Active Cell	PLACE CELL ABSOLUTE
Place Active Cell (Interactive)	PLACE CELL INTERACTIVE ABSOLUTE
Place Active Cell Matrix	MATRIX CELL (CM=)
Place Active Cell Relative	PLACE CELL RELATIVE
Place Active Cell Relative (Interactive)	PLACE CELL INTERACTIVE RELATIVE
Place Active Line Terminator	PLACE TERMINATOR
Place Active Point	PLACE POINT
Place Arc by Center	PLACE ARC CENTER
Place Arc by Edge	PLACE ARC EDGE
Place Arc by Keyed-in Radius	PLACE ARC RADIUS
Place B-spline Curve by Least Squares	MDL LOAD SPLINES; PLACE BSPLINE CURVE LEASTSQUARE
Place B-spline Curve by Points	MDL LOAD SPLINES; PLACE BSPLINE CURVE POINTS
Place B-spline Curve by Poles	MDL LOAD SPLINES; PLACE BSPLINE CURVE POLES
Place B-spline Surface by Least Squares	MDL LOAD SPLINES; PLACE BSPLINE SURFACE LEASTSQUARES
Place B-spline Surface by Points	MDL LOAD SPLINES; PLACE BSPLINE SURFACE POINTS

TOOL NAME	KEY-IN
Place B-spline Surface by Poles	MDL LOAD SPLINES; PLACE BSPLINE SURFACE POLES
Place Block	PLACE BLOCK ORTHOGONAL
Place Center Mark	DIMENSION CENTER MARK
Place Circle by Center	PLACE CIRCLE CENTER
Place Circle by Diameter	PLACE CIRCLE DIAMETER
Place Circle by Edge	PLACE CIRCLE EDGE
Place Circle by Keyed-in Radius	PLACE CIRCLE RADIUS
Place Circumscribed Polygon	PLACE POLYGON CIRCUMSCRIBED
Place Ellipse by Center and Edge	PLACE ELLIPSE CENTER
Place Ellipse by Edge Points	PLACE ELLIPSE EDGE
Place Fence Block	PLACE FENCE BLOCK
Place Fence Shape	PLACE FENCE SHAPE
Place Fitted Text	PLACE TEXT FITTED
Place Fitted View Independent Text	PLACE TEXT VI
Place Half Ellipse	PLACE ELLIPSE HALF
Place Helix	MDL LOAD SPLINES; PLACE HELIX
Place Inscribed Polygon	PLACE POLYGON INSCRIBED
Place Isometric Block	PLACE BLOCK ISOMETRIC
Place Isometric Circle	PLACE CIRCLE ISOMETRIC
Place Line	PLACE LINE
Place Line at Active Angle	PLACE LINE ANGLE
Place Line String	PLACE LSTRING POINT
Place Multi-line	PLACE MLINE
Place Note	PLACE NOTE
Place Orthogonal Shape	PLACE SHAPE ORTHOGONAL
Place Parabola by End Points	MDL LOAD SPLINES; PLACE PARABOLA ENDPOINTS

Appendix B Key-in Commands

TOOL NAME	KEY-IN
Place Point Curve	PLACE CURVE POINT
Place Polygon by Edge	PLACE POLYGON EDGE
Place Quarter Ellipse	PLACE ELLIPSE QUARTER
Place Right Cone	PLACE CONE RIGHT
Place Right Cone by Keyed-in Radius	PLACE CONE RADIUS
Place Right Cylinder	PLACE CYLINDER RIGHT
Place Right Cylinder by Keyed-in Radius	PLACE CYLINDER RADIUS
Place Rotated Block	PLACE BLOCK ROTATED
Place Shape	PLACE SHAPE
Place Skewed Cone	PLACE CONE SKEWED
Place Skewed Cylinder	PLACE CYLINDER SKEWED
Place Slab	PLACE SLAB
Place Space Curve	PLACE CURVE SPACE
Place Space Line String	PLACE LSTRING SPACE
Place Sphere	PLACE SPHERE
Place Spiral By End Points	MDL LOAD SPLINES; PLACE SPIRAL ENDPOINTS
Place Spiral By Length	MDL LOAD SPLINES; PLACE SPIRAL LENGTH
Place Spiral by Sweep Angle	MDL LOAD SPLINES; PLACE SPIRAL ANGLE
Place Stream Curve	PLACE CURVE STREAM
Place Stream Line String	PLACE LSTRING STREAM
Place Text	PLACE TEXT
Place Text Above Element	PLACE TEXT ABOVE
Place Text Along Element	PLACE TEXT ALONG
Place Text Below Element	PLACE TEXT BELOW
Place Text Node	PLACE NODE

TOOL NAME	KEY-IN
Place Text On Element	PLACE TEXT ON
Place View Independent Text	PLACE TEXT VI
Place View Independent Text Node	PLACE NODE VIEW
Polar Array	ARRAY POLAR
Polar Array Fence Contents	FENCE ARRAY POLAR
Project Active Point Onto Element	CONSTRUCT POINT PROJECT
Rectangular Array	ARRAY RECTANGULAR
Rectangular Array Fence Contents	FENCE ARRAY RECTANGULAR
Reload Reference File	REFERENCE RELOAD
Replace Cell	REPLACE CELL
Review Database Attributes of Element	REVIEW
Rotate ACS Absolute	ROTATE ACS ABSOLUTE
Rotate ACS Relative	ROTATE ACS RELATIVE
Rotate Element Active Angle Copy	ROTATE COPY
Rotate Element Active Angle Original	ROTATE ORIGINAL
Rotate Fence Contents by Active Angle (Copy)	FENCE ROTATE COPY
Rotate Fence Contents by Active Angle (Original)	FENCE ROTATE ORIGINAL
Rotate Reference File	REFERENCE ROTATE
Scale Element (Copy)	SCALE COPY
Scale Element (Original)	SCALE ORIGINAL
Scale Fence Contents (Copy)	FENCE SCALE COPY
Scale Fence Contents (Original)	FENCE SCALE ORIGINAL
Scale Reference File	REFERENCE SCALE
Select ACS	ATTACH ACS
Select and Place Cell	SELECT CELL ABSOLUTE
Select and Place Cell (Relative)	SELECT CELL RELATIVE

Appendix B Key-in Commands

TOOL NAME	KEY-IN
Set Active Depth	DEPTH ACTIVE
Show Active Depth	SHOW DEPTH ACTIVE
Show Active Entity	SHOW AE
Show Linkage Mode	ACTIVE LINKAGE
Show Pattern Attributes	SHOW PATTERN
Single Cycle Segment Linear Pattern	PATTERN LINE SINGLE
Spin Element (Copy)	SPIN COPY
Spin Element (Original)	SPIN ORIGINAL
Spin Fence Contents (Copy)	FENCE SPIN COPY
Spin Fence Contents (Original)	FENCE SPIN ORIGINAL
Symmetric Parabola (No Truncation)	PLACE PARABOLA NOMODIFY
Symmetric Parabola and Truncate Both	PLACE PARABOLA MODIFY
Thaw Element	THAW
Thaw Elements in Fence	FENCE THAW
Truncated Cycle Linear Pattern	PATTERN LINE ELEMENT
Uncut Component Lines	MDL LOAD CUTTER; UNCUT

appendix C

Alternate Key-ins

 Note: MicroStation is not case-sensitive. That is, key-ins can be uppercase or lowercase, or even a mixture of the two, and MicroStation will still understand the key-ins in the same way.

AA = ACTIVE ANGLE	set active angle
AC = ACTIVE CELL	set active cell; place absolute
AD = POINT ACSDELTA	data point—delta ACS
AE = ACTIVE ENTITY	define active entity
AM = ATTACH MENU	activate menu
AP = ACTIVE PATTERN CELL	set active pattern cell
AR = ACTIVE RCELL	set active cell; place relative
AS = ACTIVE SCALE	set active scale factors
AT = TUTORIAL	activate tutorial
AX = POINT ACSABSOLUTE	data point absolute ACS
AZ = ACTIVE ZDEPTH ABSOLUTE	set active depth

CC = CREATE CELL	create cell
CD = DELETE CELL	delete cell from cell library
CM = MATRIX CELL	place active cell matrix
CO = ACTIVE COLOR	set active color
CR = RENAME CELL	rename cell
CT = ATTACH COLORTABLE	attach color table
DA = ACTIVE DATYPE	set active displayable attribute type
DB = ACTIVE DATABASE	attach control file to design file
DD = SET DDEPTH RELATIVE	set display depth (relative)
DF = SHOW FONT	open Fonts settings box
DI = POINT DISTANCE	data point—distance, direction
DL = POINT DELTA	data point—delta coordinates
DP = DEPTH DISPLAY	set display depth
DR = TYPE	display text file
DS = SEARCH	specify fence filter
DV = VIEW	delete saved view
DX = POINT VDELTA	data point—delta view coordinates
DZ = ZDEPTH RELATIVE	set active depth (relative)
EL = ELEMENT LIST	create element list file
FF = FENCE FILE	copy fence contents to design file
FI = FIND	set database row as active entity
FT = ACTIVE FONT	set active font
GO = ACTIVE ORIGIN	Global Origin
GR = ACTIVE GRIDREF	set grid reference spacing
GU = ACTIVE GRIDUNIT	set horizontal grid spacing
KY = ACTIVE KEYPNT	set Snap divisor
LC = ACTIVE STYLE	set active line style
LD = DIMENSION LEVEL	set dimension level

Appendix C Alternate Key-ins

LL = ACTIVE LINE LENGTH	set active text line length
LS = ACTIVE LINE SPACE	set active text node line spacing
LT = ACTIVE TERMINATOR	set active terminator
LV = ACTIVE LEVEL	set active level
NN = ACTIVE NODE	set active text node number
OF = SET LEVELS <level list> OFF	set level display off
ON = SET LEVELS <level list> ON	set level display on
OX = ACTIVE INDEX	retrieve user command index
PA = ACTIVE PATTERN ANGLE	set active pattern angle
PD = ACTIVE PATTERN DELTA	set active pattern delta (distance)
PS = ACTIVE PATTERN SCALE	set active pattern scale
PT = ACTIVE POINT	set active point
PX = DELETE ACS	delete ACS
RA = ACTIVE REVIEW	set attribute review selection criteria
RC = ATTACH LIBRARY	open cell library
RD = NEWFILE	open design file
RF = REFERENCE ATTACH	attach reference file
RS = ACTIVE REPORT	name report table
RV = ROTATE VIEW	rotate view (relative)
RX = ATTACH ACS	select ACS
SD = ACTIVE STREAM DELTA	set active stream delta
SF = FENCE SEPARATE	move fence contents to design file
ST = ACTIVE STREAM TOLERANCE	set active stream tolerance
SV = SAVE VIEW	save view
SX = SAVE ACS	save auxiliary coordinate system
TB = ACTIVE TAB	set tab spacing for importing text
TH = ACTIVE TXHEIGHT	set active text height
TI = ACTIVE TAG	set copy and increment value

TS = ACTIVE TSCALE	set active terminator scale
TV = DIMENSION TOLERANCE	set dimension tolerance limits
TW = ACTIVE TXWIDTH	set active text width
TX = ACTIVE TXSIZE	set active text size (height/width)
UC = USERCOMMAND	activate user command
UCC = UCC	compile user command
UCI = UCI	user command index
UR = ACTIVE UNITROUND	set unit distance
VI = VIEW	attach named view
WO = WINDOW ORIGIN	Window Orgin
WT = ACTIVE WEIGHT	set active line weight
XD = EXCHANGEFILE	open design file; keep view config.
XS = ACTIVE XSCALE	set active X scale
XY = POINT ABSOLUTE	data point absolute coordinates
YS = ACTIVE YSCALE	set active Y scale
ZS = ACTIVE ZSCALE	set active Z scale

appendix D

Primitive Commands

 Note: MicroStation is not case-sensitive. That is, key-ins can be uppercase or lowercase, or even a mixture of the two, and MicroStation will still understand the key-ins in the same way.

NAME OF THE COMMAND	PRIMITIVE COMMAND
ACTIVE ANGLE PT2	/ACTAN2
ACTIVE ANGLE PT3	/ACTAN3
ACTIVE CAPMODE OFF	/WOCMDE
ACTIVE CAPMODE ON	/CAPMDE
ACTIVE SCALE DISTANCE	/ACTSCA
ACTIVE TNJ CB	/TJST#
ACTIVE TNJ CC	/TJST7
ACTIVE TNJ CT	/TJST6
ACTIVE TNJ LB	/TJST2

NAME OF THE COMMAND	PRIMITIVE COMMAND
ACTIVE TNJ LC	/TJST]
ACTIVE TNJ LMB	/TJST5
ACTIVE TNJ LMC	/TJST4
ACTIVE TNJ LMT	/TJST3
ACTIVE TNJ LT	/TJSTO
ACTIVE TNJ RB	/TJST14
ACTIVE TNJ RC	/TJST13
ACTIVE TNJ RMB	/TJSTl 1
ACTIVE TNJ RMC	/TJST 1 0
ACTIVE TNJ RMT	/TJST9
ACTIVE TNJ RT	/TJSTl 2
ACTIVE TXHEIGHT PT2	/TXTHGT
ACTIVE TXJ CB	/TXJS8
ACTIVE TXJ CC	/TXJS7
ACTIVE TXJ CT	/TXJS6
ACTIVE TXJ LB	/TXJSZ
ACTIVE TXJ LC	/TXJSl
ACTIVE TXJ LT	/TXJSO
ACTIVE TXJ RB	/TXJS14
ACTIVE TXJ RC	/TXJS13
ACTIVE TXJ RT	/TXJSlz
ACTIVE TXWIDTH PT2	/TXTWDT
ALIGN	/ALIGN
ATTACH AE	/ATCPTO
CHANGE COLOR	/CELECR
CHANGE STYLE	/CELELS
CHANGE SYMBOLOGY	/CELESY

Appendix D Primitive Commands

NAME OF THE COMMAND	PRIMITIVE COMMAND
CHANGE WEIGHT	/CELEWT
CONSTRUCT BISECTOR ANGLE	/ANGBIS
CONSTRUCT BISECTOR LINE	/PERBIS
CONSTRUCT LINE AA 1	/CNSAA1
CONSTRUCT LINE AA 2	/CNSAA2
CONSTRUCT LINE AA 3	/CNSAA3
CONSTRUCT LINE AA 4	/CNSAA4
CONSTRUCT LINE MINIMUM	/MDL2EL
CONSTRUCT POINT	/CNSINT
CONSTRUCT POINTALONG	/NPAE
CONSTRUCT POINT BETWEEN	/NPNTS
CONSTRUCT POINT DISTANCE	/PPAE
CONSTRUCT POINT PROJECT	/PRJPNT
CONSTRUCT TANGENT	/LNTNNR
CONSTRUCT TANGENT ARC 1	/PTARCC
CONSTRUCT TANGENT ARC 3	/ATN3EL
CONSTRUCT TANGENT BETWEEN	/LTZELP
CONSTRUCT TANGENT CIRCLE 1	/CTN1EL
CONSTRUCT TANGENT CIRCLE 3	/CTN3EL
CONSTRUCT TANGENT FROM	/PTFROM
CONSTRUCT TANGENT TO	/PITO
COPY ED	/EDCOPY
COPYELEMENT	/CPELE
COPY PARALLEL DISTANCE	/CPYPP
COPY PARALLEL KEYIN	/CPYPK
COPY VIEW	/COPY
CREATE CHAIN MANUAL	/CONNST

NAME OF THE COMMAND	PRIMITIVE COMMAND
CREATE SHAPE MANUAL	/CPXSHP
DEFINE ACS ELEMENT	/AUXELE
DEFINE ACS POINTS	/AUX3PT
DEFINE ACS VIEW	/AUXVW
DEFINE AE	/DEFPTO
DEFINE CELL ORIGIN	/DOCELL
DELETE ELEMENT	/DLELEM
DELETE PARTIAL	/DLPELE
DELETE VERTEX	/DVERTX
DEPTH ACTIVE PRIMITIVE	/ADEPTH
DEPTH DISPLAY PRIMITIVE	/DDEPTH
DIMENSION ANGLE LINES	/ANGLIN
DIMENSION ANGLE LOCATION	/PITLOC
DIMENSION ANGLE SIZE	/PTSIZ
DIMENSION ARC LOCATION	/ARCLOC
DIMENSION ARC SIZE	/ARCSIZ
DIMENSION AXIS DRAWING	/ACTAXD
DIMENSION AXIS TRUE	/ACTAXP
DIMENSION AXIS VIEW	/ACTAXV
DIMENSION DIMAETER PARALLEL	/DIAPAR
DIMENSION DIAMETER PERPENDICULAR	/DIAPER
DIMENSION DIAMETER POINT	/DIACIR
DIMENSION FILE ACTIVE	/MEAACT
DIMENSION FILE REFERENCE	/MEAREF
DIMENSION JUSTIFICATION CENTER	/JUSC
DIMENSION JUSTIFICATION LEFT	/JUSL
DIMENSION JUSTIFICATION RIGHT	/JUSR

Appendix D Primitive Commands

NAME OF THE COMMAND	PRIMITIVE COMMAND
DIMENSION LOCATION SINGLE	/LOCSNG
DIMENSION LOCATION STACKED	/LOCSTK
DIMENSION PLACEMENT AUTO	/ADMAUT
DIMENSION PLACEMENT MANUAL	/PADMAU
DIMENSION RADIUS POINT	/RADRAD
DIMENSION SIZE ARROW	/SIZARW
DIMENSION SIZE STROKE	/SIZOBL
DIMENSION UNITS DEGREES	/UNITDG
DIMENSION UNITS LENGTH	/UNITLN
DIMENSION WITNESS OFF	/WITLOF
DIMENSION WITNESS ON	/WITLON
DROP COMPLEX	/DRCMPX
EDIT AUTO	/EDAUTO
EDIT SINGLE	/EDSING
EDIT TEXT	/EDTEXT
EXTEND ELEMENT 2	/EXLIN2
EXTEND ELEMENT INTERSECTION	/EXLNIN
EXTEND LINE DISTANCE	/EXLIN
EXTEND LINE KEYIN	/EXLINK
FENCE ATTACH	/AAEFCN
FENCE CHANGE STYLE	/CFNCLS
FENCE CHANGE SYMBOLOGY	/CFNCSY
FENCE COPY	/CPFNCC
FENCE DELETE	/DLFNCC
FENCE DETACH	/RATFCN
FENCE LOCATE	/FNCLOC
FENCE MIRROR COPY HORIZONTAL	/MHCPFC

NAME OF THE COMMAND	PRIMITIVE COMMAND
FENCE MIRROR COPY LINE	/MLCPFC
FENCE MIRROR COPY VERTICAL	/MVCPFC
FENCE MIRROR ORIGINAL	/MFVERT
FENCE MIRROR ORIGINAL HORIZONTAL	/MFHRIZ
FENCE MIRROR ORIGINAL LINE	/MFLINE
FENCE MOVE	/MVFNCC
FENCE REPORT	/RPTACT
FENCE ROTATE COPY	/RTCPFC
FENCE ROTATE ORIGINAL	/RFNCC
FENCE SCALE COPY	/SCCPFC
FENCE SCALE ORIGINAL	/SCFNCC
FENCE TRANSFORM	/TRSFCC
FENCE WSET ADD	/ADWSFN
FENCE WSET COPY	/ADWSFC
FILE DESIGN	/FILDGN
FILLET MODIFY	/PFILTM
FILLET NOMODIFY	/PFILTN
FILLET SINGLE	/FILTRM
FIT ACTIVE	/FIT1
GROUP ADD	/ADDGG
GROUP DROP	/DRFGG
IDENTIFY CELL	/IDCELL
IDENTIFY TEXT	/TXNODA
INCREMENT ED	/CIDATA
INCREMENT TEXT	/CITEXT
INSERT VERTEX	/IVERTX
JUSTIFY CENTER	/EDCJST

Appendix D Primitive Commands

NAME OF THE COMMAND	PRIMITIVE COMMAND
JUSTIFY LEFT	/EDLJST
JUSTIFY RIGHT	/EDRJST
LABEL LINE	/LABLN
LABEL LINE	/LBLINE
LOCELE	/LOCELE
LOCK ACS [OFF|ON|TOGGLE]	/CPLOCK
LOCK ANGLE [OFF|ON|TOGGLE]	/ANGLLK
LOCK AXIS [OFF|ON|TOGGLE]	/AXLKFF
LOCK BORESITE [OFF|ON|TOGGLE]	/BORSIT
LOCK FENCE CLIP	/CLIP
LOCK FENCE INSIDE	/INSIDE
LOCK FENCE OVERLAP	/OVRLAP
LOCK GGROUP [OFF|ON|TOGGLE]	/GGLOCK
LOCK GRID [OFF|ON|TOGGLE]	/GRIDLK
LOCK SCALE [OFF|ON|TOGGLE]	/SCALLK
LOCK SNAP KEYPOINT	/KEYSNP
LOCK SNAP [OFF|ON]	/SNPOFF
LOCK SNAP PROJECT	/SNAPLK
LOCK TEXTNODE [OFF|ON]	/TXTNLK
LOCK UNIT [OFF|ON]	/UNITLK
MEASURE ANGLE	/LINANG
MEASURE AREA	/AREAPT
MEASURE AREA ELEMENT	/AREAEL
MEASURE DISTANCE ALONG	/MDAE
MEASURE DISTANCE PERPENDICULAR	/PRPND
MEASURE DISTANCE POINTS	/PERIM
MEASURE RADIUS	/RADIUS

NAME OF THE COMMAND	PRIMITIVE COMMAND
MIRROR COPY HORIZONTAL	/MHCPEL
MIRROR COPY LINE	/MLCPEL
MIRROR COPY VERTICAL	/MVCPEL
MIRROR ORIGINAL HORIZONTAL	/MEHRIZ
MIRROR ORIGINAL LINE	/MELINE
MIRROR ORIGINAL VERTICAL	/MEVERT
MODIFY ARC ANGLE	/MDARCA
MODIFY ARC AXIS	/MDARCX
MODIFY ARC RADIUS	/MDARCR
MODIFY ELEMENT	/MDELE
MODIFY FENCE	/MDFNC
MODIFY TEXT	/TXNODC
MOVE ACS	/AUXORC
MOVE ELEMENT	/MVELEM
MOVE FENCE	/MVFNC
NULL	/NULCMD
PLACE ARC CENTER	/PARCC
PLACE ARC EDGE	/PARCE
PLACE ARC RADIUS	/PARCR
PLACE ARC TANGENT	/PTARCC
PLACE BLOCK ORTHOGONAL	/PBLOCK
PLACE BLOCK ROTATED	/PRBLOC
PLACE CELL ABSOLUTE	/PACELL
PLACE CELL ABSOLUTE TMATRX	/PACMTX
PLACE CELL RELATIVE	/PACELR
PLACE CELL RELATIVE TMATRX	/PACRMX
PLACE CIRCLE CENTER	/PCIRC

Appendix D Primitive Commands

NAME OF THE COMMAND	PRIMITIVE COMMAND
PLACE CIRCLE DIAMETER	/PCIRD
PLACE CIRCLE EDGE	/PCIRE
PLACE CIRCLE RADIUS	/PCIRR
PLACE CONE RADIUS	/PRCONR
PLACE CONE RIGHT	/PRCONE
PLACE CONE SKEWED	/PCONE
PLACE CURVE POINT	/PPTCRV
PLACE CURVE SPACE	/PSPCRV
PLACE CURVE STREAM	/PSTCRV
PLACE CYLINDER RADIUS	/PRCYLR
PLACE CYLINDER RIGHT	/PRCYL
PLACE CYLINDER SKEWED	/PCYLIN
PLACE ELLIPSE CENTER	/PELL1
PLACE ELLIPSE EDGE	/PELL2
PLACE ELLIPSE HALF	/PPELL1
PLACE ELLIPSE QUARTER	/PPELL2
PLACE FENCE BLOCK	/PFENCB
PLACE FENCE SHAPE	/PFENCE
PLACE LINE	/PLINE
PLACE LINE ANGLE	/PLINAA
PLACE LSTRING POINT	/PPTLST
PLACE LSTRING SPACE	/PSPLST
PLACE LSTRING STREAM	/PSTLST
PLACE NODE	/PTEXTN
PLACE NODE TMATRX	/PTNMTX
PLACE NOTE VIEW	/PVITXN
PLACE PARABOLA HORIZONTAL MODIFY	/PPARMD

NAME OF THE COMMAND	PRIMITIVE COMMAND
PLACE PARABOLA HORIZONTAL NOMODIFY	/PPARNM
PLACE POINT	/PLPNT
PLACE POINT STRING	/PDPTST
PLACE POINT STRING DISJOINT	/PCPTST
PLACE SHAPE	/PSHAPE
PLACE SHAPE ORTHOGONAL	/POSHAP
PLACE TERMINATOR	/PTERM
PLACE TEXT	/PTEXT
PLACE TEXT ABOVE	/PTXTA
PLACE TEXT ALONG	/PTAE
PLACE TEXT BELOW	/PTXTB
PLACE TEXT FITTED	/PFTEXT
PLACE TEXT FVI	/PVIFTX
PLACE TEXT ON	/PTOE
PLACE TEXT TMATRX	/PTXMTX
PLACE TEXT VI	/PVITXT
REFERENCE CLIP BACK	/RFCBCK
REFERENCE CLIP BOUNDARY	/RFCBND
REFERENCE CLIP FRONT	/RFCFRO
REFERENCE DETACH	/RFDTCH
REFERENCE DISPLAY OFF	/RFDIS0
REFERENCE DISPLAY ON	/RFDIS1
REFERENCE LEVELS OFF	/RFLEV0
REFERENCE LEVELS ON	/RFLEV1
REFERENCE LOCATE OFF	/RFLOC0
REFERENCE LOCATE ON	/RFLOC1
REFERENCE MOVE	/RFMOVE

Appendix D Primitive Commands

NAME OF THE COMMAND	PRIMITIVE COMMAND
REFERENCE ROTATE	/RFROT
REFERENCE SCALE	/RFSCAL
REFERENCE SNAP OFF	/RFSNP0
REFERENCE SNAP ON	/RFSNP1
REPLACE CELL	/RPCELL
REVIEW	/RVWATR
ROTATE 3PTS	/VIEWPL
ROTATE COPY	/RTCPEL
ROTATE ORIGINAL	/ROTELE
ROTATE VIEW POINTS	/VIEWPL
ROTATE VMATRX	/VMATRX
SCALE COPY	/SCCPEL
SCALE ORIGINAL	/SCAELE
SELECT CELL ABSOLUTE	/PSCELL
SELECT CELL ABSOLUTE MATRX	/PSCMTX
SELECT CELL RELATIVE	/PSCELR
SELECT CELL RELATIVE TMATRX	/PSCRMX
SET CONSTRUCT [OFF|ON|TOGGLE]	/CONST1
SET CURVES [FAST|SLOW|OFF|ON|TOGGLE]	/FCURV1
SET DELETE [OFF|PN|TOGGLE]	/DLENSW
SET DIMENSION [OFF|PN|TOGGLE]	/DIMEN1
SET DYNAMIC [FAST|SLOW|OFF|ON|TOGGLE]	/DRAG
SET ED [OFF|ON|TOGGLE]	/UNDLI1
SSET FONT [FAST|SLOW|OFF|ON|TOGGLE]	/FFONT1
SET GRID [OFF|ON|TOGGLE]	/GRID1
SET NODES [OFF|ON|TOGGLE]	/TXNOD1
SET PATTERN [OFF|ON|TOGGLE]	/PATRN1

NAME OF THE COMMAND	PRIMITIVE COMMAND
SET TEXT [OFF|ON|TOGGLE]	/TEEXT1
SET TPMODE ACSDELTA	/AXDTEN
SET TPMODE ACSLOCATE	/AXXTEN
SET TPMODE DELTA	/MDELTA
SET TPMODE DISTANCE	/MANGL2
SET TPMODE LOCATE	/LOCATE
SET TPMODE VDELTA	/MDLTVW
SET WEIGHT [OFF|ON]	/DWGHT1
SHOW HEADER	/HEADER
SURFACE PROJECTION	/PRJELE
SURFACE REVOLUTION	/SURREV
SWAP SCREEN	/SWAP
TRANSFORM	/TRSELE
UPDATE1...UPDATE8	/UPDATE
UPDATE BOTH	/UPDBTH
UPDATE LEFT	/UPDAT2
UPDATE RIGHT	/UPDAT1
UPDATE VIEW	/UPDATV
VIEW OFF	/VIEWOF
WINDOW AREA	/WINDA1
WINDOW CENTER	/WINDC1
WINDOW VOLUME	/WINVOL
WSET ADD	/ADWSEL
WSET COPY	/ADWSEC
WSET DROP	/WSDROP
ZOOM IN 2	/HALF1
ZOOM OUT 2	/DOUBL1

Index

SYMBOLS

$ (dollar sign), pen table text with, 479
& (ampersand), filtering with, 675–676
( ) (parentheses), calculations with, 224
* (asterisk)
   dimension text with, 399
   filtering with, 675
   include files with, 677, 708
. (period), attribute key-ins with, 296–297
... (ellipses)
   buttons with, 9
   menu items with, 7, 18
: (colon)
   path with, 14
   Status bar messages with, 26
; (semi-colon), tool action strings with, 729
<< >> (angle brackets), Data Fields with, 312
= (equal sign), calculations with, 223
> (greater than), Status bar messages with, 26
\ (backslash), path with, 14
_ (underscore), Data Fields with, 307–308, 312
| (vertical bar), filtering with, 675–676
~ (tilde), shortcut characters with, 733, 734
± (plus/minus sign), tolerance values with, 438, 452–453
2 Point leaders, 611, 612
2 Points
   Rotate (sp o/c) element by, 191–193
   Rotate View by, 118–120
2D designs. *See also* Designs
   3D compared to, 747
   AccuDraw shortcuts for, 231–232
   coordinate system for, 67, 68
   design plane in, 30–31
   grid orientation for, 81
3 Point leaders, 611, 612
3 Points, Rotate (ro p o/c) element by, 193–194
3 Points scaling (sc p o/c), 186, 188–189
3D Construct tool box, 777–780
3D designs, 747–748
   AccuDraw in, 226, 775–777
   active depth in, 756–758
   cell creation/placement in, 792
   chamfer edge/surface in, 790–791
   change surface normals in, 786
   composite solids in, 781–786
   convert solid/surface status in, 775
   coordinate systems in, 67
      auxiliary, 761–767
      drawing, 759–760
      view, 760–761
   create file for, 749–750
   cut solids in, 788
   design cube in, 30, 753
   dimensioning, 792
   display depth in, 753–756
   drawing composition in, 796
   extrude along path in, 778
   fence manipulations in, 791
   fillet edge/surface in, 789–790
   grid orientation for, 81
   import/export, 601
   modify solids in, 786–787
   place 2D elements in, 781
   place primitives in, 767–775
   place text in, 791
   precision inputs for, 758–761
   projected surfaces in, 777–778
   remove faces in, 787
   rendering, 792–795
   shell solids in, 780
   surface of revolution in, 779
   text placement for, 275, 306
   thicken to solid in, 780–781
   view rotation in, 750–752
3D Modify tool box, 781–784, 786–791
3D Utility tool box, 768
3D View Control tool box, 751, 793, 795

A

AA=, 296–297
Above dimension text, 449–450
Above Element text (pla tex a), 275–277
Absolute cell placement, 503, 506–507
Absolute coordinates
   place points by, 67–68, 70–72
   set active depth by (az =), 757
   set display depth by (dp=), 754
Absolute ratio, scale reference by, 574
Absolute rotation, 752, 765
ac=, 504

Accelerator keys, setting, 735
Accept button, mouse/puck, 28
Acceptance point, pattern, 549
AccuDraw
    activating (a a), 216–217, 776
    compass for, 217–220
    coordinates box for, 220–222
    features of, 215–216
    key-in shortcuts for, 217, 231–233
    Modify Element tool with, 338
    pop-up calculator for, 223–224
    recall values with, 222–223
    settings for (a d se), 224–233
    Smart Lock for, 224
    SmartLine tool with, 241–242
    working with, 233–236
Accuracy
    coordinate display, 35, 36, 617
    design environment, 34
    dimension, 455, 457
AccuSnap. *See also* Snap Lock
    AccuDraw with, 236
    features of, 45, 94
    settings for, 63–64, 94–97
ACS (auxiliary coordinate systems)
    define, 761–762, 763–765
    move active (mov a), 766
    orient grid to, 81
    precision key-ins for, 762–763
    rotate active, 765
    save, 766–767
    select active, 766
ACS tool box, 763–766
ACS Triad, setting, 761, 762
Active Angle, rotation by
    array, 199, 200
    element (ro o/c), 190–191
    text, 204–205, 296
Active attributes, change
    dimension (cha dim), 432
    element, 107, 108, 109
Active cells
    define defaults for, 525
    placement of, 504–506
        Absolute or Relative, 506–507
        Cell Matrix for (matr c), 508–510
        Interactive mode for, 507
        line terminators in (pl t), 510–512
        point tools for (pl po), 512–519
        Select and Place (se c i), 508
    replace or update, 529–531
    select, 503–504
Active color, setting, 103–104
    ByLevel option for, 596, 597–598
    fill color and, 558
Active depth, 756–758, 792
Active level
    concept of, 98

    settings for, 99–100
        display, 103
        symbology, 596, 597–598
Active line style, setting
    ByLevel option for, 596, 597–598
    element, 104–105
Active line weight, setting
    ByLevel option for, 596, 597–598
    element, 105–106
Active Scale method (sc o/c), 185–188
AD=, 762–763
Add
    cell buttons, 526
    design files to plot, 491–492
    graphic group elements (gr a), 586–587
    menus/menu items, 736
    selection set elements, 247, 252
Adjust dimension line, 444–445
Advanced mode, IntelliTrim, 167
    Extend with, 176–177
    Trim with, 174–175
Alert box, Delete View Group, 137
Algebraic constraints, 624, 636
Alignment
    ACS with element (d a e), 763
    ACS with view (d a v), 764–765
    dimension leader, 441
    linear dimension, 395–396, 443–444, 792
Along Element text (pl tex al), 279–281
AM=, 4
Ambient lighting, 794
Angles
    AccuDraw
        recall previous, 222–223
        roundoff values for, 229, 230
        show negative, 227
    active cell placement, 505
    camera lens, 793–794
    constrain arc with, 61, 63
    dimension, format for, 457
    line, constrain (con e), 626–627
    Measure (me a), 386
    modify arc (modi a a), 342–343
    multi-line, constrain, 163
    multi-line cap, 364
    pattern, 545
    place lines at, 49
    place shapes at, 53
    readout settings for, 35, 36
    rotation by
        array, 199, 200–201, 202
        element, 190–191
        point, 513
        reference, 575
        text, 204–205, 296
        view, 751, 752
    specify points by, 69–70
    stream curve, 161, 162

Index

Angular dimensioning, 392
   Arc Location (dim ar l), 416–418
   Arc Size (dim ar s), 415–416
   for Chamfer (dim ang ch), 418
   Between Lines (dim a l), 413
   from Location (dim a lo), 411–412
   from Size (dim a s), 410–411
   tools for, 393, 409–410
   unit format for, 457
   from X Axis (dim a x), 414
   from Y Axis (dim a y), 415
Annotation tools, 607–621
   Annotate (Flag), 608–611
   Drafting, 611–616
   options for, 607
   XYZ Text, 617–621
Anti-alias rendering, 795
ap=, 504
Append. *See also* Attach
   ?fmt-raw to URL, 650
   configuration variables, 707, 708
   file, Export Coordinates and, 620
   tag sets, 328
Application window components, 17–27
Apply
   Saved Views, 142
   View Group, 137
   to Window, 115
Arbitrary dimension alignment, 395, 443
Archive files, 651, 710
Arcs
   dimensioning
      Center (dim c m), 424–425
      Diameter (dim d), 421–424
      Distance (dim ar dist), 426
      Location (dim ar l), 416–418
      Radius (dim radiu), 419–421
      Size (dim ar s), 415–416
   modify, 341–345
   multi-line caps with, 364
   Place
      By Center (pl a c), 60–61
      By Edge (pl a e), 61–63
      Point Curve tool for, 161
      SmartLine tool for, 237, 238
Arcs tool box, 60, 62, 342–344
Area
   Change Element to Active (chan a), 556
   create holes in, 555–556
   create shape from, 356–357
   fill/color options for, 240
   measurement of, 387–391
   select by property of, 594–595
   stream curve, set, 161, 162
Arrange view windows (w arr), 130–133
Arrays, construct, 198–202
Arrow Marker, Place, 615

Arrows
   Dimension Size with (dim si a), 396–397
   place notes at, 283–284
   placing, AccuDraw for, 233–234
   settings for, 447, 448–449
ASC Triad settings, 138, 139
ASCII files
   access tag reports in, 327
   export coordinates to, 619
   import coordinates from, 621
   import text from, 295, 601
Aspect ratio (Y/X), grid, 81
Assign Equation (as e), 636
Assign Variable (as v), 635–636
Association option
   Place Active Cell, 505
   Place Note, 284
Associative dimensioning
   constrain design with, 634
   enabling, 395
   Reassociate (dim reas) for, 431–432
Associative patterns, 545–546
   fence, 550
   hole area, 555, 556
Attach. *See also* Append
   ACS, 767
   cell library, 497–498
   Ellipse or circle (at c), 633–634
   Engineering Links, 659–660
   line-string or shape (at ls), 631–632
   multiple views, 796
   References (ref a), 564, 566–571
   remote design file, 651–652
   settings group file, 666
   tags, 317–318
   tool box to menu item, 734–735
Attach Elements tool box, 625
Attributes
   dimension
      extension line, 442
      line, 440
      terminator, 448
      text, 451
      tolerance, 452–453
   element
      change, 106–108, 257
      drop complex status and, 358
      key-ins for, 296–297
      Match, 109
      plot output settings for, 481, 482
      selection by, 251, 591–596
      settings for, 97–106
   Match Pattern (mat pat), 557–558
   multi-line component, 362–363, 364, 365
   print, 470–472
   tag, 316
   text
      Change (modi te), 301

Display (i t), 302
key-ins for, 296–297
Match (mat t), 300
view
   ACS Triad, 761, 762
   area fill, 559–560
   Constructions, 627, 639
   Data Field, 308
   Fast Cells, 533
   Grid, 78–79
   HTML file, 654, 656
   Patterns, 542–543
   settings box for, 138–140
   Text Node, 304–305
Attributes tool box, 22–23, 103–106, 676
Auto Fill In Enter Data Fields, 309, 310
Auto Load AccuDraw, 216, 225
Auto Point Placement, AccuDraw, 225, 226
AutoCAD drawings, 600
Automatic option
   Create Complex Chain, 347–349
   Create Complex Shape, 350–352
   dimension text placement, 444
   terminator placement, 446
Automatically Identify Elements, 63 64, 96
Auxiliary coordinate systems. *See* ACS (auxiliary coordinate systems)
AX=, 762–763
Axis. *See also* X axis; Y axis; Z axis
   constrain ellipse, 153, 155
   indexing, 220, 229, 231
   Mirror About (mi o/c h), 195–197
   Modify Arc (modi a ax), 343–345
Axis Lock, 83, 225, 226
az =, 757

B

Background settings, 138, 139
Backup files, configuration variables for, 721
Batch printing, 483–492
Below Element text (pla tex b), 278
Bentley printer driver, 468, 469, 473
Bisector snap mode, 93
Block Fence, Place (pl f b), 259
Blocks
   Drop Line String/Shape Status (dr st) for, 65–66
   placing
      AccuDraw for, 235–236
      orthogonal (pl b o), 50–51
      rotated (pl b r), 51–52
   select by, PowerSelector (pow ar b) for, 247, 248–249
Bold text, 208, 293
Boolean operations
   create composite solids with, 781–786
   create regions with, 352–356
   patterning with, 548, 550–552

Borders
   print, 472, 486
   reference, 565
      highlight, 583
      move, 572
      scale, 574
      view, customize, 737–738
Boresite lock, 758
Boundaries
   clipping, 576–578
   display settings for, 138, 139
   Hilite, 583
   print, 472, 486–487
Boundary elements, projecting, 777–778
Box text frames
   dimensions with, 451
   notes with, 283, 284
   word wrap in, 282
Break Link, 643
Browse Cell(s), 504, 546
Bspline leaders, 611, 612
Buttons
   Cell Selector, customize, 520–525, 526
   dialog box, 9
   Minimize/Maximize window, 132
   mouse, 3–4
   puck, 4–5, 28–29
By Origin text method
   3D design and, 791
   settings for, 204, 205
   Text Nodes and, 306–307
   using, 209–210
ByLevel symbology settings, 101, 596, 597–598

C

Cache Policy, 649
Calculator, AccuDraw, 223–224
Callout Leader, Place, 611
Camera settings, 138, 139, 792–794
Caps, multi-line, 360
   control placement of, 376
   define, 363–364
   Edit (ml e c), 378
Capsule frames, dimension text, 451
Capture images, 604–605
Cascade windows (w c), 127–128
CD=, 535
CEL files, 497
Cell buttons
   add, 526
   customize, 520–525
Cell library, 494–495
   3D design and, 792
   add buttons from, 526
   Attach, 497–498
   Compress, 535
   create HTML files from, 655–657

Index

create new (di ce), 495–497
create/replace cells in, 500, 502, 535–536
delete cells from, 534–535
display full path to, 525
edit buttons from, 522
edit cells in, 534
enhancements for, 45
pattern cell from, 546
remote opening of, 652
shared cells in, 536–538
Cell Selector box
  create and use files for, 526–528
  customize, 520–525
  open, 520
Cells
  3D designs with, 792
  configuration variables for, 710
  creating
    concepts for, 498–500
    new version of, 535–536
    steps for, 500–502
  Data Fields for, 307
  delete, 534–535
  dimension symbol, 437
  dimension terminator, 449
  dimension-driven design with, 639
  edit information on, 534
  enhancements for, 45
  housekeeping tools for, 528–533
    Drop Complex Status (dr e), 531–532
    Drop Fence Contents (f dr c), 532–533
    Fast Cells View, 533
    Identify Cell (i c), 528–529
    Replace Cells (rep cells e), 529–531
  pattern, 503–504, 525, 546, 547
  pen table, explode, 477
  placing active, 504–506
    Absolute/Relative modes for, 506–507
    Cell Matrix (matr c) for, 508–510
    Interactive mode for, 507
    line terminator tool (pl t) for, 510–512
    point tools (pl po) for, 513–519
    select tools for, 503–504
  print boundary for, 486
  settings group for, 668, 670–671
  shared, 536–539
  types of, 503
  uses for, 494
Center
  Place Arc by (pl a c), 60–61
  Place Circle by (pl ci c), 55–57
  Place Ellipse By (pl el ce co), 152–153, 235
  plot output, 470, 488
Center Mark, dimension from (dim c m), 424–425, 445
Center snap mode, 84, 88, 91
Center-justified text, 450
Centesimal degrees, 457

Centimeters, working units in, 31
CGM (Computer Graphics Metafile) format, 600, 601, 722
Chains
  complex
    create, 345–349
    drop status of (dr e), 358–359
  place dimensions in (dim lo s), 400–401
Chamfer
  Construct (ch), 166–167
  Dimension Angle of (dim ang ch), 418
  Dimension (dim sy), 407–408
  Edges (ch e), 790
  Between Surfaces (ch su), 790–791
Chamfered vertices
  modify, 338
  place SmartLines with, 237, 238–239, 241–242
Change
  Area (chan a), 556
  Batch Print specifications, 485, 487, 489, 491
  cell button size, 524
  constraints, 623
  Dimension to active (cha dim), 432, 459
  element attributes, 106–108
  fill type (chan f), 560–561
  highlight and pointer color, 598–599
  Normal Direction (chan n), 786
  solid/surface status (chan su c), 775
  Tags (chan t), 321–323
  text attributes (modi te), 301
Change Attributes toolbox, 106–107, 109
Characters
  Data Field, 307–308
  menu shortcut, 733, 734
  placing, point tools for, 512–519
Check boxes, 9
  AccuDraw, 216
  AccuSnap, 95
  Lock Toggles, 6, 7
  Use Shared Cell, 536–537
  View Attributes, 138–140
Child text styles, 291
choose element (cho e), 252
Circle Fence, Place (pl f c), 260–261
Circles
  arcs appearing as, 343
  construct attached (at c), 633–634
  dimensioning
    Center (dim c m), 424–425
    Diameter (dim d), 421–424
    Distance (dim ar dist), 426
    Location (dim ar l), 416–418
    Radius (dim radiu), 419–421
    Size (dim ar s), 415–416
  Place
    By Center (pl ci c), 55–57

By Diameter (pl ci d), 58–59
By Edge (pl ci e), 57–58
Place Polygon inside (pl pol in), 156–157
Place Polygon outside (pl pol c), 157–158
Circular Fillet, Construct (fill n/m/s), 164–166
Circumscribed Polygon, Place (pl pol c), 157–158
Class, element
   place flags by, 608
   selection by, 595
Clear
   cell button configuration, 523–524
   selection set, 248
Clip Back/Front, 138, 139, 569
Clip Fence, 264
Clip Reference (refe c bo), 576–577
Clip Volume, 138, 139
Clipping mask, reference, 577–579
Close
   designs (clo d), 42
   files, Undo and, 146
   line style library files, 679
   SmartLines, 239–240
   view windows, 27, 124–126
   Web browser, 648–649
Closed arrowheads, 447
Closed Cross Joint (jo cr c), 365–366
Closed Tee Joint (jo t c), 369
CO=, 103, 296–297
Color
   AccuDraw compass, 218, 227
   cell button, 521
   configuration variables for, 710–711
   dimension component, 440, 442, 448, 451
   element
      change, 106–108
      match, 109
      set, 103–104
   filter levels by, 676
   highlight/pointer, change, 598–599
   level, 101, 596–598
   multi-line, 363
   selection by, 591, 593
   SmartLine options for, 240
   text
      enable, 293
      key-ins for, 296
      select, 209
      values for, 292
   vector output, 467
Color options menu, 103–104
Columns
   active cell matrix, 509, 510
   rectangular array, 199, 200
Command button, puck, 4–5, 29
Commands
   input methods for, 28–30
   Undo/Redo for, 146–148
Comments, border, 472

Compass, AccuDraw
   color settings for, 227
   coordinate system for, 217–218
   display, 225, 226
   features of, 215–216
   indexing, 220, 229, 231
   move origin of, 225, 234
   orthogonal axes for, 218
   rotating
      key-in shortcuts for, 232
      process for, 218–219, 235
      settings for, 229, 230
Composite solids, creating, 781–786
   difference operation (constru d) for, 784–786
   intersection operation (constru i) for, 783–784
   union operation (constru u) for, 781–783
Compound line styles, 682, 689
Compress
   cell library, 535
   design, pattern deletion and, 557
Compressed files, extracting, 744–745
Cone, Place (pl co), 772–773
Configuration
   Cell Selector button, 521–522
   grid, options for, 81
   hardware, 2–5
Configuration files
   plot, save, 473–474
   user, create, 691–692
Configuration variables
   browser, 646
   by category, 710–723
   cell library, 495
   create/delete, 709
   dialog box for, 706–707
   edit, 707–708
   line style, 677
   plot file, 464
   select, 708
   uncategorized, 724–726
Confirmations, AccuDraw pop-up, 227, 228
Connect
   to Browser, 646–647
   to URLs, 660–661
Constant rendering, 795
Constants, Assign (as v), 635–636
Constrain
   coincident points (con c), 629–630
   Elements (con e), 626–627
   intersection points (con i), 628–629
   location (con l), 631
Constraints
   changing, effect of, 623
   definition of, 624
   dimensional
      Assign Variable (as v) to, 635–636
      Modify Value (mo e) of, 638
   ellipse, set, 153, 155

Index

example design with, 622–623
multi-line, apply, 163
Re-solve (up m), 638
tools for, 625
Construct
  Array, 198–202
  attached ellipse/circle (at c), 633–634
  attached line-string/shape (at ls), 631–632
  Chamfer (ch), 166–167
    edge (ch e), 790
    between surfaces (ch su), 790–791
  cut (constru cut), 788
  Difference (constru d), 784–786
  extrude along (constru e a), 778
  Fillet
    circular (fill n/m/s), 164–166
    edge (fill e), 789
    between surfaces (fill su), 789–790
  Intersection (constru i), 783–784
  Joint
    Closed Cross (jo cr c), 365–366
    Closed Tee (jo t c), 369
    Corner (jo c), 372
    Merged Cross (jo cr m), 368
    Merged Tee (jo t m), 371
    Open Cross (jo cr o), 367
    Open Tee (jo t o), 370
  Points
    Along Element (constru po a), 517–518
    Between Data Points (constru po b), 513–514
    at Distance Along (constru po d), 518–519
    at Intersection (constru po i), 516–517
    Projected Onto Element (constru po p), 515–516
  shell (constru s), 780
  surface projection (constru s p), 777–778
  surface revolution (constru s r), 779
  thicken (constru th), 780–781
  Union (constru u), 781–783
Construction elements
  constraints on, 622–623
  definition of, 624
  draw, 625–626
  view attributes for, 138, 139, 627, 639
Contents, Help menu, 42–43
Context sensitivity, AccuDraw, 225–226
Convert
  3D, 775
  image, 605
Coordinate readout, 35–36
Coordinate systems
  3D design
    AccuDraw and, 776
    drawing, 759–760
    view, 760–761
  AccuDraw
    settings for, 229–231
    switching (a m), 217–218
  auxiliary, 761–767

polar, 69–70
rectangular, 67–69
use, 70–73
Coordinates
  design plane, 30–31
  export, 619–620
  import, 621
  place labels for, 617–618
  set depth by, 754–755, 757
  tentative snap point, 96
  types of, 66
Coordinates box, AccuDraw
  features of, 215–216
  floating or docking, 217
  input fields in, 220–222
  perform calculations in, 223–224
  recall values in, 222–223
Copies
  Mirror (mi c), 194–198
  Rotate (ro c), 190–194
  Scale (sc c), 185, 187–189
Copy
  Data Fields, 311
  Element (cop el), 179–181
    AccuDraw and, 236
    parallel (cop p), 183–185
    selected element, 257
  graphic group, 588
  models, 37–38
  printer specification, 485
  reference attachments (refe c), 564, 568, 573
  tag sets, 317
  text, 209, 210, 212
  tool frames, 731
  tools, 728
Copy and paste
  cell buttons, 522–523
  into Word documents, 639–640
Copy/Increment
  Data Fields (incr e), 311–312
  Text (incr t), 302–303
Corner Joint (jo c), 372
CR=, 534
Create. *See also* Define
  3D design file (cr d), 749–750
  AccuDraw shortcuts, 229
  cell libraries (di ce), 495–497
  Cell Selector files, 526–527
  cells, 498–503, 535–536
  Complex Chain (cr ch), 345–349
  Complex Shape (cr s), 345–346, 349–352
  composite solids, 781–786
  configuration variables, 709
  dimension styles, 435–436
  dimension-driven cell, 639
  dimension-driven design, 625–637
  export files, 619
  extension lines, 442

835

function key definitions, 738–741
HTML files, 653–658
level filters, 674–676
line style library files, 679
line styles, 681, 682–689
menus/menu items, 735–736
models, 36–38
pen tables, 475
printer specification, 485
print/plot files, 472–473
Region, 352–357
settings group, 667–668
tag reports, 323–327
tag set and tags, 314–316
tag set library, 327–328
text styles, 288, 289–290
tool box, 727–728
tool frame, 730–731
user interface, 692–693
View Group, 135–136
workspace, 691–692
Create Package wizard, 744
Crosshatch area
   example of, 544
   placing, methods for, 549–554
   settings for, 545–546
   tool for, 543–544, 549
CTRL key, flood patterning with, 553
Cumulative area, measure, 388
Cumulative length, measure, 387
Cursor
   place data points with, 48–49
   pointer-type, 6
   text, moving, 8, 210
Cursor menus, 29–30
Curves
   AccuSnap to, 96–97
   Fast, enable, 138, 139
   Place Point/Stream, 160–162
   place text along, 279–281
Customizing, 663
   Cell Selector, 520–525
   dimension symbols, 437–438
   font library, 741–743
   function keys, 738–741
   level filters, 674–676
   line styles, 676–689
   linear dimensions, 403
   packaging environment in, 744–745
   scripts and macros in, 746
   settings groups, 664–674
   units, 32–33
   workspace, 690–738
      configuration variables for, 706–726
      preference settings for, 693–705
      set up active space for, 690–693
      user interface in, 726–738

Cut
   cell buttons, 523
   elements
      IntelliTrim for, 167–168
      Quick mode for, 172–173
   multi-lines
      All Component (cu a), 374
      Single Component (cu s), 373
      Uncut (un) for, 375
   Solid (constru cut), 788
Cutting elements
   Advanced Extend with, 176, 177
   Advanced Trim with, 174, 175
Cylinders
   create composite solids from, 782, 784, 785
   Place (pl cy), 771–772
Cylindrical coordinate system, 762

D

D (TriForma Document) format, 600
Dash strokes, define, 684–685
Dashed lines, depth and, 755, 757
Data buttons, mouse or puck, 3–5, 28
Data Fields
   characters for, 307–308
   Copy (cop e), 311
   Copy/Increment (incr e), 311–312
   edit text in, 312
   fill in (edi s/au), 309–310
   set justification for, 308–309
   view attributes for, 138, 139, 308
Data files, configuration variables for, 711–712
Data points. *See* Points
Database
   configuration variables for, 711
   preference settings for, 695
Date, replace pen table text string with, 479
dd=, 754–755
Decimal numbers, 31
Default settings
   AccuDraw origin, 225, 226
   cell button, 524–525
   grid unit, 78
Define. *See also* Create
   ACS, 761–765
   cell buttons, 522, 524–525
   Cell Origin (d c o), 499–500, 501
   Custom Units, 32–33
   Drafting Tool Properties, 616
   Flag Information, 609, 610
   multi-lines, 359–365
   Tag/Tag Sets, 314–316
Degrees, angle dimension, 457
Degrees of freedom, dimension-driven design, 624
Delayed Update, AccuDraw, 227
Delete
   ACS, 767
   cell buttons, 523–524

Index

cells, 534–535
configuration variables, 709
dimension style, 436
Dimension Vertex (del dimv), 430–431
Element (del), 64–65
    cut and, 173
    selected element, 256–257
    Undo and, 148
Engineering Link, 662
Fence Contents, 148, 263–266, 269
function key definitions, 740
line style, 681
Model, 38
multi-line component, 362
Partial
    element (del p), 177–178
    Multi-line (ml p d), 376
Pattern (del pat), 557
pen table items, 476, 479
printer specification, 485
reference clip mask (refe c m d), 578–579
Saved View, 144
settings group, 667
shared cell copy, 539
tag/tag set, 316
text, 211, 212
Text Styles, 290
Vertex (del v), 339–340
View Group, 137
Delta Angle, polar array, 200–201, 202
Delta distance, stream curve, 161, 162
Depth
    3D cells and, 792
    active, 756–758
    display, 753–756
    referenced model, 568
Description fields
    ACS, 767
    cell, 502, 521, 525, 534
    custom tool, 729
    reference file, 567, 569
Deselect text, 211–212
Design cube, 753
Design File Settings
    Axis, 83
    Color, 598–599
    Coordinate Readout, 35–36
    Grid, 79–81
    Working Units, 31–32
Design files
    3D, create, 749–750
    attach cell library to, 497–498
    change properties of, 38–40
    configuration variable for, 712
    convert format of, 602–603
    copy elements from, 639–640
    create fence from, 262
    create HTML files from, 653–655
    create new, 10–15
    delete shared cells from, 539
    enhancements for, 45
    fit display depth to, 756
    import and export formats in, 603
    linking and embedding in, 639–643
    opening, 16–17
    Packaging utility for, 744–745
    Place Fence from Active (pl f a), 262
    plotting
        Batch Print for, 483, 491–492
        print settings for, 472
    references to, 564–566
        attach, 566–571
        manipulate, 571–583
    remote opening of, 649
    saving and exiting, 40–42
Design models, create, 36–38. *See also* Models
Design plane, 30–31, 753. *See also* Drawing plane
Designs. *See also* 2D designs; 3D designs; Dimension-driven design
    begin new, 10–15
    history for, 45
    level display controls for, 102
    level symbology in, 596, 597–598
    plotting, process for, 464–465
    view area controls for, 111–116
Destination files, font, 743
Detach/Detach All references, 580–581
Detail Marker, Place, 615–616
DGN file format, 14, 600. *See also* Design files
DI=, 69–70, 760–761
Dialog boxes, 6–10
Diameter
    dimensioning (dim d)
        select symbols for, 438
        tools for, 421–426
    place circles by, 57, 58–59
Dictionary, Edit User, 299
Difference
    create composite solids (constru d) from, 784–786
    Create Region (cr r d) from, 355–356
    Measure Area (me ar d) of, 389–390
    patterning, 548, 552
Digitizing tablet, 4–5
Dimension lines
    adjust, 444–445
    description of, 393, 394
    relative, 445
    settings for, 438–440
Dimension styles, 396
    create and modify, 434–458
    settings group for, 664–667, 668, 671–672
Dimension text
    description of, 393, 394
    edit, 398–399

place, 419
settings for
　　location, 444–445
　　placement/appearance, 449–451
　　prefix/suffix symbol, 437–438
　　unit format, 456–458
Dimensional constraints, 624
　Assign Variable (as v) to, 635–636
　Modify Value (mo e) of, 638
Dimension-driven cells, 639
Dimension-driven design
　constraints in, 622–623
　create, 625–637
　create cells from, 639
　modify, 637–639
　terms for, 624
　tools for, 624–625
Dimensioning, 392
　3D designs, 792
　alignment controls for, 395–396
　angular, 409–418
　associative, 395, 634
　element, 408
　enhancements for, 45
　linear, 396–408
　miscellaneous tools for, 427–434
　ordinate, 427–428
　radial, 418–426
　settings for (di dims), 434–436
　　change, 432, 459
　　custom symbols, 437–438
　　dimension lines, 439–440
　　extension lines, 441–442
　　leaders, 440–441
　　matching, 433, 458–459
　　placement, 443–445
　　terminators/symbols, 446–449
　　text, 449–451
　　tolerance, 452–453
　　tool, 454
　　units/unit format, 454–458
　styles for, 396
　terminology for, 393–394
　tools for, 392–393
Dimensions
　AccuSnap to, 96–97
　add to dimension-driven design, 634
　Change (cha dim) to active, 432, 459
　Delete Vertex (del dimv) for, 430–431
　Insert Vertex (ins dimv) for, 430
　Label Lines (l l) with, 428–429
　Match Settings (mat dim) for, 433, 458–459
　Modify Location (mod dim loc) of, 431
　Reassociate (dim reas), 431–432
　view attributes for, 138, 139
Direction, Change Normal (chan n), 786
Directories, 12, 14

Disconnect from Browser, 648–649
Display. *See also* View windows; View(s)
　3D primitives, 768
　AccuDraw, 227–229
　Batch Print, specification for, 483, 489–491
　cell buttons, 521
　dimension units, 456–458
　Engineering Links, 660
　grid, 78–79
　images, 606
　levels, 98–99, 102–103
　multi-line joints, 365
　references, 568, 582–583
　saved views, 141
　by selection criteria, 594
　system requirements for, 3
　tags, 318, 321, 322
　Text Attributes (i t), 302
　view groups, 134–135
Display depth
　3D cells and, 792
　fitting, 756
　set active depth to, 758
　set (dep d), 754–755
Distance. *See also* Length
　AccuDraw
　　indexing, 222, 229, 231
　　recall previous, 222–223
　　roundoff values for, 229, 230
　　Smart Key-ins for, 226
　advanced settings for, 34
　chamfering, 166
　construct points along (constru po d), 518–519
　Dimension Arc (dim ar dist), 426
　key-in, extend lines by, 333
　measurement of, 382–385
　Move Parallel by, 183, 185
　readout settings for, 35
　specify points by, 69–70
　stream curve, 161, 162
　units for, 31
Distant light, 794
Divisors, Keypoint snap, 90
DL=, 68–69, 759–760
Docking
　AccuDraw window, 217
　tool boxes, 20–21, 129
Dots
　place, 512–519
　settings for, 448–449
Downloading files, 649–650
dp=, 754
Drafting tools, 607, 611–616
Dragging
　handles, 255–256
　selected elements, 254–255
　tool boxes, 19–21

Index

Drawing composition, 796
Drawing coordinate system, 759–760
Drawing plane
   align dimensions to, 395, 396, 443–444
   orient/rotate
      2D design and, 218–219, 232, 235
      3D design and, 776–777
Drawing properties, 38–40
Drawing tools, 77–97
   AccuSnap, 94–97
   Axis Lock, 83
   grid system, 78–83
   Snap Lock, 84–94
Drawing(s). *See also* Designs
   3D, advantages of, 747–748
   construction elements, 625–626
   import and export, 600–603
   settings group for, 664–673, 674
Drive, specifying, 14
Drop Element (dr e), 358, 531–532, 538–539
Drop Fence Contents (fe dr), 269, 359, 532–533
Drop From Graphic Group (gr d), 588
Drop Line String/Shape Status (dr st), 65–66
Drop tool box, 66
Drop-down menus. *See* Menus, drop-down
Duplicate. *See also* Copy
   line styles, 682
   multi-lines, 362
   tag sets, 317
dv=, 144
DWG file format, 600, 722
DX=, 760–761
DXF file format, 600, 722
Dynamic Display, panning with, 121–122
Dynamic highlights, flood area, 553
Dynamic update, view, 138, 139
Dynamically adjusted dimension, 444–445
dz=, 757

E

Edge
   Chamfer (ch e), 790
   Fillet (fill e), 789
   line segment, control shape of, 238–239
   Place Arc By (pl a e), 61–63
   Place Circle By (pl ci e), 57–58
   Place Ellipse By (pl el ed co), 154–155
   Place Polygon By (pl pol ed), 158–160
Edit. *See also* Modify/Modification tools
   AccuDraw shortcuts, 228–229
   cells/cell buttons, 521–522, 534
   configuration variables, 707–708
   data fields, 309–310
   dimension styles, 436
   Engineering Links, 661
   Flag, Show and (sh f), 610
   function key definitions, 739–740
   line styles (lines e), 678–689
   model properties, 38
   multi-line components, 361–362, 378
   reference URL, 582
   saved view properties, 143–144
   settings, 666–674
   tags, 317, 318–320
   text, 299–300
      Data Field, 312
      dimension, 398–399
      keyboard shortcuts for, 210, 212
   text styles, 289–290
   tool box, 727–729
   tool frame, 730–731
   view border, 737–738
   view group properties, 136–137
Edit field, dialog box, 8
Edit menu, 22, 147
Element attributes. *See* Attributes, element
Element points, place, 512–519
Element Selection tool box, 251–252
Elements
   2D, add to 3D design, 781
   active/display depths for, 756
   attach tags to, 317–318
   change area of (chan a), 556
   constrain (con e), 626–627
   construct points along (constru po a), 517–518
   coordinate labels for, 618
   create cells from, 499
   create fence from, 261
   define ACS by (d a e), 763
   dimensioning, 408–409
   embedding, 639–640
   extruding, 747, 777–778
   filling, 558–561
   grouping (*See* Groups)
   locating, 45, 63–64, 96, 179
   manipulating, tools for, 63–73, 179–202
      AccuDraw, 236
      Construct Array (ar r/p), 198–202
      Copy (cop el), 179–181
      Delete (del), 64–65
      Drop Line String/Shape Status (dr st), 65–66
      handles for, 245, 246
      Identify Automatically for, 63–64, 96
      Mirror (mi o/c), 194–198
      Move (mov e), 181–182
      Move Parallel (mov p/cop p), 182–185
      Precision Input, 66–73
      Rotate (ro o/c), 190–194
      Scale (sc o/c), 185–189
      SmartLine, 237–241
      Undo (und), 148
   measurement of
      Area (me ar e), 388
      Distance, 383–385

modifying, tools for, 164–178, 331–378
   AccuDraw with, 338
   Arc, 341–345
   Construct Chamfer (ch), 166–167
   Construct Circular Fillet (fill n/m), 164–166
   Create Complex Chain/Shape (cr ch/s), 345–352
   Create Region (cr r), 352–357
   Drop Complex Status (dr e), 358–359
   Extend Line (ext l), 331–336
   Multi-line Joints, 365–378
   Multi-lines, 359–365
   Partial Delete (del p), 177–178
   Trim (tri m), 167–177
   vertex/vertices (modi e), 336–341
patterns for, 548, 549, 555–556
Pen Table criteria for, 479–480
Place Fence from (pl f e), 261
Place Text (pl tex) in relation to, 275–281
placement tools for, 47–63, 151–163
project points onto (constru po p), 515–516
referenced
   clipping, 576–577
   highlighting, 583
selected (*See* Selection set)
selecting
   By Attributes, 591–596
   Element Selection tool for, 251–252
   pointer for, 179
   PowerSelector tool for, 245–251
snapping to
   AccuSnap for, 94, 96–97
   Tentative button for, 87–89
solid (*See* Solid models)
E-Links tool box, 646, 659–662
Ellipse
   construct attached (at c), 633–634
   Place By Center and Edge, 152–155
Ellipses tool box, 56–59, 152–154
Embedding elements in documents, 639–640
End, place point symbol at, 687, 688
End caps, multi-line, 360, 363–364
Engineering Links
   Attach, 659–660
   configuration variables for, 712–713
   Delete, 662
   Display, 660
   Edit, 661
   Follow, 660–661
English units, 31–32
ENTER key, 8, 9, 25
Equations
   add text for, 634–635
   Assign (as e), 636
Evaluate Constraints tool box, 625
Excel spreadsheets, insert, 639, 641–642
Exchange reference, 591

Exclusive selection mode, 593
Execute selection criteria, 591, 592, 593, 596
Exit
   Line Style Editor, 681
   MicroStation program (exi), 42
Explode Cells, pen table, 477
Export
   coordinates, 619–620
   drawings, formats for, 600, 601, 603
   tag sets, 327–328
Extend elements
   Advanced mode for, 176–177
   graphically (ext l), 332–333
   IntelliTrim tool for, 167–168
   to Intersection (ext l 2/in), 333–336
   by key-in (ext l k), 333
   Modify tools for, 331–332
   Quick mode for, 170–171
Extension lines, dimensioning
   description of, 393, 394
   settings for, 441–442
Extensions, configuration variables for, 713–715
Extract compressed files, 744–745
Extrude 2D elements, 777–778

F

Faces
   Locate By, 769
   relocate (str f), 786–787
   Remove (rem f), 787
Fast Cells view, 138, 139, 533
Fast Curves view, 138, 139
Fast Font view, 138, 139
Feature control frames, 433
Feel settings
   AccuSnap, 97
   Preference, 695–697
Feet, working units in, 31–32
Fence tool box, 258
Fences
   3D, 791
   Change Tags (chan t) in, 322
   coordinate labels for, 618
   create cells from, 500–501
   Drop Contents (fe dr) of, 269, 359, 532–533
   manipulate contents of, 268–269
   Measure Area (me ar f) of, 388–389
   modes for, 263–266
   modify shape (modi f) of, 266–267
   Move (mov f), 267–268
   patterns for, 548, 550, 555–556
   Place (pl f), 259–262
   print settings for, 467, 472
   references with
      clip boundary (refe c bo) for, 576–577
      clip mask (refe c m) for, 577–579
      Mirror (refe mi), 576

Index

Rotate (refe ro), 575
Scale (refe s), 574
remove, 271
replace/update cells in, 530, 531
Status bar field for, 26
Stretch (f st), 270–271
Undo delete for, 148
File formats
enhancements to, 44, 45
exchange, import and export, 600–603
printer/plotter driver, 468–469
File menu, 16–17, 22
File names
ACS, 767
cell library, 497
design, 12, 13–14
reference, 567
replace pen table text string with, 478
tag report, 323, 324
URL-style, format for, 648
Files. *See also* Design files; Reference files; *specific type*
archive, 651, 710
backup, 721
closing, Undo and, 146
extract, 744–745
Fit View (fit v e) for, 117, 756
import views from, 144–145
merge, 589–590
remote opening of, 646–652
seed, 10, 14–15, 720, 749
source, select fonts from, 742, 743
specify path to, 14
URL format for, 648
user configuration, 691–692
Fill
cell button, display, 521
element
Change to Active (chan f), 560–561
type/color options for, 558–559
view attributes for, 559–560
multi-line, set, 363
SmartLine options for, 240
view attributes for, 138, 139
Fill in
Data Fields, 309–310
Text Nodes, 306–307
Filled arrowheads, 447
Filled Hidden Line rendering, 795
Fillet
Construct Circular (fill n/m/s), 164–166
Edges (fill e), 789
Surfaces (fill su), 789–790
Filters, level, 674–676
Find
tag values, 321, 322
tools, 43–44
view windows, 133

Fit View (fit v e), 116–118, 756
Fitted text, Place (pl tex f), 274–275
Fix Point at Location (con l), 631
Flags
Place (pl fl), 608–609
Show/Edit (sh f), 610
Update (fl u), 610–611
Flashbulb lighting, 794
Floating AccuDraw window, 217
Floating origin, AccuDraw compass, 225
Flood
Create Region (cr r f) from, 356–357
Measure Area (me ar fl) of, 390–391
patterning with, 548, 552–555
Focal length, camera, 793–794
Focus, AccuDraw coordinate, 217, 221
Follow Engineering Link, 660–661
Font Installer, 741–743
Fonts. *See also* Text styles
change, 208, 299
dimension text, 451
geometric tolerance, 433
install, 741–743
key-in settings for, 296
line weight for, 209
select, 206–207, 291
stacked fractions and, 286
Fork points
create chains with, 347, 348
create shapes with, 350, 351–352
Four-button puck, 4–5
Fractions
single-character, 285–286
working units in, 31
FT=, 296–297
FTP (file transfer protocol), 648
Function keys, 738–741

G

Gap, set maximum
for chain, 347, 348
for flood pattern, 554
for region, 356, 357
for shape, 351
Gap Mode, Move Parallel, 183, 184, 185
Gap Size, cell button, 524
Gap strokes, define, 684, 686
Generate
tags report, 325–326
tags template, 323–325
tolerance, 452
Geometric constraints, 624, 625
Geometric Tolerance, 433–434
Geometry
dimensioning, set, 439, 442, 447, 453
link URLs to, 659–662
Simplify complex chain/shape, 346, 348, 350, 352

841

Global Origin. *See also* Origin
  AccuDraw settings for, 225, 226
  design cube, 753
  location of, 67
  relocate (GO=), 30–31, 68
Global Update, cell, 530
GR=, 80
Graphic cells, 502, 503
Graphic groups, 586–588
Graphically Move Parallel, 184–185
Graphically Scale, 3-Points method for, 186, 188–189
Grayscale, print vectors in, 467
Grid Lock, 78
  element manipulation and, 179
  settings for, 81–83
  tentative snaps and, 88
Grid system
  settings for, 78–83
  view attributes for, 138, 139
Groups. *See also* Selection set
  attached view, 796
  change attributes of, 257
  create complex chain/shape from, 345–352
  create (gr s), 252–253
  Create Region (cr r) from, 352–357
  Drop Complex Status (dr e) of, 358–359
  Fence manipulation tools for, 258–271
  Graphic, create and manipulate, 586–588
  Lock/Unlock elements in, 254
  selecting elements for
    Automatically Identify feature for, 63–64, 96
    Element Selection (cho e) for, 251–252
    PowerSelector (pow) for, 245–251
  Settings, 664–674
  Stretch (f st), 271
  Ungroup (ung), 253–254
GU=, 80

H

Handles, 245
  dragging, modify shape by, 255–256
  elements with, 246
  grouped elements with, 252–253
Hard copy, 463
Hard drives, 2
Hardware configuration, 2–5
Hatch area
  example of, 544
  modify settings for, 669–670
  placing, methods for, 549–554
  tool for, 543, 545–546, 549
Height settings
  arrowhead, 447
  dimension text, 451
  text, 205–206, 292
    change, 299
    key-ins for, 296
Help, 22, 42–44

Hidden levels, display, 102
Hidden Line rendering, 795
Highlight
  change color of, 598–599
  elements, 96, 179
  references, 583
  text, 8
Hint, Show Tentative, 95
History, design, enhancements for, 45
Holes, pattern, 552–553, 555–556
Horizontal axis, Mirror About (mi o/c h), 195–196, 576
Horizontal dimension text, 449–450
HTML Author, 653–658
HTML files, create
  from Basic macros, 657–658
  from cell library, 655–657
  from design file saved view, 653–654
  from design file snapshot, 654–655
HTML link, create, 659

I

Icons, display of
  Locks, 26
  snap mode, 87, 95
Identify
  Cell (i c), 528–529
  elements, 63–64, 96, 179
  text (i t), 302
Identify button, mouse/puck, 28
IEEE 64-bit floating point format, 30
IGES format, 601, 723
Image Plane Orientation, camera, 793
Images
  configuration variables for, 719–720
  flag, select, 608–609
  HTML, format for, 654, 656
  import/export, 601
  manipulate, 604–607
Import
  coordinates, 621
  drawings, formats for, 600, 601, 603
  fonts, 741, 743
  line styles, 680
  models, 38
  saved views, 144–146
  tag sets, 328–329
  text files, 295–297
  text styles, 294
In Line dimension text, 449–450
Inches, working units in, 31–32
Include Text File dialog box, 295
Inclusive selection mode, 593
Increment, Copy and
  Data Fields (incr e), 311–312
  Text (incr t), 302–303
Indents, key-ins for, 296
Indexing axis/distance, 220, 222, 229, 231

Index

Indicators, Text Node, 304
Individual selection methods
   Element Selection, 251
   PowerSelector (pow ar i), 247, 250
Input
   devices for, 3–5
   methods for, 28–30
   Preference settings for, 695
Inscribed Polygon, Place (pl pol in), 156–157
Insert
   cell buttons, 522
   Dimension Vertex (ins dimv), 430
   menu/menu item, 735–736
   multi-lines, 362
   objects in design files, 640–642
   pen table sections, 476
   pen table text, 477–479
   tools, 728–729
   Vertex (ins v), 340–341
Inside Fence modes, 263–264, 270
Inside terminator placement, 446, 447
Installing fonts, 741–743
IntelliTrim, 167–168
   Advanced Extend with, 176–177
   Advanced Trim with, 174–175
   Quick Cut with, 172–173
   Quick Extend with, 170–171
   Quick Trim with, 168–169
Interactive cell placement, 507
Interchar spacing, 205, 280
Interface
   new design, select, 10, 11
   user, customize, 726–738
   workspace, set up, 690, 691–693
Interface modification files, 726–727
Internet, 645
   access resources on, 646–649
   link geometry to, 659–662
   open files on, 649–652
   publish data to, 653–658
Internet Explorer (Microsoft), 646–647
Intersection
   Constrain Point At (con i), 628–629
   Construct Point at (constru po i), 516–517
   create composite solids from (constru i), 783–784
   Create Region from (cr r in), 352–353
   Extend Element(s) to (ext l in/2), 333–336
   Measure Area of (me ar i), 389–390
   patterns for, 548, 550–551
Intersection snap mode, 92
Invert selection set
   Element Selection mode to, 251
   PowerSelector (pow m i) mode to, 248–249, 250
Isometric view
   3D design and, 749–750
   configure grid for, 81
Italic text, 208, 293

J

Job set, save, 492
Join
   extension lines, 442
   SmartLines, 239
Joints, multi-line, 360
   modifying, 365–378
      Closed Cross (jo cr c), 365–366
      Closed Tee (jo t c), 369
      Corner (jo t c), 372
      cut lines (cu s/a) for, 373–374
      Edit Multi-line Cap (ml e c) of, 378
      Merged Cross (jo cr m), 368
      Merged Tee (jo t m), 371
      Open Cross (jo cr o), 367
      Open Tee (jo t o), 370
      uncut lines (un) for, 375
   settings for, 364–365
Justification, text
   Above Element, 277
   Along Element, 280
   data field, 308–309
   dimension, 450
   On Element, 279
   fitted, 274
   menu options for, 207–208
   note, 284
   Text Style settings for, 293
   word wrap and, 283

K

Keep Original option, Create Region, 352, 353
Keyboard, 3, 28
Key-in shortcuts
   AccuDraw, 217
      edit, 227, 228–229
      list of, 231–233
      lock fields with, 221
      move compass with, 219
   cell, 504, 522, 525
   menu name, 733, 734
   pop-up confirmations for, 228
   Text Editor window, 210–212
Key-in window, 17, 25
Key-ins
   enter commands by, 25, 28
   extend lines by (ext l k), 333
   precision coordinate input, 758–761, 762–763
   rotate view by, 752
   Smart, 225, 226
   text/element attribute, 296–297
Keypoint Sensitivity, AccuSnap, 97
Keypoint snap mode, 89–90
KY=, 90

L

Label
   coordinates, 617–618

dimensions, 455
Lines (l l), 428–429
menu, change, 732–734
Label item, dialog box, 8
Layout specification, Batch Print, 483, 487–489
LC=, 104
Leaders
  callout, place, 611
  dimension
    description of, 394
      place extended, 420–421, 422–423
      settings for, 440–441, 447
  notes with, 284
  text with, 612
Leading zeros, dimensions with, 458
Left-justified dimension text, 450
Length. *See also* Distance
  angle dimension, set, 457
  Measure (me l), 386–387
  multi-line, constrain, 163
  place arcs by, 61, 63
  place lines by, 49
  place shapes by, 53
Level Display settings box, 98, 99
Level Manager, 98, 99
Level names, key-in for, 296, 297
Level Properties dialog box, 101–102
Levels
  active cell, 506, 507
  dimension, 439
  element, 97–98
    change, 106–108
    change properties of, 101–102
    control display of, 102–103
    create new, 100–101
    manage, tools for, 98–99
    match, 109
    set active, 99–100
    set attributes for, 101
  enhancements for, 45
  filters for, 674–676
  flag, 608
  selection by, 591, 593
  Status bar field for, 26
  symbology for
    override, 439
    set and use, 596–598
    view attributes for, 138, 139
Lighting
  reference model, 569
  rendering with, 794
Limits, dimension tolerance, 452–453
Line spacing, text
  key-in for, 296
  By Origin method and, 205
  settings for, 276, 280
Line Style Editor, 678–689

Line Style menu, 105
Line styles
  custom, 676–689
    create and modify (lines e), 678–689
    select, 676–677
    settings for (lines s), 677–678
  dimension component, 440, 442, 448
  element
    change, 106–108
    match, 109
    settings for, 104–105
  filter levels by, 676
  level, setting, 101, 596–598
  reference model, 569
  selection by, 591, 593
  view attributes for, 138, 139
Line Terminator, Place Active (pl t), 510–512
Line text frames, 283, 284
Line weight
  dimension component, 440, 442, 448, 451
  element
    change, 106–108
    match, 109
    settings for, 105–106
  filter levels by, 676
  key-in for, 296
  level, setting, 101, 596–598
  selection by, 591, 593
  text, 209
  view attributes for, 138, 140
Line Weight menu, 106
Linear dimensioning
  alignment controls for, 395–396, 443–444
  arrows with (dim si a), 397–398
  for chamfer, 407–408
  custom, 403
  for element, 408–409
  example of, 392
  half, 406–407
  from location, 400–403
  perpendicular to point/line, 404–405
  with strokes, 399–400
  symmetric, 406
  tools for, 393, 396–397
Linear Elements tool box
  move and resize, 19–20
  use, 48, 160–163, 237
Lines
  constrain angle (con e) of, 626–627
  Dimension Angle Between (dim a l), 413
  Dimension Size Perpendicular to (dim si p l), 405
  display start/end caps as, 364
  Extend (ext l), 331–336
  Label (l l), 428–429
  Mirror About (mi o/c l), 195, 197–198
  multi-line
    Cut Single/All (cu s/a), 373–374

Index

define, 361–363
description of, 360
Move (ml e p), 376–377
Uncut (un), 375
Place (pl l)
coordinates for, 70–73
pointing device for, 48–49
SmartLine tool for, 237–242
Place Text (pl tex) in relation to, 275–281
select elements by, 247
settings group for, 668, 672
Line-string
Attach (at ls), 631–632
Drop status (dr st) of, 65–66
Links
Engineering, 659–662, 712–713
OLE, 639–643, 716
List attached references, 570–571
List boxes, 8–9, 25
Live Nesting, reference, 568, 569
Load
cell library, 526
scripts or macros, 746
Local files
specify location of, 650
URL format for, 648
Local storage, 649
Localhost, 648
Locate
By Faces, 769
reference models, 569
Location
Dimension, 400–403
Angle (dim a lo), 411–412
Arc (dim ar l), 416–418
Modify (mod dim loc), 431
text, 444–445
element, select by, 45, 594
fix point at, 624, 631
Lock Toggles settings box, 6, 7
Locks
AccuDraw
compass, 219
coordinate box field, 221
key-in shortcuts for, 232
SmartLock, 224
Axis, 83
Boresite, 758
Graphic Group, 587, 588
Grid, 78, 81–83, 88
selected element (chan lo), 179, 254
Snap, 84–94
Sticky Z axis, 225, 226
text height/width, 205
Text Node, 306, 307
X/Y Scale factor, 187
Locks icon, Status bar field for, 26
Locks submenu, 81, 82

Look and Feel, set preferences for, 695–697
LS=, 296–297
lt=, 504
LV=, 100, 296–297

M

Macros
Basic, create HTML files from, 657–658
load and run, 746
Main tool frame, 19–20, 106
Manage Line Style Definitions, 680
Manage View Groups, 135–137
Manipulate tool box, 180
Manipulation tools
cell, 494
element, 63–73, 179–202
AccuDraw and, 236
element selection and, 245–258
Identify Automatically feature for, 63–64, 96
SmartLine and, 237–241
Fence, 258–271
graphic group, 587–588
image, 604–607
reference, 564, 571–583
text, 8, 297–303
Manual method
Create Complex Chain (cr ch), 346–347
Create Complex Shape (cr s), 349–350
dimension text placement, 444
Margins, set dimension text, 451, 453
Mark, set and undo, 147
Mask Reference (refe c m), 577–578
Master Control menu, Pen Table, 480–481
Master file, merge references into, 579
Master Units
change settings for, 31–32
define custom, 32–33
dimension, 455
grid, 78, 80
Match
Dimension Settings (mat dim), 433, 458–459
Element Attributes, 109
Pattern Attributes (mat pat), 557–558
Text Attributes (mat t), 300
Match tool box, 459
Match/Change Element Attributes, 108
Matrix, Place Active Cell (matr c), 504, 508–510
Maximize view windows, 132–133
MDL (MicroStation Development Language),
configuration variables for, 715–716
Measure tool box, 382
Measurement tools, 381–391
Angle (me a), 386
Area (me ar), 387–391
Distance (me dist), 382–385
Length (me l), 386–387
Radius (me r), 385–386

845

Memory configuration, 2
Menu bars
  customize, 731–737
  drop-down menus in, 6–7
Menus
  cursor, 29–30
  drop-down
    customize, 69–75, 731–737
    display, 6–7
    location of, 17
    Standard tool equivalents in, 22
    use, 18–19
  tablet, 4
Merge references, 579
Merge Utility, 589–590
Merged Cross Joint (jo cr m), 368
Merged Tee Joint (jo t m), 371
Messages, Status bar, 26
Meters, working units in, 31
Metric units, 31–32, 458
MicroStation fonts, 206–207, 209
MicroStation Manager
  components of, 6–8
  Merge Utility in, 589
  open design files in, 16–17
  set up active workspace in, 690–693
  starting from, 5–6
MicroStation on Web, 44
MicroStation v8
  enhancements for, 44–45
  starting, 5–6
Midpoint snap mode, 90–91
Minimize view windows, 132–133
Minimum Distance, Measure (me dist m), 384–385
Mirror
  Element (mi o/c), 194–198
  Reference (refe mi), 575–576
Miscellaneous (Misc) dimension tools, 393, 427
  Change Dimension (cha dim), 432
  Delete Dimension Vertex (del dimv), 430–431
  Dimension Ordinates (dim o), 427–428
  Geometric Tolerance (mdl load geomtol), 433–434
  Insert Dimension Vertex (ins dimv), 430
  Label Line (l l), 428–429
  Match Dimension (mat dim), 433
  Modify Dimension Location (mod dim loc), 431
  Reassociate Dimension (dim reas), 431–432
Miter Gap mode, Move Parallel, 183, 184
MNU files, 738
Models
  3D, types of, 748
  create, 36–38
  enhancements for, 45
  references to, 564–566
    Attach (refe a), 566–571
    Detach (refe d), 580–581

Hilite modes for, 583
list attached, 570–571
rendering, methods for, 794–795
Modify Constraint tool box, 625
Modify Surfaces tool box, 775, 786
Modify tool box, 165–178
Modify/Modification tools. *See also* Edit
  Dimension Location (mod dim loc), 431
  dimension styles, 436
  dimension-driven design, 637–639
  element, 164–178, 331–378
    AccuDraw with, 338
    Arc, 341–345
    Construct Chamfer (ch), 166–167
    Construct Circular Fillet (fill n/m), 164–166
    Create Complex Chain/Shape (cr ch/s), 345–352
    Create Region (cr r) in, 352–357
    Drop Complex Status (dr e) for, 358–359
    Extend Line (ext l), 331–336
    Multi-line, 359–365
    Multi-line Joints, 365–378
    Partial Delete (del p), 177–178
    Trim (tri m), 167–177
    vertex/vertices (modi e), 336–341
  Fence (modi f/mov f), 266–268
  function key definitions, 738–741
  menus, 732–735
  pen table sections, 477–482
  plotter driver files, 465, 469
  settings group components, 668–673
  Solid (str f), 786–787
  Text (modi te), 301
  text styles, 290
  tool, 729
  tool frames, 731
  Value, dimension or variable, 638
  view border, 737–738
Monochrome, print vectors in, 467
Mosaic (NCSA), 646
Mouse
  buttons on, 3–4
  Preference settings for, 697
  tentative snap with, 84
  using, 28–29
Move
  active ACS (mov a), 766
  compass, 218–219
  compass origin, 234
  Element (mov e), 181–182
  Fence Contents (f mo), 269
  Fence Position (mov f), 267–268
  Multi-line Profile (ml e p), 376–377
  Parallel (mov p), 183–185
  References (refe m), 572
  selected elements, 257, 258
  view windows, 27, 130–131

846

Index

Movies dialog box, 607
Multi-line Joints tool box, 365–378
Multi-line text
   Copy/Increment and, 303
   display attributes for, 302
   Place (pl tex), 209
   Text Nodes for, 305
Multi-lines
   components of, 360
   Cut/Uncut lines from, 373–375
   define components of, 359–365
   edit cap (ml e c) of, 378
   mirror offsets for, 195, 196
   modify joints of, 365–378
   Move Profile (ml e p) of, 376–377
   Partial Delete (ml p d) for, 376
   Place (pl m c), 162–163
   scale offsets for, 185
   settings group for, 668, 672–673

N

Name/Rename. *See also* File names
   cells, 502, 525, 534
   level
      define, 101–102
      filter criteria for, 675–676
   line styles, 678, 680, 682
   menus, 732–734
   pen table sections, 476
   printer specification, 485
   tag sets/tags, 315, 316
   text styles, 288, 289, 290–291
   tool boxes, 727–728
   tool frames, 730–731
   views, 143
Natural fractions, stacking, 285–286
Nearest snap mode, 90
Negative angles, 227
Negative distances, 222, 226
Negative scale factors, 186
Nested attachments, reference, 568, 569
Netscape Navigator (NCSA), 646
Network files, URL format for, 648
New
   cell library (di ce), 495–497
   design, 10–15
   selection set, 248
   shortcut, AccuDraw, 229
Notes
   Dimension Radius/Diameter (dim rad/dia no), 425–426
   Geometric Tolerance, 433, 434
   Place (pl not), 283–285

O

Object linking and embedding (OLE), 639–643, 716
Offsets
   chamfer, 338
   grid, 81
   multi-line, 185, 195, 196, 361, 362
   stacked dimension, 439
On Element text (pla tex o), 278–279
Online help, 42–44
ON=/OFF= level display, 102–103
ON/OFF toggles, scroll bar, 133. *See also* Check boxes
Opaque fill, 558–559
Open
   Cell Selector box, 520
   Cell Selector files, 528
   design files, 16–17
   files, formats for, 600, 601–602
   files remotely, 646–652
   function key definition files, 740–741
   line style library files, 679
   plot configuration file, 474
   Reference tools, 564–565
   tool boxes, 18–19
   view windows, 27, 124–126
Open arrowheads, 447
Open Cross Joint (jo cr o), 367
Open File dialog box, 144–145
Open Tee Joint (jo t o), 370
Operating system, 2
Operations
   configuration variables for, 716–718
   settings for
      AccuDraw, 216, 218, 225–226
      Preference, 698–699
Option button menu, 10
Ordinate dimensions (dim o), 427–428
Orientation settings
   dimension terminator, 446
   dimension text, 449–450
   grid, 81
   Image Plane, 793
   print, 469
   reference model, 567
Origin, 67. *See also* Global Origin
   AccuDraw compass, 225, 226, 234
   ACS, 761, 767
   active cell, 506
   active cell matrix, 509, 510
   cell, define (d c o), 499–500, 501
   dimension from
      angular, 411–412, 416–418
      linear, 400–403
   place Point Symbol at, 687, 688
   set plot, 470, 488
Origin snap mode, 93
Origin terminator, 448–449
Ortho grid configuration, 81
Orthogonal axes, AccuDraw, 218
Orthogonal Block, Place (pl b o), 50–51
Orthogonal Shape, Place (pl sh o), 54–55
Orthogonal vertex, 338

Outlined fill, 558–559
Output Actions, Pen Table, 480–482
Outside terminator placement, 446, 447
Overlap Fence mode, 264
Overlined text, 293
Override
   dimension level symbology, 439
   snap mode, 87
Overwrite
   configuration variables, 707, 708
   existing file, 619, 620

P

Packaging MicroStation environment, 744–745
Pan View (pan v), 121–122
Paper size, 469, 484
Parallel
   Dimension Diameter (dim d p), 423–424
   Move/Copy (mov p/cop p), 183–185
Parallel snap mode, 94
Parameter Constraints tool box, 625
Parent text styles, 291
Partial Delete
   element (del p), 177–178
   Multi-line (ml p d), 376
Paste cell button, 522–523
Paste Special, Word, 640
Path
   cell library, display, 525
   configuration variables for, 718–719, 722
   design file, 14
   Extrude Along (constru e a), 778
Pathname, URL, 648
Pattern area
   create holes in, 555–556
   example of, 544
   modify component settings for, 669–670
   place, 549–554
   settings group for, 668
   tool for, 544, 545–547, 549
Pattern cells
   define defaults for, 525
   scale, 547
   select, 503–504, 546
Patterns
   commands for, 543–544
   control view of, 542–543
   create holes in, 555–556
   Delete (del pat), 557
   Match Attributes (mat pat) of, 557–558
   methods for, 547–548
   place, 549–554
   tool settings for, 545–547
   uses for, 541
   view attributes for, 138, 140
Patterns tool box, 543–544

Pen plotters, 464
Pen tables
   create, 475
   features of, 474–475
   modify sections of, 476–482
Perimeter, Measure Area (me ar p) of, 391
Perpendicular
   Dimension Size (dim si p p), 404–405
   Measure Distance (me dist p), 383–384
Perpendicular/Perpendicular Point snap modes, 94, 95
Phong rendering, 795
Pixels, 231, 524
Placement button, active cell, 503, 504
Place/Placement tools, 47–63, 301–313
   3D primitive
      Cone (pl co), 772–773
      Cylinder (pl cy), 771–772
      Slab (pl sl), 769–770
      Sphere (pl sp), 770–771
      Torus (pl to), 773–774
      Wedge (pl w), 774–775
   AccuDraw with, 233–234, 235–236
   Arc (pl a), 60–63
   Arrow Marker, 615
   Block (pl b), 50–52
   Callout Bubble, 612–613
   Callout Leader, 611
   Cell
      active placement, 504–510
      active point, 512–519
      active terminator, 510–512
      shared status and, 538
      types of, 503–504
   Circle (pl ci), 55–59
   Detail Marker, 615–616
   dimension, 443–445
   dimension text, 449–450
   Ellipse (pl el), 152–155
   Fence (pl f), 259–262
      modes for, 263–266
      remove fence with, 271
   Flag (pl fl), 608–609
   Leader and Text, 612
   Line (pl l)
      coordinates for, 70–73
      pointing device for, 48–49
   Multi-line (pl m c), 162–163
   Note (pl not), 283–285, 433, 434
   Pattern, 549–554
      settings for, 545–547
      types of, 543–544
   Point or Stream Curve (pl cu p/st), 160–162
   Regular Polygon (pl pol), 155–160
   Section Marker, 613–614
   Shape (pl sh), 52–55
   SmartLine (pl sm), 237–242

Index

Text Node (pla n), 305–306
Text (pl tex), 202–212
   3D design and, 791
   Above Element (pla tex a), 275–277
   Along Element (pl tex al), 279–281
   Below Element (pla tex b), 278
   dimension-driven design and, 634–635
   On Element (pla tex o), 278–279
   Fitted (pl tex f), 274–275
   geometric tolerance and, 433, 434
   By Origin, 209–210, 306–307
   settings for, 203–209
   shortcuts for, 210–212
   View Independent, 275
   Word Wrap, 282–283
Title Text, 614
Plot configuration, save, 473–474
Plot files, 464, 472–473
Plot specifications, Batch Print, 483–491
Plotter driver files, 464–465, 469
Plotters, 464, 469
Plotting
   Batch Print in, 483–492
   merge references for, 579
   overview of, 464–465
   pen tables for, 474–482
   Print settings for, 465–472
   print/create plot file in, 472–473
   save plot configuration in, 473–474
Point absolute, 67
Point cells, active
   3D designs with, 792
   characteristics of, 503
   define defaults for, 525
   place, 512–519
   select, 503–504
Point Curve, Place (pl cu p), 160–161
Point delta, 68
Point distance, 69
Point light, 794
Point On snap mode, 94
Pointer, screen
   change color of, 598–599
   select elements with, 179, 252
Pointing devices
   manipulate text with, 8
   place data points with, 48–49
   select text with, 212
   tentative snap with, 84
   using, 28–29
Points
   Auto placement of, 225, 226
   Constrain at Intersection (con i), 628–629
   Constrain Coincident (con c), 629–630
   Construct Points Between (constru po b), 513–514
   coordinate values for, 30
   Define ACS By (d a p), 764

define text box with, 282
Dimension Size Perpendicular to (dim si p p), 404–405
Fix (con l), 631
line style, create, 686–688
line style component, create, 682
Measure Area Defined By (me ar p), 391
Measure Distance Between (me dist po), 382
modify component settings for, 669
patterning area enclosed by, 548, 554
placement of, 512–519
   Active Point (pl po), 513
   Along Element (constru po a), 517–518
   coordinate systems for, 66–73, 758–763
   Between Data Points (constru po b), 513–514
   at Distance Along (constru po d), 518–519
   at Intersection (constru po i), 516–517
   pointing device for, 48–49
   Project Onto Element (constru po p), 515–516
   select type for, 512–513
   tools for, 512
Rotate References (refe ro) by, 575
Scale References (refe s) by, 574
select solids and surfaces with, 769
settings group for, 668
text justification and, 208
Polar arrays (ar p), 198, 200–202
Polar coordinates
   AccuDraw
      3D design and, 776
      indexing distance to, 222
      input fields for, 220–222
      settings for, 229–231
      switch to (a m), 217–218
   relative, 69–70, 73
Polygons
   Place Regular, 155–160
      Circumscribed (pl pol c), 157–158
      By Edge (pl pol ed), 158–160
      Inscribed (pl pol in), 156–157
      Place Shape (pl sh), 52–55
Polygons tool box, 50–55, 156–160
PopSet toggle, 24
Pop-up confirmations, AccuDraw, 227, 228
Position settings
   camera, 793
   print, 470
Positive scale factors, 186
PowerSelector (pow)
   dragging elements and, 255, 256
   select elements with, 245–251
Precision input
   3D design, 758–759
      ACS and, 762–763
      drawing coordinate system and, 759–760
      view coordinate system and, 760–761
   absolute rectangular coordinates for, 67–68, 70–72

coordinate systems for, 66–67
Grid lock and, 78
relative polar coordinates for, 69–70, 73
relative rectangular coordinates for, 68–69, 72–73
Preference settings
  solid/surface selection, 769
  user, 693–705
Prefixes
  cell button, 525
  coordinate label value, 617
  dimension, 437–438
Prepend configuration variables, 707, 708
Preview
  plot, 470, 479, 482
  print, 466
  text styles, 289
Primary search paths, configuration variables for, 718–719
Primary tool box, 23, 216
Primary units, dimension, 454, 455
Primitives, 3D, 748
  place, 769–775
  settings for, 768–769
  tools for, 767
Print Area specification, Batch Print, 483, 486–487
Print dialog box, 465–473
Print size, setting, 469–470, 488
Printer driver files, 468, 479
Printers
  Batch Print, specification for, 483, 484–485
  plotting with, 464, 468–469
Printing
  Batch Print for, 483–492
  configuration variables for, 718
  design file/to plot file, 472–473
  enhancements for, 45
  settings for, 465–472
Priority, element plotting, 481, 482
Processors, 2
Profile tool box, 625
Project
  select, 10, 11
  set up workspace for, 690, 691–692
Project Point Onto Element (constru po p), 515–516
Projected surfaces, 777–778
Prompts, command, 26
Properties
  Drafting Tool, 616
  drawing, 38–40
  Edit Saved View, 143–144
  Level, 101–102
  Select By, 591, 592, 594–595
  View Group, 136–137
Proportional Scale option, 186, 189
pt=, 504
Publishing to Internet, 653–658
Pucks
  button on, 4–5

tentative snap with, 84
using, 28–29
PZIP files, extracting, 744–745

Q

Quick mode, IntelliTrim, 167
  Cut with, 172–173
  Extend with, 170–171
  Trim with, 168–169

R

Radial dimensioning
  Arc Distance (dim ar dist), 426
  from Center Mark (dim c m), 424–425
  Diameter (dim d), 421–422
    (Extended Leader) (dim d e), 422–423
    Parallel (dim d p), 423–424
  for element (dim e), 408
  example of, 392
  Note (dim rad/dia no) with, 425–426
  Radius (dim radiu), 419–420
  Radius (Extended Leader) (dim radiu e), 420–421
  tools for, 393, 418–419
Radius
  constrain ellipse, 153, 155
  dimensioning, 419–421, 425–426
  Measure (me r), 385–386
  Modify Arc (modi a r), 15, 341–342
  place arcs with, 61, 62
  place circles with, 57, 58
  place polygons with, 156–158
  vertex rounding, 338
RAM (random access memory), 2
Raster images
  convert, 605
  display of, 568
  Fit View for, 117
  Preference settings for, 699–700
rc=, 497
RDL (Redline) files, 600
Read-only mode
  open designs in, 17
  Status bar field for, 26
  store downloaded files in, 650
Reassociate Dimension (dim reas), 431–432
Recalling values, AccuDraw feature for, 222–223
Rectangle, Capture, 605
Rectangular arrays (ar r), 198, 199–200
Rectangular coordinates, 67
  absolute, 67–68, 70–72
  AccuDraw
    3D design and, 776
    indexing distance to, 222
    input fields for, 220–222
    settings for, 229–231
    switch to (a m), 217–218
  define ACS with, 762
  relative, 68–69, 72–73

Index

Redo, 146, 148
Redundant constraint, 624
Reference File Agent dialog box, 581–582
Reference files, remote opening of, 651–652
References
   Attach (ref e), 566–571
   attach views as, 796
   enhancements for, 45
   examples of using, 565–566
   Fit View for, 117
   list attached, 570–571
   manipulation of, 571–583
      Clip (refe c bo), 576–577
      Copy (refe c), 573
      Delete Clip (refe c m d), 578–579
      Detach (refe d), 580–581
      Exchange, 581
      Hilite mode for, 583
      maintain local copies, 581–582
      Mask (refe c m), 577–578
      Merge into Master (mdl load refmerge), 579
      Mirror (refe mi), 575–576
      Move (refe m), 572
      Reload (refe r), 579–580
      Rotate (refe ro), 575
      Scale (refe s), 573–574
      select, 571
      update display sequence for, 582–583
   overview of, 564–565
   Preference settings for, 700–702
   use units from, 445
References settings box, 564–565, 570
References tool box, 564
Region, Create (cr r), 352–357
Reject button, 28
Relative cell placement, 503, 506–507
Relative coordinates
   place points with, 68–70, 72–73
   set depth by, 754–755, 757
Relative rotation, 752, 765
Reload/Reload All references (refe r), 564, 579–580
Remote files, opening
   archive file, 651
   cell library, 652
   design file, 649–650
   launch Web browser for, 646–647
   reference file, 651–652
   settings file, 650–651
   URL scheme for, 647–648
Remove
   Faces (rem f), 787
   fence, 271
   menu/menu item, 736
   selected elements, 250, 252
   tags/tag set, 316
   tool, 728
   tool frame, 731
   View Border tool, 738

Rename. *See* Name/Rename
Rendering
   configuration variables for, 719–720
   methods for, 794–795
   place light sources for, 794
   set up cameras for, 792–794
Repetitions, stroke pattern, 684
Replace
   Cells (rep cells e), 529–531
   pen table text, 477–479
   tag values, 321, 322
   text, 210, 212
Reset button
   mouse, 3–4
   puck, 4–5
   using, 28, 29
Resize
   Text Editor window, 209
   tool boxes, 20
   view windows, 27, 130–131, 132
Resolution settings
   design environment, 34
   display, 231
   HTML page, 653, 654, 656, 657
Re-solve Constraints (up m), 638
Restore
   dimension styles, 436
   saved views, 141
Reversed terminators, 446, 447
Review Tags (rev t), 320
Revolution, Construct surface of (constru s r), 779
"Right-hand rule," view rotation, 752
Right-justified dimension text, 450
Rotate
   AccuDraw compass, 218–219, 229, 230
   AccuDraw plane, 232, 235, 776–777
   active ACS, 765
   cells, 499, 505
   character/cell points, 513
   elements, 190–194
   patterns, 545
   polar array, 200–201, 202
   rectangular array, 199, 200
   References (refe ro), 575
   text, 204–205, 296
   View, 118–120, 750–752
Rotated Blocks, Place (pl b r), 51–52, 235–236
Rotation, constrain ellipse, 153, 155
Round Gap mode, Move Parallel, 183, 184
Rounded vertices
   modify, 338
   SmartLines with, 237, 238–239, 241, 242
Rows
   cell matrix, 509, 510
   rectangular array, 199, 200
RPT files, 326
RSC files, 677, 708
Rule lines, surface, 768

851

Run
    AccuDraw shortcuts, 228
    scripts or macros, 746

S

Saved views
    Apply options for, 142
    attach reference with, 567–568
    create HTML file from, 653–654
    delete, 144
    display, 141
    edit properties of, 143–144
    import, 144–146
    settings box for, 140–146
Save/Save As
    ACS, 766–767
    Cell Selector files, 527
    create text files with, 295
    design file settings, 41–42
    design files, 40–41
    files, formats for, 600, 602–603
    function key definitions, 740
    images, 604
    job set, 492
    line style library files, 680–681
    pen table, 482
    plot configuration files, 473–474
    print, 473
    print configuration files, 474
    settings groups, 664
    views, 140–146
Scale
    active cells, 505
    designs, 30
    dimensions, 456
    elements, 185–189
        3-Points method (sc p o/c) for, 188–189
        Active method (sc o/c) for, 186–188
    flags, 608
    line styles, 678
    line terminators, 511
    linked objects, 641
    pattern cells, 547
    plot, 469–470, 488
    points, 513
    references, 567–568, 569, 573–574
    settings group for, 664, 674
    text, 205–206
Scale Assistant, 469–470
Scheme, URL, 648
Screen Capture methods, 604–605
Scripts, load and run, 746
Scroll bars, 8–9, 133
Search paths, configuration variables for, 718–719, 722
Search tab, Help, 42, 43
Secondary units, dimension, 454, 456
Section Marker, Place, 613–614
Seed files
    configuration variables for, 720
    create 3D designs from, 749
    select, 10, 14–15
Segment types, SmartLine, 237, 238–239
Select
    AccuDraw coordinates field, 221
    active ACS, 766
    active workspace, 690, 691
    By Attributes, 591–596
    cells, 503–504
    configuration variable items, 708
    custom line style, 676–677
    design files to plot, 491–492
    elements
        automatic identification for, 63–64, 96
        Element Selection for, 251–252
        pointer for, 179
        PowerSelector methods for, 245–251
    files to merge, 589–590
    plot area, 467
    plot specifications, 485, 487, 489, 490
    printers, 468–469
    references, 571
    seed files, 10, 14–15
    settings group, 664–665
    solids and surfaces, 769
    source/destination font files, 742–743
    text, shortcuts for, 211–212
    text styles, 287
    view groups, 134–135
    views to import, 146
Select All mode, PowerSelector, 248
Select and Place Cell (se c i), 504, 508
Selection set. *See also* Groups
    change attributes of, 257
    change highlight color for, 598, 599
    creating
        Element Selection for, 251–252
        PowerSelector for, 245–248
        select By Attributes for, 251–252
    delete elements from, 256–257
    dragging, 254–255
    Fence manipulation for, 258–271
    Group (gr s), 252–253
    lock/unlock elements in, 254
    manipulation tools for, 257–258
    remove elements from, 250
    Status bar field for, 26
    Ungroup (ung), 253–254
Semi-auto dimension text placement, 444
Separators, coordinate label value, 617
Servername, 648
Set Active Depth (dep a), 756–758
Set Display Depth (dep d), 754–755
Set Mark (mar), 147

Settings boxes, 6–10
Settings files
  remote opening of, 650–651
  Save (fi), 41–42
Settings groups
  attach, 666
  maintain, 666–674
  select, 664–665
Shading 3D models, 792, 795
Shapes
  construct attached (at ls), 631–632
  creating complex, methods for, 345–346
    Automatic (cr s a), 350–352
    Difference (cr r d), 355–356
    Flood (cr r f), 356–357
    Intersection (cr r in), 352–353
    Manual (cr s), 349–350
    Union (cr r u), 354–355
  Drop Status (dr st/dr e) of, 65–66, 358–359
  modify, drag handles to, 255–256
  modify fence (modi f), 266–267
  patterning over, 552–553, 555
  Place Fence (pl f s), 260
  Place (pl sh), 52–55
  print boundary for, 486
  selecting, PowerSelector for, 247
Shared cells, 536–539
Sharp vertices
  modify, 338
  SmartLines with, 237, 238–239
Sheet models, create, 36–38
Shell Solid (constru s), 780
Shift stroke patterns, 678, 684
Shortcuts, key-in. *See* Key-in shortcuts
Shorten elements
  graphically, 332–333
  to intersection, 333–336
  by key-in distance, 333
Show
  Active Depth (az/dz = $), 757
  Display Depth (dp/dd = $), 755
  Engineering Links, 660
  Print Details, 466–467
Show/Edit Flag (sh f), 610
Simplify Geometry for complex chain/shape, 346, 348, 350, 352
Single-character fractions, 285–286
Slab, Place (pl sl), 769–770
Slant angle, text, 292
Smart Key-ins, AccuDraw, 225, 226
Smart Lock, AccuDraw, 224
SmartLine tool, 237–242
SmartSolids/SmartSurfaces, display, 768
Smooth rendering, 795
Snap Lock. *See also* AccuSnap
  element manipulation and, 179
  modes for
    AccuDraw shortcuts for, 233
    select, 29–30, 85–87
    Status bar field for, 26
    Tentative button for, 29, 87–89
    toggles for, 84–85
    types of, 89–94
  purpose of, 84
Snap Mode button bar, 85–86, 95
Snap Mode pop-up menu, 29–30, 86–87
Snappable line styles, 682
Snappable patterns, 546
Snapshot, create HTML file from, 654–655
Solar lighting, 794
Solid area, create holes in, 555–556
Solid models
  3D design with, 748
  Chamfer Edges (ch e) of, 790
  change status (chan su c) of, 775
  creating composite, 781–786
    difference operation (constru d) for, 784–786
    intersection operation (constru i) for, 783–784
    union operation (constru u) for, 781–783
  Cut (constru cut), 788
  Fillet Edges (fill e) of, 789
  Modify (str f), 786–787
  Remove Faces (rem f) from, 787
  Revolution, Construct (constru s r), 779
  selection of, 769
  set display method/rule lines for, 768
  Shell (constru s), 780
  Thicken To (constru th), 780–781
Solve for constraints, 624
Sound On Snap, Play, 96
Source files, font, 742, 743
Source lighting, 794
Space characters for fractions, 286
Spacebar, switch coordinate systems with, 218
Spacing, settings for
  active cell matrix, 509, 510
  grid, 79–80
  pattern, 545
  rectangular array, 199, 200
Spell Checker, 208
  Preference settings for, 702–703
  using, 297–299
Sphere, Place (pl sp), 770–771
Spherical coordinate system, 762
Spin Element (sp o/c), 191–193
Split Solid, 788
Spot lights, 794
Stack offset, dimension line, 439
Stacked Dimension Location (dim lo st), 401–403
Stacked fractions, 285–286
Standard tool box, 21–22, 147, 148
Start
  AccuDraw (a a), 216–217, 776
  MicroStation V8, 5–6
  new design, 10–15

Start Angle, place arc by, 61, 63
Start caps, multi-line, 360, 363–364
Statistics settings, Properties dialog box, 40
Status bar, 17, 26
Step translation, configuration variables for, 723
Stereo rendering, 795
STG files, 664, 667
Stickiness, AccuSnap, 97
Sticky Z Lock, AccuDraw, 225, 226
Stop AccuDraw, 216–217
Store Read Only, 650
Stream Curve, Place (pl cu st), 161–162
strech faces (str f), 786–787
Stretch Fence (f st), 270–271
Stroke, Dimension Size with (dim si s), 399–400
Stroke pattern
   associate point with, 687
   create, 682, 683–685
   settings for, 677–678
Stroke terminator, 448–449
Style library files, 677, 679–680
Sub Units
   change settings for, 31–32
   define custom, 32–33
   dimension, 455
   grid, 78, 80
Submenus
   add items to, 734
   display, 7, 18
Subtract from selection set, 247
Suffixes
   cell button, 525
   dimension, 437–438
Summary settings, Properties dialog box, 39
Suppress Rule Lines, 768
Surfaces
   3D design with, 748
   add thickness to (constru th), 780–781
   Chamfer Edges/Chamfer Between, 790–791
   Change Normal Direction (chan n) of, 786
   change status (chan su c) of, 775
   create composite solids from, 783, 786
   Fillet Edges/Fillet Between, 789–790
   projected, 777–778
   Revolution, Construct (constru s r), 779
   selection of, 769
   set display method/rule lines for, 768
sv=, 143
Sweep Angle, place arc by, 61, 63
Symbology
   configuration variables for, 720
   level
      change, 101
      set, 596–598
      view attributes for, 138, 139
   select elements by, 591, 592, 593
   tools for, 22–23

Symbols. *See also* Cells
   dimension, customize, 437–438
   geometric tolerance, 433
   point, 687, 688
   terminator, 448–449
   wild card, filter with, 675–676
Sync Find tag values, 321, 322
System environment, configuration variables for, 720–721

T

Tables, HTML page, 656
Tablet menu, digitizing, 4
Tag Increment
   Data Field, 311, 312
   text, 303
Tag report template, 314, 323–325
Tag reports, 314
   access, 327
   create, 323–327
   generate, 325–326
   uses for, 313
Tag set library, 314, 327–329
Tag sets, 314
   append, 328
   change tags in, 321–322
   create, 314–316
   export, 327–328
   import, 328–329
   maintain definitions for, 316–317
Tags
   Attach, 317–318
   attributes of, 316
   Change (chan t), 321–323
   configuration variables for, 721
   create, 314–316
   Edit (edi t), 318–320
   Edit (Internet), 661
   maintain definitions for, 316–317
   Preference settings for, 703
   Review (rev t), 320–321
   Select By, 591, 592, 595
   terminology for, 314
   uses for, 313
   view attributes for, 138, 140
Tags tool box, 317–321
Tangent edges
   chamfer (ch e), 790
   fillet (fill e), 789
Tangent/Tangent From snap modes, 93, 95
Target, setting camera, 793
Temp files, configuration variables for, 721
Template (.tmp) files, 326
Tentative button
   mouse, 3–4
   puck, 4–5
   terminator placement and, 512
   using, 29, 30, 87–89

Index

Tentative snap points
    Boresite lock and, 758
    placing
        AccuDraw for, 234
        AccuSnap for, 94
        Snap Lock for, 84, 87–89
    reference element, 568, 569
Terminator cells, active
    define defaults for, 525
    Place (pl t), 510–512
    select, 503–504
Terminators
    dimension line
        description of, 393, 394
        settings for, 441, 446–448
        specify symbols for, 448–449
    leader, specify, 611, 612
    Place Active Line (pl t), 510–512
Text. *See also* Dimension text
    3D design, 791
    AccuSnap to, 96–97
    callout, 611, 612–613
    Copy/Increment (incr t), 302–303
    Data Field, 309–312
    dimension-driven design, 634–635
    Edit (edi te), 299–300
    edit field for, 8
    enhancements for, 45
    flag, 609, 610
    geometric tolerance, 433, 434
    Import (in), 295–297, 601
    leader, 612
    manipulation tools for, 297–303
    Mirror, 195, 196
    modify component settings for, 673
    Notes with, 283–285
    patterning over, 552–553, 556
    pen table, substitutions for, 477–479
    placement of, 202–212
        Above Element (pla tex a), 275–277
        Along Element (pl tex al), 279–281
        Below Element (pla tex b), 278
        On Element (pla tex o), 278–279
        Fitted (pl tex f), 274–275
        keyboard shortcuts for, 210–212
        By Origin, 209–210, 306–307
        Place (pl tex), 203
        settings for, 204–209
        View Independent, 275
        Word Wrap for, 282–283
    Preference settings for, 703–704
    section marker, 613–614
    settings group for, 668
    spell check, 208, 297–299
    Text Node, 306–307
    Title, 614
    view attributes for, 138, 140

Text attributes. *See* Attributes, text
Text Editor
    edit text in, 300
        Data Field, 312
        dimension text, 398–399
    keyboard shortcuts for, 210–212
    place text in, 202–203, 209–210
    settings in, 208
    Stacked Fractions option in, 285–286
Text files, import or export, 295–297, 601
Text frames
    dimension text, 451
    notes with, 283, 284
    word wrap in, 282
Text Nodes
    coordinate label, 618
    definition of, 203
    display attributes of, 302
    fill in, 306–307
    function of, 304
    justification for, 293
    Place (pla n), 305–306
    view attributes for, 138, 140
    viewing, 304–305
Text size
    change, 299
    dimension, 451, 453
    key-ins for, 296
    set, 205–206, 292
Text strings
    create data fields in, 307–308
    definition of, 203
    pen table, replacing, 477–479
Text styles. *See also* Fonts
    create, 288, 289–290
    delete, 290
    import, 294
    maintain, 288–289
    modify, 290
    parent-child hierarchy for, 291
    select, 204, 287
    settings for, 288–289
        Advanced, 294
        change, 299
        General, 290–293
Text tool box
    manipulation tools in, 297–303
    placement tools in, 203, 275–285
TH=, 296–297
Thicken To Solid (constru th), 780–781
Thickness, shell, 780
Three-button mouse, 3–4
Through Point snap mode, 92
Tile windows (w t), 128–129
Time, replace pen table text string with, 479
Tips, display snap point, 96
Title bar, 6–7

Title blocks
  reference files for, 565
  Text Nodes for, 304
Title Text, Place, 614
Tolerance
  AccuSnap, 97
  dimensioning, 438, 452–453
  Geometric, 433–434
  indexing, 229, 231
  Measure Area/Length, 387, 388
  pattern, 545
  stream curve, 161, 162
Tool boxes
  access tools in, 19, 106
  Annotation, 607
  attach to menu items, 734–735
  customizing, 727–730
  Dimension, 392–393
  Dimension-Driven, 624–625
  docked, view window and, 129
  docking/undocking, 20–21
  location of, 17
  major types of, 20, 21–23
  opening, 18–19
  resizing, 20
  tool tips in, 23–24
Tool frames
  Annotation, 607
  customizing, 730–731
  Dimension-Driven, 624–625
  location of, 17
  Main, 19–20, 106
Tool Index, Help menu, 43–44
Tool Settings, dimension, 454
Tool Settings window
  location of, 17
  opening, 24
  Place Text, 203, 204–209
Tool tips, 23–24, 729
Tools
  accessing, 19
  Status bar display for, 26
Tools menu, 18, 106
Torus, Place (pl to), 773–774
Trailing zeros, dimensions with, 458
Translation, configuration variables for, 722–723
TriForma Document (D) Files, 600
Trim elements (tri m)
  Advanced Extend with, 176–177
  Advanced mode for, 174–175
  modes for, 167–168
  Quick Cut with, 172–173
  Quick Extend with, 170–171
  Quick mode for, 168–169
True dimension alignment, 395, 396, 443–444
True Scale, place cell by, 45, 505
True Type fonts, 207, 209

Truncate
  chamfer, 791
  fillet, 164–166, 790
TW=, 296–297
Two-button mouse, 3–4
TX=, 296
Types, element, selection by, 591, 593

U

UCF files, 693
Uncut Component lines (un), 375
Underconstrained construction, 624
Underline text, 208, 293, 451
Undo All (und a), 147–148
Undo Mark (und m), 147
Undo (und), 146–148
  cell deletion and, 535
  cut multi-lines, 375
  design compression and, 557
Undocking tool boxes, 21
Ungroup (ung), 253–254
Union
  create composite solids (constru u) from, 781–783
  create region (cr r u) from, 354–355
  Measure Area (me ar u) of, 389–390
  patterns for, 548, 551
Unit format, dimension text, 456–458
Units
  coordinate label, 617
  dimensioning, 454–456
  enable reference file, 445
  grid, 78, 80
  working, 31–34, 664
Unload pen table, 482
Unlock selected elements (chan u), 254
Unrotated option, Rotate View, 118, 120
Update
  cells, 529–531
  Flag (fl u), 610–611
  linked objects, 642
  reference files, 582
  Sequence, 582–583
  view, 111, 138, 139, 227
URLs (uniform resource locators)
  associate references with, 581–582
  link geometry to, 659–662
  open files remotely with, 649–652
  schemes for, 647–648
User configuration files, create, 691–692
User interface
  create, 692–693
  customizing, 726–738
    menu bar in, 731–737
    modification files for, 726–727
    tool boxes in, 727–730
    tool frames in, 730–731
    view border tools in, 737–738

Index

User workspace
  configure, 690, 691–692
  select, 10–11

V

Variables
  Assign (as v), 635–636
  Modify Value (mo e) of, 638
  place text for, 634–635
VBA (Visual Basic for Applications), 45, 723, 746
Vectors, set output color for, 467
Vertex
  Delete Dimension (del dimv), 430–431
  Insert Dimension (ins dimv), 430
  Modify Element
    Delete (del v), 339–340
    Insert (ins v), 340–341
    tools for, 336–338
  Modify Fence (modi f), 266–267
  place point symbol at, 687, 688
  SmartLine options for, 237, 238–239
Vertical axis, Mirror About (mi o/c v), 195, 196–197, 576
Vertical text, 293
vi=, 752
Video adapter, 3
View Attributes settings box, 138–140. *See also* Attributes, view
View border tools, customize, 737–738
View Control bar, 110–111
  location of, 17, 27
  Rotate View from, 751
  toggle for, 133
View Control pop-up menu, 29, 110–111
View Control tool box/tools, 109–123
  accessing, 29, 109–111
  Fit View (fit v e), 116–118
  Pan View (pan v), 121–122
  Rotate View, 118–120
  Update View (up), 111
  View Next (vi n), 122–123
  View Previous (vi p), 122
  Window area (w a e), 114–116
  Zoom In (z i e), 112–113
  Zoom Out (z o e), 113–114
View coordinate system, 760–761
View grid orientation, 81
View groups, create and use, 133–137
View Groups settings box, 124, 125, 134
View Independent text, 275, 306
View Next (vi n), 122–123
view on/off (vi on/off), 124
View Previous (vi p), 122
View Rotation settings box, 751–752
View windows. *See also* Display
  arrange multiple, 126–131
  arrange single, 131–133

Capture, 605
create/use groups of, 133–137
features of, 123–124
find, 133
open and close, 124–126
overview of, 27
scroll bar toggles for, 133
set grid display for, 79
View(s), 27. *See also* Display
  3D design, 747
    compose, 796
    default display of, 749–750
    orient drawing plane to, 776–777
    rotate, 750–752
  align dimensions parallel to, 395, 396, 443–444
  Capture, 605
  controlling design area in, 111–116
    Window Area (w a e) for, 114–116
    Zoom In/Out (z i e/z o e) for, 112–114
  current, precision input and, 761
  Define ACS with (d a v), 764–765
  Fit (fit v e), 116–118
  Pan (pan v), 121–122
  Place Fence From (pl f v), 261–262
  plot area for, 467, 486
  Preference settings for, 705
  rendering, 792–795
  Rotate, 118–120
  saving, 140–146
  Update (up), 111
  View Next (vi n), 122–123
  View Previous (vi p), 122
Visual Basic for Applications (VBA), 45, 723, 746
Void Fence modes
  cell creation and, 501
  delete contents with, 265–266
  Fence Stretch and, 270
Void-Clip Fence mode, 266
Void-Overlap Fence mode, 265
Volume, 3D fence, 791

W

Web browsers, launching, 646–647
Web pages, create, 653–658
Web sites. *See also* Internet
  MicroStation resources on, 44
  specify file names for, 648
Wedge, Place (pl w), 774–775
Well-constrained construction, 624
Width settings
  arrowhead, 447
  dimension text, 451
  text, 205–206, 292
    change, 299
    key-ins for, 296
Wild card symbols, filtering with, 675–676
Window Area, specify (w a e), 114–116

Window menu, 124, 125, 126–131
Windows system printer
   load driver file for, 468
   print with, 464, 472–473
   set parameters for, 469
Windows XP (Microsoft), 126, 132
Wireframe display mode, 768
Wiremesh rendering, 795
Witness lines. *See* Extension lines
Word documents, embed elements in, 639–640
Word processors, stacked fractions in, 286
Word Wrap text, 282–283
Work environment, packaging, 744–745
Working Areas, set units for, 34
Working line, multi-line, 361, 377
Working units
   Advanced Settings for, 34
   change, 31–32
   components of, 31
   custom, 32–33
   dimension, 445, 455
   settings group for, 664
Workspace
   characteristics of, 690
   customizing, 690–738
      configuration variables for, 706–726
      Preference settings for, 693–705
      set up active space for, 690–693
      user interface in, 726–738
   select, 11
Wrapping text, 209
WT=, 106, 296–297

X

X axis, 67, 68
   3D design and, 748
   AccuDraw
      display of, 218, 227
      rotate, 218–219
   ACS and, 761
   design cube and, 753
   design plane and, 30
   dimension angle from (dim a x), 414
   drawing coordinate system and, 759–760
   view coordinate system and, 760–761
X coordinates, locate points with, 67–69
X origin, plot/print from, 470, 488
X scale factors
   active scaling with, 185–188
   place active cell with, 505
   plot/print, 470, 488
XML:simple link, create, 659
XY=, 67–68, 759–760
XYZ Text tool box, 607, 617–621

Y

Y axis, 67, 68
   3D design and, 748
   AccuDraw
      display of, 218, 227
      rotate, 218–219
   ACS and, 761
   design cube and, 753
   design plane and, 30
   dimension angle from (dim a y), 415
   drawing coordinate system and, 759–760
   view coordinate system and, 760–761
Y coordinates, locate points with, 67–69
Y origin, plot/print from, 470, 488
Y scale factors
   active scaling with, 185–188
   place active cell with, 505
   plot/print, 470, 488

Z

Z axis, 67, 68
   3D design and, 748
   AccuDraw lock for, 225, 226
   ACS and, 761
   design cube and, 30, 753
   drawing coordinate system and, 759–760
   view coordinate system and, 760–761
Zeros, display leading/trailing, 458
Zip files, extracting, 744–745
Zoom In/Zoom Out, 79, 112–114